W9-COY-538

Experimental Behavioral Ecology and Sociobiology

In Memoriam Karl von Frisch 1886–1982

Edited by Bert Hölldobler and Martin Lindauer

Sinauer Associates, Inc. Sunderland · 1985

Unter Verantwortung der
Akademie der Wissenschaften und der Literatur, Mainz.

Gefördert durch
das Bundesministerium für Forschung und Technologie, Bonn
und das Bayerische Staatsministerium für Unterricht und Kultus, München
sowie durch die Stiftung Volkswagenwerk.

International Simposium of the Akademie der Wissenschaften und der Literatur, Mainz
October 17th–19th, 1983 at Mainz

Library of Congress Cataloging in Publication Data
Main entry under title:

Experimental behavioral ecology and sociobiology.

Includes bibliographies and index.
1. Animal behavior – Addresses, essays, lectures.
2. Animal ecology – Addresses, essays, lectures.
3. Sociobiology – Addresses, essays, lectures.
4. Zoology, Experimental – Addresses, essays, lectures.
I. Hölldobler, Bert, 1936–0000. II. Lindauer, Martin.
QL751.E96 1985 591.51 84–29807
ISBN 0-87893-460-X
ISBN 0-87893-461-8 (pbk.)

distributed in North America by Sinauer Associates Inc.,
Sunderland, Massachusetts 01375 U.S.A.

© Gustav Fischer Verlag · Stuttgart · New York · 1985
All rights reserved
Composed and printed by: Laupp & Göbel, Tübingen 3
Bound by: Großbuchbinderei Heinrich Koch, Tübingen
Printed in Germany

ISBN 0-87893-461-8 paper
ISBN 0-87893-460-X cloth

In Memoriam Karl Ritter von Frisch

Karl Ritter von Frisch (Nobel Laureate 1973). A great biologist of the twentieth century. The pioneer and doyen of sensory physiology, experimental behavioral ecology, and sociobiology. He taught us respect and wonder for all living things. His friendship extended around the world.

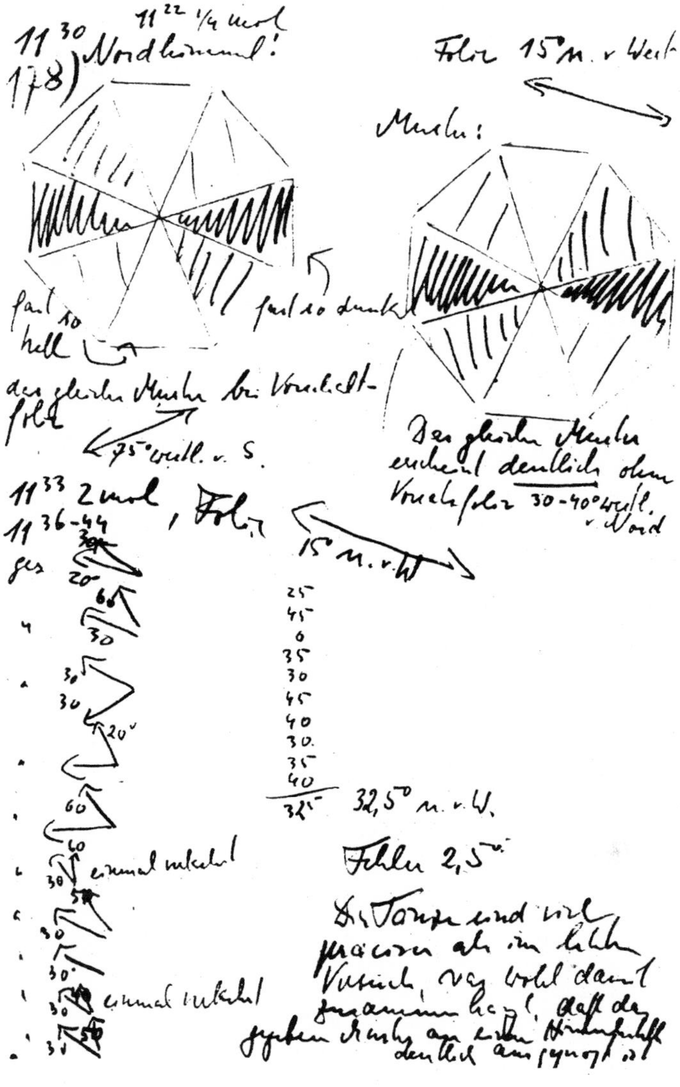

Excerpt from the notebook of Karl v. Frisch from the year 1949
Experiment No 178 of August 31, 1949

Die Bienen verkehren an einem Futterplatz im Westen.
Die Sternfolie als Modell des Bienenauges zeigt – zum Nordhimmel gerichtet – ein charakteristisches Muster. Beim Tanz auf horizontaler Wabe haben die Bienen Ausblick zu diesem Muster. Von 11.36 Uhr bis 11.44 Uhr wird über dem Tanzboden eine Polarisationsfolie so aufgelegt, daß dieses Muster 30° westlich von Nord verlagert erscheint. Die Tänze zeigen jetzt nicht mehr direkt zum Futterplatz nach Westen, sondern 32,5° nördlich von West (Fehler 2,5°).

Preface

The closely interlocked fields of behavioral ecology and sociobiology have experienced a tremendous revitalization during the past 10 to 15 years. They have been energized by the merging of ethology, population genetics, and modern evolutionary theory, in a manner that has proven effective in generating new hypotheses about the adaptiveness and evolution of behavior. We have now reached the phase where these theoretical concepts have to be put to the test by rigorous experimentation. It is here where we owe so much to KARL v. FRISCH. His way of asking questions and obtaining answers from the animals by ingenious experimentation, has been and will continue to be a major inspiration for experimental behavioral ecologists and sociobiologists.

This volume presents the papers which were read at a memorial symposium held in honor of KARL v. FRISCH and sponsored by the Mainzer Akademie der Wissenschaften und Literatur and the Stiftung Volkswagenwerk. We are most grateful to these institutions and to all contributors to this volume.

BERT HÖLLDOBLER MARTIN LINDAUER

Contents

IV. Social Organization

V. Physiology and Societies

Epilogue

List of Authors

Dr. H. St. Bartz: Museum of Comparative Zoology, Harvard University, Cambridge, Massachusetts 02138, USA.

Prof. Dr. J. Bradbury: University of California, San Diego, Department of Biology, La Jolla, California 92039, USA.

Prof. Dr. J. Beetsma: Labor. of Entomology, Agricultural University, Wageningen, Netherlands.

Dr. J. R. Cheverton: Department of Zoology, Edward Grey Institute of Field Ornithology, South Park Road, Oxford OX1 3PS, United Kingdom.

Dr. F. C. Dyer: Department of Biology, Princeton University, Princeton, N.J. 08544, USA.

Prof. Dr. St. Emlen: Cornell University, Section of Neurobiology and Behavior, Division of Biological Sciences, Ithaca, N.Y. 14850, USA.

Dr. N. Franks: School of Biological Sciences, University of Bath, Bath BA2 7AY, United Kingdom.

Prof. Dr. J. L. Gould: Department of Biology, Princeton University, Princeton, N.J. 08544, USA.

Prof. Dr. D. Griffin: Rockefeller University, New York, N.Y. 10021, USA.

Prof. Dr. B. Heinrich: The University of Vermont, Department of Zoology, Marsh Life Science Bldg. Burlington, Vermont 05405, USA.

Prof. Dr. B. Hölldobler: Museum of Comparative Zoology, Harvard University, Cambridge, Massachusetts 02138, USA.

Prof. Dr. Warren Holmes: Department of Psychology, University of Michigan, Ann Arbor, Michigan 48109, USA.

Prof. Dr. D. v. Holst: Universität Bayreuth, Zoologisches Institut, Universitätsstr. 30, 8580 Bayreuth, FRG.

Dr. A. Kacelnik: Department of Zoology, Edward Grey Institute of Field Ornithology, South Park Road, Oxford OX1 3PS, United Kingdom.

Prof. Dr. J. Krebs: Department of Zoology, Edward Grey Institute of Field Ornithology, South Park Road, Oxford OX1 3PS, United Kingdom.

Prof. Dr. M. Lindauer: Zoologisches Institut II der Universität, Röntgenring 10, 8700 Würzburg, FRG.

Prof. Dr. K. Linsenmair: Zoologisches Institut III der Universität, Röntgenring 10, 8700 Würzburg, FRG.

Prof. Dr. H. Markl: Fachbereich Biologie der Universität, Postfach 733, 7750 Konstanz, FRG.

Prof. Dr. P. Marler: Rockefeller University, Center for Field Research, Tyrrel Road, Millbrook, N.Y. 12545, USA.

Prof. Dr. R. Menzel: Zoologisches Institut II der Freien Universität Berlin, Grunewaldstr. 34, 1000 Berlin 41, FRG.

Prof. Dr. C. Michener: Department of Entomology, University of Kansas, Lawrence, Kansas 66045, USA.

Prof. Dr. P.-F. Röseler: Zoologisches Institut II der Universität, Röntgenring 10, 8700 Würzburg, FRG.

Dr. S. Rossel: Zoologisches Institut der Universität, Winterthurer Str. 190, CH-8006 Zürich, Switzerland.

Prof. Dr. F. Ruttner: Institut für Bienenkunde, Universität Frankfurt, Im Rotkopf 5, 6370 Oberursel 1, FRG.

Prof. Dr. H. SCHEICH: Zoologisches Institut der Techn. Hochschule, Schnittspahnstr. 3, 6100 Darmstadt, FRG.

Prof. Dr. T. SEELEY: Yale University, Department of Biology, Osborn Memorial Laboratory, P.O. Box 6666, New Haven, Conn. 06511, USA.

Prof. Dr. P. SHERMAN: Cornell University, Division of Biological Sciences, Section of Neurobiology and Behavior, Ithaca, N.Y. 14850, USA.

Dr. BARBARA L. THORNE: Museum of Comparative Zoology, Harvard University, Cambridge, Mass. 02138, USA.

Dr. W. F. TOWNE: Department of Biology, Princeton University, Princeton, N.J. 08544, USA.

Prof. Dr. SANDRA L. VEHRENCAMP: University of California, San Diego, Department of Biology, La Jolla, California 92039, USA.

Prof. Dr. H. VELTHUIS: Laboratorium voor vergleichende Fysiologie der Rijksuniversiteit Utrecht, Jan Van Galenstr. 40, 3572 La Utrecht, Netherlands.

Prof. Dr. CHR. VOGEL: Lehrstuhl f. Anthropologie der Universität Göttingen, Bürgerstr. 50, 3400 Göttingen, FRG.

Prof. Dr. R. WEHNER: Zoologisches Institut der Universität, Winterthurer Str. 190 Ch-8006 Zürich, Switzerland.

Prof. Dr. E. O. WILSON: Harvard University, Museum of Comparative Zoology, Cambridge, Mass. 02138, USA.

Fortschritte der Zoologie, Bd. 31 · Hölldobler/Lindauer (Hrsg.): Experimental Behavioral Ecology
G. Fischer Verlag · Stuttgart · New York · 1985

Karl von Frisch and the beginning of experimental behavioral ecology

Bert Hölldobler

Department of Organismic and Evolutionary Biology, Harvard University, Cambridge, MA 02 138, USA

Experimental behavioral ecology can be said to have begun around 1911, when the young Karl von Frisch wondered why flowers are colorful. Like other biologists previously, he supposed that color attracts bees and other pollinators. However, this assumption was starkly at variance with the results obtained by the famous professor for ophthalmology Carl von Hess, who by rigorously controlled laboratory experiments appeared to have demonstrated that honeybees as well as other invertebrates and fish, cannot perceive color. Von Hess placed honeybee workers into experimental chambers, where environmental factors could be controlled, and presented them with two light spots simultaneously, each of a different color and varying light intensity. In this situation the bees were invariably attracted to the brightest spot, whatever the color was, which seemed to prove that bees are color blind and react only to different light intensities.

Karl von Frisch reasoned that bees motivated to find a way out of a dark chamber obviously do not care for color, but aim for the brightes light spot which might indicate the exit hole. The question of whether or not bees perceive color has to be put to them when they are motivated to react to different colors, and this would be during foraging, when they visit flowers to collect nectar and pollen.

For the first time K. v. Frisch introduced the powerful method of training animals by «associative learning» in the study of sensory physiology. In a series of ingeniously simple experiments, he let the bees remain free in their natural environment, and they gave him the correct answer: yes, they can perceive colors and readily associate color with rewards during foraging. These discoveries not only became an important milestone in the field of comparative sensory physiology, they also revealed for the first time the behavioral mechanisms by which the individual foraging bees maintain their remarkable fidelity to particular flower species, a crucial necessity for achieving effective pollination of the plants. Foraging bees associate a reward (nectar or pollen) with a whole array of stimuli (color, shape, scent and others) presented by the flower, and are thereby conditioned to visit flowers of the same kind repeatedly.

This is experimental behavioral ecology at its best! Behavioral ecology and sociobiology comprise the study of the evolutionary adaptiveness of behavior in general, including the most complicated social interactions. Any difference in structure, physiological process, behavioral pattern, or complex of traits that increase the inclusive fitness of one organism as compared to another organism of the same species is an evolutionary adaptation. But since the effect of specific traits on the inclusive fitness (or reproductive success) of organisms is very difficult if not impossible to measure directly, evolutionary adaptations are usually inferred from the organisms' good fit to their environment. It is one of the major tasks of experimental behavioral ecology to investigate and critically evaluate this particular correlation.

Niko Tinbergen, the great experimental behavioral ecologist, spoke of the «survival value» while Karl von Frisch referred to the «biological significance» or «function» of traits that are subject to experimental analysis. In 1971, on the occasion of the reprinting of his original papers, Tinbergen said that it is «my growing conviction that such ‹survival value› studies, if they are to outgrow the status of speculation about correlations, require experimentation». He continued, «Karl von Frisch's work had shown me the power of simple experiments in as natural conditions as possible». We know how imaginatively and effectively Niko Tinbergen has employed experiments to understand behavioral adaptations.

An excellent naturalist from childhood, Karl von Frisch believed that phenomena such as the colors and scents of flowers, or the Weberian ossicles of catfish must have an adaptive significance and he endeavoured to find it, not by telling adaptive stories, as some characterize the work of behavioral ecologists, but by ingenious experimentation. He consistently challenged the anthropocentric bias that animals have only a very limited sensory capacity. We learned from him that each species has its own unique sensory world, often vastly different from ours. Bees cannot see red, which we can, but perceive UV as color, which we do not.

I think James Gould recently expressed this contribution of Karl von Frisch extremely well when he said: «His pioneering work inspired the discovery of several otherwise unimaginable sensory systems in animals: infrared detectors in night-hunting snakes, ultrasonic sonar in dolphins and bats, infrasonic hearing in birds, and magnetic field sensitivity in a variety of animals. Doubtless, other systems are still to be discovered. The lesson is a melancholy one: We are blind to our own blindness, and must not try to read our own disabilities into the rest of animal kingdom«.

Karl von Frisch's discoveries laid the roots for many now flourishing scientific disciplines in zoology. Modern comparative sensory physiology has been built on his pioneering work. The discovery of an internal clock and sun compass in bees (at the same time independently discovered by Gustav Kramer in birds), with their supplementary employment of the polarized sky light for navigation, has led to alsmost uncountable studies of animal orientation in time and space. His work and that of his students on perceptual preferences in honeybee learning has foreshadowed the current intensive investigations of species differences in perceptual predispositions and their ecological significance. His discovery of a «Schreckstoff» in fish was the first demonstration of an alarm pheromone. Furthermore, von Frisch elaborated extensively on the role of environmental odors and chemical signals in the honeybees' orientation and communication during foraging long before «chemical ecology» became an established field. Although these latter experiments indicated that environmental odors and pheromones play a role in attracting honeybee newcomers to a rich food source, a series of controlled experiments suggested that the frequently observed massive recruitment response must be due to other, much more effective communication mechanisms. This finally led Karl von Frisch to the ingenious decoding of the dance language of the honeybees. In the view of most biologists this remains the single most fascinating discovery in the study of animal communiation.

How could such a sophisticated communication behavior have evolved in the first place? In a long sequence of comparative investigations, von Frisch and Martin Lindauer and their associates searched for more elementary forms of communication in other bee races and species, including the stingless bees, and they concluded from their findings that the waggle dance communication behavior in honeybees is in fact a miniature and highly ritualized replica of the flight to the target area. Simple motor displays, mechanical signals, and chemical cues of some stingless bees presumably represent the more primitive mechanisms from which the waggle dance originated.

KARL VON FRISCH had a broad interest in animals. He began his scientific career investigating the ability of fish to adapt to the color of their background, and he demonstrated color vision in fish. And in later years he always returned to the fish, especially during the winter months, when the bees were not active. He demonstrated that fish can hear, provided the first experimental support for the rod-cone theory of color vision, discovered chemical communication in fish, and more. Many of his students extended this work and laid the foundations for experimental behavioral research conducted today.

But KARL VON FRISCH's true scientific love must have been the honeybees, which he once compared to a magic well: «The more you draw from it, the more there is to draw«. And this is still true today. Four academic generations of researches, all represented here today, have succeeded KARL VON FRISCH. The life of the honeybees continues to be a rich source of new scientific discoveries and an inspiration even for those of us who chose to study other organisms.

I would like to close with the first sentence of a laudatio HANS JOCHEM AUTRUM wrote on the occasion of KARL VON FRISCH's 90th birthday: «‹Simplicity is always a sign not only of truth, but of genius›. Anyone who has ever read one of KARL VON FRISCH's works, or heard him lecture or talked with him, can verify these words of SCHOPENHAUER».

Fortschritte der Zoologie, Bd. 31 · Hölldobler/Lindauer (Hrsg.): Experimental Behavioral Ecology
G. Fischer Verlag · Stuttgart · New York · 1985

Personal recollections of Karl von Frisch

MARTIN LINDAUER

Zoologisches Institut der Universität Würzburg, F. R. Germany

Many of us had the chance to meet KARL VON FRISCH in his long life, which was enriched not only by an extraordinary scientific career but also by profound personal friendship. At the beginning of this symposium let me ask the question: «What was the real basis and background for his pioneer work in sensory physiology, behavioral ecology and sociobiology?»

In my personal recollections (from 1944 – 1982) I find 3 principles which may answer this question.

1. K. v. FRISCH was inspired by a great *love of nature* and of the living organism. It was not a purely emotional love; its real origin was a *profound knowledge* of biology, morphology and physiology of the animal kingdom, something which nowadays has become a rare value. As a teacher K. v. FRISCH had delivered in Munich and in Graz two lecture series in a masterly survey: «Allgemeine Zoologie» and «Vergleichende Anatomie der Wirbeltiere». None of his students will forget his fascinating demonstrations on the function of the fish swimbladder, of the cow-stomach etc. Teaching demands careful preparation, but the sympathy for the animals you have to deal with grows with the knowledge acquired.

K. v. FRISCH's comprehensive knowledge of taxonomy, biology and ecology is demonstrated in an unique way by his *Museum in Brunnwinkl,* a private collection of the local fauna. It was started in 1910, and it has become a historical document, particularly in view of the fact that many species of butterflies, beetles, solitary bees, amphibia etc. have become extinct.

In his love of nature K. v. FRISCH also enthusiastically supported the idea of «Naturschutz», the preservation of natural environment:

If a visitor comes to Brunnwinkl these days and can still drink in front of the Mühlhaus the clear cool water from the private spring, if he still can walk the romantic «Führberger Weg» along the Wolfgangsee, then this is only possible because K. v. FRISCH was the advocate who fought against the project of a highway on the east side of the Wolfgangsee.

His *love of bees* is evident to all who have read his books or who assisted him in his so-called simple experiments in free nature (Fig. 1).

I will never forget that day in October, when the weather turned cold and the last forager bee on the feeding table 300 m away from the hive became stiff, so that she could not fly off:

K. v. FRISCH warmed her up in the hollow of his hand and took her back into the hive.

Fig. 1: KARL v. FRISCH marking bees on the feeding table. In this quiet countryside of Brunnwinkl – preserved as natural environment – the most exciting discoveries were made.

2. Next to his love for animals was his overwhelming *enthusiasm for research*, an enthusiasm that has created many new ideas. K. v. FRISCH undertook his research not in

the hope of acquiring awards, but in order to elucidate and analyse the many mysteries of animal life. This enthusiasm was inspired by a powerful impetus to find the *truth*.

And as far as this was concerned he was inconsiderate towards himself, his co-workers and his students. «Die Wissenschaft ist eine strenge Göttin, die nichts neben sich duldet» he said.

In his attempts to find the truth in all details K. v. FRISCH always had a weakness for *little things*, for unexpected results, for so called *«singularities»*, which nowadays receive increasing attention in biophysics, in biochemistry and in biology.

«Nature never lies» he said and he always impressed on his students: «auch kleine Dinge muß man in der Biologie ernst nehmen» (in biology you also have to consider little things seriously).

Our publications «Himmel und Erde in Konkurrenz bei der Orientierung der Bienen» or «Richtungsweisung der Bienen bei Seitenwind» etc. originated from so called «unimportant observations». And the «Missweisung» in the beedance – small deviations in direction indication – was the starting point for his pioneer work concerning orientation by polarized light; last but not least the «Restmissweisung» gave me and my co-worker H. MARTIN the first inkling that the earth's magnetic field may be involved in the orientation of bees.

A personal comment may be added here:

The enthusiasm of K. v. FRISCH for research was not limited to his own work and his own discoveries. I remember when he reported to us in Graz immediately after World War II on GRIFFIN's work on echolocation in bats, or Pardi's sun compass orientation in *Talitrus* and later in Munich, when he reported MICHENER's manifold work on solitary bees, or WILSON's and HÖLLDOBLER's fine works on ants: he was as happy as if he had done all this research himself.

It was the same with BALTZER's, HADORN's, SCHARRER's, MÖHRES's work; and it was one of the high points in my life when I had given him the very first report on the debate of the scouts when they announce a new nesting site in the swarm cluster. With a smile on his face he said: «Congratulations! You have witnessed an ideal parliamentary debate; your bees can evidently change their decision when other scouts have to announce a better nesting site.»

3. A third principle that made the research work of K. v. FRISCH so successful was his unshakeable *optimism*.

When you start an experiment with bees you first have to look at the sky to see whether the sun will be shining that day. Even during the ill-famed «Salzburger Schnürlregen» K. v. FRISCH always saw a small bright spot somewhere between the dense clouds; «let's try again» he said, and we had to climb up the Schafberg. This went on day after day; even if it was raining and misty at the time; however (after 2 weeks) we were able to finish our experiments on the top of the mountain in bright sunshine. This incident is expressive of his typical optimism, which continued to prevail even when his hard work was not rewarded by clear or expected results. For example in a series of protracted experiments we wanted to find out immediately whether bees have included special information for a goal *up* and *down* in their dance. What was the result? «There is no vocabulary in the bee dance for up and down – which means – in an objective biological sense – it is of no use – the language of the bees is free of gossip.»

Let us cherish his love of nature
his enthusiasm for research
his optimism in all situations
as the legacy of a great biologist.

I. Orientation, Learning, Foraging

Fortschritte der Zoologie, Bd. 31 · Hölldobler/Lindauer (Hrsg.): Experimental Behavioral Ecology
G. Fischer Verlag · Stuttgart · New York · 1985

The bee's celestial compass – A case study in behavioural neurobiology

RÜDIGER WEHNER and SAMUEL ROSSEL

Department of Zoology, University of Zürich, Switzerland

Abstract

The scattering of sunlight within the earth's atmosphere produces an extensive pattern of polarization (Figs. 1 and 2) as well as coarse spectral gradients.

In 1949 KARL VON FRISCH showed that bees can use the celestial pattern of polarization as a compass, but how they manage to accomplish this task has remained elusive. In theory, there are a number of geometrical and non-geometrical ways of computing the exact position of the sun from single patches of polarized light in the sky. As shown in this paper, bees do not seem to resort to any of these methods. Instead, they use a very simple strategy in assuming that any particular e-vector orientation occurs at a fixed azimuthal position with respect to the solar and anti-solar meridian. In the real sky, these e-vector positions vary with the elevation of both the sun and the point observed, but in the bee's internal model of the sky they do not (Figs. 6, 7 a and 10). Thus, orientation errors must occur whenever e-vector positions in the sky do not coincide with the e-vector positions predicted by the bee's e-vector compass. It is on the basis of these orientation errors exhibited by the bees when viewing a single e-vector in the natural sky or a spot of artificially polarized light that the bee's e-vector compass has been derived (see e. g. Figs. 4 and 5).

Even when large parts of the natural blue sky are available, bees make mistakes exactly as large as predicted from their simple e-vector compass (Fig. 12). The prediction implies that the orientation error induced by any pattern of e-vectors is the mean of the orientation errors induced by individual «pixels» of sky within each of which the e-vector is approximately constant. Each e-vector is equally important, irrespective of its degree of polarization. Orientation errors do not occur when the patch of sky is positioned symmetrically with respect to the anti-solar meridian. Then the errors induced by the left and right halves of the pattern are equal in magnitude, but opposite in sign, and thus cancel each other out.

The bee's e-vector compass can be derived not only experimentally (from the bees) but also theoretically (from the sky) by referring exclusively to the most prominent feature of skylight polarization, the arc of maximum polarization. As it turns out, the bee's e-vector compass mimics the mean distribution of the maximally polarized e-vectors as they occur in the upper part of the sky (Figs. 7 c and 10 b).

Where does the e-vector compass reside within the bee's visual system? A crucial step in answering this question is our discovery that the ultraviolet receptors of a set of specialized ommatidia located at the dorsal margin of the bee's eye are necessary for the detection of polarized skylight. Over the entire region of this «POL area» the microvillar axes of the ultraviolet receptors vary in a fan-shaped array that resembles the distribution of the maximally polarized e-vectors in the sky. Thus, the geometry of the bee's e-vector compass as derived by behavioural experiments seems to reside in the utmost periphery of the bee's visual system – in the spatial arrangement of the retinal analysers (ultraviolet receptors). In addition to the fan-shaped array of the retinal analysers, the POL area exhibits a number of anatomical and physiological specialisations which can be expected to have functional consequences for the detection of polarized skylight.

How is the POL area used in deriving compass information from the celestial e-vector pattern? On the basis of both the neurophysiological and behavioural evidence mentioned above we propose the following model (Fig. 16): In the bee's eye the array of retinal analysers mimics the array of maximally polarized e-vectors in the sky. Equipped with such an array of analysers the bee can use a very simple strategy in determining the position of the solar or anti-solar meridian from any particular e-vector in the sky. If the bee scans the sky, i. e. rotates about its vertical body axis and thus sweeps its array of analysers across the celestial e-vector pattern, peak responses will occur whenever the bee faces the solar or anti-solar meridian. (There is a simple rule by which the bee can distinguish between these two possibilities.) To be exact, the former statement only holds when the e-vector observed is in its state of maximum polarization, because the design of the receptor array is based on the array of maximally polarized e-vectors in the sky. For other e-vectors peak responses will occur when the bee is not aligned exactly with the solar (or anti-solar) meridian. What then results are the orientation errors we have observed.

An important consequence of our model is that bees can derive compass information only from e-vector patterns that are designed according to the rules of Rayleigh scattering, because the retinal array of analysers is designed according to the celestial array of e-vectors. Furthermore, bees do not perceive polarized skylight as an extra quality of light. While scanning the sky, they translate e-vector orientations into modulations of perceived ultraviolet intensity. When peak responses occur, the bees are aligned with the solar or anti-solar meridian.

Bees use spectral information in addition to e-vector information when determining the position of the solar (or anti-solar) meridian. They interpret small unpolarized green and ultraviolet stimuli as lying along the solar meridian and within the anti-solar half of the sky, respectively (Figs. 13 and 14). The non-POL parts of the bee's dorsal retina are sufficient for the orientation by means of spectral cues: Bees in which the POL areas of both eyes have been painted out are able to derive compass information from unpolarized green and ultraviolet stimuli exactly as the untreated control bees do. When they view a polarized spot of ultraviolet light, they interpret it as if it were an unpolarized spot (Fig. 15). Furthermore, when large parts of the sky are available, they orient nearly as precisely (by referring to spectral cues) as the untreated controls do which can use the celestial e-vector pattern in addition. Possibly the polarization-sensitive ultraviolet channels interact with polarization-insensitive channels somewhere along the visual pathway.

The hasty reader is invited to take the figures and their legends as a short story of our hypotheses, results and conclusions.

1 Introduction: History and perspectives

No event did more to establish the fame and prestige of KARL VON FRISCH's work in the field of sensory physiology than his discovery that bees can perceive the polarized light in the sky and use it as a celestial compass. Actually, two decades earlier the myrmecologist FELIX SANTSCHI had already observed that ants could find their way even when they could see nothing but a small patch of blue sky. In an interesting but unfortunately neglected work almost forgotten in the Mémoires de la Société Vaudoise des Sciences Naturelles, SANTSCHI (1923) literally asked the question «What is it in this small patch of sky that guides the ants back home?» SANTSCHI, practising as a physician in North Africa, could not tell. In one experiment he had even used a ground glass disk (which depolarized the light from the sky) and put it above a homing ant. The ant instantly stopped and searched around at random, but SANTSCHI did not draw the right inferences from this important observation. Even if he had, at the time of his writing it would have been exceedingly difficult to perform behavioural experiments involving polarized light, simply because large polarizers such as sheets of polaroid filters were not yet available[1]. After a quarter of a century had passed, in 1947, VON FRISCH did an experiment with bees almost identical to the one SANTSCHI had performed with ants. He got the same result, asked the same question, and – *horribile dictu* – could not tell either. However, VON FRISCH, then Head of the Department of Zoology at the University of Graz, was in a better position to answer such questions than SANTSCHI had ever been. At one of the next Faculty Meetings he told the story to his colleague of the Physics Department, HANS BENNDORF. The physicist advised him to check for polarized light. Next summer, in 1948, VON FRISCH did the crucial experiment: He placed a polarizer above a bee which performed its recruitment dances on a horizontal comb, and as he rotated the polarizer, the direction of the bee's dances changed correspondingly (VON FRISCH, 1949). This was the first demonstration that an animal used skylight polarization for adjusting the direction of its course.

With this stirring discovery VON FRISCH has left to us two fascinating questions: How does the insect's polarization compass work, and what does it mean to the insect? These are the questions of neurobiology and behavioural ecology, respectively. Even though both questions call for quite different lines of experimentation, they are closely intertwined. In evolutionary terms, it is the role the celestial compass plays in the insect's overall system of navigation that has shaped the insect's visual system by deciding which information the insect's nervous system must extract from the sky, and which it can discard. In this chapter, we shall concentrate on the first question, but the answer we shall provide on how bees use polarized skylight will make full sense only when discussed in the context of other visual capacities and navigational strategies as well.

Are we to conclude from SANTSCHI's und VON FRISCH's early findings that insects can infer the position of the sun correctly from any particular patch of sky? Or in other words, does a bee place, in its mind, any patch of polarized light in the correct celestial position with respect to the sun? This would imply that the bee is informed exactly about all

---

[1] Nicol prisms were used by the first investigators who studied the responses of animals towards polarized light (CROZIER and MANGELSDORF, 1924; VERKHOVSKAYA, 1940). In these experiments a number of arthropods such as daphnids, blowfly and beetle larvae as well as drosophilid flies had to make phototactic choices between a polarized and an unpolarized beam of light with both beams adjusted for equal total intensity. The first study yielded negative, the second positive results.

geometrical properties of the extensive pattern of celestial polarization as well as about how this pattern varies during the course of the day. It has been assumed by VON FRISCH and many later workers in this field[2] that these questions can be answered in the affirmative, but the evidence for this conjecture is rather slim.

We tried to take a fresh look at the whole problem by starting with a series of rather straightforward experiments. Under the full blue sky bees were trained to a given compass course leading them to an artificial food source. Immediately thereafter they had to select this compass course solely by means of a small patch of blue sky. As was already apparent from these very first experiments, bees made mistakes in all but very few stimulus conditions. If this had not been surprising enough, the situation became more astounding when we presented the bees with large patches of sky. Even then consistent navigational errors occurred. Obviously, the bees were not informed exactly about where any particular patch of polarized skylight was positioned with respect to the sun. This was perplexing at first, but led us finally to the formulation of an extremely simple hypothesis of how bees could use skylight patterns as a compass. This hypothesis as first proposed by behavioural work was later substantiated by neuroanatomical and neurophysiological studies of the bee's peripheral visual system. According to this hypothesis bees do not take advantage of all visual information present in the sky, but resort to some principal features of their celestial world. Even though the skylight patterns are indeed rather complex, the bee's strategy is simple.

When first reported (ROSSEL, WEHNER and LINDAUER, 1978), our observations and conclusions were met with considerable doubt, and our hypothesis that the bee's model of the sky was a simplified rather than correct copy of the outside world, was treated with disbelief (WATERMAN, 1981, 1984; DYER and GOULD, 1983). In the meantime, a host of additional experiments has led to a confirmation of this apparently preposterous idea. We shall present in this article our principal conclusions and most of the critical evidence. Detailed documentation has already been provided (ROSSEL and WEHNER, 1982, 1984; WEHNER and ROSSEL, 1983; WEHNER, 1982, 1983) or will be published elsewhere (ROSSEL and WEHNER, in press; WEHNER and STRASSER, in prep.).

The following is organized so as to let the reader experience step by step how we arrived at our present understanding of the bee's celestial compass. We start with a short inquiry about the skylight cues used by the bees (Section 2). Then we derive the bee's celestial compass from behavioural experiments (Section 3). After the experimental techniques have been introduced (Section 3.1.1), the basic one-spot experiment, and its astonishing result, is described in detail (Section 3.1.2): Bees make mistakes. It is from these mistakes that we develop the hypothesis of the bee's stereotyped e-vector compass (Section 3.1.3). The spatial design of the bee's e-vector compass is derived first from the bees (by analyzing their navigational errors: Section 3.1.3., first part) and then from the sky (by referring to the most prominent feature of skylight polarization: Section 3.1.3., second part). Our hypothesis is later tested, and vindicated, by presenting the bees with more than one pixel of sky, artificial beams of polarized light, or large parts of the natural blue sky (Section 3.1.4). Finally, spectral information used by the bees alongside e-vector information is incorporated into the bee's celestial compass (Section 3.2). A discussion comparing our data with those of other authors (Section 3.3) rounds off the behavioural analysis of the bee's celestial compass. Next comes the crucial question: How does the bee's celestial compass work physiologically? Section 4 deals with this question. After a brief survey of

---

[2] One of us has recently summarized many aspects of skylight navigation in insects (WEHNER, 1982, 1984). The reader is referred to the extensive lists of references provided by these summaries.

some classical views (Section 4.1), a new approach is introduced (Section 4.2). Drawing upon neurobiological data (Section 4.2.1) and behavioural evidence (Section 4.2.2), a model is proposed according to which the bee adopts a global scanning strategy to determine the position of the solar and anti-solar meridian (Section 4.2.3). We end by inquiring about some of the more general implications of our results and conclusions (Section 5).

2 Celestial patterns: What skylight cues are used by the bees?

(i) *The scattering of sunlight within the earth's atmosphere creates e-vector patterns and spectral gradients.*

As *skylight polarization* arises from the scattering of sunlight by the air molecules within the earth's atmosphere (Fig. 1), the whole pattern of polarization is fixed relative to the sun. From the way it is portrayed in Fig. 2 a its main features are immediately apparent:

First, the *e-vector orientations (angles of polarization)*[3] form concentric circles around the sun. This simply follows from the laws of Rayleigh (primary) scattering, according to which light is always polarized perpendicularly to the plane of the scattering angle, i. e. the plane containing the sun, the observer, and the point observed (Fig. 1 a).

Second, the *degree of polarization* (per cent of polarized light relative to the total amount of light emanating from a given point in the sky) increases from the sun, where it is zero, to an angular distance of 90° from the sun. From there, it decreases towards the anti-solar point, where it is zero again (Fig. 2 a). In more detail, the laws of Rayleigh scattering imply that the degree of polarization is proportional to some sine function of the scattering angle, i. e. the angle which the earthbound observer forms with the direction of the sun and the celestial point observed (Fig. 1 a). Thus, direct light from the sun is unpolarized, while light emanating from celestial points that lie at an angular distance of 90° from the sun, are polarized maximally. As a consequence, the great circle[4] positioned 90° off the sun forms what can be called the *arc of maximum polarization*. During the course of the day, when the sun is above the horizon (see Fig. 1 a), this arc of maximum polarization passes across the anti-solar part of the sky. Generally speaking, a highly polarized (anti-solar) half of the sky can be distinguished from a less polarized (solar) half of the sky.

Third, there is a global *plane of symmetry* passing across the entire celestial hemisphere. It contains the solar and anti-solar meridian. The e-vector orientations at corresponding points on both sides of the solar and anti-solar meridian are mirror images of each other.

---

[3] Light can be described as a transverse electromagnetic wave characterized by a magnetic component (h-vector) and an electric component (e-vector). These vectors are perpendicular to each other and to the direction of propagation. It is the electric component that interacts with the electrons of atoms and molecules and thus is involved in optical phenomena such as the scattering of light. In a beam of unpolarized light, there is no preferred plane within which the e-vector vibrates, but in a beam of linearly polarized light, the e-vector vibrates in a fixed plane. The orientation of this plane with respect to the vertical is called the angle of polarization (or e-vector orientation, denoted by χ). For conventions see Figs. 1 b and 2 b.

[4] Great circles are defined as the largest possible circles which can be drawn on the surface of a sphere. The radius of a great circle is identical with the radius of the sphere. This implies that in the celestial sphere the plane formed by any great circle always includes the observer in the centre of the sphere.

of each other. Along this arc of symmetry, and nowhere else in the sky, the e-vectors are oriented parallel to the horizon.

Having described the static picture, let us now set it in motion. During the course of the day, the sun moves across the sky, and the e-vector pattern moves along with it. Of course, as mentioned earlier, the pattern remains fixed within a sun-related system of coordinates,

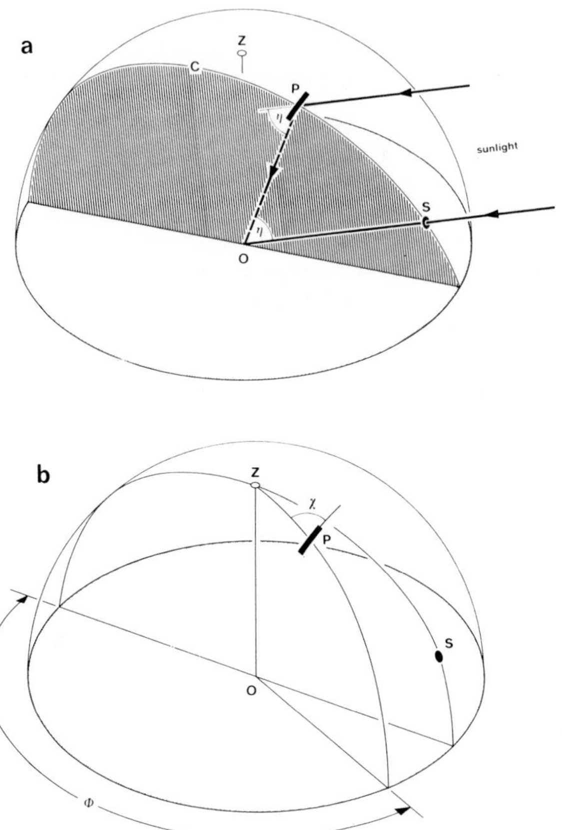

Fig. 1: Skylight optics: The scattering angle (η) and the angle of polarization (e-vector orientation, χ).
a An observer (O) receives direct sunlight from *S* (the apparent position of the sun in the celestial hemisphere) and scattered light from the sky as shown here for the celestial point *P*. Scattered skylight is partially linearly polarized with the e-vector oriented perpendicularly to the plane of the scattering angle η. This plane *(hatched)* includes the direct rays from the sun (*heavy solid lines*) and the observer's line of sight (*heavy dashed line*). The circle C in which the scattering plane intersects the celestial sphere is a meridian in a sun-related system of coordinates, in which the sun and the anti-solar point form the poles of the sphere (see Fig. 2 a). In the more conventional horizon-system of coordinates the zenith (*Z*) is the pole of the celestial hemisphere.
b The orientation of the e-vector (χ) is defined as the angle the e-vector forms with the vertical (see also Fig. 2 b). Throughout this paper the following convention is used: positive values of χ refer to angles measured clockwise from the vertical (as seen by the observer in the centre of the celestial hemisphere). Φ denotes the azimuth position of *P* relative to the anti-solar meridian.

but due to the sun's change in elevation, it changes within the horizon-system of coordinates. Just look at what happens to the arc of maximum polarization. As the sun moves up along the solar meridian, this arc tilts down across the anti-solar half of the sky. Concomitantly, all e-vectors change their positions with respect to the axis of symmetry of the pattern. A special situation exists at sunrise and sunset, when the sun is on the horizon. Then, the arc of maximum polarization passes through the zenith and forms a second axis of symmetry in the sky.[5]

A patch of sky is characterized not only by its state of polarization, but also by its spectral composition. Both parameters depend largely on the angular distance from the sun. Thus, the celestial straylight patterns provide *spectral gradients* alongside polarization gradients. As the radiant intensity of scattered light is inversely proportional to the fourth power of the wavelength, the relative intensity of short-wavelength (e.g. ultraviolet) radiation is higher in scattered skylight than in direct sunlight. More specifically, those parts of the sky that exhibit maximum polarization also exhibit the highest relative content of short-wavelength radiation. To sum up, all characteristics of sky radiation – intensity (radiance), colour, degree of polarization, and angle of polarization (e-vector orientation) – depend largely on the angular distance from the sun and thus can be used as cues for inferring the position of the sun from isolated patches of sky. If bees were able to detect and process all these parameters independently, they could determine the position of the sun from the skylight information available in a single patch of sky[6]. However, before we inquire about how this could be done in theory, and what the bee does in practice, let us have a look at the real sky.

(ii) *The real world is not so neat and tidy.*

All information presented above about skylight patterns applies to the simple case of the primary scattering of light within an ideal atmosphere. In such an atmosphere, the incoming unpolarized sunlight is scattered by molecules that are smaller than the wavelength of light, isotropic, and of low concentration, so that sunlight is scattered only once before it reaches the eye of the observer. In reality, however, these conditions are rarely met. Instead, the ideal Rayleigh pattern gets distorted by various kinds of pollution, haze or clouds as well as by reflections from the ground. Factors such as higher-order scattering, reflection, absorption and molecular anisotropy markedly affect the polarization and spectral gradients of the sky. What results is a celestial pattern in which the sun and the pattern of e-vector orientations provide the most reliable cues, whereas the degree of polarization, the radiant intensity and the spectral gradients often become subject to dramatic disturbances and variations. This is clearly borne out by the skylight measurements performed by Brines and Gould (1982) under a variety of atmospheric conditions.

---

[5] In a horizon system of coordinates, the pattern of polarization rotates about the zenith, due to the westward movement of the sun. Note, however, that the whole pattern does not merely rotate. Due to the sun's change in elevation mentioned above the pattern changes its intrinsic properties as it rotates about the zenith.

[6] This holds for all but one class of exceptions: At each elevation above the horizon there are, in general, two optically equivalent points in the sky, which coincide in all aspects of skylight radiation and thus are indistinguishable to any detector system. These are the points characterized by vertical e-vectors (see Figs. 2 c and 2 d).

(iii) *Bees cope amazingly well with the unreliabilities of their celestial world.*

If bees cannot see the sun, the most important skylight parameter on which they rely is the e-vector orientation. The degree of polarization is not used as a compass cue (VON FRISCH, 1965: 397; ZOLOTOV and FRANTSEVICH, 1973). It is important merely in the sense

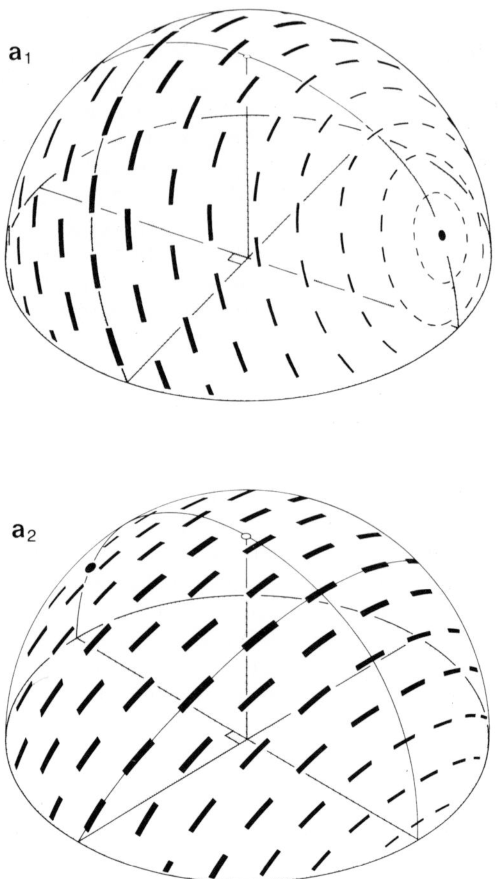

Fig. 2: The pattern of polarized light in the sky (e-vector pattern).
a Three dimensional representation of the celestial e-vector pattern. The two hemispheres are rotated relative to each other so that the reader faces either the solar (a_1) or the anti-solar meridian (a_2). The sun is shown as a black disk positioned at an elevation of 24° above the horizon. The electric vectors, or e-vectors, of light (*black bars*) form concentric circles around the sun. The e-vector pattern is further characterized by a global gradient of the degree (percentage) of polarization as indicated by the varying widths of the black bars. The direct light from the sun is unpolarized. Maximum polarization occurs at an angular distance of 90° from the sun. The arc of maximum polarization is depicted by the semicircle combining the widest bars. The other semicircle shown in either figure represents the symmetry plane of the e-vector pattern. It comprises the solar meridian and the anti-solar meridian. The zenith is marked by an *open circle*.

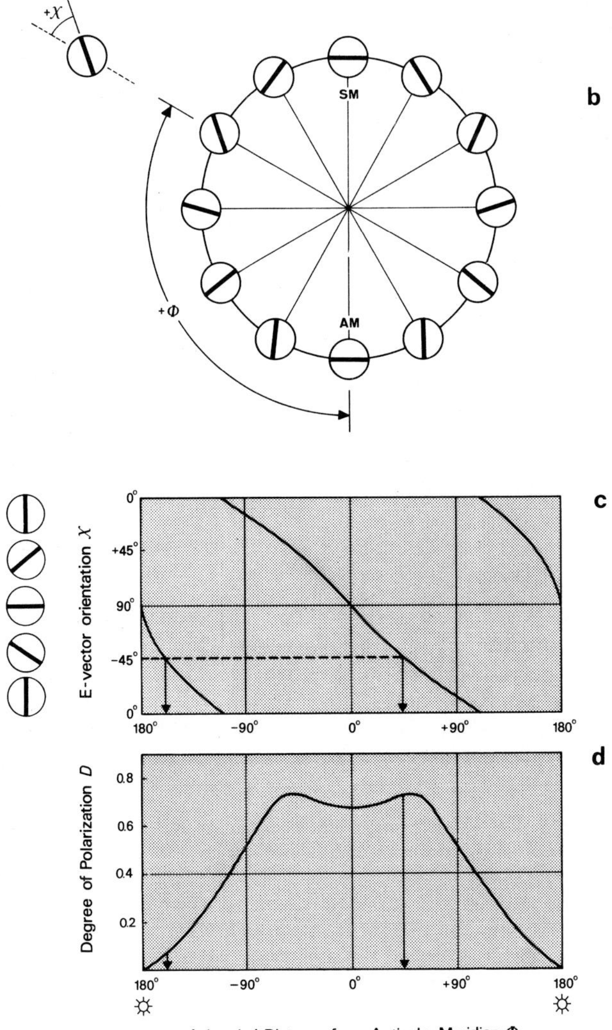

b E-vector orientations as they occur at an elevation of 60° when the sun is at an elevation of 24° above the horizon. To use e-vector patterns as a compass means to associate e-vector orientations (χ) with their corresponding azimuth positions (Φ, φ). Note definitions and sign conventions for χ and φ. *AM* anti-solar meridian, *SM* solar meridian. The correlation between χ and φ varies with both the parallel of altitude (here 60°) and the elevation of the sun (here 24°). It is a long-held view that bees are informed about all possible χ/φ correlations, so that they can assign the correct azimuth position to any particular e-vector occurring in the sky. This paper shows otherwise. Bees use a stereotyped rule of correlating χ and φ, no matter at what elevation and at what time of day an e-vector is presented.

c Correlation between e-vector orientation (χ) and azimuth direction (Φ, φ) as outlined schematically in Fig. 2 b. In the following this correlation is called the χ/φ function. As in Fig. 2 b it is depicted here for an elevation of 60° above the horizon (elevation of sun 24°). Note that each e-vector orientation that it must exceed a certain minimum value (about 0.1), before the bee is able to detect the orientation of an e-vector. Spectral cues are used to some extent (VAN DER GLAS, 1976, 1980), at least to discriminate between sun and sky (EDRICH, NEUMEYER and VON HELVERSEN, 1979; BRINES and GOULD, 1979). Direct sunlight contains less than 10 per cent of ultraviolet radiation, while in scattered skylight the relative amount of ultraviolet radiation is larger by a factor of up to four. Bees use this difference in the relative content of ultraviolet radiation to make the distinction between sun and sky. In this context, it is interesting to note that within the bee's visual system information about skylight polarization is mediated exclusively by the ultraviolet receptors[7]. Recall that those parts of the (anti-solar) sky that exhibit the most saturated ultraviolet tinge are also the ones that exhibit maximum polarization. Apparently, this important physical property of skylight patterns has been incorporated into the bee's visual system. In this context, it is attractive to assume that bees are able to distinguish between sun and sky by exploiting spectral gradients within the celestial hemisphere, and that they make finer spatial distinctions by using polarization gradients within the anti-solar half of the sky. We shall take up this idea after we have presented our experimental data. At present, it suffices to mention that the bee's celestial compass is based on the most reliable sources of celestial information – e-vector orientation and hue of colour.

(iv) *To understand the bee's celestial compass a less conventional model than hitherto suggested is required.*

Now that we have sketched out the skylight patterns, and discussed the skylight cues on which bees rely, let us again ask the crucial question: How do bees use these patterns as a compass? To what geometrical relationship between various parts of these patterns do they refer? This is the main topic of this paper. In more operational terms, the question can be phrased as follows: How do bees deduce the azimuthal position of the solar or anti-solar meridian from the information provided, for example, by a single patch of sky?[8] If

(e. g. χ = -45°, *dashed line*) is represented at two azimuth positions: one is closer to the solar meridian, the other closer to the anti-solar meridian.

d Correlation between degree (percentage) of polarization (D) and azimuth direction (Φ, φ), shown here for ultraviolet light. Note that any two points in the sky that are identical in e-vector orientation vary in degree of polarization. The member of a pair of identical e-vectors that is positioned farther from the sun is polarized more strongly than the member positioned closer to the sun. This difference holds for all but the vertical e-vectors (χ = 0°). As experimental evidence shows that bees do not rely on the degree of polarization, the intriguing question arises as to how they resolve the ambiguity inherent in the χ/φ function. The following figures provide the answer.

---

[7] Almost twenty years ago, VON FRISCH (1965: 408) showed that the e-vector orientation of bees depended upon the wavelength of light. He concluded from his experiments that only short-wavelength radiation (ultraviolet and blue lieght) was used in the detection of polarized skylight. More detailed studies later refined this conclusion (DUELLI and WEHNER, 1973; VON HELVERSEN and EDRICH, 1974): The ultraviolet receptors alone are necessary and sufficient for mediating e-vector orientation in both bees and ants. This is not to say, however, that the insect's celestial compass relies exclusively on ultraviolet input channels. As we shall see later, spectral cues are used as well, and the processing of such information requires some kind of interaction between ultraviolet and other spectral channels.

[8] There is a second problem not treated in this paper, the problem of time-compensation. As the azimuthal position of the solar and anti-solar meridian changes during the course of the day, the celestial compass must first be calibrated against a non-celestial system of references. For recent data and hypotheses see GOULD (1980) and WEHNER and LANFRANCONI (1981).

you stop and think about it, this is a rather tricky problem. Bees could remember the e-vector pattern last seen, and later try to match the memorized celestial image with whatever they experience in the sky. In this case it is not necessary to assume that bees know anything general about how these patterns are designed. They could form memory images of arbitrary celestial patterns just as they have been shown to form memory images of landmark panoramas which are unpredictable from one location to the other (WEHNER, 1981, 1983; CARTWRIGHT and COLLETT, 1983). Another possibility is that bees are informed about the spatial design of the celestial e-vector patterns. They could know, for example, that the e-vectors form concentric circles around the sun, or in other words: that the sun is positioned on the great circle oriented perpendicularly to the e-vector present in any given patch of sky. In following this rule, they could determine the orientations of at least two e-vectors, trace out the corresponding great circles, and compute the position of the sun as the point where the great circles intersect. However, a little careful thinking reveals the obvious fact that ambiguities arise when skylight vision is restricted to a single e-vector in the sky. None of these have been observed[9].

The two hypotheses outlined above are challenged by our experiments described below. Certainly, a less conventional model is required if one tries to understand how bees deduce compass information from skylight patterns.

3 Behavioural analysis: The bee's stereotyped celestial compass

3.1 Use of e-vector information

(i) *Experimental design: Bees perform their recruitment dances on a horizontal plane surrounded by an artificial sky vault.*

The design of the basic experiment is simple. Individually marked bees are trained to an artificial food source several hundred meters away from the hive. During training they are allowed to view the complete e-vector pattern in the sky, but in the subsequent test their view of the sky is restricted to a single e-vector, or a limited number of e-vectors. The direction which the bees now select, if they select any direction at all, should inform us about what knowledge they have of the celestial e-vector pattern.

All this is easier said than done. If a walking insect like an ant sets out for a foraging trip, it can easily be followed and its path recorded by the human observer. In addition, its visual environment can be manipulated experimentally by moving various kinds of optical equipment along with the insect as it walks (WEHNER, 1982). In the foraging bee, however, such experiments are not feasible. One cannot follow a flying bee, let alone put optical screens and filters on top of it as it flies. Bees could be trained to reach a food source by walking, but this is difficult to achieve experimentally (BISETZKY, 1957). Fortunately enough, when the bee has returned from a foraging flight, it indicates its foraging direction by the direction of its recruitment dance, the so-called waggle dance, by which it communicates information about the location of the food source to its nest mates.

---

9 Most previous authors have agreed upon the assumption that bees determine individual e-vector orientations locally, but it is a little difficult to judge what they have assumed the bees to do next. Most workers including STOCKHAMMER (1959) and VON FRISCH (1965: 398) have certainly leaned towards the first hypothesis outlined above (that bees perform some kind of image matching), but KIRSCHFELD, MARTIN and LINDAUER (1975) have favoured the second (that bees possess some more general knowledge about the geometry of skylight patterns).

If the hive is put horizontally, and if the bee dancing on the horizontal comb has full view of the sky, or of a large circular patch of sky centred about the zenith (Fig. 3 b), the bee's waggle runs point exactly in the direction towards the food source. We can conclude from this kind of control experiment that the device used in our investigations does not disturb or misdirect the bees. In another control experiment, we surround the dancing bee by a translucent Plexiglass hemisphere which depolarizes the sky completely and provides the bee with nothing but homogeneously lit visual surroundings (Fig. 3 a). In this case, the bee's waggle runs are oriented randomly, as they are under a fully overcast sky, with no

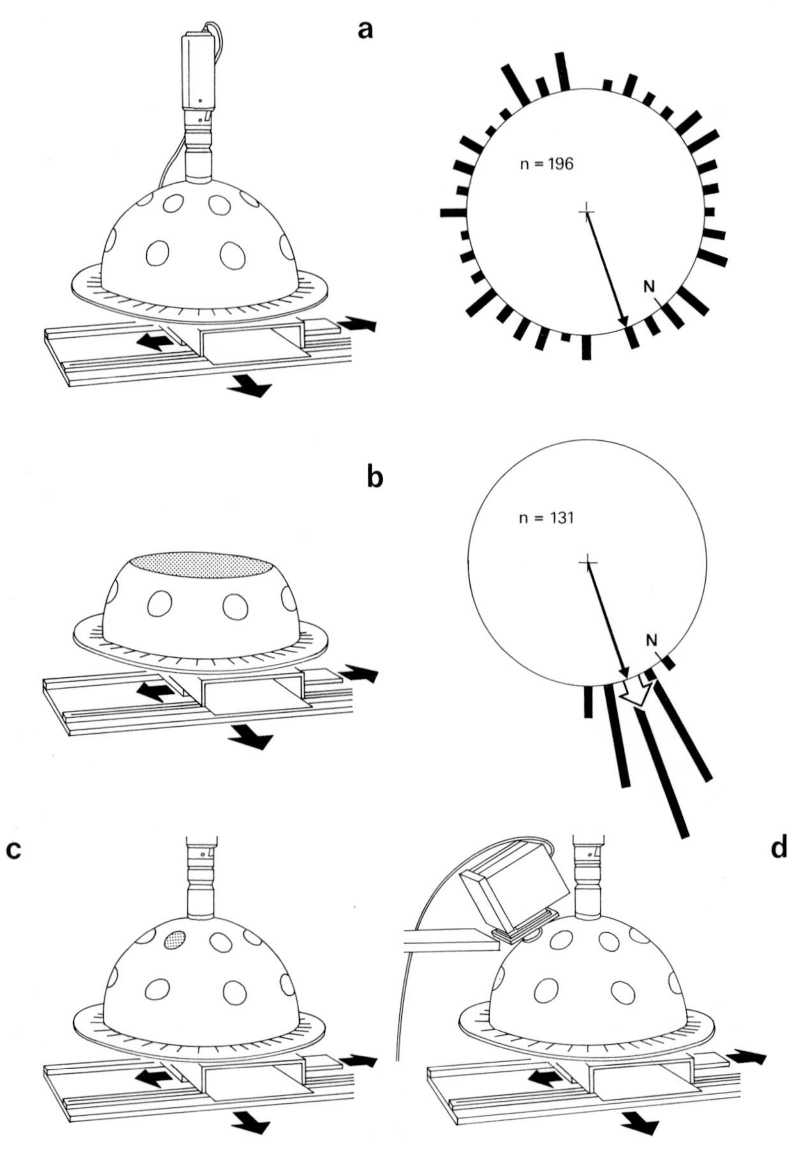

other visual cues available. This assures us that the diffuse illumination prevailing inside the Plexiglass dome does not offer visual cues that could guide the bees as they dance.

Now comes the crucial experiment. One of the apertures built into the Plexiglass hemisphere is opened, so that the bee can view the unobscured blue sky through a single small circular window (Fig. 3 c).

Before we describe in the next section how the bees behave in this potentially ambiguous situation (see p. 19), we must add some technical remarks. The diameter of the circular aperture through which the bees can see the sky is a mere 10°, sometimes even 5°. In additional experiments, artificially polarized light is displayed to the bees through one of the apertures (Fig. 3 d). In this case, the light entering the aperture is provided by a Xenon arc source (400 W). Before it reaches the eye of the dancing bee, it passes through a heat-absorbing filter (KG 2, Schott), a diffuser, spectral cutoff or broad-band interference filters (e.g. GG 400, UG 11, Schott), and a polarizer (HNP'B, Polaroid). In addition, two or more patches of artificially polarized light can be presented simultaneously by using more than one light source and more than one set of filters. This arrangement enables us not only to mimic stimulus conditions that occur in the natural sky, but also to display spatial configurations of e-vectors that do not occur in the natural sky and thus have never been experienced by any bee throughout its entire lifetime. Clearly, it is this kind of experiment which will finally provide a touchstone of any hypothesis about the bee's celestial compass. First, however, let us perform the basic experiment announced above and derive from it the hypothesis that will later be tested in more detail.

(ii) *The basic experiment: The bee's internal model of the sky is not a memory image of the e-vector pattern last seen.*

After the bees have been trained under the full blue sky (Fig. 4 a), they perform their recruitment dances underneath the translucent Plexiglass dome. In the first experiment, their skylight vision is restricted to a small patch of sky characterized by an e-vector orientation of -45°. The stimulus condition is sketched out in Fig. 4 b. In this and the following figures a heavy black arrow marks the foraging direction, which in this case points 15° to the left of the anti-solar meridian and 96° to the left of the skylight window opened within the Plexiglass dome. Much to our surprise, the mean direction of the bee's

---

Fig. 3: Experimental design. Bees trained to an artificial food source perform their recruitment dances (waggle runs) on a horizontal comb which is surrounded by a translucent Plexiglass hemisphere. The hive including the comb can be moved in x and y directions so as to centre the dancing bee with respect to the hemisphere *(black arrows)*. The dances are recorded by a video camera mounted in the zenith. In the data plots the directions of the recruitment dances are indicated by *black bars*. The *big open arrow* and the *small black arrow* mark the mean direction of the recruitment dances and the direction of the food source, respectively. N north, *n* number of recruitment dances recorded.

a Control experiment no. 1: Neither natural skylight nor artificially polarized light is displayed to the bees. As underneath a completely overcast sky, the dances of the bees are oriented at random.

b Control experiment no. 2: The bees view the entire upper half of the natural blue sky (elevation > 45°), but the elevation of the sun is < 45°. The dances of the bees point exactly in the direction of the food source.

c One of the apertures of the Plexiglass dome is opened enabling the bees to view a small patch of blue sky (diameter 10°). The result of one of these experiments is described in Fig. 4.

d A precisely controlled artificial source of light (dimaeter 10°) is displayed to the dancing bees. E-vector orientation, degree of polarization, spectral composition, and radiant intensity can be varied independently. In some experiments more than one patch of natural sky or more than one artificial source of light have been presented.

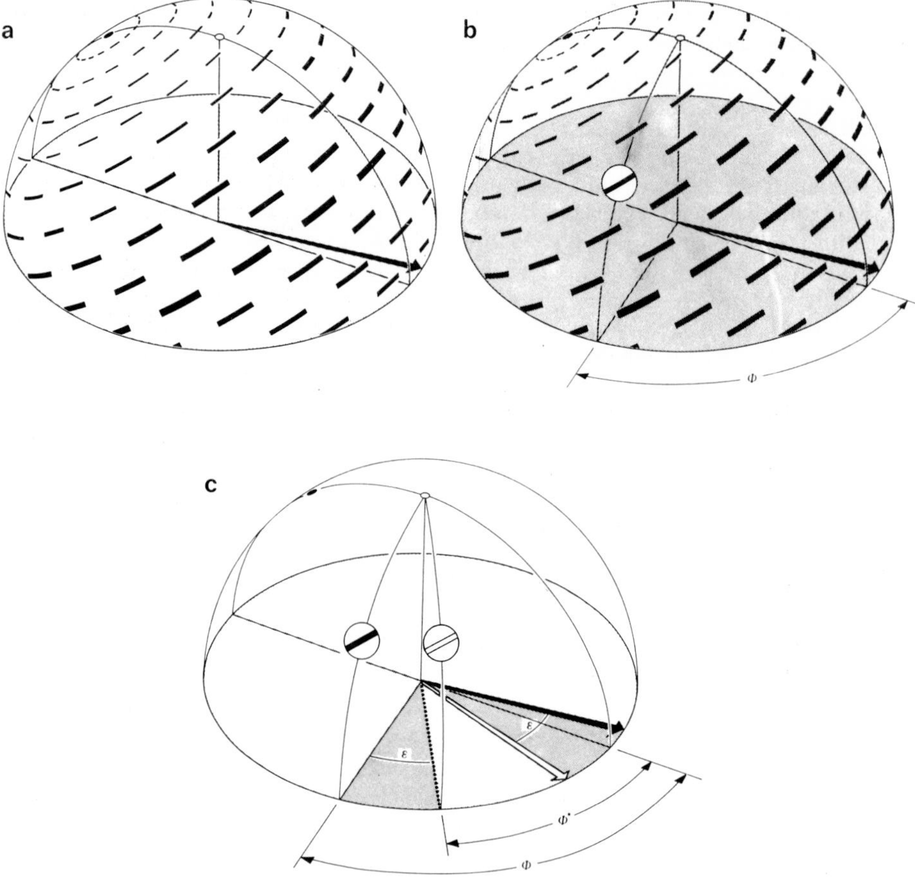

Fig. 4: The basic experiment. Bees are trained to fly under the full blue sky to a distant food source (*a*). The direction of the food source is marked by a *black arrow*. In the experiment described here the black arrow points 15° to the left of the anti-solar meridian. Having returned from the food source, the bees perform their recruitment dances underneath a translucent Plexiglass dome, under which their skylight vision is restricted to a small patch of sky (*b*). In the present experiment the orientation of the e-vector available in the patch of sky is -45° (elevation of patch 45°, elevation of sun 53°). The result (*c*) is astounding: The bees deviate considerably (*white arrow*) to the right of the foraging direction. The error angle ε is about 30°. Conclusion: Bees expect the -45° e-vector to occur nearer to the anti-solar meridian (*white bar,* azimuth position Φ^*) than it actually does (*black bar,* azimuth position Φ). Note that $\Phi^* = \Phi - \varepsilon$. When the -45° e-vector is presented at other times of day (when the sun is at other elevations), the values of Φ are different from the one shown here, but Φ^* turns out to be always the same. Obviously, bees expect the -45° e-vector to occur at a fixed azimuth position (Φ^*) relative to the symmetry line of the e-vector pattern. The same value of Φ^* holds for the -45° e-vector that is positioned closer to the sun (recall that at any given elevation each e-vector usually occurs twice; see Fig. 2 b, c). Hence, the e-vector closer to the sun is interpreted by the bees as if it is the one farther from the sun. This «misinterpretation» results in extremely large error angles ε.

recruitment dances as indicated by the open arrow deviates by as much as 30° to the right of the foraging direction.

Astounding as this result may appear, its tentative explanation is simple. Let us first assume that the bee while foraging has learned in what direction the food source is located with respect, say, to the anti-solar meridian[10]. Let us further assume that later, while dancing, the bee tries to infer the azimuth position of the anti-solar meridian from whatever information is available in the sky, in this case from a single e-vector (-45°) presented at a given elevation. To accomplish this task, the bee must somehow be informed about the spatial layout of the e-vector pattern. As noted earlier, it could have acquired this information during its previous foraging flight by building up in its mind a detailed image of the e-vector pattern in the sky as seen when heading for the food source. However, the errors made by the bees in the experiment described above show otherwise (Fig. 4 c). Obviously, bees use an internal model of the sky in which the -45° e-vector is positioned closer to the anti-solar meridian (see open e-vector symbol) than it is in the real sky – closer by exactly that amount (ε) by which the bees deviate from their foraging direction towards the e-vector.

The experiment is outlined in more detail in Fig. 5. The main difference with respect to the experiment described in the former figure is that in the present case artificially polarized ultraviolet light rather than natural sky-light is presented. Nevertheless, as described in the next paragraph, the result is the same.

By now the reader will be so familiar with the design of the experiment that we will not lose him by switching from the three-dimensional representation used so far to the more convenient two-dimensional representation adopted in this and the following figures. The stimulus condition and the bee's dance directions are depicted in Figs. 5 a and 5 b, respectively. When the -45° e-vector is presented at an elevation of 60° above the horizon at a time of day when the sun is also at an elevation of 60° above the horizon, the bees deviate by 30° to the right of the true foraging direction. The conclusion drawn from this result is sketched out in Fig. 5 c: In the bee's internal model of the sky (inner circle) the -45° e-vector lies 30° closer to the anti-solar meridian than it does in the outside world (outer circle). Now imagine that the bee turns until it has accomplished a match between its internal representation of the celestial e-vector pattern and what it actually experiences in our experimental set-up. To visualize what happens, the reader is invited to rotate – in his or her mind – the inner circle with respect to the outer one until both 45° e-vectors are aligned (lower graph of Fig. 5 c). Then, the white arrow (indicating the direction in which the bee assumes the food source to occur) deviates from the black arrow (indicating the actual direction of the food source) by the error angle that has been observed under the experimental conditions mentioned above.

In conclusion, the bee's internal model of the sky cannot be a memory image of the actual e-vector pattern as experienced during the bee's previous foraging flight. The bee assumes the -45° e-vector to occur at a position which does not coincide with the real position of this e-vector in the sky. Another model is required whose precise formulation must await further evidence. Such evidence is presented in the following sections.

---

[10] In this and the following considerations we take the anti-solar meridian as the reference azimuth position. This is merely for the sake of convenience; we could have equally justifiably selected the solar meridian.

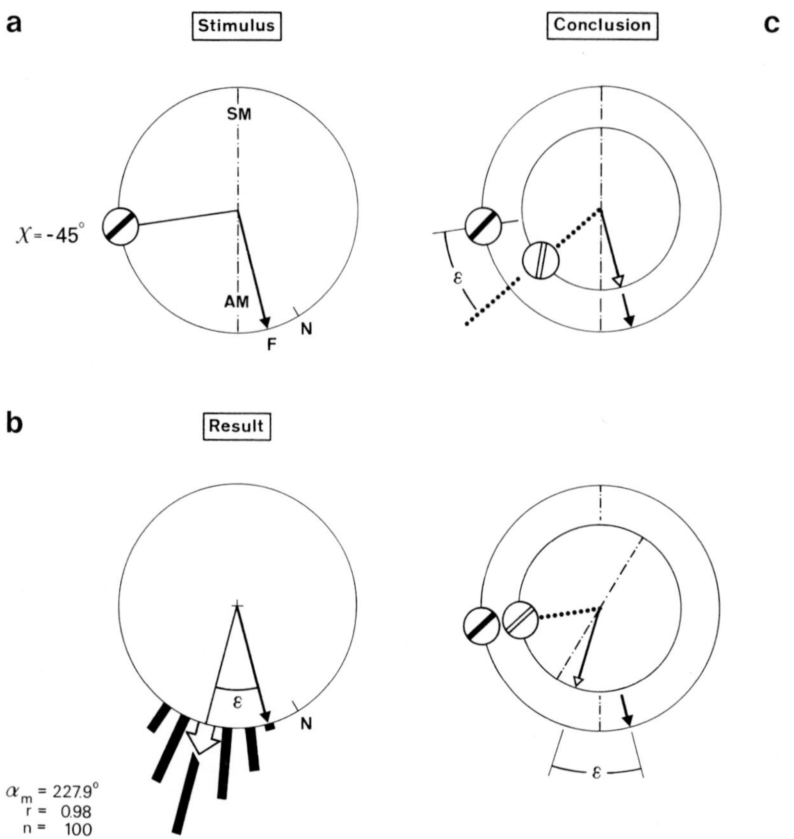

Fig. 5: Same experiment as in Fig. 4, but with the -45° e-vector presented by means of a beam of artificially polarized ultraviolet light (Fig. 3 d) rather than natural skylight (Fig. 3 c).
a The *black arrow* points towards the food source (*F*). *N* denotes north. *SM* solar meridian, *AM* anti-solar meridian. The e-vector is presented at an elevation of 60° above the horizon (elevation of sun 60°).
b The bees' waggle runs (*black bars,* mean value α_m indicated by *big white arrow*) deviate by about 30° (error angle ε) to the right of the foraging direction.
c The outer and the inner circles depict the position of the -45° e-vector in the real sky *(black bar)* and in the bee's model of the sky *(white bar)*, respectively. The latter position is derived from the error angle ε observed in the experiment (see *b*). The *black arrow* marks the direction of the food source (as in *a*). Imagine that during its preceding foraging flight performed underneath the full blue sky, the bee has learned the direction of its food source with respect to the solar and anti-solar meridian (*white arrow*). Further imagine that the bee when confronted with a -45° e-vector alone tries to align the -45° e-vector of its internal compass with the -45° e-vector in the sky. Metaphorically, this can be done by rotating the inner circle relative to the outer one by the error angle ε. Then the *white arrow* points in the direction in which the bees actually dance (compare *b* and *c, lower figure*). See Appendix for definition of symbols.

(iii) *Outlining a hypothesis: The bee's e-vector compass reflects the mean azimuth positions of the maximally polarized e-vectors in the sky.*

The next logical step is to display to the bees different e-vectors presented at different elevations and at different times of day, to record the error angles exhibited by the bees, and to derive from these error angles where in the sky the bees assume any particular e-vector to occur. What has emerged from these tedious investigations during which more than ten thousand waggle runs have been recorded on videotape, is the bee's internal representation of the celestial e-vector pattern – the bee's e-vector compass. It is depicted at the inner circle of Fig. 6.

Let us just highlight the main features of this compass:

(a) The compass does not depend on the elevation at which a given e-vector is presented above the horizon. Each point of the compass is characterized invariably by its own e-vector orientation. In the real sky, at each azimuthal distance from the solar or anti-solar meridian the e-vector orientation varies with both the elevation of the point observed and the elevation of the sun (Fig. 6, outer circle), but in the bee's compass it does not. Referring back to Fig. 4 c, we can rephrase this important statement as follows: The actual azimuth position Φ of a given e-vector varies with the elevation of both the e-vector and the sun, but the azimuth position the bee assumes (Φ^*) stays put. In conclusion, in the bee's e-vector compass e-vectors remain at fixed azimuth positions with respect to the solar and anti-solar meridian (see also Fig. 10 a). This holds at least for elevations larger than 30° above the horizon.

(b) This stereotyped e-vector compass is used by the bee under all conditions tested so far. The following situation is especially intriguing. Depending on the elevation of the sun, and thus the time of day, certain e-vectors do not occur at a given elevation above the horizon. However, the bees are not disturbed when such an e-vector is presented by means of a beam of artificially polarized light. They expect this e-vector to occur exactly where it should according to their internal compass (see e.g. Fig. 9).

(c) The compass is restricted to the highly polarized half of the sky, that is the half centred around the anti-solar meridian. Recall that at each elevation above the horizon any particular e-vector usually occurs twice: once in the less polarized half of the sky (close to the sun) and another time in the highly polarized half of the sky (farther away from the sun). To assure oneself of this crucial point, one should consult again Figs. 2 b–d. However, the bee is not informed about this ambiguity. It indiscriminately expects all e-vectors to occur in the highly polarized half of the sky. This is one of the strongest arguments in favour of the previous statement that bees do not pay attention to the degree of polarization, which as noted earlier is highly variable in the natural sky. It further implies that the bee's e-vector compass does not cover all 360°, but only 180° of the points of the compass.

(d) In detail, the full range of e-vector orientations is spread along the points of the bee's e-vector compass in the following way:
Horizontal e-vectors are positioned along the anti-solar meridian ($\varphi = 0°$) forming the symmetry line of the bee's «demi-compass». This is the only case in which the bee's compass coincides with the real sky. Thus, no errors occur, if a patch of horizontally polarized skylight is displayed (Fig. 8).
Vertical e-vectors are positioned at right angles to the left and right of the symmetry line (at $\varphi = +90°$ and $\varphi = -90°$). When a vertical e-vector is presented, the bees usually exhibit bimodal orientation expecting the vertical e-vector at two positions lying opposite to each other and at right angles to the solar and anti-solar meridian (Fig. 9)[11].
All other e-vectors are spread out between the horizontal and vertical e-vectors in such

a way that the compass exhibits mirror-symmetry around the anti-solar meridian. Their exact positions can be inferred from Fig. 6 (inner circle) and in more detail from the *standard* χ/φ *function* shown in Fig. 7 a.

In conclusion, bees are not informed exactly about all intricacies of the celestial e-vector patterns. They follow a rather simple rule in assigning azimuth positions to e-vector orientations (Figs. 6 and 7 a).

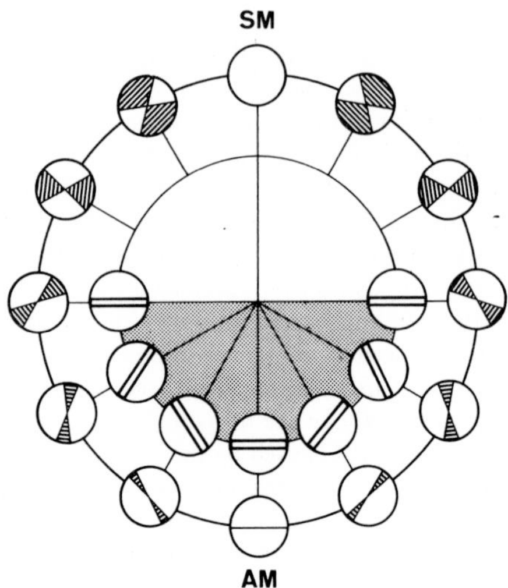

Fig. 6: The bee's e-vector compass *(inner circle)*.
SM solar meridian, *AM* anti-solar meridian. The two most striking features of the bee's e-vector compass are the following: (1) E-vector orientations occur at fixed azimuth positions as measured with respect to the symmetry plane of the skylight pattern *(SM-AM)*. (2) The bee's e-vector compass is restricted to the anti-solar half of the sky *(shaded semicircle)*. The *outer circle* depicts the azimuth positions at which the e-vectors displayed to the bees during the course of the experiments actually occurred in the sky. At each azimuth position marked by a small circle the range of e-vector orientations displayed in the experiments is indicated by the *hatched sectors*.

---

[11] While presenting vertical e-vectors, Brines and Gould (1979) usually observed unimodal dance orientation with a distinct preference for the e-vector to the right of the sun. This was thought to represent a dance convention used by the bees to avoid ambiguities. In contrast, we observed bimodal dance orientation throughout the day. However, when the sun was close to the horizon, one direction was sometimes selected more frequently than the other, but preference for either vertical e-vector could change from one day to the other. A close analysis of the dance directions revealed that in these cases, when there was a preference for one direction, the e-vectors presented were interpreted as being inclined by a few degrees relative to vertical. Thus we are not able to confirm the "right"-rule observed by Brines and Gould. Further work will be necessary to clarify this point. Note, however, that beyond this "right"-rule the major discrepancy between our results and those of Gould and his students is as follows. The latter workers have found that bees expect a vertical e-vector to occur exactly where it should according to its position in the sky. We find that bees expect vertical e-vectors to occur invariably at right angles to the solar and anti-solar meridian.

Now comes the poignant question we have pondered for quite some time: What properties of the actual e-vector patterns are hidden behind the bee's e-vector compass? Is there a rule by which the bee's compass strategy can be derived theoretically from the e-vector patterns in the sky? In evolutionary terms, the bee must have adopted such a rule, and so it

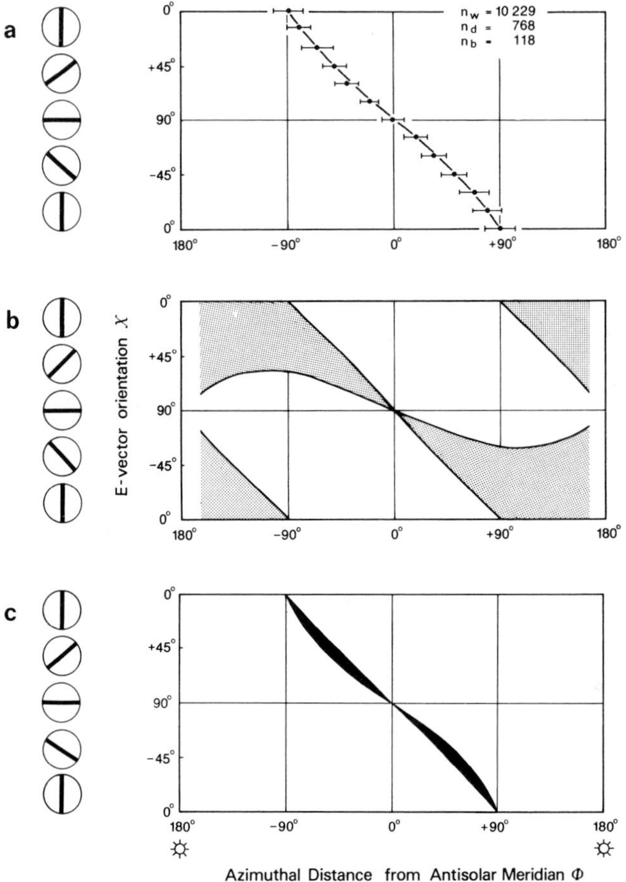

Fig. 7: *a* The bee's e-vector compass in detail: The *bee's standard χ/φ function* as derived from single-spot experiments (see Fig. 5). The figure demonstrates at which azimuth position (Φ, φ) the bee expects any particular e-vector orientation (χ) to occur. Mean values and standard deviations are given. For a more vivid representation see Fig. 6 (inner circle) and Fig. 10 a. n_w number of waggle runs recorded, n_d number of recruitment dances, n_b number of bees tested.
b Azimuth positions at which all e-vector orientations tested in *a* occurred in the sky during the time of the experiments *(celestial χ/φ functions)*. The elevations of the e-vectors displayed to the bees varied from 30° through 66°, and the elevation of the sun from 15° through 60°. In the figure the latter range is extended down to 0° (sun on horizon).
c Azimuth positions of the maximally polarized e-vectors in the sky (*celestial* χ_{max}/φ *functions*). The range of elevations of both e-vector and sun for which the χ_{max}/φ functions have been determined is identical with the one shown in *b*. Note the close resemblance between the mean celestial χ_{max}/φ function *(c)* and the bee's standard χ/φ function *(a)*.

is a challenge to the human observer to inquire what the properties of the skylight patterns are that the bee has discovered and incorporated into its visual system.

The answer to this question is rather simple. The bee's e-vector compass can be derived theoretically by referring exclusively to the most prominent feature of skylight polarization, the arc of maximum polarization (see Fig. 2a). More specifically, the bee's e-vector compass reflects the mean distribution of the maximally polarized e-vectors as they occur across the sky during the course of the day.

What does this mean? The arguement can be understood easily by consulting Fig. 10b. As we have seen before, the arc of maximum polarization passes across the sky vault at an angular distance of 90^0 from the sun. When the sun is on the horizon, the arc passes through the zenith. Consequently, at sunrise and sunset, the vertical e-vectors are part of the arc of maximum polarization. When this is the case, they occur at two azimuth positions which lie opposite to each other and at right angles to the solar and anti-solar

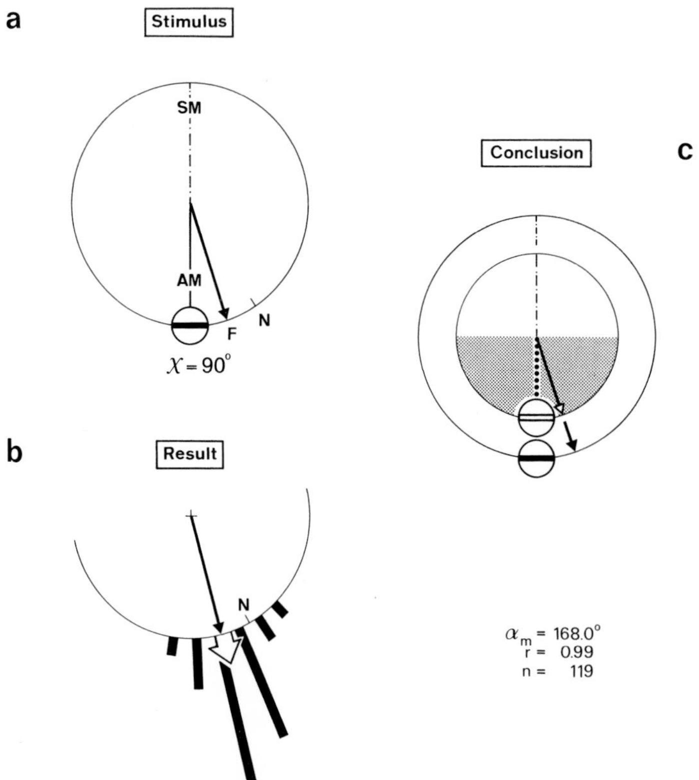

Fig. 8: In the sky, horizontal e-vectors ($\chi = 90°$) are positioned along both the solar and the anti-solar meridian. However, as borne out by the present experiment, bees expect horizontal e-vectors to occur exclusively along the anti-solar meridian. The experiment is as follows: An artificial beam of horizontally polarized ultraviolet light is presented to the bees at an elevation of 60°. The elevation of the sun is also 60°. The bees dance correctly in the direction of the food source *F* (see *b*) when the horizontal e-vector is presented at the azimuth position of the anti-solar meridian (see *a*). For explanation of symbols see Fig. 5 and Appendix.

meridian, exactly as described above for the bee's e-vector compass. When the sun moves up, the arc of maximum polarization tilts down. In Fig. 10 b the distribution of maximally polarized e-vectors as caused by the downward tilt of the arc of maximum polarization is shown for an elevation of 45°, but in fact it is very similar for all elevations > 30° above the horizon (Fig. 7 c). Furthermore, if one takes the mean of all these distributions, one arrives at the bee's celestial compass (compare Figs. 7 a and 7 c).

The similarity between what the bee uses as a compass (Fig. 7 a) and what is displayed in the sky by the maximally polarized e-vectors (Fig. 7 c) is striking. The bee's e-vector

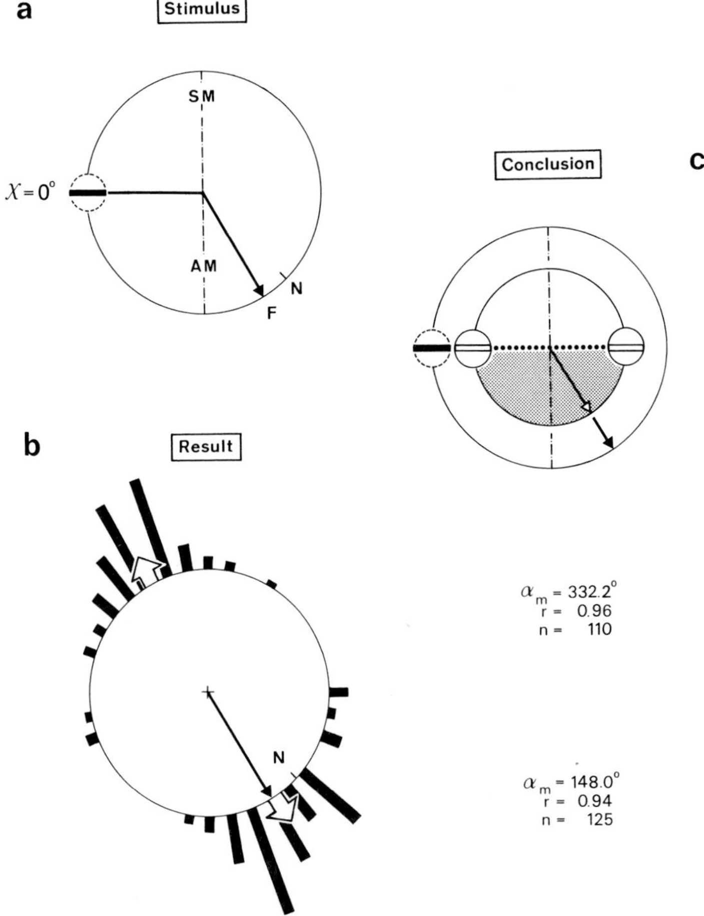

Fig. 9: Vertical e-vectors ($\chi = 0°$) are expected by the bees to occur at right angles to the left and right of the solar or anti-solar meridian. This follows from the experiment described in this figure. An artificially produced vertical e-vector is displayed to the bees at an elevation (60°) at which it does not occur in the sky at the time at which the experiment is performed (elevation of sun 60°). It is presented 90° to the right of the solar (and anti-solar) meridian (*a*). The bees dance bimodally in the direction of the food source and in the diametrically opposite direction (*b*). Note that in *c* there are two orientations of the inner circle in which the azimuth positions of the vertical e-vectors of both circles coincide.

compass and the maximally polarized e-vectors are restricted to the anti-solar half of the sky. With the only exception of the vertical e-vectors each maximally polarized e-vector is positioned at only one point of the compass. No ambiguities occur. Finally, the azimuth position of each maximally polarized e-vector in the sky is almost independent of the elevation of the e-vector above the horizon. This does not hold for all other (less polarized) e-vectors present in the sky. Their azimuth positions vary with the elevation above the horizon and, at each particular elevation, with the elevation of the sun (Fig. 7 b).

In conclusion, the bee's celestial compass can be derived not only experimentally as described in the former section, but also theoretically by referring to the arc of maximum

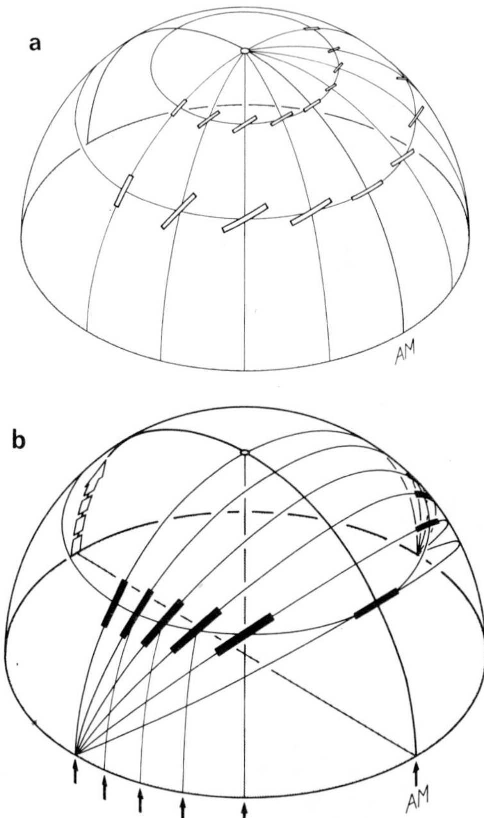

Fig. 10: *a* Three-dimensional representation of the bee's e-vector compass (compare Fig. 6, inner circle). The same distribution of e-vectors (the bee's standard χ/φ function, see Fig. 7 a) holds for all elevations > 30°. *AM* anti-solar meridian.
b The maximally polarized e-vectors shown for an elevation of 45° above the horizon. When the sun moves up *(white broken arrow),* the arc of maximum polarization which is always positioned at an angular distance of 90° to the sun (Fig. 2 a) tilts down. As a consequence, the maximally polarized e-vectors are confined to the anti-solar half of the sky. This reflects an important aspect of the bee's e-vector compass. Moreover, the spatial distribution of the maximally polarized e-vectors is very similar for all elevations higher than about 30° above the horizon (Fig. 7 c), and is very similar to the spatial distribution of the e-vectors as assumed by the bees (Fig. 7 a).

polarization in the sky. What the bee has incorporated into its e-vector compass are the average azimuth positions of the maximally polarized e-vectors as they occur in the upper part of the sky.

(iv) *Testing and vindicating the hypothesis: Bees apply their simple compass rule even when large parts of the natural blue sky are available.*

The hypothesis outlined above implies that all the bee knows about skylight polarization is the simple compass rule depicted in Fig. 7a. We have derived this compass empirically (from the bees) and theoretically (from the sky). Nevertheless, in all experiments described so far only a single e-vector, either natural or artificial, has been displayed to the bees. One may argue that the bees resort to their simple model of the sky only when such restrictive stimulus conditions prevail. Whenever they see large parts of the sky, they might exhibit more detailed knowledge of the natural e-vector patterns.

To cut a long story short, this is not the case. In the following we shall describe a number of experiments in which bees are confronted with more than one e-vector or with large parts of the natural blue sky. Even under such conditions the bees invariably apply their simple compass rule. Just as in the single spot experiments, each e-vector induces an orientation error, and the mean of these errors shows up as an angular deviation of the bees from the correct compass course.

Two-spot experiments: Only the most intriguing of a series of experiments will be described. In this experiment the horizontal and the vertical e-vector are presented simultaneously at an angular distance at which they occur in the natural sky (Fig. 11a).

One might expect that in this situation no orientation errors would occur, because as described above (Fig. 8), the horizontal e-vector alone suffices for establishing the correct compass course. So it comes as a real surprise that the bees again make mistakes. They deviate by 21° from the correct course (Fig. 11b). This is a really counter-intuitive observation: additional information is presented – a vertical e-vector in addition to the horizontal one – and orientation deteriorates. However, this is exactly what our hypothesis predicts.

A moment's thought helps to explain why this is so. Assume that the bee can do nothing but apply its simple compass rule even when there is more than one e-vector available. Then the bee should select a direction that is the mean of the directions it would take when confronted with each e-vector alone. This is exactly what we observe in the two-spot experiment described above. The error exhibited by the bees is the mean of the errors exhibited in each of the two one-spot experiments. When a horizontal or a vertical e-vector is presented alone, the error angles are 0° and 45°, respectively. These numbers can be read off Fig. 11c by comparing where the two e-vectors occur in the bee's compass (inner circle) and the sky (outer circle). From these numbers a mean error angle of 22.5° is predicted for the two-spot experiment. The error angle actually observed (20.7°) is statistically indistinguishable from the predicted value.

What happens if the horizontal and vertical e-vector are presented at an angular distance of 90° rather than 135°, i.e. at an angular distance that is not found in the sky at the time of the experiment? In this case, the positions of the two stimuli exactly match the bee's celestial map. No orientation errors should occur, and none have been observed. This seems to contradict common sense: When the stimulus configuration coincides with the outside world, the bees make mistakes, but when an artificial configuration is presented which is not found in the natural sky, the bees are oriented correctly. However counter-intuitive this notion might appear, it is fully in accord with the assumption that the bee is inevitably bound to the simple compass strategy outlined in the previous section.

In conclusion, when any particular e-vector is presented in addition to a horizontal e-vector (which guarantees correct orientation when presented alone), the bees do not refer selectively to the latter. In their celestial compass, all e-vectors seem to be equally important. The orientation errors observed in the two-spot experiments are the mean of the errors which the bees exhibit when they are confronted with the two e-vectors separately.

Large-field experiments: Now let us go one step further and display to the dancing bees large parts of the natural blue sky rather than sets of isolated e-vectors. Only two experiments of this kind will be described in Fig. 12, but a large number of similar experiments have all led to the same conclusion (Wehner, 1983; Rossel and Wehner, 1984). Even when large skylight windows are opened in the translucent Plexiglass dome, bees make mistakes exactly as large as predicted from their e-vector map. This holds irrespective of the size and position of the skylight window.

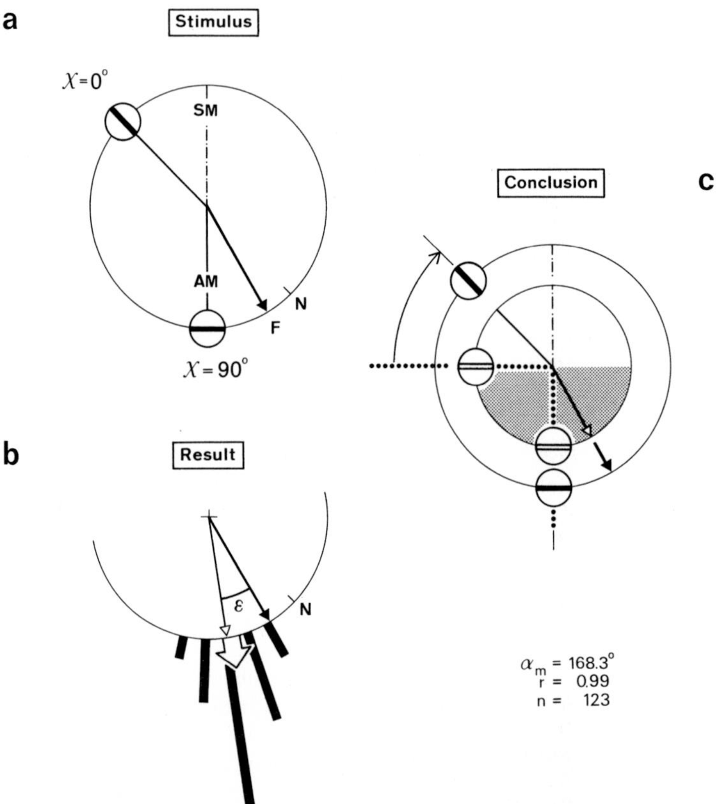

Fig. 11: Two-spot experiment. Horizontal and vertical e-vectors are presented simultaneously by two beams of ultraviolet light. The angular distance between the two e-vectors corresponds to the angular distance at which the two e-vectors occur in the sky at the time of day at which the experiment is performed. In spite of this correspondence, the directions of the bee's dances deviate from the direction of the food source *(b)*. The error angle observed can be predicted by assuming that the bees try to match their internal e-vector compass *(c: inner circle)* as closely as possible to the two stimuli displayed to them *(c: outer circle)*.

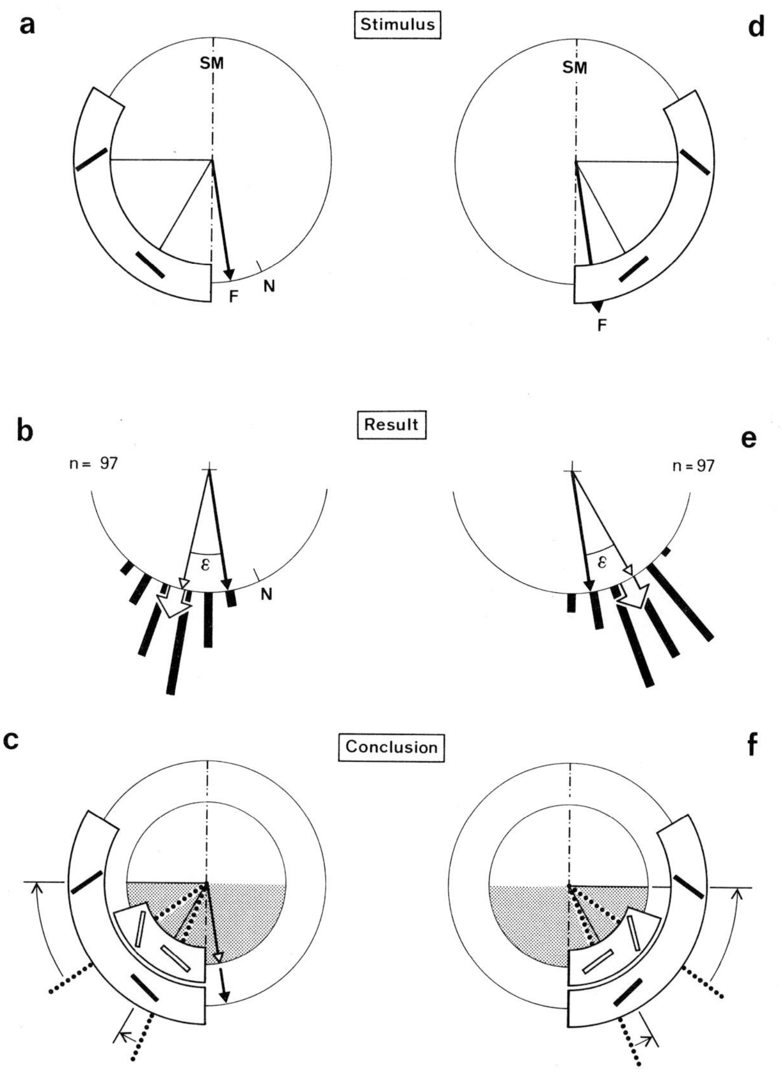

Fig. 12: Large-field experiments. The bees view a strip-like celestial window positioned 60° above the horizon either to the right *(a)* or to the left *(d)* of the anti-solar meridian. The window opened in the translucent Plexiglass dome is 135° wide and 10° high. The bees deviate to the right *(b)* and to the left *(e)* of the correct direction, respectively. As in the two-spot experiment of Fig. 11 the error angles observed *(large white arrows)* are statistically indistinguishable from the prediction *(small white arrows)* that the error induced by an e-vector pattern is the mean of the errors induced by each e-vector alone. In computing the mean error, the positions of individual e-vectors have been compared between the inner and outer circle in *c* and *f* at azimuthal distances of 10°, but only two such comparisons are shown in the figures. For the bees, each e-vector seems to be equally important, irrespective of its degree of polarization which decreases from the anti-solar side towards the solar side of the window.

A short description of one of the experiments will make the point. In this experiment (Fig. 12 a–c) the bees view a strip of blue skylight positioned to the right of the anti-solar meridian, at an elevation of 60° above the horizon (Fig. 12 a). Again, their waggle runs do not point exactly towards the food source, but deviate by 23.5° to the right (Fig. 12 b), and again this is to be expected if the bees apply the rule outlined above, i. e. calculate the mean of the error angles induced by each individual e-vector alone. For the sake of simplicity, only two of these e-vectors are shown in Fig. 12 c, but for the computation of the mean error angle e-vectors have been taken every 10°. The mean error angle computed this way is 22.1° (small open arrow in Fig. 12 b). It nicely coincides with the error angle actually observed in the experiment (23.5°, large open arrow in Fig. 12 b).

This coincidence finally disposes of the possibility that bees use the degree of polarization as an additional compass cue. The point is easy to grasp. The degree of polarization varies dramatically within the strip-like skylight window. In our experiment, it reaches its maximum near the anti-solar side of the strip and then decreases steadily towards the solar side of the strip. There, it is just high enough for the e-vectors to be detected by the bees (see p. 20). If the bees referred to the degree of polarization in addition to the e-vector orientation, the e-vectors should dwindle in importance on the way from the anti-solar to the solar side of the strip. This is not the case. As shown above, each e-vector is equally important for the bee, irrespective of its degree of polarization.

The error angles observed in the large-field experiments depend crucially on the position of the skylight window. In the experiment described above (Fig. 12 a–c), the window is positioned to the right of the anti-solar meridian. If it is positioned to the left, with all other things being equal (Fig. 12 d–f), the bees deviate to the left from the correct compass course by the same angular amount by which they deviated to the right in the former case (Figs. 12 a–c). It is only when the skylight window is positioned symmetrically with respect to the anti-solar meridian that no orientation errors occur. Then, the errors induced by the left and the right half of the pattern are equal in magnitude, but opposite in sign, and thus cancel each other out. As a consequence, the bees' waggle runs point towards the food source. Never have systematic errors been observed in such situations. This kind of experiment is not documented here, but described at length in WEHNER (1983) and ROSSEL and WEHNER (1984).

In conclusion, even when large parts of the natural blue sky are available, bees make mistakes exactly as large as predicted from their simple compass rule. The prediction implies that the orientation error induced by any e-vector pattern is the mean of the orientation errors induced by each e-vector alone. Each e-vector is equally important, irrespective of its degree of polarization (provided that the degree of polarization is high enough for the e-vector to be detected). Orientation errors do not occur when the patch of sky is positioned symmetrically to the anti-solar meridian, because then the errors induced by the left and the right half of the pattern neutralize each other.

3.2 Use of spectral information

How do bees behave if they view a patch of *unpolarized* light? In the full blue sky the only patch that is unpolarized is the sun and the area immediately around the sun. Therefore we would expect that bees take an unpolarized beam of light for the sun. However, bees behave according to this expectation only if the beam of unpolarized light is rich in long-wavelength radiation, so that the green receptors are stimulated predominantly (Fig. 13 b). If a beam of unpolarized ultraviolet light is presented instead, the bees expect this spot of light to lie within the anti-solar half of the sky (Fig. 13 c). The scatter within the data is much larger than in the former case (compare Figs. 13 b and 13 c).

Clearly, the mean vector of the directions plotted in Fig. 13 c points toward the anti-solar meridian. This is to be expected whenever the data are arranged symmetrically with respect to the anti-solar meridian, but it would be rash to conclude from this result that in the bee's internal image of the sky an unpolarized spot of ultraviolet light lies just along the anti-solar meridian rather than anywhere within the anti-solar half of the sky.

This is borne out by another set of experiments (Fig. 14). A polarized beam of ultraviolet light (e-vector orientation -45°) is presented in addition to an unpolarized beam of ultraviolet light (UV). The latter is positioned either 90° to the left (Fig. 14 a) or 90° to the right (Fig. 14 d) of the polarized beam. Even though the two stimulus conditions are completely symmetrical, the results are not (Figs. 14 b, e). In the first experiment, with the beam of unpolarized light to the left of the e-vector, the bees hehave as if the e-vector had been presented alone (compare Figs. 14 b and 5 b). They seem to disregard completely the

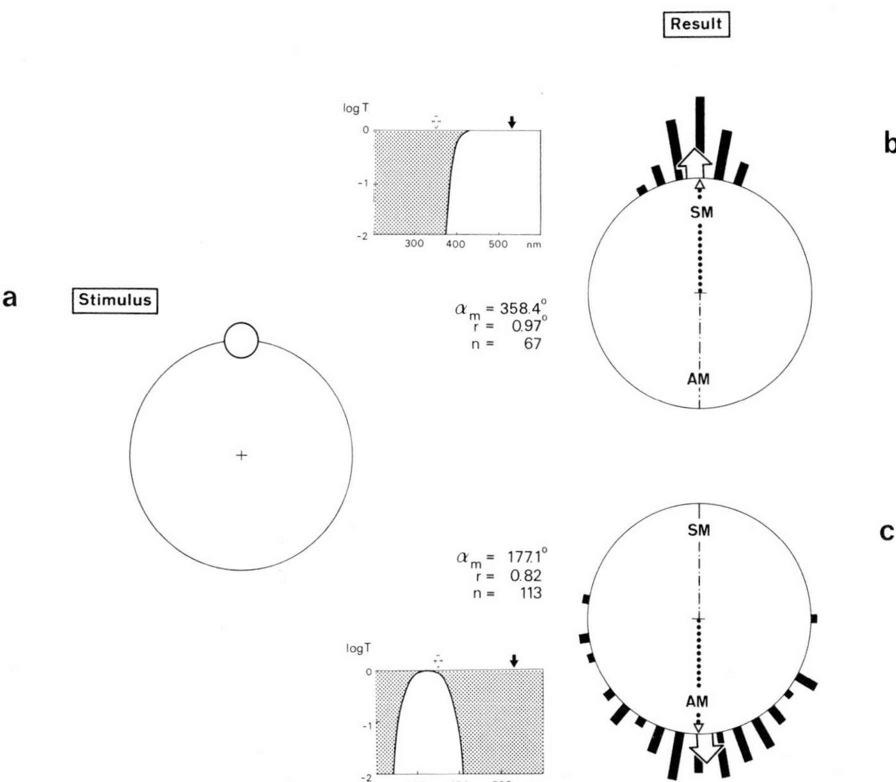

Fig. 13: Bees are presented with an unpolarized beam of green *(b)* or ultraviolet *(c)* light (diameter 10°, elevation 60°). In the first case they take the unpolarized stimulus for the sun, in the second for a celestial point lying anywhere within the anti-solar half of the sky. Note that in plotting the data the convention used in this and the following figures differs from the one used in the preceding figures. The *dotted line* with the *small white arrow* marks the direction in which the bees are expected to dance if they assume the stimulus to lie along the solar *(b)* or anti-solar *(c)* meridian. *Insets:* Spectral transmission *(T)* of the filters used. The black and white arrows point to the sensitivity peaks of the bee's green and ultraviolet receptors, respectively.

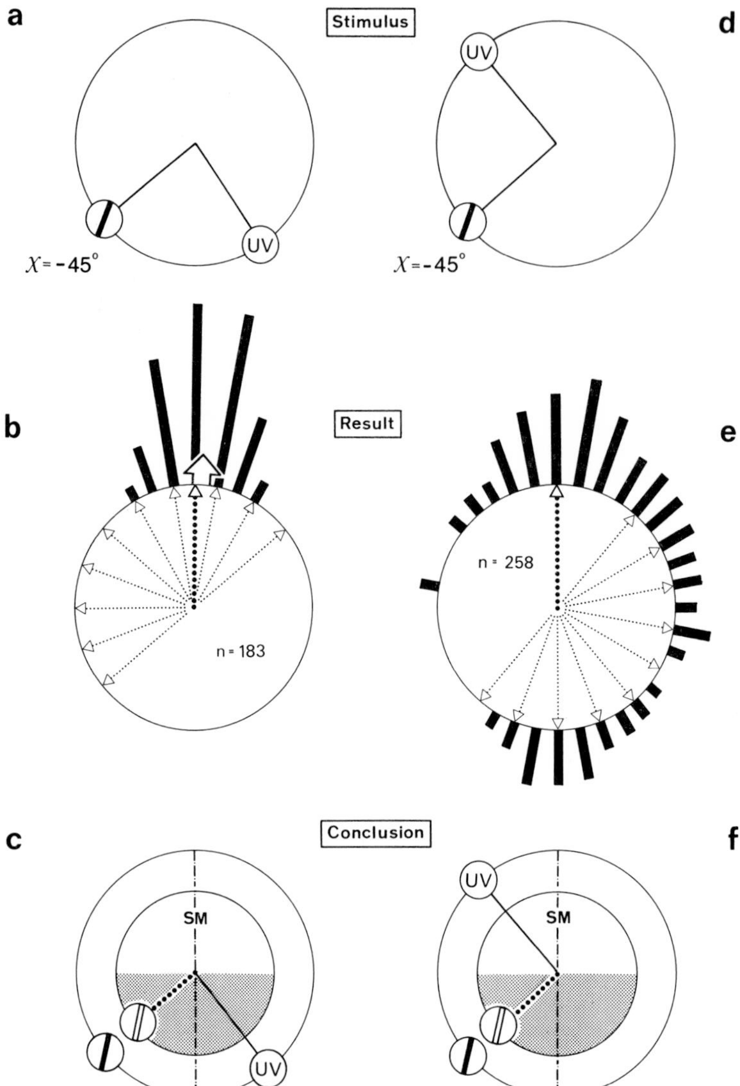

Fig. 14: Polarized and unpolarized beams of ultraviolet light are presented simultaneously. The unpolarized beam denoted by *UV* is positioned either to the left *(a)* or to the right *(d)* of the polarized beam (χ = -45°). In the former case the bees behave as if the unpolarized beam were not present at all *(b)*, whereas in the latter case the directions of their dances scatter considerably *(e)*. In *b* and *e* the *heavily dotted line (with arrow)* indicates the direction in which the bees should dance when they refer to the -45° e-vector alone. The *lightly dotted lines (with arrows)* mark the range of dance directions expected when the bees use exclusively the unpolarized spot of ultraviolet light. Figures *c* and *f* show why the unpolarized stimulus greatly worsens the orientation towards the polarized stimulus when it is to the right *(f)* but not when it is to the left *(c)* of the polarized beam. This becomes immediately apparent if one recalls that bees expect an unpolarized patch of ultraviolet light to lie within the antisolar rather than solar half of the sky (see Fig. 13).

unpolarized beam of ultraviolet light. This contrasts greatly to what happens in the second experiment, in which the unpolarized beam is presented to the right rather than the left of the polarized beam. As it appears, the bees are very disturbed when confronted with the latter stimulus configuration. They circle around hesitantly for long periods of time before they finally perform a waggle run. Successive waggle runs often point in different directions, and many times the course is changed even during a single waggle run. The casual observer might well get the impression that the bees are completely disoriented. When the directions of the waggle runs are finally plotted (Fig. 14 e), a peak shows up, roughly in the direction it should be in if the bee had used the -45° e-vector alone, but the angular distribution of the data is very broad covering nearly all points of the compass.

This result is perplexing at first, but once understood extremely rewarding. Do bees realize that there is any difference between the two stimulus conditions other than the mirror image symmetry? A comparison of Figs. 14 c and 14 f provides the answer. Imagine that the bees use the -45° e-vector to set their compass, i. e. to determine where the anti-solar meridian is in the sky. In the first experiment (Fig. 14c) they then would realize that the unpolarized beam of ultraviolet light lies within the anti-solar half of the sky where it really should (see Fig. 13 c). As a consequence, the unpolarized light adds nothing but redundant information. In the second experiment (Fig. 14 f), however, the stimulus condition is at variance with what the bees expect according to their celestial compass. The unpolarized beam of ultraviolet light lies outside the anti-solar half of the sky as defined by the polarized beam of light. For the bees, something must be wrong, and this explains the certain uneasiness we have observed in their behaviour. Taken at face value, the data of Fig. 14 e suggest that the bees respond alternatively to the polarized and the unpolarized beam of light, but, on closer scrutiny, one can see that more intricate interactions between the two responses are likely to occur.

One feature of Fig. 14 b deserves further comment. This is the close correspondence between this figure and Fig. 5 b. The only difference between the experiments described in the two figures is that in the one experiment (Fig. 14 a–c) an unpolarized beam of ultraviolet light is presented in addition to the -45° e-vector, whereas in the other (Fig. 5) the -45° e-vector is presented alone. The correspondence between the results of the two experiments finally assures us that bees do not interpret an unpolarized patch of ultraviolet light as lying exclusively along the anti-solar meridian. If they did, the stimulus condition shown in Fig. 14 a should introduce a certain ambiguity, because then the position of the anti-solar meridian as defined by the unpolarized beam of light would not coincide with the position of the anti-solar meridian as defined by the polarized beam of light. However, as mentioned above, no such ambiguities are observed.

In conclusion, when steering compass courses bees use spectral information in addition to e-vector information. A green spot of unpolarized light is taken for the sun, whereas an ultraviolet spot of unpolarized light is interpreted as lying anywhere whithin the anti-solar half of the sky.

3.3 Conclusion and implications

The most striking feature of the bee's celestial compass as described in this paper is its simplicity. Generally speaking, the compass consists of a green sun and an ultraviolet anti-solar half of the sky in which any particular point of the compass is further characterized by its own private e-vector orientation. The design of this compass is based on a few principal skylight features: a coarse spectral gradient extending over large parts of the sky and the more finely tuned spatial distribution of the maximally polarized e-vectors.

It is worth digressing here for the sake of perspective, to see how our data compare with those of other authors who have done similar experiments. This is particularly important since what we are finding runs counter to the prevailing view of how bees perceive and use polarized skylight. Since our first report (ROSSEL, WEHNER and LINDAUER, 1978) at least one feature of the bee's stereotyped *e-vector compass* has been confirmed by BRINES and GOULD (1979): At a given elevation above the horizon each e-vector orientation usually occurs at two azimuth positions, but bees invariably interpret either member of any such pair of e-vectors as the one farther from the sun. Beyond that rule BRINES and GOULD (1979) assume that the bees know exactly where a particular e-vector is positioned in the sky. The authors do not report that small navigational errors occur when individual e-vectors are displayed to the bees. These errors, however, are at the heart of the simple compass strategy we shall propose in the next section.

In fact, the errors exhibited by the bees under certain stimulus conditions are usually small and can be easily overlooked. We ourselves were convinced of them only after we had discovered their systematic nature by repeatedly varying the azimuthal positions of both food source and sun, and – most convincingly – by alternately presenting small patches of sky that were positioned symmetrically to the left and right of the anti-solar meridian. The latter set of experiments is especially convincing, because the error angles observed under these stimulus conditions are equal in magnitude but opposite in sign. It follows that the directions of the bees' dances induced by the two alternately presented patches of sky differ by an angle that is exactly twice as large as the individual error angle, and thus can be recognized easily even by the naked eye[12].

Apart from the difference between GOULD's and our data there is, more importantly, a major discrepancy between his and our point of view in interpreting the data. GOULD assumes that the «e-vector rule» mentioned above is an arbitrary convention applied by the bees only in the context of their dance language system of communication to ensure that both sender and receiver use the same reference system. The rule is «not necessary or even desirable for any of the vast number of social and nonsocial animals that .... lack a symbolic language» (BRINES and GOULD, 1979: 573).

In contrast, we assume that the «e-vector rule» reflects an integral and hard-wired property of the bee's navigational system used in steering compass courses in the field as well as performing recruitment dances in the hive. This is in accord with the hypothesis to be proposed shortly, that principal features of the bee's e-vector map are already laid down in the geometry of the receptor array. It is further borne out by the observation that desert ants, *Cataglyphis*, use essentially the same e-vector compass (work in preparation), but do not recruit.

With respect to *spectral cues*, EDRICH, NEUMEYER and VON HELVERSEN (1979) as well as BRINES and GOULD (1979) have reported that unpolarized spots of green or ultraviolet light are taken as the sun or the anti-solar point, respectively. Our data confirm the «green-rule», but in addition show that an unpolarized spot of ultraviolet light is expected by the bees to lie anywhere within the anti-solar half of the sky rather than exclusively along the anti-solar meridian. For statistical reasons, of course, the mean

---

[12] Long ago, VON FRISCH himself noticed that the mean dance directions of the bees sometimes did not coincide completely with the direction of the food source. However, he attributed such deviations to his "own inaccuracies in reading the directions of the dances rather than to the bees themselves" (personal communication). The verbal descriptions provided by him in his final summary (VON FRISCH, 1965: 393–397) are not detailed enough to allow any worthwhile generalization.

vector of the directions in which bees place an unpolarized spot of ultraviolet light would always point towards the anti-solar meridian.

Finally, let us return from this digression to the main story and inquire about some of the more general implications of our results. Certain overall impressions are inescapable. Bees operate underneath a complex celestial canopy, but do so by relying on some very simple rules. They adopt what can be called short-cut strategies or rules of thumb. Such short-cut strategies imply that under certain stimulus conditions navigational errors must occur – and, indeed, they do. Actually, it was from these errors that our concept of the bee's stereotyped celestial map was derived in the first place.

In the open, one usually does not observe such errors: first, because the errors only show up under certain stimulus conditions, and second, because the insect has at its disposal a number of back-up systems drawing upon non-celestial cues such as landmarks and including an efficient searching strategy when getting lost. What appears to us as unerring precision in an insect's foraging or homing path is in fact the result of some rather simple navigational rules, each relying on approximate rather than exact solutions of the underlying navigational problem and thus sacrificing absolute precision for a workable neural strategy (WEHNER, 1983). With respect to the bee's celestial compass, what is this neural strategy?

4 Model and neurobiological evidence: Scanning the sky

Until now we have been concerned with the question of where in the sky the bee expects any particular e-vector to occur. What resulted was the bee's stereotyped celestial compass. Next we ask, what does this compass mean in physiological terms. Where does it reside within the hardware of the bee's visual system, and how is it read and used?

4.1 The long-held view

It has been proposed by VON FRISCH (1965) and many later workers in this field that bees first determine the orientations of individual e-vectors available in the sky, and then use some knowledge about the celestial e-vector patterns to infer their compass course from the individually determined e-vector orientations. BRINES and GOULD (1979), for example, argue that a bee dancing on a horizontal comb and being provided with a single e-vector as the only navigational cue would first determine the orientation of the e-vector and then resort to memory information about the celestial pattern last seen as well as to a set of dance language conventions to avoid potential ambiguities. According to this concept, two separate questions have usually been asked. First, how does the bee determine the orientation of individual e-vectors in the sky, and second, how does it infer the position of the sun from a given e-vector or set of e-vectors?

By far the lion's share of recent publicity concerning the bee's e-vector compass has focused on the first question. Therefore it may seem strange that this fundamental topic should be discussed so late in this chapter, but at the time of this writing no agreed explanation is available on how bees solve this problem.

Nevertheless, some bits of information are at hand. We know from behavioural work done many years ago in both bees and ants (DUELLI and WEHNER, 1973; VON HELVERSEN and EDRICH, 1974) that it is only the ultraviolet type of receptor that mediates the analysis of polarized skylight. In the bee's eye the polarization sensitivity of such ultraviolet receptors reaches values as high as 10 (LABHART, 1980; WEHNER and BERNARD, 1981). We further know, as especially shown in this paper, that in a given patch of sky the orientation of the e-vector is detected independently of variations in intensity and degree of polarization. To accomplish this task three variables must be taken into account: e-vector orientation, degree of polarization, and radiant intensity. As a consequence, a set of three analysers (polarization-sensitive ultraviolet receptors), each characterized by its own analyser (microvillar) direction, is necessary to determine the orientation of an e-vector instantaneously. Alternatively, if the analysers are allowed to rotate about their optical axes, so that sequential readings can be taken, one analyser suffices for unambiguous e-vector detection (KIRSCHFELD, 1972; BERNARD and WEHNER, 1977).

In the strict sense mentioned above neither condition seems to be met in the bee's eye. Sets of ultraviolet receptors exhibiting three different microvillar directions have not yet been found either in individual ommatidia or in groups of adjacent ommatidia[13]. On the other hand, photoreceptors do not rotate within the eye. Nevertheless, as the foregoing suggests, bees are indeed able to rely on the orientation of the e-vector independently of the dedegree of polarization and the radiant intensity. Where do we go from here? In such complicated circumstances there is bound to be a certain uneasiness about the traditional way of posing the principal question. Thus, let us challenge the long-held view outlined above and entertain a new approach.

4.2 A new approach

The inconclusive nature of the neuroanatomical data presented at the end of the last paragraph drives home a major point: For all we know, there may be no *local* e-vector detectors[14] within the bee's visual system. However, there are *global* arrays of polarization-sensitive photoreceptors. Each individual photoreceptor is endowed with its own unique analyser direction[15], and the analyser directions of adjacent photoreceptors vary systematically. Even though a single photoreceptor ist not able to determine the orientation of an e-vector unambiguously, the whole array of photoreceptors certainly is. Just imagine what happens when the bee scans a small patch of polarized sky, i. e. turns about its vertical body axis and thus sweeps its array of retinal analysers through the patch of sky. While the bee turns, peak responses occur in that photoreceptor in which the analyser

---

13 In all but the few specialized ommatidia located at the dorsal margin of the bee's eye there is only one polarization-sensitive photoreceptor per ommatidium. This polarization-sensitive photoreceptor is the short ninth cell. Referring to the three-channel hypothesis according to which the orientation of an e-vector is determined by the interaction of three polarization-sensitive photoreceptors (KIRSCHFELD, 1972), RIBI (1980) has claimed that the short ninth cells of adjacent ommatidia form a regular pattern of triplets each characterized by three separate microvillar directions (0° [= vertical], +40°, −40°). Other workers, however, have not been able to confirm these findings (MENZEL and SNYDER, 1974; WEHNER and MEYER, 1981). Furthermore, as will be shown later (p. 44), the parts of the eye for which such triplets of analysers have been described by RIBI (1980), are not at all able to detect the polarization of skylight.

14 A *local e-vector detector* is defined as a local set of polarization-sensitive photoreceptors (e-vector analysers) that allows for detecting instantaneously and unambiguously the orientation of an e-vector.

15 The *analyser direction* of a polarization-sensitive photoreceptor is the direction of maximum sensitivity for linearly polarized light: Maximum responses occur within the photoreceptor when light is polarized parallel to its analyser direction. In all rhabdomeric photoreceptors studied so far the analyser direction coincides with the microvillar direction of the rhabdomere (unless the photoreceptor is twisted).

(microvillar) direction coincides with the orientation of the e-vector present in the patch of sky. This strategy sounds very simple indeed, but it has snags. For the model to work the bee must be informed about the analyser direction of the photoreceptor in which the maximal response occurs. Therefore, the ultimate usefulness of any such scanning model depends largely on how the analyser directions of adjacent photoreceptors are spatially arranged within the retina, and how well certain geometrical aspects of the skylight patterns are reflected by the array of the retinal analysers. Until the geometrical properties of the receptor array are well documented it seems premature to attempt to formulate specific hypotheses of how the scanning model proposed above actually works. Thus let us defer the full formulation of our hypothesis until later, and first have a closer look at the bee's eye.

(i) *Neurobiology: The ommatidia at the dorsal margin of the bee's eye exhibit striking structural specialisations which can be expected to have functional consequences for the detection of polarized light.*

At the uppermost dorsal margin of the bee's eye, the photoreceptors are arranged in a way that betokens specialization. All nine photoreceptors of each ommatidium are straight and of equal length. This is in stark contrast to what occurs in the remainder of the eye, where the photoreceptors are twisted and each ommatidium contains eight long photoreceptors and a short ninth cell (for details see WEHNER, BERNARD and GEIGER, 1975; WEHNER and MEYER, 1981). All ultraviolet receptors positioned a the dorsal margin of the eye are highly sensitive to polarized light. Measured values of polarization sensitivity[16] range up to 10, or even more. In all other ommatidia of the bee's eye the only photoreceptors that are sensitive to polarized light are the short ninth cells.

The most important thing of all to grasp from an anatomical reconstruction of this specialized area of the eye is the orderly, albeit somewhat curious, layout of microvillar directions. First, in each individual ommatidium the microvilli of the three ultraviolet receptors (cells R1, R5 and R9) are arranged in a mutually perpendicular way. The directions of the microvilli of R1 and R5 coincide and run at right angles to the microvilli of R9. Second, within the retina the ommatidial sets of orthogonally arranged photoreceptors form a striking fan-shaped pattern that spreads across the entire dorsal margin of the bee's eye (for details see WEHNER, 1982; MEYER, 1984).

When this pattern is projected into visual space, it reveals some gross geometrical properties of the bee's e-vector compass. Furthermore, as we have seen before, the bee's e-vector compass is based on the most prominent feature of skylight polarization – the distribution of the maximally polarized e-vectors. Taken together, these findings are extremely interesting, because they might imply that the bee's retina shares some important geometrical properties with the celestial patterns of polarization.

The structural and functional specilizations described so far[17] raise the intriguing question whether the ommatidia at the dorsal margin of the bee's eye are specialized for the

---

[16] Polarization sensitivity is defined as the sensitivity to light polarized parallel to the microvilli divided by the sensitivity to light polarized perpendicularly to the microvilli.

[17] In addition, there are other specializations that characterize the dorsal margin of the bee's eye. They refer to both the optics overlying the photoreceptors (MEYER and LABHART, 1981; WEHNER, 1982) and the visual neuropiles to which the photoreceptors project (MEYER, in press). Structural specializations similar to the ones described here for bees have been reported for ants (WEHNER, 1982), flies (WUNDERER and SMOLA, 1982; HARDIE, 1984), moths (MEINECKE, 1981) and crickets (BURGHAUSE, 1979; AEPLI et al., in press). A comparison between flies and bees is especially interesting because these two groups of insects differ strikingly in the structure of their rhabdoms,

detection of polarized skylight and, more specifically, whether the geometry of the fan-shaped receptor array tells us something about how this is done. Our curiosity was aroused even more by the fact that all specializations mentioned above were confined to an extremely small fraction of the bee's retina. The dorsal marginal ommatidia contribute 2.5 percent of the in total 5′600 ommatidia of the worker bee's eye. One may question whether such a small number of photoreceptors is really sufficient to handle and use all e-vector information available in the sky. On the other hand, it seems difficult to escape the conclusion that the geometrical similarities that exist between receptor array and skylight pattern are more than sheer coincidence. Furthermore, the fact that at the dorsal margin of the bee's eye each ommatidium contains two mutually perpendicular (crossed) analysers is clearly in favour of the view that this small marginal part of the eye has been shaped by natural selection to provide the bee with high-contrast e-vector signals. The reason behind this argument is that polarization-sensitivity can be enhanced by antagonistically comparing the outputs of two mutually perpendicular analysers (WATERMAN, 1981; WEHNER, 1982). Recent intracellular recordings from specialized marginal photoreceptors of the fly have shown that negative electrical interactions between orthogonally arranged ultraviolet receptors are likely to occur (HARDIE, 1984). Nevertheless, neurophysiological data and theoretical considerations are good as far as they go, but convincing evidence can be provided only by investigating the bee's behaviour. This is the topic to be discussed next.

(ii) *Behaviour: The ommatidia at the dorsal margin of the bee's eye are necessary for the detection of the polarized light in the sky.*

To test the hypothesis outlined in the preceding section the first crucial experiment to be performed is to anaesthetize a bee, paint out the specialized ommatidia at the dorsal margins of both its eyes, and let it dance again within the translucent Plexiglass dome. In the experiments described in Fig. 15 a a single beam of polarized ultraviolet light is displayed. When the specialized ommatidia of the two eyes have been occluded, the bees are no longer able to detect the e-vector. Instead, they behave as if the stimulus were an *unpolarized* beam of ultraviolet light, and place it anywhere within the anti-solar half of the sky (Fig. 15 c). This is immediately apparent from a comparison of Fig. 15 c and Fig. 13 c. In the latter case the beam of ultraviolet light is actually unpolarized, but the results of the two experiments are virtually the same. A number of additional experiments including those in which bees can view the sky only with the dorsal marginal ommatidia of their eyes (WEHNER, 1982; WEHNER and STRASSER, in prep.) are all in accord with the conclusion that these specialized ommatidia are indeed necessary for the detection of polarized skylight and thus comprise what can readily be called the POL area of the bee's eye.

---

which are open or fused, respectively. Yet the specializations of the rhabdoms that occur at the dorsal margin of the eye are strikingly similar. To make the point, let us briefly refer to the fly's eye. At the dorsal margin of the eye, the rhabdomeres of receptor cells R7 and R8 are greatly enlarged in diameter, reduced in length, and do not twist. Thus, high polarization sensitivity is to be expected, and has indeed been found (HARDIE, 1984). Furthermore, in contrast to the remainder of the eye in which R7 is the only type of ultraviolet receptor, at the dorsal margin of the fly's eye R7 and R8 are ultraviolet receptors. Due to this additional property R7 and R8 are similar in all respects except for their (mutually perpendicular) analyser directions. This makes such pairs of cells ideal input channels to any system used in the analysis of polarized light.

Finally, one example should be mentioned that looks like the exception to the rule. It is the water bug *Notonecta*. In this insect mutually perpendicular ultraviolet receptors occur in the ventral rather than dorsal part of the eye. However, when flying from one pond to another, *Notonecta* uses polarized light produced by reflections from water surfaces to decide where to descend and plunge into the water (SCHWIND, 1984).

Given this crucial role that the uppermost dorsal part of the eye plays in skylight vision, it may come as a surprise that bees are well oriented even when the POL areas of their eyes have been painted out – provided that large parts of the sky are available. We have already seen that bees treated this way can use the hue of colour to decide whether a small unpolarized stimulus belongs to the solar meridian or the anti-solar half of the sky (Fig. 13). Now, if we display to the bees the entire upper half of the sky (Fig. 3 b), the bees are oriented nearly as precisely as the controls in which the POL areas have been left open. In another experiment, a strip-like window, 90° wide and 10° high, is opened within the wall of the Plexiglass dome. In this case, the bees select the correct compass course when the

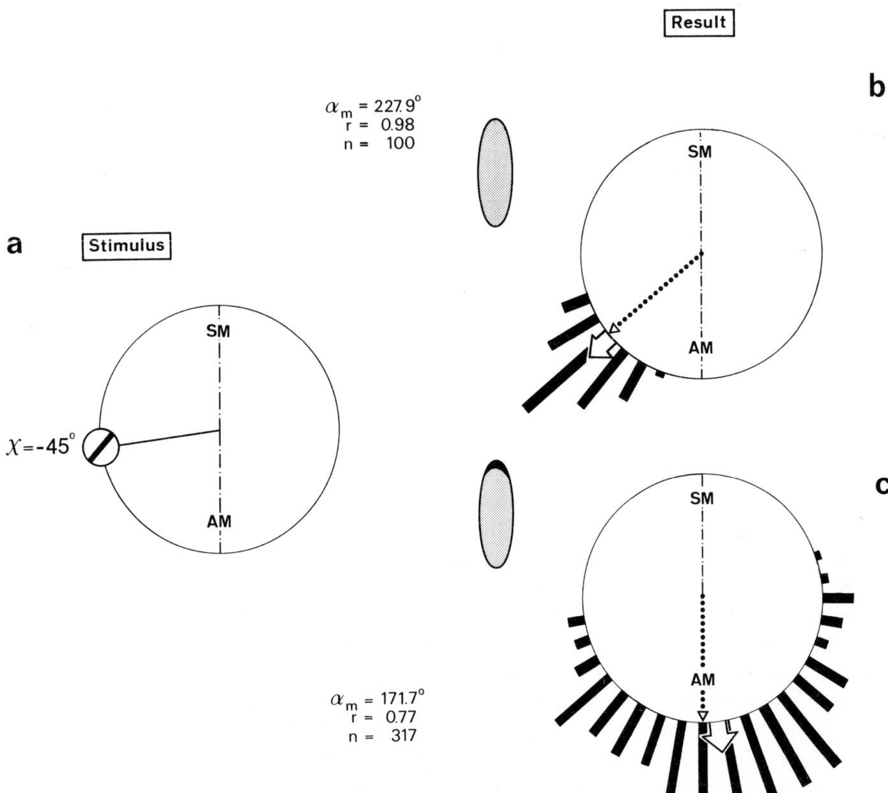

Fig. 15: Specialized ommatidia at the dorsal margin of the bee's eye are necessary for the detection of polarized skylight. In the present experiment, the stimulus is a polarized beam of ultraviolet light *(a)*. The experimental data *(black bars)* plotted in *b* and *c* show where the bees expect the stimulus *(a)* to occur in the sky. The controls *(b)* behave as is to be expected if they use their stereotyped e-vector compass and place the −45° e-vector accordingly. Actually, the data are the same as the ones shown in Fig. 5. On the other hand, the bees in which the specialized ommatidia at the dorsal margins of the eyes have been painted out (*c*) place the stimulus anywhere within the anti-solar half of the sky. This means that they take it for an unpolarized patch of ultraviolet light (compare Fig. 13 c). Conclusion: The ommatidia outside the uppermost dorsal part of the eye do not detect the polarization of light. They are used to interpret a celestial stimulus on the basis of its hue of colour.

skylight window is centred about the anti-solar meridian, but deviate by 45° to the left or right, when the sky-light window is shifted by 45° to the left or right, respectively. Obviously, they resort to spectral cues in the sky. If they had used the e-vector pattern available in the strip-like window, their orientation errors would have been much smaller than 45° (compare e.g. Fig. 12). In the mean, the bees with the uppermost dorsal ommatidia painted out take the centre of the patch of sky positioned to the left or right of the anti-solar meridian for the anti-solar meridian itself.

These and similar experiments are described and discussed in detail by Rossel and Wehner (in press). What they show is that the uppermost dorsal part of the eye is necessary for the detection of polarized skylight (POL area). The remainder of the dorsal part of the eye is able to exploit spectral cues in the sky. The latter cues, however, are conveniently reliable only when large patches of the sky are available and when such large patches include the symmetry line of the skylight patterns.

(iii) *Model: Bees determine the position of the solar or anti-solar meridian by sweeping the POL areas of their eyes through the celestial e-vector pattern.*

Armed with the neurobiological and behavioural data just described, we now propose a model in which the POL area of the bee's eye is used in a rather direct way to determine the position of the solar or anti-solar meridian in the sky.

What information does the POL area of the eye extract from the celestial e-vector patterns? We had started with the conjecture, however vague, that understanding the receptor array within the POL area will go part of the way of understanding how the bee's celestial compass might work. Now that we have shown that this part of the eye is indeed necessary for analyzing polarized skylight, it is time to consider this possibility in greater detail.

As proposed earlier (p. 42), the bee could determine the orientation of an e-vector by sweeping its array of retinal analysers through the e-vector and detecting in which analyser (photoreceptor) the peak response occurs. Recall, however, that there is a problem. In order to read the orientation of the e-vector from the peak response, the bee must somehow know what the orientation of the analyser is within which the peak response occurs. It must further know where this e-vector is positioned in the sky with respect to the solar meridian. The whole problem could be solved most elegantly, if the geometry of the array of analysers mimicked the geometry of the celestial e-vector pattern to be analyzed. In the following, we shall propose a model eye in which this condition is met.

In the model eye, the spatial layout of the receptor array is assumed to match the spatial layout of the maximally polarized e-vectors in the sky. This is not a completely arbitrary assumption. We had seen that bees expect any e-vector to occur in its position of maximum polarization, and beyond that we had argued that within the POL area of the bee's eye the retinal array of ultraviolet receptors[18] shares some principal geometrical properties with the celestial array of the maximally polarized e-vectors. For our model to work, the bee must scan the sky with its model array of photoreceptors, and associate the occurrence

---

[18] For the sake of simplicity, we refer only to one analyser direction per ommatidium. Actually there are two (see p. 43), but the model works even if there were only one. Two mutually perpendicular analysers that interact antagonistically merely enhance the modulation of the signal that is mediated by a single analyser. Of course, such antagonistic interactions are an important improvement of the system, but they are not essential for understanding the rationale behind our scanning model.

of peak responses with being aligned with the solar or anti-solar meridian. To exemplify this point, let us select three stimulus conditions, which have been used before in the behavioural experiments, and discuss the bee's behaviour in terms of the scanning strategy proposed above.

First, the bee views a small patch of sky in which the e-vector is in its position of maximum polarization. In this stimulus condition, peak responses occur whenever the bee is aligned exactly with the solar or anti-solar meridian. This is because the array of analysers matches the celestial distribution of the maximally polarized e-vectors, and in the experiment described the e-vector is actually presented in its position of maximum polarization. Therefore, the bee can correctly assume that it faces the solar or anti-solar meridian whenever it experiences a peak response. No navigational errors should occur. Indeed, none have been observed.

Why do peak responses occur when the bee faces either the solar or the anti-solar meridian rather than only one of the two meridians? The reason for this ambiguity is that within the bee's receptor array each analyser direction occurs twice: once in the left (right) eye and once again in the diametrically opposite position in the right (left) eye. Owing to this peculiarity two peak responses are elicited whenever the bee performs a full turn

Celestial e-vector gradient

Spatial distribution of maximally polarized e-vectors in the sky. *AM* anti-solar meridian, *SM* solar meridian. Compare Figs. 7c and 10b.

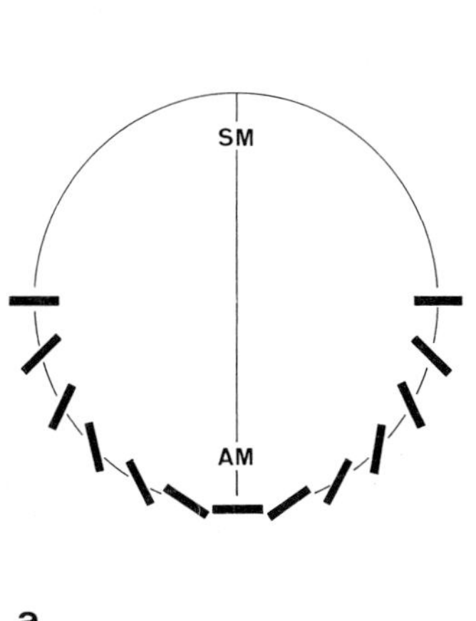

Retinal array of analysers (model)

Spatial distribution of analyser (microvillar) directions within the retina. At the dorsal margin of the bee's eye (POL area) the microvilli of the ultraviolet receptors are arranged in a way similar to that shown here. *L*, *R* left and right visual field, respectively.

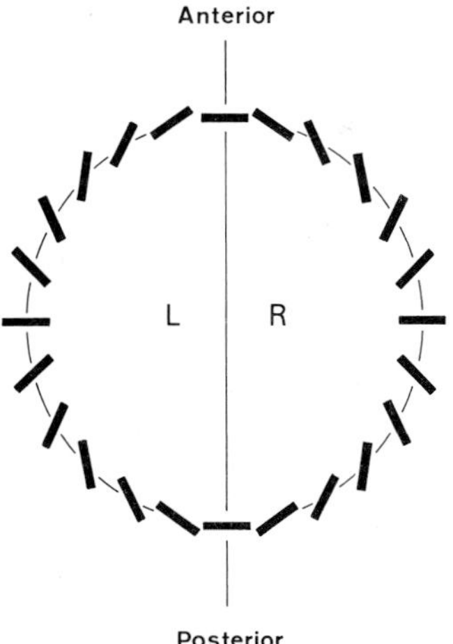

Fig. 16: Scanning model. When the model bee rotates about its vertical body axis and thus sweeps its array of analysers *(b)* through the sky *(a)*, peak responses occur whenever the bee is aligned with the solar or anti-solar meridian. A simple rule helps solve the ambiguity: When the peak response occurs in the anterior or posterior part of the visual field, the bee faces the anti-solar or solar meridian, respectively. How such a scanning system can be used as a celestial compass is described in detail in the text.

(scannning cycle). As our behavioural data suggest, bees use a very simple rule to solve this ambiguity. They associate peak responses invariably with the anti-solar meridian. Thus, when they experience a peak response in the anterior half of the visual field, they assume themselves to be facing the antisolar meridian, but when the peak response occurs in the posterior half of the visual field, they assume that they have the anti-solar meridian in their back and thus face the solar meridian. This simple rule applies to all but the vertical e-vectors. As the vertical analyser directions occur in the bee's lateral field of view, exactly at right angles to the forward and backward direction, the bee cannot distinguish the solar from the anti-solar meridian when it is confronted with a vertical e-vector alone. As a consequence, it orients in two preferred directions with one lying diametrically opposite to the other (Fig. 9).

Second, the bee views a small patch of sky in which the e-vector is not in its position of maximum polarization. Again, the bee while scanning the sky assumes that it is aligned with the solar or anti-solar meridian whenever a peak response occurs, but in this case the bee is mistaken. As the bee's receptor array reflects the celestial pattern of e-vectors only when the e-vectors are in their positions of maximum polarization, errors must occur whenever the e-vectors are not in this position, as is usually the case. Thus, the position of the solar or anti-solar meridian as assumed by the dancing bee deviates from the actual position of the corresponding meridian in the sky, and the directions of the bee's dances deviate by exactly that amount from the direction of the food source (Fig. 5). It was on the basis of these orientation errors that the bee's stereotyped e-vector compass was derived originally.

Third, the bee views a large part of the blue sky. Our hypothesis implies that even in this situation, with many e-vectors being provided, a bee is bound to use its simple scanning strategy. Of course, of the many e-vectors present in the patch of sky only a few will be in their positions of maximum polarization and only these few will simultaneously match the bee's receptor array whenever the bee faces the solar or anti-solar meridian. Any other e-vector will induce a peak response when the bee is not aligned exactly with the solar or anti-solar meridian, and thus will indicate an incorrect position of the corresponding reference meridian. How does the bee cope with this potentially ambiguous situation caused by the multitude of possible solar or anti-solar meridians? Our model proposes that the bee has no means of sorting out what peak response is induced by a maximally polarized e-vector and what by another one. A simple rule to apply is to take the mean of all possible positions of the solar or anti-solar meridian, and this is what the bee seems to do. In all our experiments in which bees were presented simultaneously with a number of e-vectors, the error angle exhibited by the bees coincided with the arithmetical mean of the error angles induced by each individual e-vector alone (see Figs. 11 and 12).

To sum up, in the three examples described above, as well as in all other experiments performed so far, the bee's behaviour is consonant with the scanning model. Although this model has been sketched out here rather hastily, it should have been possible for the reader to grasp the gist of the argument and to discover the rationale behind it. At the heart of the hypothesis is the assumption that while scanning the sky the bee translates e-vector orientations into temporal modulations of receptor responses. All that the bee derives from these responses is *when* during a scanning cycle it faces the solar (or anti-solar) meridian. A simple way of accomplishing this task is to use an array of analysers that, to a certain degree, reflects the spatial layout of the celestial e-vector patterns. In our model we have assumed that the array of analysers mimics the celestial distribution of the maximally polarized e-vectors. The gross features of the peculiar fan-shaped receptor array within the POL area of the bee's eye support this assumption.

In this context, a further peculiarity of the POL area must be mentioned, which might put the final piece of evidence in place. The visual fields of the photoreceptors at the dorsal margin of the eye are exceedingly wide but the polarization sensitivities of the photoreceptors do not decrease with off-axis stimulation (LABHART, 1980). The wide visual fields are mainly due to structural specializations of the corneal lenses, which contain dense arrays of rugged-walled pore canals widely scattering the incoming light (MEYER and LABHART, 1981; AEPLI, LABHART and MEYER, in press.). Consequently, the ultraviolet receptors of the POL area act as wide-field e-vector analysers. At first glance, the low acuity of this part of the eye seems to be a rather severe constraint on the general utility of the POL area as a means of deriving spatial information from the sky. On closer scrutiny, however, this may be just an interesting example of a constraint being turned into a virtue. First, the large visual fields of photoreceptors considerably increase the area of the sky that can be scanned by the rather limited number of POL ommatidia at the dorsal margin of the eye. Furthermore, wide-field sampling may be advantageous not only in rendering e-vector signals less susceptible to disturbances caused, for example, by clouds or the canopy of vegetation, but also in averaging out e-vector orientations. Finally, if wide-field sampling is accomplished by a corresponding degree of neural pooling, spatial irregularities within the array of receptors dwindle in importance. In any event, a wide-field, low-acuity detector system seems to be well suited for deriving *global* spatial information from the sky – such as information concerning the position of the solar or anti-solar meridian.

Once the bee has aligned itself with the solar or anti-solar meridian, it must turn about its vertical body axis until it faces the direction of the compass course intended. To accomplish this task it could resort to a number of strategies. Only one possibility shall be mentioned and discussed briefly: The bee could use any distant visual cue to select its final compass course.

To illustrate this point, take our basic experiment (Figs. 4 and 5) in which a single spot of polarized light is displayed within the otherwise featureless surroundings of the Plexiglass dome. The compass course intended by the bee is 15° to the left of the anti-solar meridian. First, the bee would align itself with, say, the anti-solar meridian by turning around and tuning in for peak responses to occur within the POL areas of its eyes. It is here that errors will occur (unless the e-vector presented to the bee is in its position of maximum polarization). Then, having aligned itself with the anti-solar meridian – or what it thinks is the anti-solar meridian – the bee must turn so as to shift the retinal image of the spot by 15° to the right, and lock on to it. Of course, outside the sterile visual world of our Plexiglass dome, there are usually many more visual cues available for selecting the final compass course than the single beam of light displayed in our basic experiment: Clouds, isolated patches of sky, or distant landmarks may be used. Setting its compass course by referring to such cues is very similar to what a bee would usually do when it uses the sun as a reference cue.

As we have seen before, there is no need for fine grain vision in accomplishing the first step of the task. Indeed, the POL area exhibits low visual acuity. However, fine grain vision is certainly required for accomplishing the second step of the task. Such fine grain vision is provided only by ommatidia lying outside the POL area. As the POL area of each eye looks contralaterally, it shares a common field of view with parts of the contralateral eye that lie outside the POL area and thus exhibit the high visual acuity afforded. Note, however, that in principle for setting the final compass course any non-POL part of the eye could be used, irrespective of whether it belongs to the monocular or binocular field of view.

A final point should be mentioned. Once the bee has selected its proper compass course, it again could use the pattern of polarized light in the sky for maintaining this course. The reason is that marked response modulations occur within the retinal image of the celestial e-vector pattern whenever the bee deviates from a straight course (see the sequence of false colour images in WEHNER, 1982).

5 Epilogue: Speculation built on fact

What we have proposed in this paper is a rather simple solution to the long-standing puzzle of how bees analyse the polarized light in the sky. The most thought-provoking feature of our hypothesis is that bees are not able to determine e-vector orientations *locally* and *instantaneously* in any particular part of the sky. Instead, bees seem to use the celestial polarization merely to determine, in a rather direct way, the position of the solar (or anti-solar) meridian, and seem to do so by applying some kind of *global scanning* strategy.

More specifically, our hypothesis implies that in deriving compass information from the sky, bees use an array of photoreceptors within which the microvillar orientations vary systematically, and sweep this array of photoreceptors through the celestial e-vector pattern. This hypothesis is substantiated by our finding that a specialized small area at the dorsal margin of the bee's eye, termed POL area, is necessary for detecting polarized skylight, and that within this POL area the receptor array shares some principal geometrical properties with the e-vector patterns in the sky. Based on these observations, a model bee is designed, which uses e-vector patterns as a compass by following a rather simple rule: turn around, and whenever peak responses occur in the POL area of the eye, know that the solar (or anti-solar) meridian is straight ahead. While stationary, the bee would not know *where* the solar (or anti-solar) meridian is positioned in the sky, but while turning it would be able to determine *when* it faces the solar (or anti-solar) meridian.

Of course, our model owes its simplicity to its limitations. Unlike physicists, bees cannot determine the orientation of any arbitrary e-vector. They can derive useful information only from e-vector patterns that are designed according to the rules of Rayleigh scattering. This is because some principal design features of their receptor arrays have been adopted from these patterns. As receptor arrays are fixed, but celestial e-vector patterns are not (they vary with the elevation of the sun), navigational errors must occur under a variety of stimulus conditions – and they do exactly as predicted. In our model bee, however, they are minimized since we have allowed evolution to be intelligent: the retinal array of analysers mimics the most prominent feature of skylight polarization, that is the spatial distribution of the maximally polarized e-vectors. All behavioural data we have obtained so far under different experimental conditions are in full accord with the behaviour of the model bee.

We note in passing that our hypothesis disposes of the possibility that bees perceive polarized skylight as an extra quality of light. It is sufficient to assume that bees while scanning the sky translate e-vector orientations into modulations of perceived ultraviolet intensity (or saturation). Besides their polarization-sensitive ultraviolet channels bees use polarization-insensitive channels to read spectral information from the sky. The intriguing question remains as to whether and where along the visual pathway the polarization-sensitive and the polarization-insensitive channels interact.

Finally, a cautionary note must be added. The hypothesis outlined above is included in this chapter as the current best bet about how bees deduce compass information from the sky, not as received fact. Further research over the next few years will certainly shape and perhaps modify our present ideas. For example, the extent to which the array of receptors at the dorsal margin of the bee's eye really matches the bee's e-vector compass that we had deduced from our behavioural data, has still to be determined. For the proposed model to work, however, a complete match is not required. Another question relates to possible interactions between the dorsal marginal receptors (POL area) and the remainder of the dorsal retina. Of special interest is the contribution of those parts of the dorsal retina that share a common field of view with the POL area of the contralateral eye.

Notwithstanding many pieces of evidence that are still missing, the hypothesis meets the primary criterion of a successful explanation. It coordinates a series of observations that would otherwise remain unconnected and, in many cases, downright peculiar. Finally, the hypothesis is simple rather than complex and contrived. It implies that the bee's behaviour results from a set of simple rules deduced from and operating within a complex world.

For obvious reasons we have entitled our analysis of the bee's celestial compass a case study in behavioural neurobiology. It was only by keeping in line with VON FRISCH's early approach of combining behavioural studies alongside neurobiology that the bee's compass strategy could have been unravelled. In fact, we had discovered many optical, anatomical, and neurophysiological details of the ommatidia at the uppermost dorsal margin of the bee's eye long before the behavioural data enabled us to interpret these neurological idiosyncracies in functional terms and to establish the concept of the bee's «POL area». Therefore let us conclude by referring to how the neurobiologist GORDON SHEPHERD (1983) has recently paraphrased to the point the well-known aphorism of the evolutionary biologist TTHEODOSIUS DOBZHANSKI: «Nothing in neurobiology makes sense – except in the light of behaviour».

Appendix

Symbols and definitions used in the figures of this chapter.

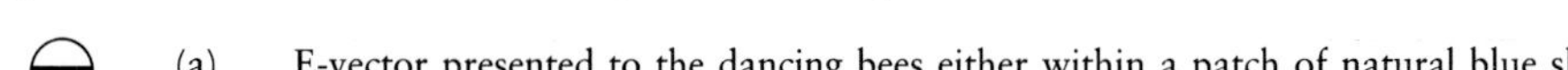

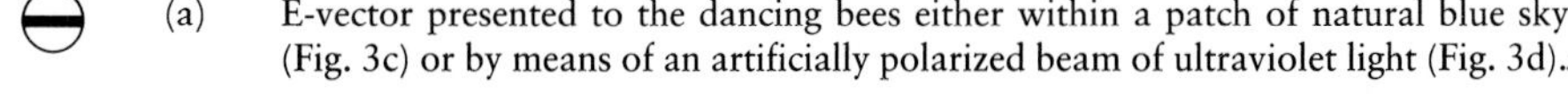

(a) E-vector presented to the dancing bees either within a patch of natural blue sky (Fig. 3c) or by means of an artificially polarized beam of ultraviolet light (Fig. 3d).

(b) Artificially produced e-vector presented at an elevation at which it does not occur in the natural sky at the time of day at which the experiment is performed.

(c) E-vector shown at the azimuth position at which it occurs within the bee's e-vector compass (Fig. 6, inner circle; Fig. 7a).

(d) Direction of food source (*F*).

(e) Direction in which the bee expects the food source to occur when using the compass strategy outlined in this paper. This direction is not denoted by a separate symbol when (d) and (e) coincide.

(f) Mean direction of orientation as determined experimentally (α_m).

(g) Error angle (ε) observed in the experiments.

(h) Direction in which the bees expect a given celestial cue (e. g. a particular e-vector orientation, the solar or the anti-solar meridian) to occur in the sky. In Figs. 13–15 the directions of the bees' dances *(black bars)* are plotted so as to coincide with the *dotted line* when the bees behave according to the expectation, e. g. assume the stimulus to lie along the solar or anti-solar meridian, or in any other direction as predicted from the bee's celestial compass.

–·–·–·– (i) Symmetry plane of the celestial patterns. It includes the solar and anti-solar meridian.

AM anti-solar meridian, *F* direction of food source, *N* north, *n* number of waggle runs recorded, *r* length of mean orientation vector (for statistical significance see BATSCHELET, 1981), *SM* solar meridian, α_m direction of mean orientation vector, ε error angle, η scattering angle, χ e-vector orientation (for conventions see Figs. 1b and 2b).

References

Aepli, F., Labhart, T., Meyer, E. P.: Structural specializations of the cornea and retina at the dorsal rim of the compound eyes in hymenopteran insects. In press.

Batschelet, E. (1981): Circular Statistics in Biology. Academic Press, London, New York.

Bernard, G. D., Wehner, R. (1977): Functional similarities between polarization vision and color vision. Vision Res., *17,* 1019–1028.

Bisetzky, A. R. (1957): Die Tänze der Bienen nach einem Fußweg zum Futterplatz. Z. vergl. Physiol., *40,* 264–288.

Brines, M. L., Gould, J. L. (1979): Bees have rules. Science, *206,* 571–573.

Brines, M. L., Gould, J. L. (1982): Skylight polarization patterns and animal orientation. J. exp. Biol., *96,* 69–91.

Burghause, F. (1979): Die strukturelle Spezialisierung des dorsalen Augenteils der Grillen (Orthoptera, Grylloidea). Zool. Jb. Physiol., *83,* 502–525.

Cartwright, B. A., Collett, T. S. (1983): Landmark learning in bees. Experiments and models. J. Comp. Physiol., *151,* 521–543.

Crozier, W. J., Mangelsdorf, A. F. (1924): A note on the relative photosensory effect of polarized light. J. gen. Physiol., *6,* 703–709.

Duelli, P., Wehner, R. (1973): The spectral sensitivity of polarized light orientation in *Cataglyphis bicolor* (Formicidae, Hymenoptera). J. Comp. Physiol., *86,* 37–53.

Dyer, F. C., Gould, J. L. (1983): Honey bee navigation. Amer. Scientist, *71,* 587–597.

Edrich, W., Neumeyer, C., Helversen, O. von (1979): «Anti-sun orientation» of bees with regard to a field of ultraviolet light. J. Comp. Physiol., *134,* 151–157.

Frisch, K. von (1949): Die Polarisation des Himmelslichts als orientierender Faktor bei den Tänzen der Bienen. Experientia, *5,* 142–148.

Frisch, K. von (1965): Tanzsprache und Orientierung der Bienen. Springer, Berlin, Heidelberg, New York.

Glas, H. W. van der (1976): Polarization induced colour patterns: A model of the perception of the polarized skylight by insects. II. Experiments with direction trained dancing bees, *Apis mellifera.* Neth. J. Zool., *26,* 383–413.

Glas, H. W. van der (1980): Orientation of bees, *Apis mellifera,* to unpolarized colour patterns, simulating the polarized zenith skylight pattern. J. Comp. Physiol., *139,* 225–241.

Gould, J. L. (1980): Sun compensation in bees. Science, *207,* 545–547.

Hardie, R. C. (1984): Properties of photoreceptors R7 and R8 in dorsal marginal ommatidia in the compound eyes of *Musca* and *Calliphora.* J. Comp. Physiol. A, *154,* 157–165.

Helversen, O. von, Edrich, W. (1974): Der Polarisationsempfänger im Bienenauge: ein Ultraviolettrezeptor. J. Comp. Physiol., *94,* 33–47.

Kirschfeld, K. (1972): Die notwendige Anzahl von Rezeptoren zur Bestimmung der Richtung des elektrischen Vektors linear polarisierten Lichtes. Z. Naturforsch., *27c,* 578–579.

Kirschfeld, K., Lindauer, M., Martin, H. (1975): Problems of menotactic orientation according to the polarized light of the sky. Z. Naturforsch., *30c,* 88–90.

Labhart, T. (1980): Specialized photoreceptors at the dorsal rim of the honeybee's compound eye: polarizational and angular sensitivity. J. Comp. Physiol., *141,* 19–30.

Meinecke, C. C. (1981): The fine structure of the compound eye of the African armyworm moth, *Spodoptera exempta* (Lepidoptera: Noctuidae). Cell Tissue Res., *216*, 333–347.

Menzel, R., Snyder, A. W. (1974): Polarized light detection in the bee, *Apis mellifera*. J. Comp. Physiol., *88*, 247–270.

Meyer, E. P.: Retrograde labeling of photoreceptors in different regions of the compound eyes of bees and ants. J. Neurocytol. In press.

Meyer, E. P., Labhart, T. (1981): Pore canals in the cornea of a functionally specialized area of the honey bee's compound eye. Cell Tissue Res., *216*, 491–501.

Ribi, W. A. (1980): New aspects of polarized light detection in the bee in view of non-twisting rhabdomeric structures. J. Comp. Physiol., *137*, 281–285.

Rossel, S., Wehner, R. (1982): The bee's map of the e-vector pattern in the sky. Proc. Natl. Acad. Sci. USA, *79*, 4451–4455.

Rossel, S., Wehner, R. (1984): How bees analyse the polarization patterns in the sky. Experiments and model. J. Comp. Physiol. A, *154*, 607–615.

Rossel, S., Wehner, R.: Celestial orientation in bees: The use of spectral cues. J. Comp. Physiol. A. In press.

Rossel, S., Wehner, R., Lindauer, M. (1978): E-vector orientation in bees. J. Comp. Physiol., *125*, 1–12.

Santschi, F. (1923): L'orientation sidérale des fourmis, et quelques considérations sur leurs différentes possibilités d'orientation. I. Classification des diverses possibilités d'orientation chez les fourmis. Mém. Soc. Vaud. Sci. Nat., *4*, 137–175.

Schwind, R. (1984): Evidence for true polarization vision based on a two-channel analyser system in the eye of the water bug, *Notonecta glauca*. J. Comp. Physiol. A, *154*, 53–57.

Shepherd, G. M. (1983): Neurobiology. Oxford University Press, Oxford, New York.

Stockhammer, K. (1959): Die Orientierung nach der Schwingungsrichtung linear polarisierten Lichtes und ihre sinnesphysiologischen Grundlagen. Erg. Biol., *21*, 23–56.

Verkhovskaya, I. N. (1940): L'influence de la lumière polarisée sur la phototaxis de quelques organismes. Bull. Soc. Nat. Moscou, Sect. Biol., *49*, 101–113 [in Russian, French summary].

Waterman, T. H. (1981): Polarization sensitivity. In: Handbook of Sensory Physiology, Vol. VII/6B, ed. H. Autrum, pp. 281–469. Springer, Berlin, Heidelberg, New York.

Waterman, T. H. (1984): Natural polarized light and vision. In: Photoreception and Vision in Invertebrates, ed. M. A. Ali, pp. 63–114. Plenum Press, New York.

Wehner, R. (1981): Spatial vision in arthropods. In: Handbook of Sensory Physiology, Vol. VII/6C, ed. H. Autrum, pp. 287–616. Springer, Berlin, Heidelberg, New York.

Wehner, R. (1982): Himmelsnavigation bei Insekten. Neurophysiologie und Verhalten. Neujahrsbl. Naturforsch. Ges. Zürich, *184*, 1–132.

Wehner, R. (1983): Celestial and terrestrial navigation: human strategies – insect strategies. In: Neuroethology and Behavioral Physiology, eds. F. Huber and H. Markl, pp. 366–381. Springer, Berlin, Heidelberg, New York.

Wehner, R. (1984): Astronavigation in insects. Ann. Rev. Entomol., *29*, 277–298.

Wehner, R., Bernard, G. D. (1980): Intracellular optical physiology of the bee's eye. II. Polarizational sensitivity. J. Comp. Physiol., *137*, 205–214.

Wehner, R., Lanfranconi, B. (1981): What do the ants know about the rotation of the sky? Nature (Lond.), *293*, 731–733.

Wehner, R., Meyer, E. P. (1981): Rhabdomeric twist in bees – artefact or in vivo structure? J. Comp. Physiol., *142*, 1–17.

Wehner, R., Rossel, S. (1983): Polarized light navigation in bees: use of zenith and off-zenith e-vectors. Experientia, *39*, 642.

Wehner, R., Strasser, S.: The POL area of insect eyes: behavioural experiments with bees. In prep.

Wehner, R., Bernard, G. D., Geiger, E. (1975): Twisted and non-twisted rhabdoms and their significance for polarization detection in the bee. J. Comp. Physiol., *104*, 225–245.

Wunderer, H., Smola, U. (1982): Morphological differentiation of the central visual cells R 7/8 in various regions of the blowfly eye. Tissue and Cell, *14*, 341–358.

Zolotov, V. V., Frantsevich, L. I. (1973): Orientation of bees by the polarized light of a limited area of the sky. J. Comp. Physiol., *85*, 25–36.

Fortschritte der Zoologie, Bd. 31 · Hölldobler/Lindauer (Hrsg.): Experimental Behavioral Ecology
G. Fischer Verlag · Stuttgart · New York · 1985

Randolf Menzel

Learning in honey bees in an ecological and behavioral context

Freie Universität Berlin, Fachbereich Biologie, F. R. Germany

Abstract

The search strategy of a generalist pollinator such as the honey bee is controlled to a great extent by individual experience at food sources. However, both learning of the features of a food source and choice behavior of potential food sources prior to any learning is guided by «prepared search images», which result from the coevolutionary adaptations between the signals of angiosperm plants and the sensory-neural capacity of the general pollinator. «Prepared search images» are hard to prove. Spontaneous choice behavior excluding prior learning has not yet been adequately analysed. More information comes from experiments which measure associative predispositions, signal preparedness and selectivity of associations in appetitive learning experiments. Useful behavioral measures are e. g. rate of acquisition, precision of learning or correctness of learned behavior, retention, sensitivity to extinction. The first example discussed in this paper is color learning. Bees learn any color, but violet is the color learned fastest and bluish-green the one learned slowest. The differential preparedness for color learning is not related to receptor or peripheral visual integration processes, but represents a central nervous weighting function. This function is not context specific, but applies both to learning at the feeding station and at the hive entrance. Other examples discussed in this paper are: learning of color patterns and learning of odors.

Another question relates to the species-specificity of color learning. Three hymenopteran species *(Apis mellifera, Melipona quadrifasciata, Osmia bicornis)* are compared. We find small but significant differences. Again, the general learning ability in all three species is the major effect and preparedness appears as an additional tuning factor.

Two other aspects of the honey bee's learning behavior are studied with respect to the neural basis of learning and memory and their ecological constraints. Reversal learning experiments demonstrate that bees are prepared to switch to new food signals after a certain number of initial trials. Short-term memory in bees has been found to be a neural strategy of fast adaptations to changing profitability of food sources, and not as a necessary mechanism to bridge time between the learning trial and the termination of slow processes establishing long-term memory. These two examples are used to demonstrate that learning both in the short and long term range is adapted to temporal and spatial features of natural food sources. «Prepared search images» include these features as well as the external signals for recognition.

It is concluded that the neural mechanisms of learning in the bee are controlled both by a strong inherent learning ability, and by species-specific preparedness, which result from the species evolutionary history.

1 Introduction

The two sources of information controlling behavior are the phylogenetic experience encoded in the genetically controlled wiring of the nervous system and the individual experience which modulates, alters, and adds neural pathways. It is common sense for the ethologist, but seemingly a quite recent revolutionary insight for the experimental psychologist that the effect of individual experience through learning is constrained by preprogrammed («prepared») combinations of stimuli and responses. As a result of selective pressure, certain adaptive responses to commonly encountered stimuli from the environment are genetically favored. This applies both to appetitve behavior – (responses to stimuli that in the history of the species have been associated with food, water, mating, safety, or relief of discomfort), and to aversive behavior (– responses to threatening or noxious stimuli).

In insects, the balance between phylogenetic and individual experience has long been thought to be on the side of phylogenetic experience; although insects are equipped with an astonishingly complex behavioral repertoire, they have been thought to act like robots following step by their instinctive guidance. Indeed, genetically programmed behavior can reach an unbelievably high degree of sophistication in insects, but in social hymenoptera indivual learning plays an important role. When a forager bee searches for food, for example, it may fly hundreds of meters in an irregular way until finding nectar or pollen producing flowers. In its flight back to the hive, the bee takes a straight path along the shortest distance, thereby integrating all the turns and using landmarks and the sun compass for orientation (v. Frisch, 1965). Obviously, the bee must learn the position of the landmarks, its relation to the wind, and its alignment to celestal cues, but programming tells the bee which signal should be attended to and stored at which time during its flight. The sun and the polarized light pattern are instinctively used as guides by the bee, but their changes in time have to be learned (Lindauer, 1959). Stimuli immediately surrounding the food source, such as colors, odors and shapes, are learned by the individual bee, and, as we shall see, the stimuli and the associations are ranked in a programmed hierarchy. I shall call those stimuli and those associations, which rank high in this hierachy «prepared stimuli» and «prepared associations». Preparedness, then, is used in this context with a wide ethological meaning, wider than the strict sense of preparedness as used in learning theory of experimental psychology (see Menzel, et al. in press).

Color and odor learning in bees at the feeding place has been a central theme in Karl von Frisch's work from the very beginning and was so through out his long and productive life. In his first and leading paper published in 1914 («Der Farbensinn und Formensinn der Biene, Zoologische Jahrbücher) he stressed the point that all the amazing sensory and learning capacities of bees have to be understood against the background of their biological adaptations. A good part of these adaptations are the result of a 80 million years long co-evolution of bees and angiosperm plants from which they obtain their carbohydrate (nectar) and protein (pollen) diet, and for which they, in turn function as species-specific fertilizing machines. As has been pointed out many times (e.g. Free, 1970, McGregor, 1976. Faegri, van der Pijl, 1978, Kevan and Baker, 1983, Proctor, Yeo, 1983) it is the discrepancy between the foraging behavior optimal for the plant and that optimal for the pollinator that has determined the evolution of features of plants and the sensory-neural capacities of the pollinators (see table 1 for a short summary).

On the side of the bee, color and odor perception and learning ability play a crucial role in the process of mutual adaptations. It should be advantageous for the bee not only to perceive the signal sent off by the flowers but also to «know» in advance what a potential food source looks like and how it may smell. In other words, the bee may have an «*innate search image*» for flowers (KEVAN and BAKER, 1983), which, through experience, develops into a precise *learned search image.*

Honey bees flying from flower to flower follow the search strategy of a generalist pollinator, which is not genetically bound to one or a few plants, but optimizes its profit by taking advantage of its social life style and relatively long individual life time. This means, that the individual bee changes its nectar and pollen sources frequently (see SEELEY, this Vol.) always trying to get the best out of the offerings available. Too strong «innate search images» are dangerous sources of information for the guidance of choices, since they may prevent the bee from finding a potentially better food source and from learning the features of that food source for later guidance. The bee, therefore, should be capable of both selecting a flower prior to any experience because it might be a potential food source *and* it should be able to learn very quickly the features fo that flower, if it indeed turns out to be a profitable food source.

The plant is involved in this ciritcal interplay of genetically prepared and individually experienced information too, because the plant wants to be both attractive to a pollinator prior to any individual experience *and* it wants to ensure a highly preferential choice of conspecific flowers after a visit of a pollinator. It should support the bee's behavior by both using common flower features and species specific features. As many flower species compete for adaptive *and* selective pollinators, each flower species should be mainly interested in being different from other flower species and still being recognized as a potential food source before any individual learning. This strategy should then put strong selective pressure on adaptive pollinators to improve their learning capacity within a framework of expected flower features (Tab. 1). In any case, we might expect a sensitive balance between «innate search images» and general learning capacity in such a general pollinator as the honey bee.

An «innate search image» not only allows the bee to select an object as a potential food source prior to learning, but also should facilitate learning of flower-like features and help to make the choice behavior more precise after learning.

The term «search image» is used here in a pragmatic sense without too many theoretical implications about its neural correlate. The term will be used as a short expression for an inferred neural representation of a compound of stimuli in a given behavioral context which has the potential of being compared with external stimuli. The addition «innate» has to be used in an even looser sense, since there is no direct evidence for the innateness of any of the «search images» which are thought to exist prior to a learning experiment. However, there is suggestive evidence in many experiments that what will be called «innate» exists prior to individual learning. The strongest evidence is that the results are independent of the age of the experimental bees, independent of the different experience of the bees and independent of the ecological conditions (e. g. seasons), at which the experiments were carried out.

When we ask the question of whether bees apply «innate search images» in their learning behavior of food sources, we look for associative predispositions, signal preparedness and selective associations in their appetitive learning. Do bees preferentially choose certain stimuli when they are searching for a food source? Are certain stimuli learned faster as food signals and do they acquire better control over the choice behavior? If there are such prepared stimuli, are they selective with respect to the behavioral context, e. g. prepared as food signals but not as signals for the hive entrance? Are such stimuli general features of flower visiting insects or are they species specific? We shall see that learning in bees is embedded in a contextual frame work which can be well understood against the background of the ecological adaptations mentioned above. However, compared with the powerful general learning ability both at the food source and at the hive

Table 1: Summary of factors in plant – pollinator coevolution

Bee	Plant
gain as much nectar and pollen with as little investment as possible; calculate net profit; adjust changes; communicate with comrades	gain as much specific pollen transfer with as little costs as possible. Keep pollinator long on the flower to increase chance of pollen deposit
The «Generalist – strategy»: Social bees potentially active all year around; They adapt to changes on the nectar/pollen market and use a strategy to optimize net profit; Flowers compete for individually adaptive pollinators;	
Perceive the signals	Mark the flowers
– location – color, color pattern – odor, odor pattern – shape – timing of food supply	– conspicuous from background (color, odor, shape, location) – different from other competing flowers especially if plants grow in sparse population
Use innate search images	Support innate search images
– to locate flowers prior to learning – to fascillitate learning – to enhance the learning effect by making it more precise – to reduce handling time at each new flower	of pollinators by – large, contrasting inflorescence – apply common «flower features» color, odor, shape, nectar guides
Learn new search images effectively and adjust quickly to changes by	Support learned search images
– rapid acquisition – short-term memory and long-term memory – low sensitivity to extinction	– make your flower different from competing flowers – offer enough reward (more on rare plants) – signal drying up of nectar/pollen supply

entrance this framework gives guidance but does not restrict it to a preprogrammed machinery.

2 How can «innate search images» and prepared associations be proved

The simplest way to demonstrate the existence of «innate search images» might be to ask a bee on her first flight out, what signals she is choosing *spontaneously* in search of a food source. Although such experiments have been attempted (OETTINGEN-SPIELBERG, 1949, BUTLER, 1951, LINDAUER, 1952, LUDWIG, cited in v. FRISCH, 1965, p. 250), they are practically not feasible, because (1) too few data are collected and (2) it can not be ruled out with certainty that a given bee did not have any experience in the field before or that she is not guided on her flight by information from other hive members. The results are thus quite inconsistent. For example, v. OETTINGEN-SPIELBERG (1949) claims that scout bees prefer yellow, BUTTLER (1951) maintains that blue and yellow are equally attractive, LUDWIG (cited in v. FRISCH, 1965) found blue to be most effective. The question of choice behavior prior to any experience and communicated guidance is a very important one, but as far as I can see it has not been adequately addressed yet.

Many data can be collected in learning experiments with experienced bees, and we are, therefore, bound to use indirect behavioral measures to demonstrate preparedness of learning. Two measures have been used successfully: (1) learning rate or acquisition of learning and (2) precision of learning or correctness of learned behavior, because learning should be quicker, more effective and more accurate for prepared signals than for others. We might even expect to find extreme cases of preparedness in so far as certain signals may not be learned at all, or that only certain stimuli are associated with each other and not others (selective association). But these are by no means the only useful behavioral measures. Others are: retention function, sensitivity to extinction, interference between consecutive learning events, multireversal effects, consolidation into long-term memory, compound vs single stimulus learning, specific differences in the association of different sensory modalities, etc. Some of these measures are used in the experiments described below.

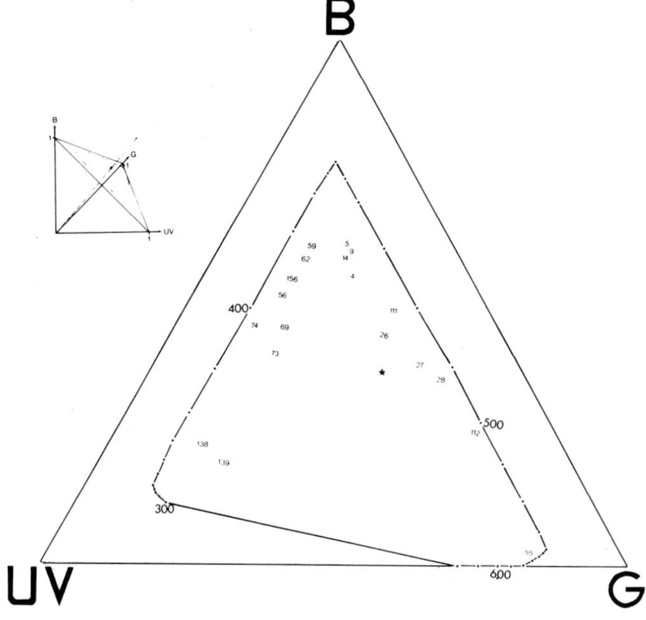

Fig. 1: The color vision of the honey bee can be appropriately described with a three-dimensional input model (see MENZEL, 1979, MENZEL and LIEKE, 1983). The three dimensions are the 3 color receptors UV, blue (B) and green (G). Their contributions define a space (upper left), in which each color is represented by a color locus. For simplicity, we refer to a two dimensional representation in a color triangle, which neglects the brightness of a color stimulus. All monochromatic colors lie along the dark line with dots showing the spectral colors at 10 nm intervals from 300 to 630 nm. The line connecting 300 and 630 nm is the purple line. The color loci of all the colors used in the experiments given in Fig. 6 and 7 are shown as numbers.

3 Bees show preparedness for learning «violet» as a flower color

Investigating preparedness for colors at the feeding place is an interesting approach to the question (v. FRISCH, 1914, MENZEL, 1967), since a lot is known about color vision in bees (AUTRUM and v. ZWEHL, 1964, DAUMER, 1956, MENZEL and BLAKERS, 1976, NEUMAYER, 1980, 1981, MENZEL, 1977, KIEN and MENZEL, 1977 a, b, HERTEL, 1980), and since the mutual adaptation of the bees' color vision and the coloration of flowers has attracted much attention (e. g. KEVAN, 1978). Color vision in bees is best described by means of a model calculation of the *color space* (Fig. 1, see MENZEL, 1977, 1979), which matches all our knowledge quite well. Bees have 3 color receptors with maxima of their spectral sensitivity functions at 350 nm, 440 nm, 540nm, respectively (Fig. 2 a). UV color

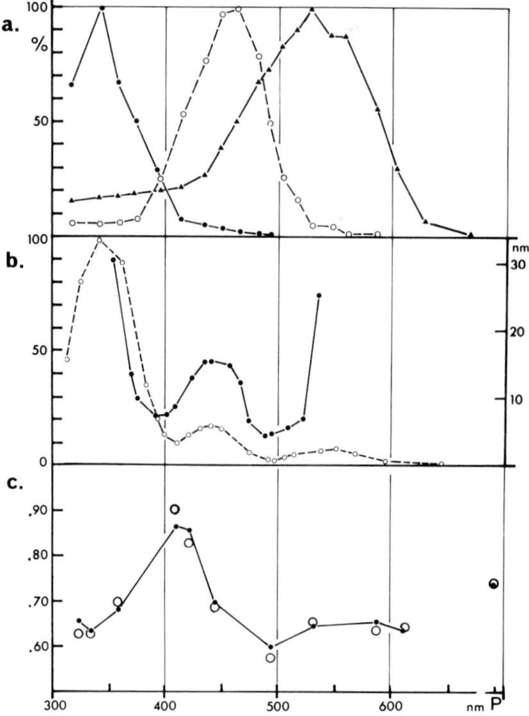

Fig. 2: Comparison of the spectral properties of color vision and color learning in bees. a. Spectral sensitivity of the 3 color receptor types (MENZEL, BLAKERS, 1976). b. Spectral threshold (dotted line) and spectral discrimination ($\Delta\lambda$) of bees in training experiment to spectral colors (v. Helversen, 1972). The left ordinate shows the relative spectral sensititity expressed for the maximal sensitivities in UV. The right ordinate shows the wavelength intervals which are separated by the bee by a criterion (70 %) of correct choices. c. Choice behavior after one reward on a spectral color λ_+) which is indicated at the abscissa. The ordinate shows the probability of choices of the trained color in a dual choice test (MENZEL, 1967). Two kinds of pretraining are used: ● three trials on an unilluminated ground glass, ○ three trials on a white illuminated ground glass. The two colors in the test situations are adjusted for each individually tested bee to induce equal choice probability (0.5) prior to training (from MENZEL, 1967).

around 360 nm appear brightest to medium-light adapted bees in a UV-free sorrounding (Fig. 2 b). Color discrimination is best for violet (around 440 nm) and bluish-green (490 nm) (Fig. 2 b). The mixture of the two ends of the visual spectrum (UV-Yellow) produces a unique color – bee purple – which is different from any other color (DAUMER, 1956, MENZEL, 1967). UV – bluish-green and blue-purple are complementary colors.

Color learning has been studied in a dual choice situation with free flying bees trained to spectral colors under constant background light conditions (MENZEL, 1967). It turns out that violet (400–420 nm) is learned fastest and is chosen most accurately. Bluish-green is learned most slowly and is chosen with least precision (Fig. 2 c, 3 c). Learning acquisition and level of correctness is a function only of the color learned (λ_+), not of the alternative color (λ_-) presented in the dual choice test (Fig. 3 a), provided that the colors can be discriminated. Neither is color learning a function of brightness (Fig. 3 b), if the color is sufficiently bright to be easily detected by the bee. Furthermore, color learning appears

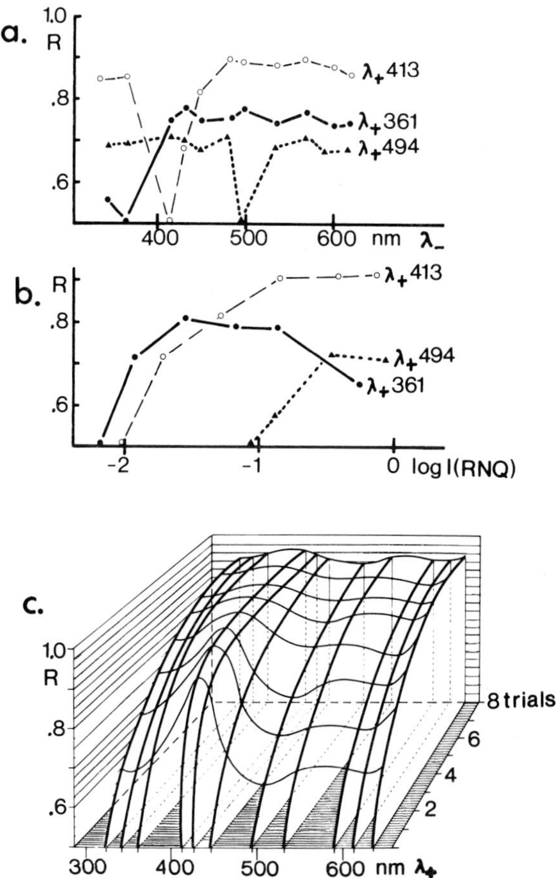

Fig. 3: *a.* Dependence of the alternative color (λ_-) on the learning of λ_+ and *b.* of the brightness of the color λ_+ on the learning of λ_+. In both cases the choice probability is given after two trials on λ_+ (see text). *c.* Acquisition functions for 10 different spectral colors and purple (P) (from MENZEL, 1967).

independent of the kind of pretraining which was necessary to introduce bees to the experimental training situation (Fig. 2 c). The sole dependence on λ_+, the color learned, is further supported by our finding that bees trained to white light choose white light equally well with various spectral lights as alternatives (LIEKE, unpubl.). Furthermore, there is evidence for an independence of the color-specific learning rate of the age of the bees and of the weather conditions (MENZEL, 1967).

We conclude from these results that color learning in the context of feeding follows a rank order of signal preparedness with violet ranking highest, followed by blue and purple, then green and UV and finally bluish-green. This rank order parallels the frequency of colors found in hymenopteran flowers, especially of those which grow in sparse populations (KEVAN, 1978, MENZEL and LIEKE, unpubl.). It well be pointed out, however, that proper spectral measurements have to be made of the flower colors as well as the calculations of their color loci in the bees' color space. Violet, the color learned best, is distinctive to the bee because it appears relatively bright *and* is highly discriminable. Furthermore, it contrasts very well with the greenish background of the foliage leaves and is less sensitive to changing UV-illumination than e. g. purple and UV in exposed versus shaded locations.

Although these results have shown that preparedness for the color of a foraging place exists, it is still a small effect compared with the strenght of the bees' learning ability, and can be demonstrated only in well-controlled, quantitative behavioral experiments. FOREL (1910) stated this correctly: «Die Farbe bildet ein Merkzeichen, aber keine Anziehung an und für sich für das Insekt» (p. 194) («Color is a learning signal but not an attractive signal in itself for this insect»).

4 Do bees show preparedness for particular color patterns?

VON FRISCH (1914) and WEHNER (1972) used a simple pattern of two vertically arranged half discs to study pattern recognition. Such a pattern has the advantage that the effect of the spatial arrangement of the pattern components can be tested easily. We used these discs to address the question of whether bees apply search images consisting of color patterns at the food source (MENZEL, LIEKE, 1983). The results we obtained are in partial agreement with the assumption of specific search images, but not all results fit into such an interpretation (Fig. 4).

Let us suppose that bees have a search image in which the UV component of the pattern is expected to appear in the upper portion and bluish-green in the lower portion of the visual field. Discrimination values should be high if the trained pattern has more overlap with the search image and low if the alternative test pattern shows more overlap. Indeed, this hypothesis is supported by experiments in which patterns with horizontal or vertical contrast lines have been trained (Fig. 4, upper left part, dark columns). If the trained pattern have either a 45 degree tilted contrast line between a UV and bluish-green sector or between a UV bright-dark sector, discrimination values contradict the hypothesis. The same is true for a search image in which the upper part of the visual field is orange and the lower part is blue (Fig. 4, second row). There are two ways of explaining these results (see MENZEL, LIEKE, 1983) by assuming either a selective change of the search image after training to tilted contrast line or by hypothesizing an additional color-selective position effect. In any case, if there is a search image for color patterns it cannot be very strong, because it is changed by a seemingly small difference in the trained pattern and/or is affected by additional parameters.

One might argue that search images for color patterns should show up more strongly in flower-like patterns e. g. in concentric UV-non-UV reflecting patterns, or purple-yellow configuration or star-like patterns with outer-inner color contrast and so on. We find it very difficult to interpret discrimination experiments using such patterns. At any rate, preliminary experiments do not support strong search images for such intuitively «flower-like» color patterns.

We have trained bees to the same half-disc color patterns at the hive entrance and – to our great surprise – found no discrimination with respect to the pattern component. Any flaws in the procedure can be excluded because the same bee colony discriminated colors at the hive entrance very well (see below). If this finding is confirmed in additional experiments, we have to conclude that learning of color patterns is context specific – a result very different from those obtained for color learning (see below). Furthermore, this result is one of the rare cases of selective association in bees. A signal, here the color pattern, is learned when sucrose solution is provided as a reward, but not learned when the reward is return into the hive.

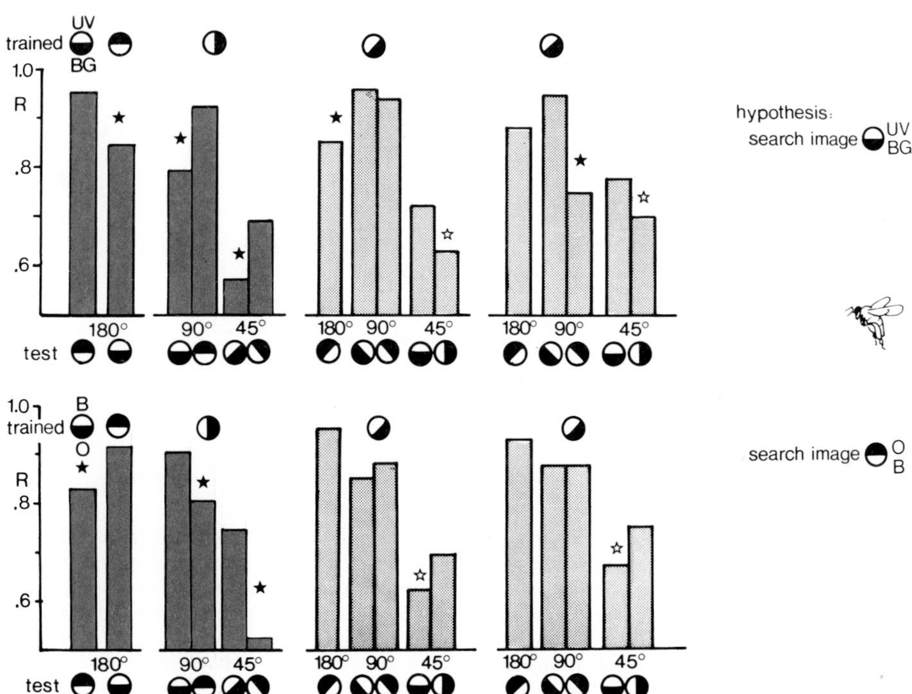

Fig. 4: Choice behavior of honey bees after training to patterns of two half circles of different colors (UV: ultra violet, No. 2 in Fig. 1, BG: bluish-green Nr. 14 B: blue No. 8 and O orange Nr. 17). The trained color pattern is given in the upper row of each set of experiments the alternative color pattern in the dual choice test is given below each column. The angular number of degrees indicate the angular deviation between test pattern and training pattern. The columns show the probability of correct choices for the respective pairs of color patterns. Heavy stars indicate highly significant differences between the two columns ($p \leq 0{,}01$, χ^2-test), light stars indicate weak statistical differences ($0{,}01 \leq p \leq 0.05$). The results given in dark columns are those which support the hypothesis of a search image shown at the right side, the grey columns do not support this hypothesis (see text) (from MENZEL, LIEKE, 1983).

5 Bees show preparedness for flower-like odors, but will learn any odor

Flower-like odors are learned faster than other odors (v. FRISCH, 1919, LAUER, LINDAUER, 1971, SCHWARZ, 1955, KRISTON, 1971, 1973). Fig. 5 c,f shows a few examples. The odor learned fastest is geraniol, a component of the Nassanov gland which is used by the bee to mark productive food sources as well as the hive entrance. VON FRISCH (1965 p. 506) found it surprising that bees can be conditioned to odors which cause obvious signs of discomfort. Lysol, for example, is used by bee keepers to prevent bees from approaching

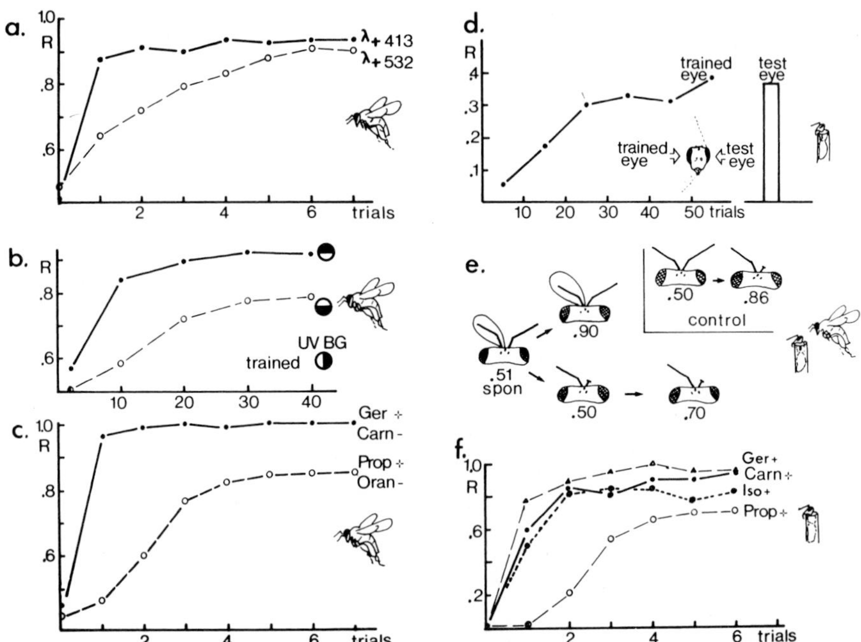

Fig. 5: *a–c:* Average learning rate of free flying bees to different food signals. Choice behavior (ordinate) at trial O (left side of abscissa) indicates the probability of choices after pretraining (usually three trials on a neutral target). a. Acquisition functions for two spectral colors $\lambda_+ = 413$ nm, $\lambda_+ = 532$ nm (see Fig. 3). b. Acquisition functions for two color patterns of the kind described in Fig. 4. Note the different scale on the abscissa (number of trials). c. Acquisition functions for 2 different pairs of odors. The bees were pretrained by three trials to an unscented target (cardboard box). Ger + Carn −: Geraniol was the rewarded odor, Carnation the alternative odor; Prop + Oran −: Propanol trained, Orange alternative odor.
d–f: Conditioning of the proboscis-extension reflex of fixed bees to color and odor stimuli. d. Acquisition function to a flash of blue light to the right eye (trained eye). Note the different scale on the abscissa. Column: response probability. e. Side specificity of odor learning in free flying bees and fixed bees. The probability for the two kinds of experiments were so similar that only the numbers of the free flying bees are given (see text). The line around the antenna indicates covering of that antenna during learning in free flying bees or exclusion from odor stimulation during proboscis conditioning in fixed bees (MASUHR, MENZEL, 1972). f. Acquisition functions for various odors during proboscis extension conditioning in fixed bees. Ger: geraniol, Carn: carnation, Iso: isoamylacetate, Prop: propanol (see text) (Experiments with Ger +, Carn +, Prop + from BITTERMANN et al., 1983, with Iso + from SALISBURY, unpublished).

a location, and yet bees choose a feeding place marked with lysol preferentially when they have been rewarded there. Propanol is another example of an odor which has a repellent effect on bees and can still be conditioned (Fig. 5 c, f). During proboscis conditioning, bees may even retract the proboscis at the beginning of traning but later respond with conditioned proboscis extension. KRISTON (1971) found that bees even prefer butyric acid to thyme after pretraining to an unscented feeding place. Flower-like odors are more salient signals and, therefore, bees choose less salient odors after training to *no* odor. There is no evidence for an odor which cannot become a conditioned stimulus to the bee. Even CO_2, water vapor and the bee's own alarm pheromone can be conditioned. The latter is very interesting (Fig. 5 f). Isoamylacetate, the main component of the alarm pheromone, is even learned as fast as flower-like odors in proboscis extension conditioning, although a few bees after training put out their sting reflexively while extending their proboscis at the same time.

We learn from these experiments that as far as odor learning is concerned, bees are innately prepared for expected odor signals at the feeding place, but that the general learning ability is powerful enough to enable the bee to learn any odor as a food signal. Here again, «innate search images» seem to be relatively weak cues when compared with the general learning potential.

6 Comparison of color, color pattern and odor learning

In comparing the most salient visual and olfactory signals, we find that odors are learned fastest and are chosen more accurately than colors and color patterns (Fig. 5 a, b, c). However, a comparison of this kind is complicated by the fact that the same behavioral measures have been used to quantify choice behavior (approaches to the target or sitting down on the target), but visual and olfactory cues are known to guide the flight path of the bee at different distances from the target (v. FRISCH, 1965). In my view, the question of acquisition rate for different learning signals is open for more appropriate experimentation, which selects more appropriate behavioral measures for different learning signals.

In contrast to free bees, large differences between learning of visual and olfactory stimuli are found in the proboscis extension conditioning paradigm (Fig. 5 d, f). Flashes of colored light are turned into conditioned stimuli only with great difficulty (MASUHR, MENZEL, 1972). A moving-stripe pattern ist not much more effective as a conditioned stimulus (ERBER, SCHILDBERGER, 1980). Olfactory stimuli, however, are condtioned as quickly as in the free-flying paradigm (Fig. 5 c, f). These are strong hints for the existence of selective associations in the bee brain between certain stimuli and a well defined motor programm (proboscis extension). The biological background for such selective associations is obvious. Olfactory stimuli are always closely related to proboscis extension both at the flower and within the hive. They are used by the running and standing bee as food signals and not only in flight.

We found another difference in the processing of visual and olfactory stimuli during conditioning in bees which favors the notion of selective association. (Fig. 5 d, e): Color signals trained with one eye are responded to at the same rate when testing with the other eye. This is different in olfactory stimuli. Here stimulation of the unconditioned antenna does not elicit the conditioned response. We may suspect that the memory trace is limited to the side of the brain ipsilateral to the conditioned antenna and is not accessible by the contralateral antenna. This is, however, an inadequate explanation, because a conditioning by the initially covered antenna to the *same* odor gives a significantly *lower* response level than conditioning without pretraining of the contralateral antenna (Fig. 5 e). This result reminds us of Forel's (1910) concept of the «topochemical sense» in bees (see v. Frisch, 1965, p. 513). An odor perceived and learned with the *right* antenna is different from an odor perceived and learned with the left antenna. Therefore, training the initially covered antenna to the same odor is a reversal learning, which has to result in lower response levels. If this interpretation is correct, the side-specific effect is not related to the representation of the memory trace in the bee brain but is a result of the side-specific processing of odor information.

7 Color learning in different behavioral context

Search images have to be specific with respect to the behavioral context. Different behavioral contexts may work with different search images. The preparedness of an animal is, therefore, not only related to certain stimuli but also to the meaning of these stimuli in a given behavioral context. Bees are color-blind in certain behaviors (flight control through optomotor response, Kaiser, Liske, 1974; polarized light orientation, v. Helversen, Edrich, 1974; phototaxis, Menzel, Greggers, unpubl.), but perceive and learn colors both for localising a food source and the hive entrance. The color signals marking a food source and the hive entrance are typically very different, although we should keep in mind that bees do feed also on greenish-brownish locations (e. g. aphids on

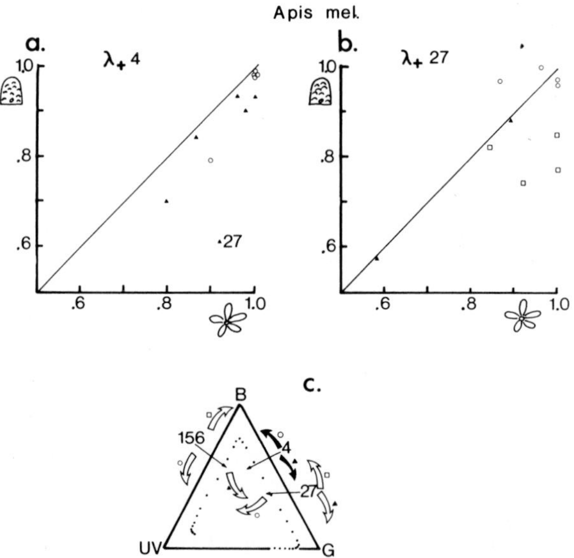

Fig. 6: Comparison of color learning of *Apis mellifera* at the feeding place and at the hive entrance. In a. the bees were trained to blue (No. 4, see c.), in b. to bluish-green (No. 27). The ordinates show the choice probability of the trained color at the hive entrance, the abscissa the choice probability of the same trained color at the feeding place. Each dot indicates identical dual choise tests in the two situations. If color discrimination were the same in the two situations all points would lie around the diagonal line (crossing through the graph). Triangles indicate color discrimination towards the bluish-green, open circles towards the UV and violet, and open squares towards the blue. c. gives the locations of the trained colors in the color triangle and explains the symbols used in a. and b.

leaves, extra-floral nectaries, phloem liquid squeezed out by leaves, and decomposing fruits). We may expect, therefore, different «innate search images» for color marks in the context of orientation at the hive entrance when compared with that of the foraging flight.

We have used identical test situations for color training at the feeding place and at the hive entrance (dual choice test, see MENZEL, LIEKE, 1983 and above). Two colors were trained: No. 4 (blue) and No. 27 (bluish-green). Discrimination was tested with alternative colors throughout the bee's color space. The results are presented in such a way that all points should lie around the line crossing the diagram (Fig. 6) if color learning is similar in both behavioral contexts. If bees learn colors better at the feeding place than at the hive entrance, the data points should lie below this line. Bees learn colors at the hive entrance very well, but most point lie somewhat below the crossing line, indicating a choice behavior at the hive entrance which is slighly less precisely controlled by learned color signals than at the feeding place. Most importantly, there is no selective effect with respect to color discrimination in the two behavioral contexts as we may have expected (see above). Some of the results are even contra-intuitive. Blusih-green is much better discriminated from blue at the feeding place than at the hive (Fig. 6 a, No. 27), and ultraviolet is equally well discriminated both from blue and bluish-green in the two test situations.

So far we have been unable to quantify the acquisitions process during hive-entrance training, but qualitative estimates indicate similar dependencies of acquisition on different colors as in appettitive training.

It is an intuitively attractive idea to expect context-specific preparedness for signal-reward aussociations, and, indeed, we have found such preparedness in the proboscis-extension-conditioning (see above). However, no selective preparedness has been found in color learning of the free-flying bees. This provides further evidence for the strength of the general learning ability in both behavioral contexts tested (feeding place and hive entrance).

8 Species-specificity of search images in color learning of hymenopterans

Flower visiting hymenopterans face the same problems in reliably indentifing and localising potential food sources and in relocating their nest sites. The specific adaptations may still differ according to the ecological conditions in which they collect nectar and pollen and relocate their nest sites. The overall conditions of light climate of the species-specific niche, for example, may influence the evolution of color perception and color learning. A comparative study may help us to identify the general properties of color vision and color learning of flower visiting hymenopterans and separate species-specific adaptations.

Such a study has been started in our laboratory, but is still in its infancy. The crucial point is that identical methods have to be used to characterize color discrimination. So far we have studied *Apis mellifera* (unpublished experiments by A. WERNER, S. MÜLLER, F. RRANK, S. MENZEL, K. WEBER and students of the course Neural Systems and Behavior at MBL in Woods Hol, Mass. 1983), *Melipona quadrifasciata* (in co-laboration with D. VENTURA, Sao Paulo) and *Osmia bicornis* (in co-laboration with E. STEINMANN, Chur) using the same methods of hive entrance training (see above).

Melipona discriminates colors in the bluish-green, green and yellow region better than *Apis*, and *Apis* is superior to *Melipona* in discriminating longer wavelength colors (bluish-green, green) from violet and UV (Fig. 7 a, b). Unfortunately, we have not yet trained *Apis* to violet (filter No. 156, Fig. 6 c) at the hive entrance, but let us compare color discrimination of *Melipona* after training to violet at the hive entrance with that of *Apis* trained to violet at the feeding place (Fig. 7 e). We may proceed in such a way, since we did not find color specific differences between discrimination of *Apis* at the food source and the hive entrance (Fig. 6 a, b). Here, too, we find *Melipona's* color discrimination to be best when bluish-green color are presented as alternatives and relatively weak when UV colors are the alternatives.

One possible explanation for these results may be, that the eye of *Melipona* quadrifasciata contains different color receptors. However, this is not the case. We have recorded intracellularly from many photoreceptors in the compound eye of Melipona (UV: λ_{max} = 350 nm, blue: λ_{max} = 440 nm, green: λ_{max} = 540 nm) and found the same three color receptor types as in *Apis mellifera* (MENZEL, VENTURA, unpubl.). Our conclusion then is that the processing of color information in the *Melipona* brain is adapted to an improvement of color discrimination in the bluish-green region and to a reduction of color discrimination in the violet-UV region. This may be a specific adaptation to the light conditions prevailing in the dense tropical rain forest. *Melipona* houses in hollow trees in the tropical forest and searches for nectar and pollen on flowering trees which are often overshadowed by the canopy of higher trees.

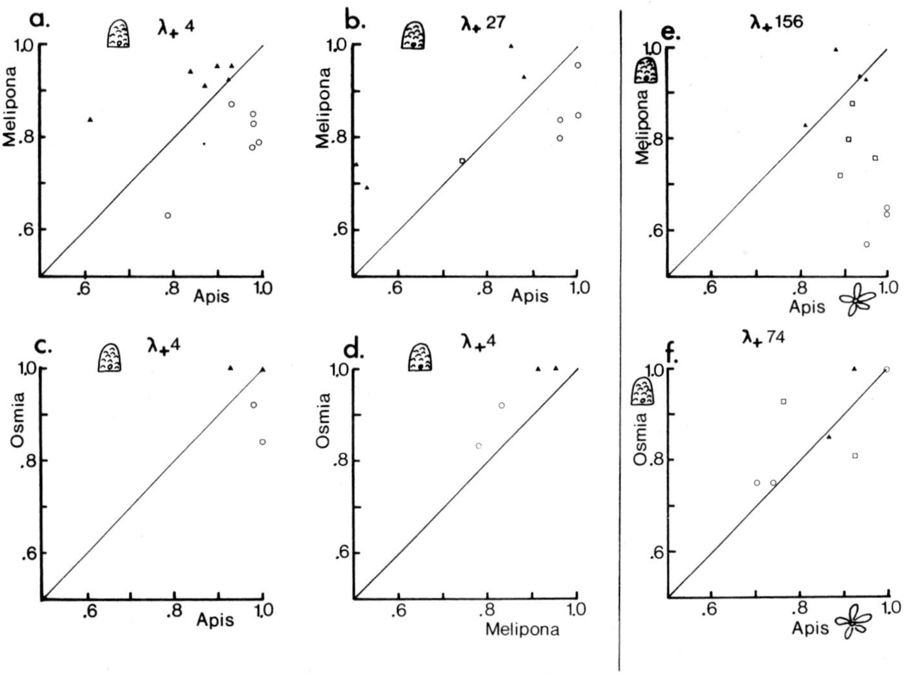

Fig. 7: A comparison of color discrimination of 3 hymenopteran species, the honey bee *Apis mellifera*, the stingless bee *Melipona quadrifasciata* and the solitary bee *Osmia bicornis*. a. and b.: Color discrimination of *Melipona* (ordinate) and *Apis* (abscissa) at the hive entrance after training to blue (No. 4) and bluish-green (No. 27). The same kind of plot and the same symbols are used as in Fig. 6. c. *Apis* compared with *Osmia* after training to blue (No. 4), d. *Melipona* compared with *Osmia* after training to blue (No. 4). e and f: A comparison between species and in different behavioral context. These plots are presented because most data we have collected so far are on *Apis* color discrimination at the feeding place, whereas *Melipona* and *Osmia* were tested only at the hive entrance. e. trained color is Violet (No. 156 see Fig. 1 and 6 c), f. trained color is Violet (No. 74 see Fig. 1) (see text).

The solitary bee *Osmia bicornis* has so far been tested with a few colors only and is found to discriminate colors in the bluish-green better or as well as *Apis* and not as well as *Apis* in the Violet-UV region (Fig. 7 c, f). It is, therefore, not surprising that *Osmia* is superior to *Melipona* (Fig. 7 d).

«Innate search images» are a product of the phyologenetic history of the species. We expect, therefore, species-specific preparedness for the association of certain stimuli with reward or reinforced behavior, and we expect more precisely controlled behavior for prepared stimuli. The species we have compared so far are phyolgenetically quite separate but relatively closely related with respect to their search behavior at the food source and at the nest-site. It is, therefore, not surprising that we find only marginal quantitative differences. Again, the general learning ability in all three species is the major effect and preparedness appears as an additional tuning factor.

9 Bees are prepared to switch to new food signals

The search strategy of a generalist is characterized by the ability to switch to new food sources when the old one becomes less attractive. As the typical food source of a generalist has a shorter duration than the life time of the animal, we might expect to find growing readiness of the animal to switch to new food sources when the old one has been visited throughout an extended period of time. Such a preparedness should be advantageous in so far as it tunes the animal to give up a food source when it exceeds an expected life time. From the learning point of view such a preparedness reflects a paradox, because ongoing

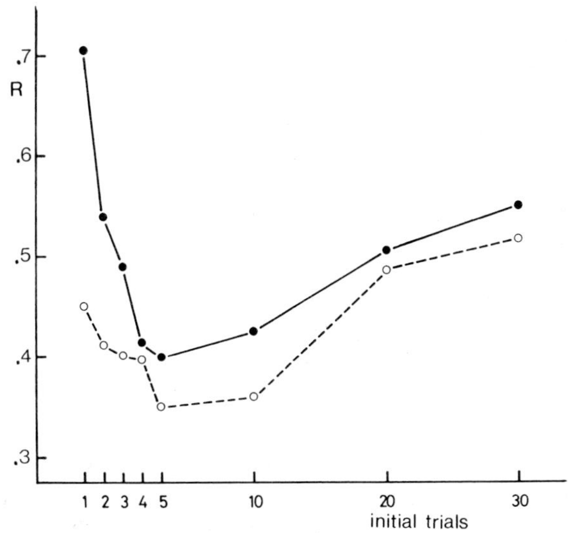

Fig. 8: Over-learning reversal effect in the honey bee (Menzel, 1969). Bees were trained to a variing number of initial trials to one of two test colors (blue-yellow) and then rewarded once at the other color. The abscissa gives the number of initial trails on blue (○) or yellow (●), and the ordinate gives the choice probability after one reversal trial on blue (●) or on yellow (○). Choice probability below. 5 means preferred choice of the initially trained color, above. 5 preferred probability of the second trained color (see text) (from Menzel, 1969).

reinforcement would not strengthen the learned behavior but, on the contrary, would actually weaken it.

We have found such an effect in the honey bee (Fig. 8). Bees are trained to one color out of a pair of colors by a variing number of rewards, each bee of a group of individually marked bees in consecutive experiments to a given number of initial rewards (ranging from 1–20 initial rewards). Then the alternative color is rewarded on several trials. The curves show the choice probability on a dual choice test after one reversal trial. Decreasing probability values give a decreasing choice of the second learned odor, increasing probability values reflect an increasing choice of the second learned color. Reversal learning is first slowed down with the number of initial trials as we would expect from the effect of strengthening a learned behavior by reinforcement. But ongoing reinforcement results in an increasing readiness to switch to the second color beyound 10 initial trials.

Such an effect is not unknown in learning psychology («overlearning reversal effect»), but there is little knowledge about what it may mean for animals under natural conditions. Our experimental conditions are not natural in a strict sense, since the bees could suck their fill during each landing, and the food source was always located at the same place. It is still tempting to interpret these results as an adaptation of a generalist pollinator to the expected temporal limits of its food sources and, therefore, as a temporal search image.

This effect is not related to any rhythm of learned behavior or time-linking of learning (KOLTERMANN, 1971, BOGDANY, 1978), because it is independent of the timing of the initial learning trials. We have looked into time-linked learning not only in this experiment but also in several other series of experiments (MENZEL, 1968, 1969, MENZEL, 1979, MENZEL, ERBER, MASUHR, 1974), and have not found any evidence for a periodicity of learning or a time-linking of learning and retrieval. Time-liking is the effect of an extended period of differential conditioning. During this long procedure timing of a signal-reward association is established through differential reinforcement (e. g. BOGDANY, 1978). The positive outcome of such a training procedure proves that bees are able to learn that certain signal-reward associations are restricted to some time in the day and other signal-reward associations to another time of the day. These experiments do, however, not demonstrate a time-linking of signal-reward associations per se.

10 Short-term memory as a mechanism of fast adaptations to changing profitability of the food sources

In honey bees short-term memory (STM) traces are thought to be coded in an orderly neural activity, whereas long-term memory (LTM) appears to have some stable structural or biochemical substrate (MENZEL, ERBER, MASUHR, 1974). STM preceeds LTM, the time course of the STM-LTM transfer can be measured (ERBER, 1975 a, b, 1976) by erasing STM with experimental procedures like narcosis, cooling or weak electronconvulsive shocks. At the beginning of our studies on STM in bees we thought that the time course of STM-LTM transfer with a half-life time of about 5 min is determined by slow biochemical processes transcribing neural activity into biochemical changes and viewed this process as an internal machinery of the nervous system necessary to establish LTM. This picture collapsed when ERBER (1975 b) found that LTM can be established within less than a minute if several very short rewards follow each other quickly (Fig. 9). The time course of STM-LTM transfer, which we measured initially (Fig. 9 a, curve 1), may thus be an active process which prevents a memory trace from reaching LTM before it has been confirmed by additional associations (MENZEL, 1979, 1983). If this concept is correct one might expect to find an adaptation of the actual time course of STM-LTM transfer on the expected interval between successive approaches to natural food sources. The underlying neural processes would then be under the control of phylogenetic adaptations which prepare these neural processes for expected time frequencies of associative events.

Evidence for this view comes from the following observations. (1) If two learning trials with two *different* conditioned stimuli follow each other within 30 sec, the memory trace for *both* is eliminated by subsequent eletroconvulsive shock (MENZEL, unpubl.). This result supports our assumption that only a confirming event establishes LTM quickly; (2) Unrewarded approaches to a trained visual target have a very weak extinction effect if the memory trace is in its longterm form (MENZEL, 1967). However, extinction trials are very effective in STM (MENZEL, 1968); (3) Reversal learning experiments after one initial trial on one color and a second trial on another color within a time interval ranging from 0,5 to 10 min demonstrate a biphasic time course (MENZEL, 1979, Fig. 9 b). This means that a second learning trial with competing information interferes with the memory trace of the first learning trial and erases the first memory trace effectively at two temporal windows: immediately after the initial learning trial and around 3 min after the initial learning trial.

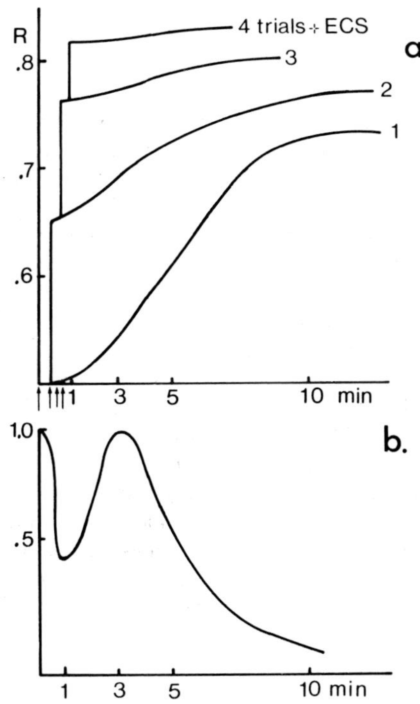

Fig. 9: Short-term memory and choice behavior in the honey bee. a. Bees were trained with 1 or 2 or 4 trials on blue and then treated with electroconvulsive shocks. The time course after one trial (lowest function) is supported by several points, those of 2–3 trials only by the beginning and endpoints (from ERBER, 1975 a, b). b. Thime course of the sensitivity of an initial learning trial to a reversal trial following the initial trial at a varying time interval (abscissa) (from MENZEL, 1979). The ordinate gives a normalized measure of the relative sensitivity of the initial memory trace to conflicting information coming from the new trial (reversal trial) (see text).

There are not enough data yet to allow any functional interpretation of this effect (see MENZEL, 1982, 1983 for discussion). Nevertheless, the biological meaning is that the establishment of a LTM trace is highly sensitive to the temporal sequence of associative experience. The temporal sequence of learning trials under natural conditions is a function of flight speed, sucking time (or pollen collecting time) and density of flowers, the latter being the most important factor. It is tempting to speculate that the time course of STM-LTM transfer is adapted to two categories of flower distribution: densely growing flowers (e. g. trees) and sparse distributed flowers. In the first case a just-established STM-trace is erased if a different experience is made within about 30 sec. Indeed, I have found an average sucking time per flower of 3 sec and an average landing interval of 10 sec in densely flowering orange trees (MENZEL, 1968). In sparsely growing flowers a just-established STM-trace should be sensitive to corrections at a longer time interval. An interval of three minutes seems quite long, and indeed we measured in widely distributed autumn flowers an average of 1 min interval between landings. However, many of these landings were unrewarded, so a time window of 3 mins for rewarded landings on sparsely growing flowers may be reasonable. These speculations have no further experimental support, but they can stimulate the design of a number of promising experiments.

11 Conclusion

Although all available evidence is indirect, it seems beyond doubt that bees have «innate search images» which guide their orientation and shape their learning processes. Does this mean that bees are little robots which carry prewired circuits in their brains with all possible subroutines ready to be switched on? My believe is that this view is too narrow and not supported by the experimental evidence at hand. In addition, such a mechanism would be maladaptive for a social insect following the search strategy of a generalist pollinator. Like FOREL (1910) and v. FRISCH (1965), I am overwhelmed by the power of the general learning ability of the bee, which allows her to learn pratically all perceived stimuli as food markers, at least under natural conditions. Comparative studies of learning in traditional paradigms have revealed great similarities between laboratory mammals and the bee (BITTERMANN at al., 1983, COUVILLON, BITTERMANN, 1980, 1982, 1983; MENZEL, 1983). Such findings support the notion of across-species rules of learning.

I conclude that the neural mechanisms of learning in the bee are controlled both by a strong inherent general learning ability following rules of interspecific applicability, and by species-specific preparedness for selective associations, which are the result of the species' evolutionary history. This concept advocates against an either – or of «general rules of learning» versus «selective associations of a preprogrammed switch board» and in favor of an integration of theories of learning psychology and ethology.

References

Autrum, H., Zwehl, V. von (1964): Spektrale Empfindlichkeit einzelner Sehzellen des Bienenauges. Z. vergl. Physiol. *48*, 357–384.

Bitterman, M. E., Menzel, R., Fietz, A., Schäfer, S. (1983): Classical conditioning of proboscis extension in honeybees *(Apis mellifera)*. J comp. Physiol. 97, 107–119.

Bogdany, F. J. (1978): Linking of learning signals in honey bee orientation. Behav. Ecol. Sociobiol. *3*, 323–336.

Butler, C. G. (1951): The importance of perfume in the discovery of food by the worker honeybee. Proc. Roy. Soc. B. *138,* 403–413.
Daumer, K. (1956): Reizmetrische Untersuchungen des Farbensehens der Bienen. Z. vergl. Physiol. *38,* 413–478.
Erber, J. (1976): Retrograde amnesia in honey bees *(Apis mellifera carnica).* J. comp. Physiol. Psychol. *90,* 41–46.
Erber, J., Schildberger, K. (1980): Conditioning of an antennal reflex to visual stimuli in bees *(Apis mellifera L:).* J. comp. Physiol. *135,* 217–225.
Faegri, K. van der Pijl, L. (1978): The Principles of Pollination Ecology. Oxford, New York, Toronto, Sidney, Paris, Frankfurt: Pergamon, pp. 244, 3rd Ed.
Free, JB. (1970): Insect pollination of crops. New York, Academic, pp. 54.
Forel, A. (1910): Das Sinnesleben der Insekten. München.
Frisch, K. von (1914/1915): Der Farbensinn und Formensinn der Bienen. Zool. Jb. Physiol. *35,* 1–188.
Frisch, K. von (1965): Tanzsprache und Orientierung der Bienen. Springer Verlag, Berlin-Heidelberg-New York, pp. 1–567.
Helversen, O. von (1972): Zur spektralen Unterschiedsempfindlichkeit der Honigbiene. J. comp. Physiol. *80,* 372–439.
Helversen, O. von, Edrich, W. (1974): Der Polarisationsempfänger im Bienenauge: Ein Ultraviolett-rezeptor. J. comp. Physiol. *94,* 33–47.
Hertel, H. (1980–: Chromatic properties of identified interneurons in the optic lobes of the bee. J. comp. Physiol. *137,* 215–231.
Kaiser, W., Liske, E. (1974): Optomotor reactions of stationary flying bees during stimulation with spectral light. J. comp. Physiol. *89,* 391–408.
Kevan, P. G. (1978): Floral coleration, its colorimetric analysis and significance in anthecology. See Ref 24, pp. 51–78.
Kevan, P.G., Baker, H. G. (1983): Insects as flower visitors and pollinators. Ann. Rev. Entomol. *28,* 407–453.
Kien, J., Menzel, R. (1977): Chromatic properties of interneurons in the optic lobes of the bee. I. Broad band neurons. J. comp. Physiol. *113,* 17–34.
Kien, J., Menzel, R. (1977): Chromatic properties of interneurons in the optic lobes of the bees. II. Narrow band and color opponent neurons. J. comp. Physiol. *113,* 35–53.
Kolterman, R. (1971): 24-std-Periodik in der Langzeiterinnerung an Duft- und Farbsignale bei der Honigbiene. Z. vergl. Physiol. *75,* 49–68.
Kriston, I. (1971): Zum Problem des Lernverhaltens von *Apis mellifica* gegenüber verschiedenen Duftstoffen. Z. vergl. Physiol. *74,* 169–189.
Kriston, I. (1973): Zum Zusammenhang zwischen Signalbewertung und Lernprozess: Die Bewertung von Duft- und Farbsignalen an der Futterquelle durch *Apis* mellifica. J. compl Physiol. *84,* 77–94.
Lauer, J., Lindauer, M. (1971): Genetisch fixierte Lerndispositionen bei der Honigbiene. Akad. d. Wiss. u. Lit. Mainz, Wiesbaden, Fr. Steiner.
Lindauer, M. (1952): Ein Beitrag zur Frage der Arbeitsteilung im Bienenstaat. Z. vergl. Physiol. *34,* 299–345.
Lindauer, M. (1959): Angeborene und erlernte Komponenten in der Sonnenorientierung der Bienen. Z. vergl. Physiol. *42,* 43–62.
Masuhr, T., Menzel, R. (1972): Learning experiments on the use of side-specific information in the olfactory and visual system in the honey bee *(Apis mellifica)* In: Information Processing in the Visual Systems of Arthropods. (Ed. R. Wehner) pp 315–322, Berlin–Heidelberg–New York, Springer Verlag.
McGregor, S. E. (1976): Insect pollination of cultivated crop plants. Agriculture Handbook 496. ARS-USDA, Washington DC pp 411.
Menzel, R. (1967): Das Erlernen von Spektralfarben durch die Honigbiene. Z. vergl. Physiol. *56,* 22–62.
Menzel, R. (1968): Das Gedächtnis der Honigbiene für Spektralfarben. I. Kurzzeitiges und langzeitiges Behalten. Z. vergl. Physiol. *60,* 82–102.
Menzel, R. (1969): Das Gedächtnis der Honigbiene für Spektralfarben. II. Umlernen und Mehrfachlernen. Z. vergl. Physiol. *63,* 290–309.

Menzel, R. (1977): Farbensehen bei Insekten – ein rezeptorphysiologischer und neurophysiologischer Problemkreis. Verh. Dtsch. Zool. Ges. pp 28–40.
Menzel, R. (1979): Behavioral access to short-term memory in bees. Nature *28,* 368–369.
Menzel, R. (1979). Spectral sensitivity and color vision in invertebrates. In: Handbook of Sensory Physiology Vol. VII 6A (Ed. H. Autrum) pp 504–580, Berlin–Heidelberg–New York, Springer Verlag.
Menzel, R. (1982): Neurobiologie and behavior of social insects: An Introduction. In: The Biology of Social Insects (Eds. D. Breed, Ch. D. Michener, H. E. Evand), pp 335–337.
Menzel, R. (1983): Neurobiology of learning and memory: «The honey bee as a model system». Naturwiss. *70,* 504–511.
Menzel, R. (in press): Report on: Biology of invertebrate learning. In: The Biology of Learning. (Eds. P. Marler, H. S. Terrace) Dahlem Konferenzen, Chemie Verlag Weinheim.
Menzel, R., Blakers, M. (1976): Colour rezeptors in the bee eye – morphology and spectral sensitivity. J. comp. Physiol. *108,* 11–33.
Menzel, R., Erber, J., Masuhr, Th. (1974): Learning and memory in the honeybee. In: Experimental Analysis of Insect Behavior. (Ed. L. Barton-Browne) pp 195–217. Berlin–Heidelberg–New York, Springer Verlag.
Menzel, R., Lieke, E. (1983): Antagonistic color effects in spatial vision of honey bee. J. comp. Physiol. *151,* 441–448.
Neumeyer, Ch. (1980): Simultaneous color contrast in the honey bee. J. comp. Physiol. *139,* 165–176.
Neumeyer, Ch. (1981): Chromatic adaptation in the honey bee: Successive color contrast and color constancy. J. comp. Physiol. *144,* 543–553.
Oettingen-Spielberg, Th. zu (1949): Über das Wesen der Suchbiene. Z. vergl. Physiol. *31,* 454–489.
Proctor, M., Yeo, P. (1973): The pollination of flowers. London, Collins, pp 418.
Schwarz, R. (1955): Über die Richschärfe der Honigbiene. Z. vergl. Physiol. *37,* 180–210.
Wehner, R. (1972): Dorso-ventral asymmetry in the visual field of the bee, *Apis mellifica.* J. comp. Physiol. *77,* 256–277.

Fortschritte der Zoologie, Bd. 31 · Hölldobler/Lindauer (Hrsg.): Experimental Behavioral Ecology
G. Fischer Verlag · Stuttgart · New York · 1985

The information-center strategy of honeybee foraging

Thomas D. Seeley

Biology Department, Yale University, New Haven, CT 06511 USA

Abstract

Honeybee colonies pursue an information-center strategy of foraging in which the thousands of foragers in a colony function together as an integrated whole. This strategy enables a colony to sample widely among forage patches, focus foragers on top-quality patches, and quickly readjust the distribution of foragers on patches when the profitability ranking of the patches changes. This system of social foraging is made possible by three features of honeybee social organization. First is the dance-language system of recruitment communication, which enables individual bees to direct nestmates to forage patches. Second is the division of labor between scouts and recruits such that most foragers locate forage patches by following recruitment dances. And third is the array of mechanisms which enables each forager to estimate the relative profitability of her forage patch. Evidently the average rate of food collection per forager is higher with than without this elaborate social organization, especially when there is wide variation in forage patch quality and top-quality patches are widely scattered in the environment.

1 Introduction

The knowledge that honeybees can share information about food sources extends back into the mists of antiquity. Aristotle, for example, observed more than 2000 years ago that once one foraging bee finds a profitable patch of flowers, other bees soon begin to arrive. Evidently the first bee recruits her nestmates. Precisely how bees communicate with one another about food sources remained a mystery until Karl von Frisch discovered the honeybee's dance language early in this century.

To date, studies of the honeybee's system of recruitment communication have emphasized its physiological mechanisms, as opposed to its ecological significance. Thus we now understand in fair detail how bees can orient to flowers several kilometers from their nest, then steer their way back home, and finally provide nestmates with precise directions to the flowers. Solving these puzzles in insect communication and orientation was itself a major scientific challenge. However, an equal challenge arises in trying to understand how the honeybee's skill in recruitment communication fits together with other aspects of their social organization to form an integrated strategy of social foraging. Also, we still do not precisely understand how this intricate social organization enhances a colony's foraging efficiency.

Partial answers to these two ecological questions are already available in the body of literature produced by von Frisch and his co-workers, most notably LINDAUER (1948, 1952, 1954), BOCH (1956), OETTINGEN-SPIELBERG (1949), and NÚÑEZ (1966, 1970). In his masterwork on the dance language, v. FRISCH (1967) addresses the topic of the honeybee's strategy of social foraging in terms of «regulation of supply and demand on the flower market». He points out that the dance language operates together with other social mechanisms «to grade the mobilization of the foraging groups in accordance with the profitability of each of the several sources of food available at the time» (p. 235). Later studies confirm this idea (VISSCHER and SEELEY 1982). However, much work remains to be done before we will understand precisely how and why honeybee colonies track rich forage patches.

The time is ripe to examine the social foraging strategy of honeybee colonies fully and directly. In this discussion I will try to demonstrate that honeybee foraging offers a powerful system with which to integrate optimal foraging theory (PYKE et al. 1977, KREBS 1978, KAMIL and SARGENT 1981) with studies of social foraging. Furthermore, understanding the adaptive basis for a honeybee colony's foraging patterns may prove valuable in enabling humans to predict the pollination patterns of honeybee colonies in agricultural settings. And finally, this avenue of research should ultimately identify the ecological significance of the honeybee's dance language, thereby greatly enriching our understanding of this classical example of animal communication.

2 Foraging behavior of a whole colony

Although natural selection acts primarily at the level of individuals (WILLIAMS 1966, ALEXANDER 1974, MAYNARD SMITH 1976), it is often useful to view the behavior of an individual social insect as contributing to the good of her colony (OSTER and WILSON 1978). This is especially true for those social insect species, such as honeybees, in which the workers do not lay eggs. In such species, each worker maximizes her inclusive fitness by helping her colony as a whole achieve high efficiency in such things as temperature regulation, food collection, and reproduction. Given that each worker honeybee has been selected to be an efficient part in a larger whole, it is appropriate to analyze honeybee foraging in terms of a colony-level foraging strategy.

A logical starting point for a colony-level analysis of honeybee foraging is a description of the foraging patterns of a colony living in nature. VISSCHER and SEELEY (1982) recently accomplished this by reading the recruitment dances of a honeybee colony to map out, day by day, the forage patches being worked by the colony. We worked with a full-size colony living in an observation hive located in a forest. Recruitment dances were randomly sampled throughout each observation day and translated into forage sites using standard techniques (VON FRISCH 1967).

A small sample of the forage maps produced using these techniques is shown in Fig. 1. These maps yielded the discovery that the overall pattern of a colony's foraging is a rapidly changing mosaic of food source patches. This pattern of rapid change in forager allocation to patches is more easily visualized when the forage map data are arranged as in Fig. 2. To make this figure, we defined forage patches as obvious groupings of same-colored points on the forage maps, and then calculated for each day the percentage of a colony's foraging which occurred at each patch. Fig. 2 indicates that consecutive days never share the same pattern of labor allocation to forage patches. This suggests that colonies steadily readjust their distributions for foragers across patches. Fig. 2 also emphasizes that on any given day the vast majority of a colony's foragers are distributed over a small number of patches. We

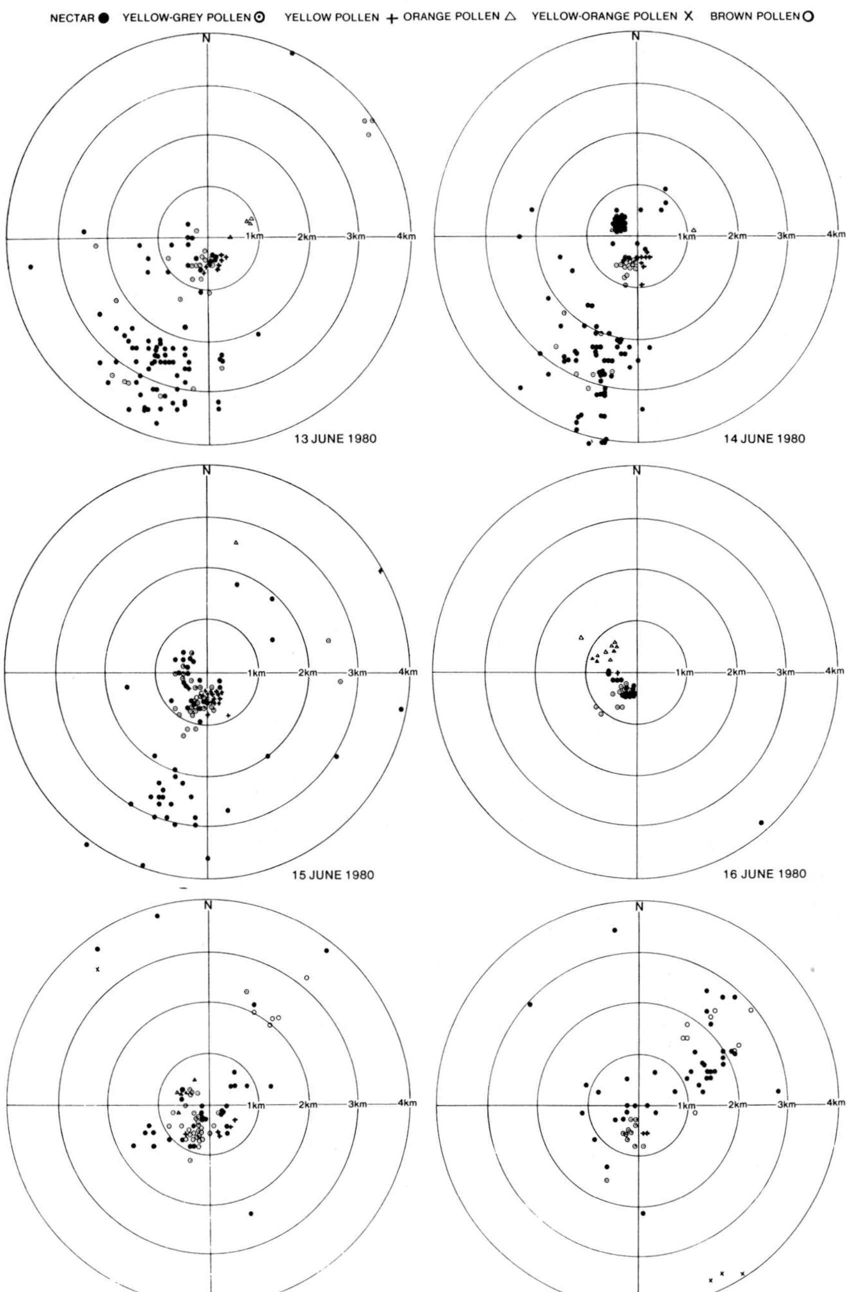

Fig. 1: Maps of the foraging locations inferred from honeybee recruitment dances. Different symbols code the color of pollen, if any, borne by the dancing bee. Locations beyond the edges of the maps (4 km) are not shown.

calculated the number of patches needed to account for 90 % of the colony's recruitment activity (and by inference, foraging activity). This yielded 9.7 ± 4.9 patches/day, averaging over all 36 days of observation. Besides revealing the patterns of a colony's forage-patch dynamics, the forage maps also provide a quantitative picture of the distances over which a honeybee colony forages in nature. As is shown in Fig. 3, these distances are remarkably large: range, 50 to 10100 m; mean ± one standard deviation, 2260 ± 1890 m; 95th percentile, 6000 m.

The results of this forage map study can be summarized in the following generalizations about the foraging behavior of colonies:

1. regular foraging at sources up to several kilometers from the nest,
2. daily readjustment of the distribution of foragers across patches,
3. relatively few patches used at any one time, and each for only a few days.

These patterns, when combined with the knowledge that honeybees possess recruitment communication, suggest that the foraging strategy of a honeybee colony involves monitoring a vast area around the nest for food source patches, pooling the reconnaissance of the foragers, and somehow using this information to focus a colony's forager force on a few high-quality patches within its foraging area. In short, a foraging honeybee colony apparently operates like an information center (Heinrich 1978, Visscher and Seeley 1982).

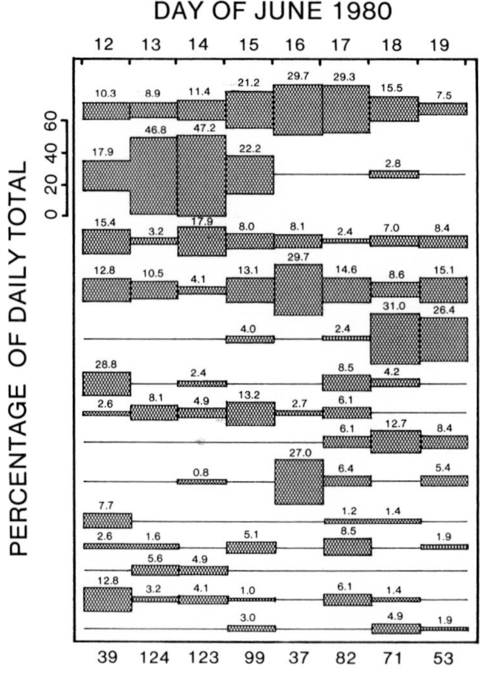

Fig. 2: Summary of the turnover of food source patches from day to day. Each line represents a patch of forage of a given pollen color. The widths of the lines denote the percentage each patch represents of the total number of sites plotted each day. Only those patches comprising at least 1 % of the record for the 9-day observation period are presented.

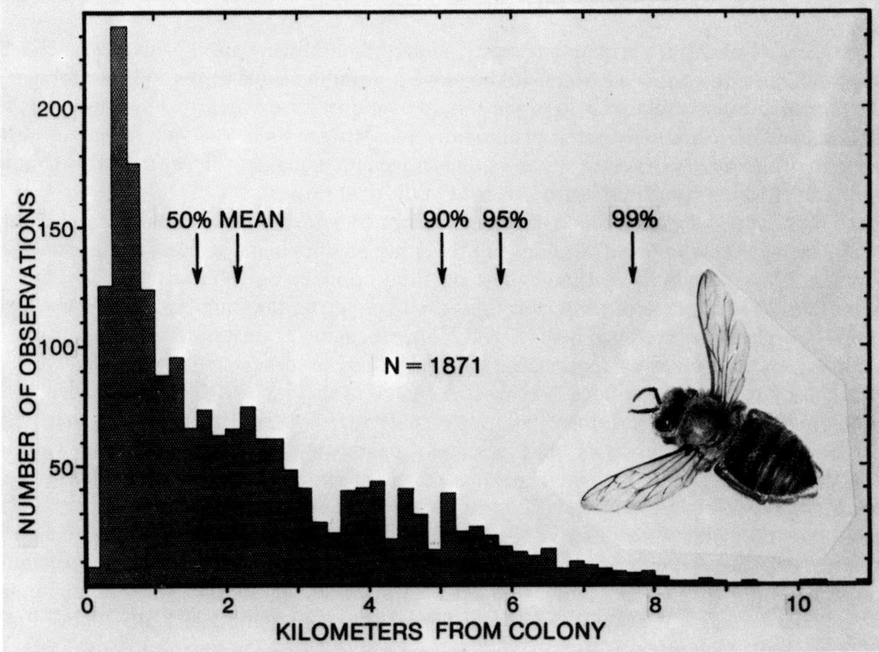

Fig. 3: Distribution of distances between a honeybee colony's nest and its forage patches.

The heart of this information-center hypothesis for the foraging strategy of honeybee colonies is the idea colonies organize their foraging so that on any given day a colony's foragers are focussed on the few best forage patches that the colony has found. This hypothesis is supported by the experiments of BOCH (1956). His studies demonstrate that a honeybee colony offered two feeding stations in various combinations of richness (concentration of sugar solution) and distance from the hive can evaluate the relative profitability of the two, allocate its forager labor accordingly, and shift its forager force from one to the other within hours of a reversal of their profitability. Further support for the hypothesis of information-center foraging comes from the observations by BUTLER (1945) and WEAVER (1979) of rapid shifts of foragers on to and later off of patches of flowers.

3 Mechanisms of forage patch selection

Obviously the mechanisms of patch choice which underlie a colony's shifts in forage sites are central to the foraging strategy of honeybee colonies. One possibility for the way that this decision-making works is that it reflects the joint operation of the following three processes:

1. workers abandon forage patches of relatively low profitability,
2. workers usually locate new patches by following recruitment dances,
3. workers only recruit to patches of relatively high profitability.

The net effect of these three processes working in concert is probably that workers are constantly leaving poor patches and moving onto good patches.

3.1 Assessing the relative profitability of forage patches

Processes 1 and 3 are probably correctly viewed as complementary processes, each one appropriate to an opposite extreme in the range of foraging experiences. When a forager's patch profitability is relatively low, she should abandon it to switch to a superior one. In contrast, when a forager's patch profitability is relatively high, she should recruit nest-mates to it. In the intermediate situation of average patch quality, the bee probably should continue working the patch but not recruit additional foragers.

The fundamental puzzle here is the mechanisms by which an individual bee can know the quality of her forage patch relative to the other patches being worked by her colony's foragers. One possibility is that honeybees, like bumblebees (Heinrich 1976), travel about and sample other patches. But this is unlikely given the immense distances which commonly occur between patches (see Fig. 1). Also arguing against the hypothesis of patch sampling are the numerous reports that labelled honeybee foragers will steadfastly work a particular patch for many days, leaving it only late in the day or when fully loaded with forage (Ribbands 1949, Singh 1950, Weaver 1957). A second possibility is that each worker possesses some sort of built-in scale of patch quality. By comparing her current patch profitability with the values on this scale, a forager could conclude whether her patch is relatively good or poor. (Built-in scales of quality are used by honeybees when evaluating nest sites [Lindauer 1955, 1961, Seeley 1977, 1982]. They enable bees with no prior experience in househunting to identify high-quality sites.) However, this possibility also is unlikely because large week-to-week or even day-to-day changes in forage availability (see Fig. 4) prevent bees from having a fixed definition of a top-quality food source. Thus, whereas a dish of 1.0 mol sugar syrup located 500 m from a hive attracts hordes of bees in New England in mid-August, a time when forage is poor, the same dish of sugar syrup is ignored by bees a few weeks later when the goldenrod plants (*Solidago* spp.) come into bloom.

One promising approach to solving the mystery of how honeybees assess the relative profitability of their forage patches is to explore certain ideas from optimal foraging theory (reviewed by Pyke et al. 1977, Krebs 1978, and various authors in Kamil and Sargent 1981). Within this literature, the ideas of central-place foraging theory (Orians and Pearson 1979, Orians 1980) are especially relevant to honeybee foraging. In this

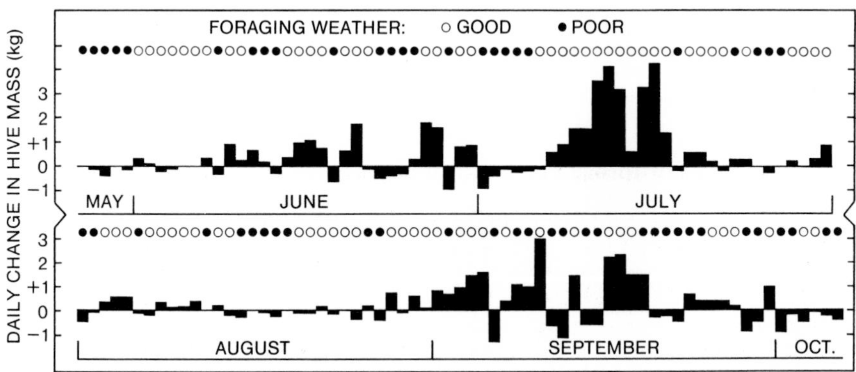

Fig. 4: Daily changes in mass of a hive of honeybees throughout a summer. «Good» foraging weather was defined here as a mean temperature above 14 °C during the hours 0600–1800 and no rain during the day.

discussion, I will apply these ideas to just one situation: bees foraging from patches of flowers producing just nectar.

It seems reasonable to assume that what a honeybee foraging for nectar would like to maximize is her rate of energy intake. PYKE (1981) reviews the validity of this assumption for nectarivores in general. Some of the justifications which apply specifically to honeybees are as follows. First, because the many foragers from a honeybee colony pool their forage, bees can specialize on either nectar or pollen collection and need not divide their time between foraging for these two types of food. Honeybee foragers do, in fact, frequently specialize in either nectar or pollen collection (reviewed by RIBBANDS 1953 and FREE 1970). Second, because of the division of labor between scouts and recruits in honeybee colonies (see below), nectar foragers working one patch need not sample alternative patches to track changes in patch profitability. Such sampling is performed by the colony's scout bees. Third, the risk of predation may be negligible for foraging honeybees and at least is probably not affected by the profitability of a bee's forage patch. And fourth, the ability of honeybee colonies to grow, reproduce, and survive winters is probably closely correlated with the total energy collected by a colony's foragers.

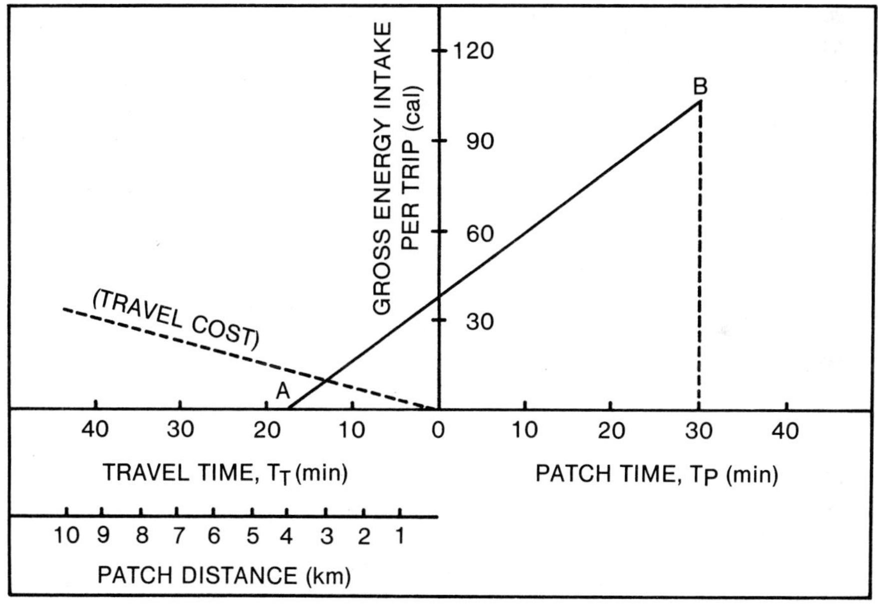

Fig. 5: Graphical representation of the calculation of a honeybee forager's gross rate of energy intake (indicated by the slope of the line AB) from a forage patch which is 4 km from the forager's nest, offers a 1.5 mol (= 43 %) sucrose solution, and requires 30 min of time at the patch for the forager to collect a full, 50 ul load of nectar. The dashed, sloping line on the left side denotes the cost, in calories, of a roundtrip flight between nest and forage patch for various distances, assuming foragers fly 7.7 m/sec (VON FRISCH and LINDAUER 1955, BOCH 1956), weigh 100 mg (Otis 1982), and have a metabolic rate while flying of 1.42 ml $0_2 \cdot g^{-1} \cdot min^{-1}$ (HEINRICH 1979 b). This line indicates that for forage patches up to about 5 km from the nest, the flight cost of a forage trip is only a tiny fraction of the trip's gross profit. In other words, for most forage flights by honeybees, the gross rate of energy intake closely approximates the net rate of energy intake.

I suggest that is is useful to think of a nectar-yielding patch as being characterized by 3 basic properties: (1) the time required to fly to and from the patch (travel time, T_T), (2) the time required to load up with nectar when in a patch (patch time, T_P), and (3) the sugar concentration of the nectar in the patch. Because all workers in a honeybee colony are roughly equivalent in size (WADDINGTON 1981), and probably load similar volumes of nectar of a given sugar concentration (VON FRISCH 1967, WELLS and GIACCHINO 1968, NUNEZ 1966, 1970)[1], it is probably also fair to characterize a patch by the gross energy acquired by a forager during a trip to the patch. Gross energy = (volume of nectar loaded) x (sugar concentration) x (4 calories/mg sugar).

One way of thinking about the relationship between the patch properties just mentioned and a patch's overall profitability to a foraging honeybee is shown in Fig. 5. The slope of the line AB denotes the gross rate of energy intake from the patch. According to this analysis, what a nectar forager needs to know in order to judge the profitability of her forage patch relative to other foragers' patches is the relative energy yield per trip to the patch and the relative duration ($T_T + T_P$) of her forage trips.

How might a bee perceive the relative energy intake per trip to her patch? If, as was argued above, all bees load equal volumes of nectar of a given sugar concentration, then the gross energy intake per trip to a patch can be deduced given knowledge of the value of just one variable – the sugar concentration of a patch's nectar. Thus a forager's problem of estimating the relative energy yield per trip to her patch may simplify to being able to perceive the relative sugar concentration of her patch's nectar.

LINDAUER (1948, 1952, 1954) has already shown that honeybees apparently have a special communication system through which foragers exchange information about nectar concentration. Information transfer occurs during a forager's interactions with the bees in the nest which first receive her nectar, the so-called «receiver bees». These bees sample widely among in-coming foragers and thus know the range of sugar concentration in the available nectars (NOWOGRODZKI 1981). The receiver bees also scale the vigor of their reception of a forager in relation to the sugar concentration of the forager's nectar. Foragers bearing relatively concentrated nectar are quickly and vigorously unloaded whereas foragers returning with relatively dilute nectar are only slowly and seemingly unenthusiastically unloaded. By noting either the delivery time, or the intensity of antennal contacts, or some other variable aspect of the reception by receiver bees, a forager can apparently estimate the relative sugar concentration of her patch's nectar, and thus the relative energy yield per trip to her patch.

Recent studies (SEELEY, unpublished) have revealed that this communication system enables colonies to preferentially allocate foragers to superior patches with high precision. If one offers a colony two food sources in opposite directions 500 m from the hive, with the two feeders identical except for the concentrations of their sugar solutions, and one compares the colony's recruitment rates to the two patches, one observes the pattern shown in Fig. 6. The recruitment rate to the poorer food source drops off exponentially as the difference in sugar concentration between the two feeders increases linearly. A difference in sugar concentration between the feeders as small as 0.125 mol elicits a significant ($p < 0.004$), approximately 30 %, difference in recruitment rate to the two food sources.

How might a bee perceive the relative duration of her forage trips? Unfortunately, there seems to be no way for foragers to exchange information about forage trip times the way they do about their patches' nectar concentrations. One possible alternative is that all

---

[1] When a bee works a nearby (≤ 600 m away) patch, the volume of her nectar load appears to be influenced by patch distance (NÚÑEZ, 1982).

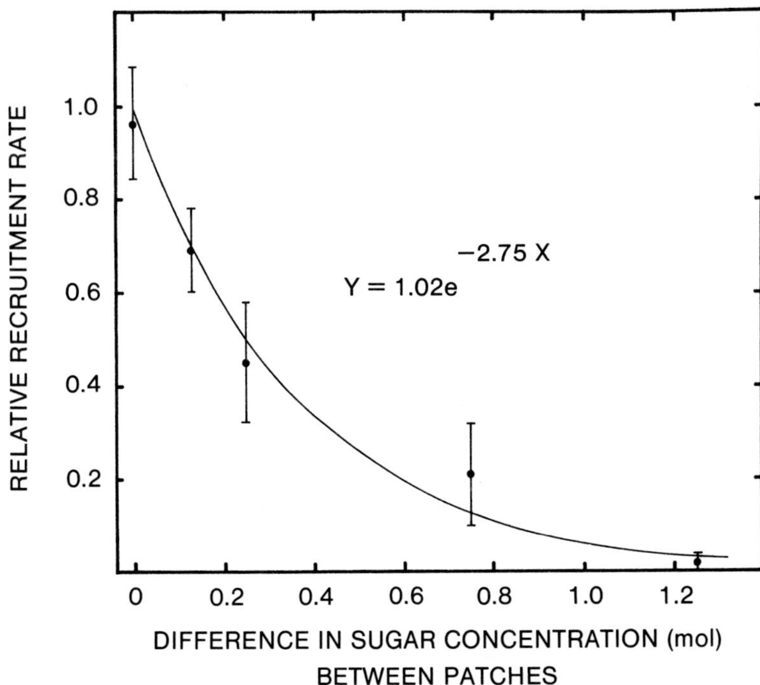

Fig. 6: Sensitivity of a honeybee colony to differences in sugar concentration between forage patches. Relative recruitment rate denotes the rate of recruitment to the patch with less concentrated nectar divided by the rate of recruitment to the patch with more concentrated nectar. A negative exponential describes a colony's response to linear increases in difference in sugar concentration between patches.

foragers possess a shared, innate scale of forage-trip times and that each forager compares the average of her own forage-trip times with the values on this scale to determine whether she is making relatively long, average, or short forage trips. Such a system would only work if the mean and variance of forage-trip durations do not vary widely throughout a summer. To my knowledge, no one has ever collected data on forage-trip times across a summer. Such data would provide a test of the plausibility of this hypothesis.

In summary, the hypothesis for how nectar foragers perceive the relative profitability of a patch is that a forager can estimate both the relative energy yield per trip to her patch and the relative duration of a trip to her patch, and can combine these two pieces of information to estimate the relative rate of energy intake available from the patch. Presumably it is this last piece of information which a nectar forager uses to decide whether to continue working or to abandon her patch. Although these ideas have yet to be tested, this situation should not last for long. Using artificial feeders, one can create forage patches whose sugar concentration, trip time, and patch time can be precisely controlled. Furthermore, by working with foragers from a colony living in an observation hive, one can easily observe the effects of changing the properties of a forage patch on a forager's behavior. Thus is should be possible to test the quantitative predictions outlined above about the relationship between patch properties and forager behavior.

3.2 Division of labor between scouts and recruits

In order for colonies to select forage patches by the method described earlier, most workers must locate new forage patches by following recruitment dances (as opposed to searching randomly on their own, i. e., scouting). Is this the case? It seems clear *a priori* that a colony needs some scouts among its foragers since it is the scout bees which discover new patches. Recruits are directed to known patches and sooner or later all of these known patches will cease yielding food. In other words, a colony needs some explorers as well as exploiters in order to achieve long-term foraging efficiency. But what sort of scout: recruit ratio prevails?

Oettingen-Spielberg (1949) made a preliminary measurement of the percent scouts among novice foragers by adding marked, newly-emerged bees to a hive kept in a flight room and observing what fraction of these bees began foraging by arriving at a feeding station, to which there was recruitment, or to a bouquet of flowers, to which no recruitment was allowed. Of 1062 bees, only 53 (5 %) found the flowers and so presumably were scouts. Lindauer (1952) repeated this measurement using a far superior experimental technique. He observed which individuals within a large cohort of bees living in an observation hive started foraging without following any dances. He observed 13 % scouts (data analyzed in Seeley 1983). I have recently extended these measurements to estimate the percent scouts among experienced foragers (Seeley 1983). My basic experimental procedure was as follows: (1) train a small group of labelled foragers from an observation hive to a feeding station, (2) monitor the foraging tempo of each forager for 2 days, (3) shut off the feeding station and watch the labelled bees back in the observation hive to determine how each individual finds her next forage patch (whether by scouting or by following recruitment dances). This procedure was repeated 7 times to get data for 78 foragers, of which 18 (23 %) were scouts. Clearly, only a small minority of honeybee foragers, both novice and experienced, find patches without following recruitment dances.

The figure of 23 % scouts among experienced foragers is by no means a fixed percentage. Rather, the proportion of scouts apparently depends upon the foraging conditions. In trials when little forage was available, approximately 36 % of the bees found forage by scouting. In contrast, in trials when nectar-bearing plants were blooming profusely, only 5 % of the foragers scouted. Evidently the proportion of scouts is adjusted to the colony's needs. Probably a colony benefits by allocating more bees to scouting when forage is poor, and its foragers are underempolyed, than when forage is rich, and the colony cannot fully exploit the many attractive sources discovered by its scouts.

What determines whether a forager serves as a scout or recruit, and how might this determination mechanism operate so that colonies can adjust the percent scouts in relation to changes in foraging conditions? The key to answering this question may be variation among individual foragers in tendency to perform as a scout or a recruit. According to this view, some bees always scout to find a new forage patch, other always follow recruitment dances, and still others can use either technique (see Fig. 7). What determines whether or not these intermediate, versatile foragers follow recruitment dances might be simply the ease of finding dancing bees to follow. If so, then as the availability of forage decreases, and the number of bees advertising underexploited forage patches dwindles, a colony will automatically allocate more bees to scouting. This in turn should boost the colony's probability of discovering new food sources. Because an exhaustive exploration of a colony's forage range probably requires only a fraction of a colony's forager force, it seems probable that there is an upper limit to the percent scouts among a colony's foragers.

At the heart of this hypothesis is the idea that foragers are not all equivalent, that there are scout and recruit specialists, and so instances of scouting and recruitment are distributed non-randomly among a group of foragers. Alternatively, all foragers are essentially equivalent, and random events determine whether a particular forager, at a given time, scouts or follows dances to find a forage patch. Careful observers of honeybee behavior have repeatedly concluded that individual bees differ markedly in foraging behavior (VON FRISCH 1923, 1967, 1969, LINDAUER 1952, OPFINGER 1949, RIBBANDS 1955, SCHMID 1964). One piece of quantitative evidence supporting the hypothesis of non-equivalent foragers emerged from my measurement of the scout: recruit ratio among experienced foragers (SEELEY 1983). A comparison of the foraging tempos of scouts and recruits during their two days of foraging from the feeding station revealed a significant difference. The scouts' foraging rates averaged higher than those of recruits (scouts, 95 ± 13; recruits, 75 ± 12 foraging trips · 8 hr^{-1} · bee^{-1}; $p < 0.01$). Moreover, individual bees were moderately consistent from day to day in their foraging rates ($r = 0.65$ for the correlation between days in the number of forage trips made daily by each individual).

The origin of these differences among individuals remains unclear. Although foragers of a wide age range can become scouts (OETTINGEN-SPIELBERG 1949, VON FRISCH 1967), foraging experience may have some effect on a bee's scouting tendency since, as described above, experienced foragers seem more likely to scout than do beginning foragers. Genetic differences and differences in developmental conditions between individuals could, of course, also shape the differentiation of scouts and recruits, but there is no solid evidence for or against the role of these two factors.

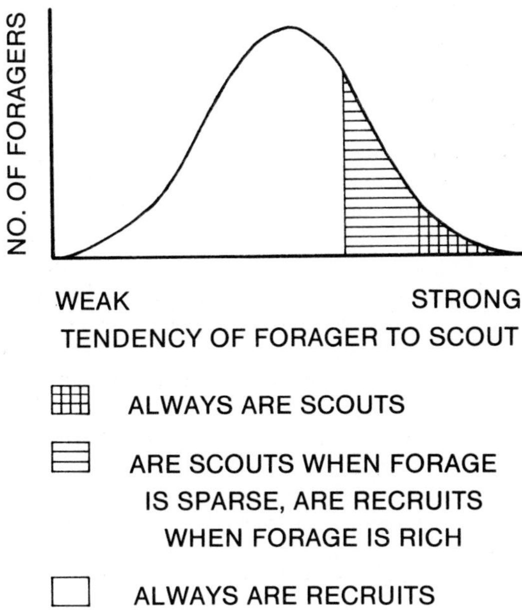

Fig. 7: Hypothetical distribution of foragers across a gradient of tendency to scout. By this scheme, the large majority of a colony's foragers have little inclination to scout and so are always recruits, a tiny minority have a strong tendency to scout and so are always scouts, and the remainder function either as scouts or recruits depending on whether the available forage is sparse or rich, respectively.

4 Foraging energetics

What is the impact on a colony's foraging energetics of the honeybee's elaborate foraging strategy, with its intricate intertwining of recruitment communication, scout-recruit division of labor, and selectivity in recruitment? A preliminary answer to this question has been made by comparing the cost to scouts and the cost to recruits of finding a new food source when the feeding station they had worked for the previous three days was shut off. Search cost comparisons were made by comparing the search times of scouts and recruits. Each bee's «search time» was the sum of the time periods spent outside the nest when searching for a new forage patch (see SEELEY 1983 for details).

Initially, I thought that this comparison would reveal that a major payoff of the honeybee's foraging strategy is a reduced cost of finding forage patches. My reasoning was that because it probably is easier for bees to find forage patches by following recruitment dances than by independently scouting, a colony with recruitment communication and a scout-recruit division of labor probably has lower search costs than one without this social organization of foraging.

Contrary to expectations, the recruits spent more time (and presumably energy) searching for a forage patch than did the scouts: recruits 138 ± 76 min; scouts, 85 ± 58 min ($p < 0.087$). The basis for this surprise is probably the fact that recruitment in the real world is much less automatic than previous studies under experimental conditions (VON FRISCH 1967, ESCH and BASTIAN 1970, MAUTZ 1970, GOULD 1976) had suggested. Whereas honeybee researchers usually study recruitment to conspicuous feeders a few hundred meters from a hive, in nature a typical situation for honeybee recruitment probably involves a small patch of flowers a few thousand meters from a nest. I observed that only rarely does a recruit locate her new patch on the first trip outside the nest after following a dancing bee. Recruits required on average 4.8 trips (range, 1–12 trips). Scouts, in contrast, seek any available forage patch, not particular patches, and so evidently expend less time and energy locating new food sources.

Apparently the payoff to recruits for their greater investment in finding a forage patch is that the patches found are unusually high in quality. This idea is supported by a comparison between scouts and recruits in their behavior following the first trip to a new forage patch. The probability of again bringing back forage was higher for recruits than scouts: 95 % and 51 %, respectively. Clearly, the patches found by recruits had a higher probability of being worth returning to than did the patches found by scouts.

Although not enough data are yet available to provide a detailed picture of how the honeybee's social organization affects a colony's foraging energetics, the following picture is emerging. First, there is a spatial effect. The average rate of food collection per forager is raised because colonies, foraging as an integrated whole, can sample widely through the environment and focus their foragers on the best patches. In insect societies without recruitment communication, such as those of bumblebees (HEINRICH 1979 a), each individual is limited to the information she herself collects when she chooses a work site. Inevitably, the area over which an individual bee can search is far smaller than the 100 or more square kilometers monitored by an entire honeybee colony. Secondly, there is a temporal effect. The honeybee's social organization also enhances foraging efficiency by facilitating the tracking of high-quality forage patches. Whenever a scout bee discovers a new rich food source, or there are changes in the relative rankings of known sources, foragers shift off the less profitable sources onto more rewarding ones. Such shifting of work sites does not depend on each forager making direct comparisons between her current and possible alternative food sources, but instead on the exchange of information among a colony's foragers. In contrast, the foragers in a bumblebee colony share little, if any, information about food sources. Thus each forager must periodically inspect alternative food sources in order to stay informed of changes in foraging opportunities.

5 Further studies

Future research on the social foraging strategy of honeybees will probably expand in two directions. One concerns foraging decisions other than patch choice, the second involves comparative studies within the genus *Apis*.

5.1 Foraging decisions besides patch choice

In collecting its food, a honeybee colony faces the following nested hierarchy of foraging decisions. First, what fraction of a colony's total work force should be devoted to food collection? Second, given a certain size forager force, how should a colony partition it between scouts and recruits? Third, given a certain number of recruits, how should they be divided between the tasks of pollen and nectar collection? And fourth, given a certain number of recruits for either pollen or nectar collection, how should they be distributed among the known patches yielding either pollen or nectar?

Anecdotal evidence suggests that colonies do make decisions at all four of these levels in the foraging process. The fraction of a colony's work force which is devoted to foraging apparently reflects forage availability. When forage is abundant, workers start foraging at an earlier age than when little forage is available (LINDAUER 1952, SEKIGUCHI and SAKAGAMI 1967), thereby creating a larger proportion of foragers in a colony during good times than bad times. Similarly, as we have seen above, the proportion of scout bees in a colony's forager force reflects forage availability. It may be affected by forage patch predictability as well. Current foraging theory (ORIANS 1981) predicts that animals should devote more effort to reconnaissance when the turnover rate of forage patches is high than when it is low. Labor allocation to pollen collection apparently also varies, depending on the relative magnitudes of pollen supply in the environment and demand in the colony. For example, colonies seem to devote a much larger fraction of their foragers to pollen collection in the early spring, when colonies need much pollen for brood rearing but still few flowers are available, than in the summer, when it is easy to balance a colony's collection and consumption of pollen (personal observations). It is also well known that equipping a beehive with a pollen trap (which scrapes the pollen loads from a proportion of the pollen collectors as they enter the hive) stimulates pollen collection (LINDAUER 1952). Finally, as was discussed above, colonies carefully distribute their foragers among patches in relation to profitability differences among patches.

Of these four decision-making processes, only one – forage patch selection – is understood in any detail.

5.2 Comparative studies of honeybee foraging strategies

One special attraction of honeybees for research on social foraging is that they provide a perfect opportunity for using the comparative approach to study the relationship between social organization and foraging ecology (KREBS and DAVIS 1981, SEELEY et al. 1982). On the one hand, the three honeybee species of southern Asia – *Apis florea, A. cerana,* and *A. dorsata* – are closely related and so share many features of their biology, such as their diet of nectar and pollen, their basic morphology, and the dance language system of recruitment communication. On the other hand, there are striking differences among the three species in such aspects of their foraging ecologies as habitat, worker size, flight range, aggressiveness in food source defense, and species diversity of food plants (Lindauer 1956, Singh 1962, Morse and Laigo 1969, Akratanakul 1977, Koeniger and Vorwohl 1979, Seeley et al. 1982). Because the three species are so closely related, differences among them probably represent a recent adaptive radiation within a monophyletic group rather than historical differences among phylogenetically distant species.

By carefully documenting these patterns of divergence in foraging behavior and ecology, it should be possible to identify the ecological pressures which have propelled the differentiation of the three species' foraging behaviors. So little is known about the foraging biologies of the Asian honeybees that it is hard to predict what general aspects of animal foraging behavior and ecology will be illuminated through study of the Asian *Apis*. One possibility is that the three species will display three distinct strategies of social foraging, all of which are based on recruitment communication, but which differ markedly in search area, forage patch selectivity, recruitment rate, vigor of patch defense, and total energy budget. Whatever story ultimately emerges, it is likely to provide a major contribution to our understanding of the ecology and evolution of cooperative foraging in animal societies.

References

Akratanakul, P. (1977): The natural history of the dwarf honey bee, *Apis florea* F., in Thailand. Ph. D. Thesis, Cornell Univ., Ithaca, New York, 1–91.

Alexander, R. D. (1974): The evolution of social behavior. Ann. Rev. Ecol. Syst. *5,* 325–383.

Aristotle (1956): Historia Animalium. Vol. IX, chapt. 40.

Boch, R. (1956): Die Tänze der Bienen bei nahen und fernen Trachtquellen. Z. vergl. Physiol. *38,* 136–167.

Butler, C. G. (1945): The influence of various physical and biological factors of the environment on honeybee activity and nectar concentration and abundance. J. exp. Biol. *21,* 5–12.

Esch, H. and J. A. Bastian (1970): How do newly recruited honeybees approach a food site? Z. vergl. Physiol. *68,* 175–181.

Free, J. B. (1970): Insect Pollination of Crops. Academic Press, New York.

Frisch, K. von (1923): Über die «Sprache» der Bienen, eine tierpsychologische Untersuchung. Zool. Jb. (Physiol.) *40,* 1–186.

Frisch, K. von (1967): The Dance Language and Orientation of Bees, Harvard Univ. Press, Cambridge, Ma.

Frisch, K. von (1969): The foraging bee. Bee World *50,* 141–152.

Frisch, K. von and M. Lindauer (1955): Über die Fluggeschwindigkeit der Bienen und ihre Richtungsweisung bei Seitenwind. Naturwissenschaften *42,* 377–385.

Gould, J. L. (1976): The dance-language controversy. Q. Rev. Biol. *51,* 211–244.

Heinrich, B. (1976): The foraging specializations of individual bumblebees. Ecol. Monogr. *46,* 105–128.

Heinrich, B. (1978): The economics of insect sociality. pp. 97–128. In: J. R. Krebs and N. B. Davies (Eds.): Behavioural Ecology. An Evolutionary Approach. Sinauer Assoc., Sunderland, Ma.

Heinrich, B. (1979 a): Bumblebee Economics. Harvard Univ. Press, Cambridge, Ma.

Heinrich, B. (1979 b): Keeping a cool head: honeybee thermoregulation. Science *205,* 1269–1271.

Kamil, A. C. and T. D. Sargent (1981): Foraging Behavior. Ecological, Ethological and Psychological Approaches. Garland STPM, New York.

Koeniger, N. and G. Vorwohl (1979): Competition for food among four sympatric species of Apini in Sri Lanka *(Apis dorsata, Apis cerana, Apis florea,* and *Trigona irridipennis).* J. Apic. Res. *18,* 95–109.

Krebs, J. R. (1978): Optimal foraging: decision rules for predators. pp. 23–63. In: J. R. Krebs and N. B. Davies (Eds.): Behavioural Ecology. An Evolutionary Approach. Sinauer Assoc., Sunderland, Ma.

Krebs, J. R. and N. B. Davies (1981): An Introduction to Behavioural Ecology. Sinauer Assoc., Sunderland, Ma.

Lindauer, M. (1948): Über die Einwirkung von Duft- und Geschmacksstoffen sowie anderen Faktoren auf die Tänze der Bienen. Z. vergl. Physiol. *31,* 348–412.

Lindauer, M. (1952): Ein Beitrag zur Frage der Arbeitsteilung im Bienenstaat. Z. vergl. Physiol. *34,* 299–345.

Lindauer, M. (1954): Temperaturregulierung und Wasserhaushalt im Bienenstaat. Z. vergl. Physiol. *36,* 391–432.

Lindauer, M. (1955): Schwarmbienen auf Wohnungssuche. Z. vergl. Physiol. *37,* 263–324.

Lindauer, M. (1956): Über die Verständigung bei indischen Bienen. Z. vergl. Physiol. *38,* 521–557.

Lindauer, M. (1961): Communication Among Social Bees. Harvard Univ. Press. Cambridge, Ma.

Mautz, D. (1971): Der Kommunikationseffekt der Schwanzeltänze bei *Apis mellifica carnica* (Pollm). Z. vergl. Physiol. *72,* 197–220.

Maynard Smith, J. (1976): Group selection. Q. Rev. Biol. *51,* 277–283.

Morse, R. A. and F. M. Laigo (1969): *Apis dorsata* in the Philippines. Monogr. Philippine Assoc. of Entomol. *1,* 1–96.

Nowogrodzki, R. (1981): Regulation of the number of foragers on a constant food source by honey bee colonies. M. S. thesis, Cornell Univ., Ithaca, New York, 1–131.

Núñez, J. A. (1966): Quantitative Beziehungen zwischen den Eigenschaften von Futterquellen und den Verhalten von Sammelbienen. Z. vergl. Physiol. *53,* 142–164.

Núñez, J. A. (1970): The relationship between sugar flow and foraging and recruiting behaviour of honey bees (*Apis mellifera* L.). Anim. Behav. *18,* 527–528.

Núñez, J. A. (1982): Honeybee foraging strategies at a food source in relation to its distance from the hive and the rate of sugar flow. J. Apic. Res. *21,* 139–150.

Oettingen-Spielberg, T. zu (1949): Über das Wesen der Suchbiene. Z. vergl. Physiol. *31,* 454–489.

Opfinger, E. (1949): Zur Psychologie der Duftdressuren bei Bienen. Z. vergl. Physiol. *31,* 441–453.

Orians, G. H. (1980): Some Adaptations of Marsh-nesting Blackbirds. Princeton Univ. Press, Princeton, New Jersey.

Orians, G. H. (1981): Foraging behavior and the evolution of discriminating abilities. pp. 389–405. In: A. C. Kamil and T. D. Sargent (Eds.): Foraging Behavior. Ecological, Ethological, and Psychological Approaches. Garland STPM, New York.

Orians, G. H. and N. E. Pearson (1979): On the theory of central place foraging. pp. 155–177. In: D. J. Horn, R. D. Mitchell, and G. R. Stairs (Eds.): Analysis of Ecological Systems. Ohio State Univ. Press, Columbus, Ohio.

Oster, G. F. and E. O. Wilson (1978): Caste and Ecology in the Social Insects. Princeton Univ. Press, Princeton, New Jersey.

Otis, G. W. (1982): Weights of worker honeybees in swarms. J. Apic. Res. *21,* 88–92.

Pyke, G. H. (1981): Optimal foraging in nectar-feeding animals and coevolution with their plants. pp. 19–38. In: A. C. Kamil and T. D. Sargent (Eds.): Foraging Behavior. Ecological, Ethological, and Psychological Approaches. Garland STPM, New York.

Pyke, G. H., H. R. Pulliam and E. L. Charnow (1977): Optimal foraging: a selective review of theory and tests. Q. Rev. Biol. *52,* 137–154.

Ribbands, C. R. (1949): The foraging method of individual honeybees. J. Anim. Ecol. *18,* 47–66.

Ribbands, C. R. (1953): The Behaviour and Social Life of Honeybees. Bee Research Assoc., Ltd., London.

Ribbands, C. R. (1955): The scent perception of the honeybee. Proc. Roy. Soc. London (B) *143,* 367–379.

Schmid, J. (1964): Zur Frage der Störung des Bienengedächtnisses durch Narkosemittel, zugleich ein Beitrag zur Störung der sozialen Bindung durch Narkose. Z. vergl. Physiol. *47,* 559–595.

Seeley, T. D. (1977): Measurement of nest cavity volume by the honey bee *(Apis mellifera).* Behav. Ecol. Sociobiol. *2,* 201–227.

Seeley, T. D. (1982): How honeybees choose a home. Scien. Amer. *247,* 158–168.

Seeley, T. D. (1983): Division of labor between scouts and recruits in honeybee foraging. Behav. Ecol. Sociobiol. *12,* 253–259.

Seeley, T. D., R. H. Seeley and P. Akratanakul (1982): Colony defense strategies of the honeybees in Thailand. Ecol. Monogr. *52,* 43–63.

Sekiguchi, K. and S. F. Sakagami (1966): Structure of foraging population and related problems in the honeybee, with considerations on the division of labor in bee colonies. Hokkaido Nat. Agric. Expt. Sta. Rep. *69,* 1–65.

Singh, S. (1950): Behavior studies of honeybees in gathering nectar and pollen. Cornell Univ. Agric. Expt. Sta. Mem. *288,* 1–34.

Singh, S. (1962): Beekeeping in India. Indian Council of Agric. Res. Publ., New Delhi.

Visscher, P. K. and T. D. Seeley (1982): Foraging strategy of honeybee colonies in a temperate deciduous forest. Ecology *63,* 1790–1801.

Waddington, K. D. (1981): Patterns of size variation in bees and evolution of communication systems. Evolution *35,* 813–814.

Weaver, N. (1957): The foraging behavior of honeybees on hairy vetch. II. The foraging area and foraging speed. Insectes Soc. *4,* 43–57.

Weaver, N. (1979): Possible recruitment of foraging honeybees to high-reward areas of the same plant species. J. Apic. Res. *18,* 179–183.

Wells, P. H. and J. Giacchino (1968): Relationship between the volume and the sugar concentration of loads carried by honeybees. J. Apic. Res. *7,* 77–82.

Williams, G. C. (1966): Adaptation and Natural Selection. Princeton Univ. Press, Princeton, New Jersey.

Fortschritte der Zoologie, Bd. 31 · Hölldobler/Lindauer (Hrsg.): Experimental Behavioral Ecology
G. Fischer Verlag · Stuttgart · New York · 1985

Reproduction, foraging efficiency and worker polymorphism in army ants

Nigel R. Franks

School of Biological Sciences, University of Bath, BA2 7AY England

Abstract

The size and growth rate of *Eciton burchelli*, army ant colonies has been determined by filming colony emigrations and by collecting the discarded pupal cases of their broods of new workers. Colonies of *Eciton burchelli* on Barro Colorado Island, Panama, have between 300,000 and 650,000 workers and they raise between 40,000 and 60,000 workers per 35 days. At one time of year the largest colonies produce a single brood of a few queens and about 4,000 males. These sexually reproductive colonies then undergo a process of binary fission to produce two daughter colonies. Parental colonies reproduce at approximately the size at which the combined growth rate of their daughter colonies just exceeds their own growth rate. This should minimize colony generation time.

There are four distinct physical castes of worker in *Eciton burchelli*, making this one of the most polymorphic of known ant species. I have evaluated the performance of these different worker castes during raids, because foraging efficiency is likely to be one of the key factors limiting colony growth rate. Of the four types of worker, submajors seem to specialise as porters; they are only 3 % of the worker population but they represent 26 % of the workers carrying prey items. Due to an unusual allometry submajors have the longest legs in proportion to their body size of any *Eciton burchelli* workers and they can run faster than any of their sisters and can carry disproportionately large items. Submajors probably also transport prey at a relatively low cost because they are the largest members of their colonies who retrieve prey items and transport costs decrease with increasing body size. The division of labour among *Eciton burchelli* workers may therefore serve to increase foraging efficiency and this should increase colony growth rate, decrease colony generation time and thereby contribute to the inclusive fitness of all members of the army ant colony.

1 Introduction

One of Darwin's (1859) most penetrating insights was his suggestion that sterile forms evolved in social insects because they are «profitable to the community» and «selection may be applied to the family, as well as to the individual». He further suggested that once this colony-level selection had begun, the sterile forms could be molded into distinct castes: «Thus in *Eciton,* there are working and soldier neuters, with jaws and instincts extraordinarily different» (Darwin, 1859). To evaluate this division of labour, which is a key to understanding the limits of social evolution (Oster and Wilson, 1978, Wilson 1980 a, b), it is necessary to relate the size, shape and behaviour of individual workers to the growth and reproductive success of their colonies. Army ants of the genus *Eciton* are particularly suitable for this analysis because their workers are exceptionally polymorphic and yet their colony life-history strategies are relatively simple.

In this paper I examine one way in which the growth and reproductive rate of *Eciton burchelli* colonies is determined by their foraging efficiency and how this in turn is related to the distribution of labour and performance among their workers. *Eciton burchelli* is a good subject for this kind of study for two reasons. First, colonies of this species can be accurately sampled and their size determined, as they nest and raid above ground: and second, the production of both workers and new colonies follow extremely stereotyped patterns.

Colonies of *E. burchelli* alternate between 15 day nomadic phases when they raid during every day and emigrate at night, and 20 day statary phases when they raid less frequently and stay at the same nest site (Schneirla, 1971; Willis, 1967). This cyclical behaviour is correlated the production of discrete broods of eggs, larvae and pupae. These broods are generally of workers; however, at one time of year the largest colonies while maintaining their activity cycles produce a single brood of a few thousand males and about six queens. After such a sexual brood has completed its development the parental colony divides to produce two daughter colonies (Rettenmeyer, 1963; Schneirla, 1971). Because workers are produced in discrete broods it is relatively easy to estimate the growth rate of undisturbed colonies in the field. Furthermore, since new colonies are formed by binary fission it is possible to relate worker production and colony growth directly to reproductive success.

Recent studies (Franks and Fletcher, 1983; Franks and Bossert, 1983) have shown that *Eciton burchelli* colonies have a distinct foraging strategy. In the statary phase these army ants separate neighbouring raids by using a pattern similar to that used by many plants in spiral phyllotaxis; and in the nomadic phase colonies navigate by raiding and emigrating in much the same direction from one day to the nest. This navigation lowers the probability that the nomadic raid path will cross itself and also separates the areas foraged in successive statary phases. These swarm raiding patterns will contribute to colony growth rate by increasing foraging efficiency. In this paper I examine the behaviour and morphology of individual workers within the raid system to determine how their design contributes to foraging efficiency. This paper also reports the first quantitative study of colony size, brood production, caste ratios, and reproduction in these army ants.

2 Methods

The study was conducted on Barro Colorado Island, Panama (Leigh et al., 1982) during the dry season, January to May of 1979. The size of seven *Eciton burchelli* colonies was estimated by filming their emigrations. Colonies emigrate on almost every night of their nomadic phase and during these migrations the entire colony moves between bivouac sites where their nests take the form of hanging baskets of living workers. The route taken by an emigration can be predicted very accurately because the emigration column follows the exact path of the principal trail of the same day's raiding (Rettenmeyer, 1963). This is the trail that the ants used to go to the raiding swarm front and along which they returned with prey items. An emigration is signaled when the ants at the swarm raid no longer return to the old bivouac, but start to form a new bivouac towards the end of that day's raid system. When this is occurring return traffic to the old bivouac quickly diminishes, as does outward traffic to the swarm: then quite suddenly with the onset of dusk, movement becomes almost entirely outward and rapidly increases as the emigration begins. This moment was taken as the cue to begin filming the emigrations of the selected colonies.

The emigration from bivouac to bivouac is so orderly that it resembles a ribbon of ants unwinding from one spool and winding-up on another. The number of ants in a bivouac can be estimated from the following information: (1) the average velocity of the column, (2) the average number of ants per unit length of the trail, and (3) the duration of the emigration. The first was estimated by timing the progress of 20 burdened and unburdened ants of various sizes over a one metre section of trail, at various times during the emigration. The second was assessed by taking a picture of a known length of about one metre of trail every ¼ of an hour throughout the emigration. Finally the duration of the emigration, to the nearest ¼ of an hour, could also be estimated from the same series of pictures. The number of ants in the bivouac (B) was estimated from the relation B = T.V.N. where T is the duration of the emigration, V is the mean velocity of the column, and N is the average number of ants per unit length of the emigration column.

The photographs were taken automatically by a 35 mm single lens reflex camera fitted with a timer, an auto-winder and a flash. A fine grain, Kodak Pan-X monochrome film was used. The number of ants per unit length of trail, as sampled every 15 minutes throughout the emigration was counted directly from the film negatives. Members of the two largest worker castes could also be distinguished and counted from the films, to estimate possible changes in caste ratios with colony size (Table 1). The camera was stationed near the old bivouac site, so only the ants in the bivouac and none of the ants in the raid system were photographed.

The filmed colonies were all either forming or leaving a statary bivouac and I estimated the size of the brood each colony produced by collecting the pupal cases of either its worker brood or its predominantly male sexual brood. Pupal cases are sometimes also deposited at the first nomadic bivouac (Rettenmeyer, 1963) and these were also collected. A D-Vac suction sampler was used to collect all the pupal cases from even the most inaccessible recesses of the hollow trees in which the colonies had been bivouacing. The cases and the debris that were collected with them were later separated, after preliminary drying, by blowing them apart. The cases were then dried to constant weight in a 50 °C oven and weighed, as was a counted subsample, to provide an estimate of the number of workers or males produced by the filmed colonies.

To estimate the size of the workers that had emerged from the empty pupal cases, callow workers still in their pupal cases were collected from one of the first nomadic emigrations. I removed these callow workers from their cocoons and recorded both the dry weight of each ant and the width of its pupal case. Thus the size distribution in each worker brood was estimated by measuring the width of 100 pupal cases taken at random from those discarded by each colony.

In a separate series of measurements the raid lengths of the filmed colonies were also recorded by laying a 100 m tape along the principal trail of some of their nomadic phase foraging systems. These distances were measured when the raids were at their maximum length just before night fall.

Morphological measurements of ants, made by using a dissection microscope fitted with an eyepiece graticule, suggest that there are four distinct castes in *E. burchelli* (see results section). With practice, individual workers can be allocated to one of these four castes by visual sorting, and this can be done with a high degree of success as checked by subsequent measurement.

The relative abundance of these four worker castes in the foraging and bivouac populations were also estimated by using the suction sampler to make unbiased collections of raiding ants and by pressing a beaker containing 70 % ethanol into the side of a pendulous bivouac. In addition, to investigate the role of the different castes as porters of prey items, just those ants engaged in this task were collected from raid trails. To ensure unbiased sampling, a flag was placed by a trail and all the ants carrying prey past this marker were collected individually. All samples of adult workers were taken from a series of different *E. burchelli* colonies, which were not filmed, but which were judged to be of a wide range of different sizes.

To analyse one aspect of foraging efficiency I measured the speed of workers running in raid trails and the size of the prey items they carried. The velocity of various sizes of workers was recorded as they ran in raid trails that passed along the length of fallen tree trunks and other such obstacle-free courses. I followed and timed individual workers as they ran over measured distances of 10 m and more and collected both the ants and any prey items they were carrying. Burdened ants were only timed if they were carrying prey items without the aid of other ants. Morphological measurements and dry weights of ants and prey items were then obtained using the standard methods outlined above.

3 Results

3.1 Colony Size

Four of the filmed colonies had worker broods and three had sexual broods. The colonies with sexual broods divided evenly at the end of their statary phase. This process is described in detail by SCHNEIRLA (1956). One of these colonies (E.b. 11) was filmed before

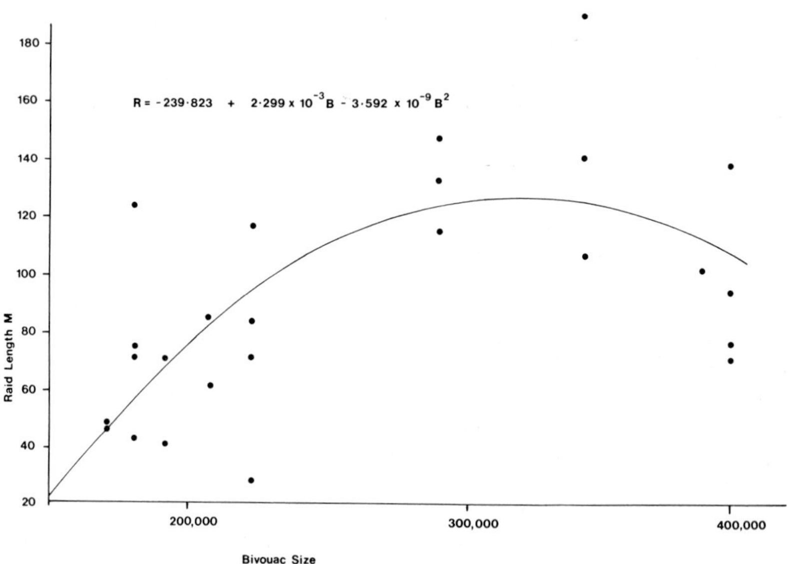

Fig. 1: Daily raid length (R) in metres as a function of bivouac size (B), with polynomial least squares curve fit ($r = 0.694$, $N = 25$, $p < 0.001$).

it entered its last statary phase, whereas one daughter of each of the other sexual colonies (E.b. 14 and 16) was filmed and the size of their parental bivouac was estimated by doubling the estimated size of the filmed daughter bivouac (Table 1). The bivouacs studied ranged in size from about 180,000 to 450,000 ants and as the largest of these have swarm raids containing 200,000 ants (WILLIS, 1967) the maximum colony size of *Eciton burchelli* on Barro Colorado Island is about 650,000.

The larger bivouacs produced bigger and longer raids as described by $R = -239.82 + 2.29 \times 10^{-3} B - 3.59 \times 10^{-9} B^2$ ($r = 0.694$, $N = 25$, $p < 0.001$) where R is the raid length in metres and B is the number of ants in the bivouac (Fig. 1).

3.2 Castes

Four distinct castes can be recognized among *E. burchelli* workers, as can be seen in Figures 2 and 3 which depict aspects of their allometry. These graphs recorded data from a large sample of workers from a bivouac that was not filmed. A random sub-sample of 573 of these bivouac ants was used to obtain size-frequency data; (Figure 4) this shows the quadrimodal distribution of worker sizes in *Eciton burchelli*. From these data, the dry weight ranges of the different worker castes were estimated as follow; minims less than 1.5 mg, medium workers 1.6 to 5.5 mg, submajors 5.6 to 9.5 mg, majors 9.6 to 11.5 mg.

Majors and large submajors (those with heads of a lighter shade) could be distinguished from all other workers in the films of the emigrating colonies. This technique was used to obtain estimates of the relative abundance of these workers in the bivouacs of the filmed colonies (Table 1). These data suggest that the caste ratios of adult workers change little, or not at all, with increasing colony size.

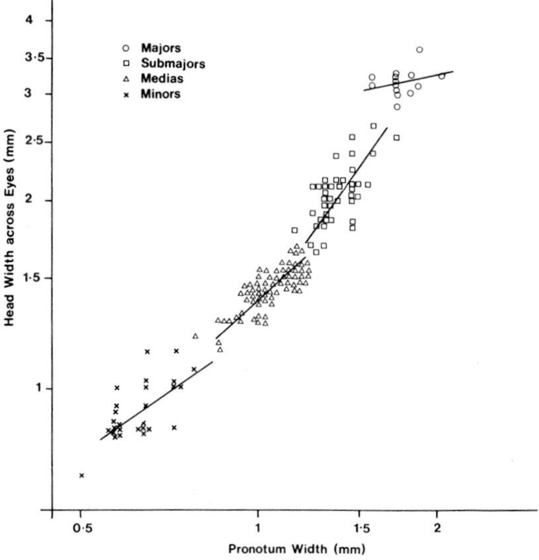

Fig. 2: Head width-pronotum width allometry, based on a sample of 150 ants. The ants were allocated to four castes before measurements were taken. Crosses indicate minims, triangles medium workers, squares submajors and circles majors. Curves have been fitted to each of the four castes by the least squares method.

In samples of dead workers, individuals can be accurately assigned to castes purely by visual sorting. This method was used to obtain caste ratios of raiding workers and those carrying prey (Table 2). It is clear from these data that the proportions of the different castes, present in raids and carrying prey are not the same as those in the bivouac as a whole.

Caste ratios of callow workers change as a function of colony size as shown in Figure 5, where the dry weights of 100 callows for each filmed colony, have been estimated from C = 0.019 $W^{3.67}$ ($r = 0.99$, $N = 10$ $p < 0.001$) where Cmg is callow dry weight and Wmm is the maximum width of the ant's pupal case.

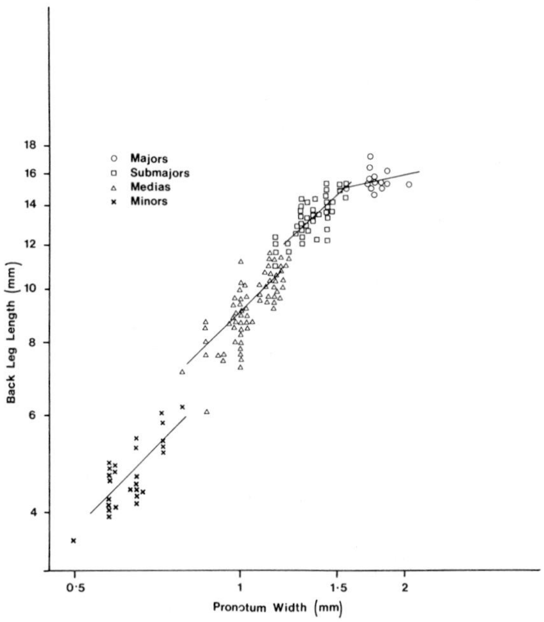

Fig. 3: Back leg length-pronotum width allometry (N = 150). The ants were allocated to four castes before measurements were taken. Crosses indicate minims, triangles medium workers, squares submajors and circles majors. Curves have been fitted to each of the four castes by the least squares method.

Table 1

Colony	Ants/M	Emigration data Velocity M/min	Duration min	Caste % Large Submajors	Majors	Bivouac size	Number of pupae
E.b.20	162	3.1	360	1.74%	0.99%	180,792	43,366 workers
E.b.19	145	3.2	450	1.80%	1.17%	208,800	44,982 workers
E.b.10	189	3.2	480	1.90%	1.48%	290,304	57,230 workers
E.b.18	219	3.0	600	1.68%	0.73%	394,200	60,112 workers
E.b.16*	154	2.9	390	1.53%	0.88%	348,348	3,798 males
E.b.14*	150	3.1	480	1.80%	1.20%	446,400	4,112 males
E.b.11	265	3.3	390	1.91%	1.70%	341,055	4,143 males

* Daughter colonies filmed

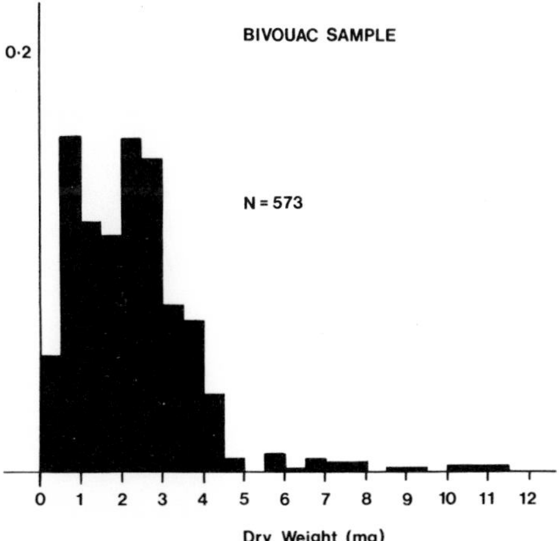

Fig. 4: The frequency-dry weight distribution of 573 bivouac workers, showing four modes at approximately 0.75 mg, 2.5 mg, 7 mg and 10.5 mg, corresponding to the mean weights of the four worker castes.

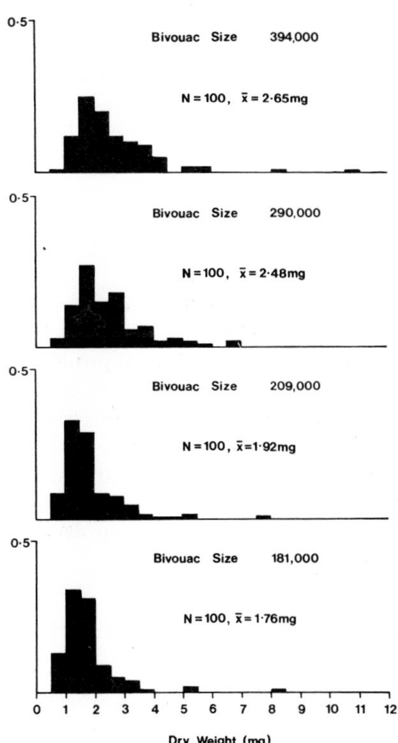

Fig. 5: Frequency – dry weight data of callow workers produced by four colonies. 100 pupal cases were measured from those produced by each colony and the weight of callows was estimated from these data (see text). The average dry weight of each colony's callows is given, as is the size of the bivouac.

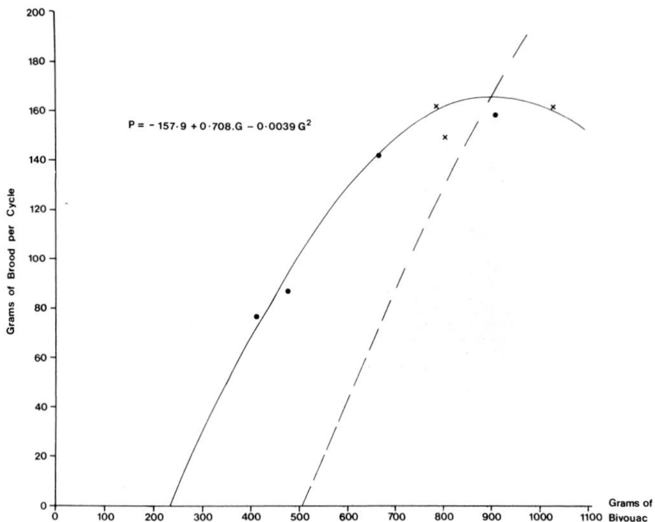

Fig. 6: Grams of brood produced per 35 days (P) as a function of bivouac dry weight (G grams). Circles indicate colonies that produced worker broods and crosses indicate colonies that produced sexual broods (see Table 1). The solid line shows a polynomial, least squares, curve fit based on all 7 colonies ($r = 0.99, N = 7, p < 0.001$). The broken line was derived from the formula of the solid line and represents the combined growth rate of two identical daughter colonies as a function of the size of their parental bivouac. The intersection of the solid and broken lines indicates the optimum size for colony division.

3.3 Colony Productivity

These data relating callow caste ratios to colony size have been used to calculate colony productivity, by estimating the dry weight of each colony's brood from its unique callow size distribution (Fig. 5) and the number of callows it produced (Table 1). The dry weight of sexual broods, which are more than 99.8 % male have been calculated from the average dry weight of males which is 39 mg. Bivouac dry weights have been estimated on the basis of the average dry weight of bivouac workers which is 2.32 mg (Fig. 4). The relationship between bivouac weight and the weight of brood produced per 35 day cycle is shown in Figure 6, which also shows a best fit line described by $P = -157.9 + 0.708\ G - 0.0039\ G^2$ ($r = 0.99, N = 7, p < 0.001$) where P is the dry weight of brood in grams and G is the dry weight of the bivouac in grams.

3.4 Foraging efficiency

Colony productivity is probably partly limited by foraging efficiency and one important aspect of raiding ergonomics is the ability of the workers to retrieve large prey items quickly and at low energetic cost. *Eciton burchelli* seems to have a specialist porter caste: submajors are engaged in the task of carrying prey items in much larger numbers than expected from their overall abundance in raid systems (Table 2). Furthermore, submajor morphology seems to be specially adapted for their role as porters. Submajors have the longest legs in proportion to their body size of any *E. burchelli* workers: the ratio of back

Table 2

	Caste populations Minim	Medium	Submajor	Major
Bivouac (N = 573)	33.68%	62.13%	3.14%	1.05%
Brood (N = 400)	29.75%	68.00%	2.00%	0.25%
Raid (N = 3314)	22.00%	74.55%	3.15%	0.30%
Carrying prey (N = 244)	7.38%	66.39%	26.23%	0%

leg length to body length (sting to occiput) is less than 1 for minims, between 1 and 1.3 for medium workers, between 1.4 and 1.5 for submajors and between 1.3 and 1.4 for majors (see also Fig. 3). Possibly as a consequence of their disproportionately long legs, submajors can run faster than any of their sisters. The velocity of unburdened running ants, excluding majors, is an exponential function of their leg length as described by: velocity = 0.038 exp 0.058L ($r = 0.79$, $N = 12$, $p < 0.005$) where velocity is in metres per second and L is back leg length in mm (Fig. 7). However, when minims, medium workers and submajors are carrying prey items they all run at much the same speed, as described by: velocity = 0.054–0.001 C ($r = 0.19$, $N = 24$ $P > 0.1$) where C is ant dry weight in mg and velocity is in metres per second. The reason the burdened ants run at the same speed is probably that the larger ants carry disproportionately large prey items, with the exception of majors who

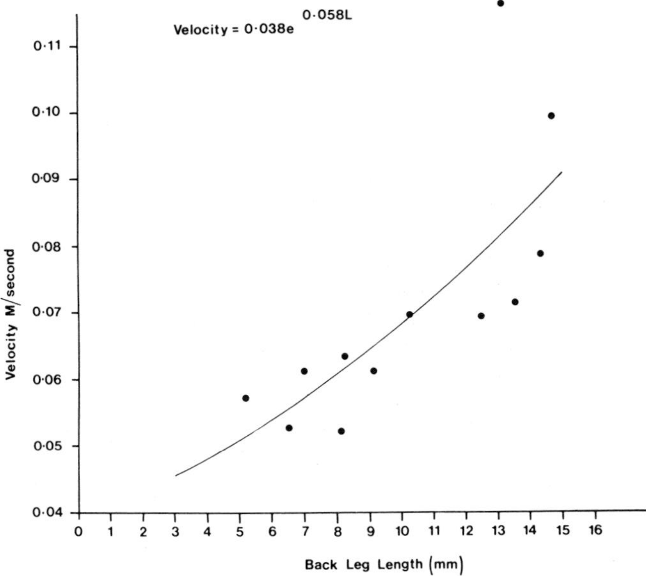

Fig. 7: The velocity of individual unburdened workers in metres per second as a function of their back leg length (Lmm). The curve was fitted by the least squares method ($r = 0.79$, $N = 12$, $p < 0.005$).

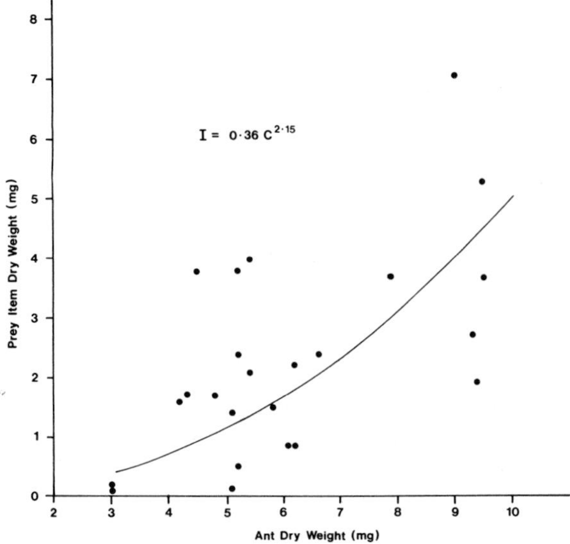

Fig. 8: The dry weight of prey items (Img) carried by individual workers as a function of worker dry weight (Cmg). Least squares curve fit ($r = 0.66$, $N = 24$, $p < 0.005$).

never carry anything (Rettenmeyer, 1963). Not only do submajors carry the largest items, but their burdens are so large that they also transport the greatest weight per unit of their own body weight: the relation between prey item dry weight I mg and ant dry weight C mg is $I = 0.36\,C^{2.15}$ ($r = 0.66$, $N = 24$ $p < 0.005$) (Fig. 8).

4. Discussion

To relate worker design to colony efficiency it is first necessary to consider the factors that limit the fitness of all the individuals within the colony. In army ants this problem is greatly simplified because new colonies are formed by the division of old ones (Schneirla, 1956; Schneirla and Brown, 1952). Hence the fitness of all members of an army ant colony will be largely determined by the rate at which the colony can divide and this will depend upon colony growth rate. Natural selection should act to minimize the generation time (Lewontin, 1965), by maximizing the rate of successive colony divisions. Colonies should therefore grow as quickly as possible and then divide at such a size that the average growth rate of their daughters is also maximized. For this reason we might expect army ant workers to be designed and allocated to castes in such a way as to maximize colony growth rate. To evaluate this hypothesis I will first test a simple model for the reproductive strategy of *Eciton burchelli.* This model highlights the importance of colony growth rate which is probably most limited by foraging costs. I will then go on to consider how worker design minimizes these foraging costs and contributes to colony productivity.

How and when should a colony divide? First, if the survival probabilities of the daughter colonies are the same, which is likely because at division the workers can chose the fittest queens (Franks and Hölldobler, in prep.), then the parental colony should divide its worker pouplation equally in two, as this will minimize the average generation time of its daughters (CALOW et al., 1979). Second, the timing of division should be determined by the growth rate of both parental and daughter colonies. Consider the growth rate of an established colony to be given by $y = f(S)$ which decelerates with colony size as in Figure 9. Now let each colony divide at $S_0 = 2.S_1$ to give two daughters of size S_1. So the generation time is given by:

$$T = \int_0^T dt = \int_{S_1}^{2S_1} \frac{dS}{S} = \int_{S_1}^{2S_1} \frac{dS}{f(S)}$$

which is a function of S_1. For a minimum $\frac{dT}{dS_1} = 0$ but $\frac{dT}{dS_1} = \frac{2}{f(2S_1)} - \frac{1}{f(S_1)}$. So $f(2S_1) = 2f(S_1)$ and $f(S_0) = 2f(S_1)$. Thus to maximize its fitness each colony should divide equally in two and at such a size that the combined growth rate of its daughters is equal to its own growth rate.

This seems to be the case in *Eciton burchelli*. Figure 6 shows the biomass of brood raised by different biomasses of bivouac in 35 days: the solid line in this figure can be taken to represent average colony growth rate as a function of colony size because both overall colony size and worker mortalities are likely to be proportional to bivouac size (FRANKS, 1982 a). The broken line in Figure 6 has been derived from the formula of the solid line, and represents the combined growth rate of two identical daughter colonies as a function of the size of their parental bivouac. For this reason the intersection of the solid and broken lines in this figure indicates the optimum bivouac size for division, which will be the best compromise between maximizing parental and daughter colony growth rates to minimize the generation time. Parental bivouacs quite closely approximate this optimum size (Fig. 6). Colony division occurs only in the dry season in Panama (SCHNEIRLA, 1971) and this seasonality must place an additional constraint on reproductive colony size. The decision whether to divide this year or next will inevitably mean that parental colony size can only be approximately optimized and this may account for the small scatter of parental colonies about the optimum size for division.

From this analysis it is clear that the shape of the colony growth function ultimately determines colony fitness. One of the most likely reasons for the fall in growth rate with

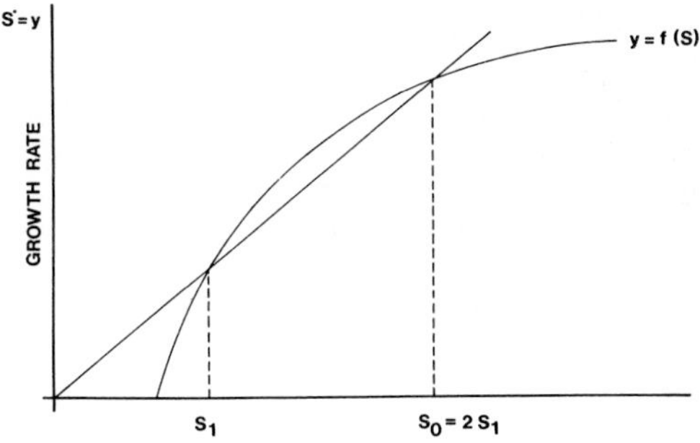

Fig. 9: A hypothetical model of colony growth rate as a function of colony size (see text for explanation). To minimize the generation time a colony should divide into two daughter colonies of equal size (S_1) whose combined growth rate is just equal to its own growth rate.

colony size is that foraging costs will increase for bigger colonies. A common assumption is that the size of social insect colonies is limited by the costs of central place foraging (Hölldobler and Lumsden, 1980; Orians and Pearson, 1979). These costs result from the local depletion of resources which means that foragers must go further and further, at ever increasing costs, to retrieve less energy per unit time. Paradoxically these costs are perhaps most limiting for nomadic army ants because their foraging time is also limited by their need to make frequent emigrations which employ the entire colony (Mirenda and Topoff, 1980). For example, nomadic *Eciton burchelli* colonies commonly spend 12 hours raiding and 8 hours emigrating each day. As large colonies have only the same time in which to forage as small colonies, but must produce more extensive raids (Fig. 1), it is reasonable to assume that raiding velocity is one of the key factors limiting colony growth. During the course of an average raid 30,000 prey items, which have been dismembered at the swarm front, are carried back to the bivouac (Franks, 1982 a). The ants carrying such prey items have to run considerable distances as the average raid length at the termination of foraging is 105 m (Franks, 1982 b) and only when these ants have delivered their burdens are they free to turn around and run back to join the swarm. Sometimes the army ants store prey items in booty caches situated along the raid system (Rettenmeyer, 1963), and these may help to reduce transport costs. Nevertheless, the progress of the swarm will be limited by the time the ants spend running from the raid to such prey stores and the bivouac, and back again. For this reason, foraging efficiency and colony growth rate will be both determined in part, by the ability of the ants to run fast and carry large prey items.

What design criteria should apply to workers that specialize in the role of prey transport? First, such workers should be relatively large, because the net cost of transport,

Fig. 10: A major on the right rears above a submajor who is carrying a fragment of a roach with the assistance of a medium worker.

defined as the quantity of energy used to transport one unit of body weight over one unit distance, is inversely proportional to body weight. For vertebrates the rapidly diminishing cost of transport with body size can be expressed by a negative linear relationship on a log log plot of cost versus weight (FEDAK and SEEHERMAN, 1979). Furthermore, JENSEN and HOLM-JENSEN (1980) have shown that ants have values for the cost of transport that are only slightly higher than those expected by extrapolation from vertebrates (ALEXANDER, 1982).

In addition to transporting items at low energetic cost the army ants also need to move their brood and booty quickly. This suggests that the large porter caste should have disproportionately long legs. The rationale for this prediction comes from an analysis of terrestrial locomotion by ALEXANDER (1976). He suggests that where inertia and gravity interact as with animals moving on land, the Froude number provides a good criterion for assessing the physical similarity of their movements. The Froude number is $u^2/g\,L$ where u is velocity, g the acceleration of free fall, and L is a characteristic dimension such as leg length. The theory of physical similarity suggests that relative stride length will be a function of the Froude number, even when animals are not geometrically similar and this implies that velocity is likely to be a positive function of leg lenght. These considerations of the cost and rate of transport therefore suggest that to be efficient porters workers should be both relatively large and disproportionately long legged. Remarkably, *Eciton burchelli* has a specialist porter caste with exactly these morphological adaptations.

Submajors clearly specialize in the role of prey transport as they are only 3 % of the workers population of the colony but represent 26 % of the ants carrying prey items (Table 2). Submajors are the largest *E. burchelli* workers that carry brood or booty: they are only marginally smaller than majors who never act as porters (RETTENMEYER, 1963). In addition to their relatively large size, submajors can run much faster than any of the other workers. The greater speed of submajors is unlikely to be purely a consequence of their larger size, because if all the workers were the same shape but of different sizes they would be expected to move at the same speed, since the power available and the power needed to run would both be proportional to body mass (ALEXANDER, 1971). As the Froude number analysis suggests, to move faster submajors should have relatively longer legs than their sisters. This is the case. Leg length is determined by a positive allometry for minims up to and including submajors, but in majors leg length is determined by a negative allometry (Fig. 3). A similar complex allometry of leg length is also found in *Eciton hamatum* (TOPOFF, 1971), but not in *Atta* (WILSON pers. comm.). This difference is expected because leaf cutting ants can forage from their sedentary nest for their immobile food for 24 hours a day. So in *Atta* running velocity will not limit foraging to the extent that it does in *Eciton*. Furthermore, *Eciton burchelli* workers run at much the same speed when they are carrying burdens; this may be due to the larger ants and submajors in particular carrying disproportionately large loads. This is itself may be an adaptation to prevent traffic congestion in the columns of ants returning to the bivouac. The speeds of burdened *Atta* workers seem to be similarly adjusted (WILSON, 1980 b). In *Eciton burchelli* the unladen workers move out to the swarm raid by running at great velocity down the sides of the foraging columns. Army ants are also unusual in that they carry prey items beneath their bodies, which enables teams of ants to co-operate efficiently when they carry very large prey items. An observation by RETTENMEYER (1963) further emphasizes the importance of this division of labour and the cost of transport. He observed that the frequency with which adult workers are carried varies inversely with their size, and increases near the ends of emigrations. It is probably less expensive for the colony as a whole, for a submajor to carry a minim than to have both ants move under their own power.

Submajors are specialists and so apparently are two of the other three castes in *Eciton burchelli.* Minim workers are disproportionately common in the bivouacs (Table 2) where they are believed to act as brood nurses (Schneirla, 1971; Topoff, 1971; de Silva, 1972) and majors are equipped with mandibles like ice-tongs and seem to be a specialist defensive caste against would-be vertebrate predators (Schneirla 1971). By contrast medium workers are a generalist caste and they seem to be equally abundant in all roles within the colony (Table 2).

The value of a particular worker caste will be determined not only by current performance but also by manufacturing and maintenance costs and by the rate at which the workers will need to be replaced as determined by their longevity. The manufacturing costs of workers are likely to be proprotional to adult dry weight (Wilson, 1980) and hence submajors with a dry weight of 7 mg will be much more expensive for the colony to grow than medium or minor workers with dry weight of 2.5 mg and 0.75 mg respectively. However, maintenance costs may not be much higher for larger workers as it is well known that in ants metabolic rate per gram of body tissue tends to fall rapidly with increasing body size (Peakin and Josens, 1977). Furthermore, majors and submajors may live longer than their smaller sister so their replacement rate will be less and their relative cost to the colony will also be reduced.

In army ants the importance of worker longevity is highlighted again by the phenomenon of reproduction by colony division. Worker longevity will not only affect colony growth rate directly but it will also determine the value of workers to the daughter colonies that receive them. The minimum average longevity of *Eciton burchelli* workers can be estimated from their rate of production in a colony of average size. Such a colony has about 400,000 workers and produces about 50,000 new workers everv 35 days, thus each worker must live at least 280 days for the colony to be able to grow at all.

Data are not available on caste specific death rates in *E. burchelli*, but differential mortality rates can be tentatively inferred from the rate of production of the various types of worker, and this can be estimated from the size of the pupal cases that the colonies deposited (Fig. 5). The most surprising thing about these data is that brood caste ratios change as a function of colony size, with larger colonies producing larger workers. Such shifts in the caste ratios of adult workers are known for many other ant species (Oster and Wilson, 1978), but marked changes in the relative abundance of the four worker castes in *Eciton burchelli* are unlikely to occur for the following reasons. First, parental colonies provision their daughter colonies with their own workers; hence caste ratios must be very similar in mother and daughter colonies. Second, all colonies maintain similar nomadic and foraging life styles (Franks and Fletcher, 1983; Franks and Bossert, 1983) and lastly, as a consequence of reproduction by binary fission the largest colonies are only twice as large as the smallest, whereas in many other non-doryline species mature colonies are thousands of times more populous than incipient ones (Wilson, 1971).

Submajors are the one caste that might be more useful in greater relative abundance in larger colonies, because bigger colonies forage further and will incur greater transport costs (Fig. 1). However, the proportion of even submajors does not appear to change markedly in the adult worker populations of the growing colonies (Table 1). For these reasons the data on callow worker production suggests that larger colonies are provisioning their daughter colonies with all, or most, of the submajors and majors they will need for a considerable period of time. This follows from the observation that small colonies are either not producing these larger worker castes or producing them at a very low rate. Furthermore, summed over all colony sizes the larger workers are produced at a much lower rate than their smaller sisters (Table 2). These findings suggest that submajors and majors live longer than their smaller sisters. This is important because if these larger castes do live longer then the relative costs of these workers to the colony will also be reduced.

Colony level selection is likely to determine the longevity of sterile workers and for this reason the demographic statistics of sterile workers are of great theoretical interest not least because workers are unlikely to suffer from conventional senescence (WILSON, 1975). This prediction stems from the hypothesis that conventional senescence is due to the accumulation of genes that are beneficial before, but detrimental after, the last age of reproduction (MEDAWAR, 1957; WILLIAMS, 1957). Since sterile workers do not reproduce themselves they are unlikely to senesce in the conventional way.

Clearly, the realized longevity of workers will be determined in part by the risks they take, and colony level selection is likely to have caused the well known temporal division of labour, where young, and hence more valuable, workers are employed on such safe tasks as nursing brood while older workers take the hazardous roles of foraging and colony defense (WILSON, 1971). The much more speculative idea that I wish to discuss here is that colony level selection can alter the fundamental longevity of workers simply by altering their size.

If the same types of correlation apply to social insect workers as to a wide variety of other organisms then larger workers will tend to have inherently greater longevities. For a wide variety of oragnisms from *Escherichia coli* to *Sequoia* generation time is proportional to organism length to the power 0.9 (BONNER, 1965) which implies that generation time is proportional to body mass to the power 0.3. An equally all encompassing correlation is that within broad categories of organisms metabolic rate is proportional to body mass to the power 0.75 (KLEIBER, 1961) although within more closely related groupings the relationship may be closer to the more theoretically satisfying proportionality of mass to the power 0.67 (HEUSNER, 1982). Furthermore, PEARL (1928) has suggested, in his hypothesis of a ‹rate of living› that these correlations are causally related, and that larger organisms live longer because they have a lower metabolic rate per unit of body mass. It should be noted that PEARL's hypothesis and MEDAWAR's (1957) explanation for conventional senescence may be complimentary if for example the rate of metabolic activity per gram of body tissue is associated with the accumulation of gene replication errors that are only deleterious after the last age of reproduction (GRANT, 1978).

The important point that I wish to draw from these hypotheses is that larger workers are likely to have greater physiological longevities than their smaller sisters if they have a lower metabolic rate per unit body weight. Most intriguingly, large ant workers do have lower respiratory rates per unit body weight than smaller workers (PEAKING and JOSENS, 1977; WILSON, 1980 b), and in the harvester ant *Pogonomyrmex badius* this decline is so rapid that majors and minors have the same resting rate of oxygen consumption even though the larger caste is ten times heavier than the smaller (GOLLY and GENTRY, 1964). Hence larger workers may have both relatively low maintenance costs and inherently greater longevities. Similar factors may contribute to the low replacement rate of submajors and majors in *Eciton burchelli*. Clearly worker longevity is not only of direct importance in social evolution but it may also shed light on the very nature of biological senescence (WILSON, 1975).

From DARWIN onwards, studies of biological adaptation have been conducted by the comparative method. Social insects provide an unrivalled opportunity to refine this technique because colonies with polymorphic workers allow comparisons to be made not only with other species but also among the individuals within the same society. In species such as *Eciton burchelli* adaptations in behaviour, morphology, physiology and even longevity can be analysed in the context of the ecology of the society and the inclusive fitness of its individual members.

Acknowledgements

I wish to thank Mark Hodson for the anaylsis of allometry described in this paper. N. F. Britton for the optimum life history model and Stuart Reynolds for his criticism of an earlier draft of this manuscript. R. McNeill Alexander, E. Broadhead, B. Hölldobler, S. D. Porter, S. E. Reynolds and E. O. Wilson provided helpful advice and encouragement. Facilities were generously provided by the Smithsonian Tropical Research Institute, Republic of Panama. This research was supported in part by a Natural Environment Research Council Studentship and by a Fellowship from the Royal Commission for the Exhibition of 1851.

References

Alexander, R. McN. (1971): Size and shape. Arnold London.
Alexander, R. McN. (1976): Estimates of speeds of dinosaurs. Nature. *261,* 129–130.
Alexander, R. McN. (1982): Locomotion of Animals. Blackie Glasgow.
Bonner, J. T. (1965): Size and cycles: and essay on the structure of biology. Princeton University Press, Princeton.
Calow, P. Beveridge, M., Sibly, R. (1979): Heads and tails: adaptational aspects of asexual reproduction in freshwater triclands. Amer. Zool. *19,* 715–727.
Darwin, C. (1859): On the origin of species. Murray, London.
Fedak, M. A. Seeherman. H. J. (1979): Reappraisal of energetics of locomotion shows identical cost in bipeds and quadrupeds including ostrich and horse. Nature. *282,* 713–716.
Franks, N. R. (1982 a): Ecology and population regulation in the army ant *Eciton burchelli.* In: Leigh, E. G., Rand. A. S., Windsor, D. W. (eds) The ecology of a tropical forest: Seasonal rhythms and long-term changes. Smithsonian Institute Press. Washington DC pp. 389–395.
Franks, N. R. (1982 b): A new method for censusing animal populations: the number of *Eciton burchelli* army ant colonies on Barro Colorado Island. Panama. Oecologia (Berlin) 52, 266–268.
Franks, N. R., Bossert, W. H. (1983): The influence of swarm raiding army ants on the patchiness and diversity of a tropical leaf litter ant community. In Sutton, S. L., Whitmore, T. C., Chadwick, A. C. (eds) Tropical rain forest: ecology and management. Blackwell, Oxford pp. 151–163.
Franks, N. R., Feltcher, C. R. (1983): Spatial patterns in army ant foraging and migration: *Eciton burchelli* on Barro Colorado Island, Panama. Behav. Ecol. Sociobiol. *12,* 261–270.
Golley, F. B., Gentry, J. B. (1964): Bioenergetics of the southern harvester ant *Pogonomyrmex badius.* Ecology *45,* 217–225.
Grant, P. (1978): Biology of Developing Systems. Holt, Rinehart and Winston, New York.
Heusner, A. A. (1982): Energy metabolism and body size. Is the 0.75 mass exponent of Kleiber's Equation a Statistical Artifact? Respiration Physiol. *48,* 1–12.
Hölldobler, B., Lumsden, C. J. (1970): Territorial strategies in ants. Science *210,* 732–739.
Jensen, T. F., Holm-Jensen, I. (1980): Energetic cost of running in workers of three ant species, *Formica fusca* L., *Formica rufa* L. and *Componotus herculeanus* L. (Hymenoptera, *Formicidae*). J. Comp. Physiol. B *137,* 151–156.
Kleiber, M. (1961): The fire of life. An introduction to animal energetics. John Wiley, New York.
Leigh, E. G., Rand, A. S., Windsor, D. W. (eds) (1982): The ecology of a tropical forest: Seasonal rhythms and long-term changes. Smithsonian Institute Press, Washington DC.
Lewontin, R. C. (1965): Selection for colonizing ability. In: Barker, H. G., Stebbins, G. L. (eds) The genetics of colonizing species. Academic Press, New York 79–94.
Medawar, P. B. (1957): The uniqueness of the individual. Methuen, London.
Mirenda, J. T., Topoff, H. (1980): Nomadic behaviour of army ants in a desert-grassland habitat. Behav. Ecol. Sociobiol. *7,* 129–135.
Orians, G. H., Pearson, N. E. (1979): On the theory of central place foraging. In: Horn, D.J., Mitchell, R. D., Stairs, G. R. (eds) Analysis of Ecological Systems. Ohio State University Press, Ohio pp. 155–177.

Oster, G. F., Wilson, E. O. (1978): Caste and ecology in the social insects. Princeton University Press, Princeton.

Peakings, G. J., Josens, G. (1978): Respiration and energy flow. In: Brian, M. V. (ed) Production ecology of ants and termites. Cambridge University Press, Cambridge pp. 111–163.

Pearl, R. (1928): The rate of living. Knopf, New York.

Rettenmeyer, C. W. (1963): Behavioural studies of army ants. Univ. Kans. Sci. Bull. *44,* 281–465.

Schneirla, R. C. (1956): A preliminary survey of colony division and related processes in two species of terrestrial army ants. Insectes Sociaux *3,* 49–69.

Schneirla, T. C. (1971): Army ants: a study in social organization. Topoff, H. R. (ed) Freeman, San Francisco.

Schneirla, T. C., Brown, R. Z. (1952): Sexual broods and the production of young queens in two species of army ants. *Zoologica 37,* 5–32.

Silva, M. M. T. G. da (1972): Contribucao ao estudo da biologia de *Eciton burchelli* Westwood (Hymenoptera, Formicidae) S. D. (Doutor em Ciencias) thesis. University of Sao Paulo, Brazil.

Topoff, H. (1971): Polymorphism in army ants related to the division of labour and colony cyclic behaviour. Am. Nat. *105,* 529–548.

Williams, G. C. (1957): Pleiotropy, natural selection, and the evolution of sensecence. Evolution *11,* 398–411.

Willis, E. O. (1967): The behaviour of bicolored antbirds. Univ. Calif. Publ. Zool. *79,* 1–127.

Wilson, E. O. (1971): The insect societies. Belknap Press of Harvard University Press, Cambridge (Massachusetts).

Wilson, E. O. (1975): Sociobiology: The new synthesis. Belknap Press of Harvard University Press, Cambridge (Massachusetts).

Wilson, E. O. (1980 a): Caste and division of labor in leaf-cutter ants (Hymenoptera: Formicidae: *Atta*) 1. The overall pattern in *A. sexdens* Behav. Ecol. Sociobiol. *7,* 143–156.

Wilson, E. O. (1980 b): Caste and division of labor in leaf-cutter ants (Hymenoptera: Formicidae: *Atta*) II. The ergonomic optimization of leaf cutting. Behav. Ecol. Sociobiol. *7,* 157–165.

Fortschritte der Zoologie, Bd. 31 · Hölldobler/Lindauer (Hrsg.): Experimental Behavioral Ecology
G. Fischer Verlag · Stuttgart · New York · 1985

Optimal foraging: constraints and currencies

JOHN CHEVERTON, ALEJANDRO KACELNIK, JOHN R. KREBS

Edward Grey Institute of Field Ornithology, South Parks Road, Oxford, OXl 3PS England

Abstract

In order to construct an optimality model of behaviour, one has to identify constraints, and to identify constraints one has to know something about the mechanisms controlling behaviour. Thus optimality modelling implies a knowledge of mechanisms. By adding more constraints one modifies the model's predictions, but this does not mean that constrained animals are «suboptimal». We illustrate the idea of constraints and mechanisms with reference to the within-inflorescence movement rules of bumble bees. The bees' rule is dependent on both body posture and visibility of the next flower.

A second component of optimality models which may be modified in the light of experimental evidence is the currency. As an example of how currency assumptions affect predictions we consider the crop load of honey bees. If bees maximise gross energy gain they should fill the crop before flying home. If they maximise net gain (Gain – costs) or efficiency (Gain/costs) they should return to the hive from short distances with partially empty crops. The data are closer to the predictions of the second and third possibilities than to the first.

1 Introduction

To judge from its growth in the literature, optimal foraging theory (OFT), although not without its vigorous critics (e.g. GOULD and LEWONTIN 1979, LEWONTIN 1983, MYERS 1983), is a research topic of exponentially increasing interest (KREBS et. al. 1983). It is also a much reviewed area (e.g. PYKE et. al. 1977, WERNER and MITTELBACH 1981, KREBS et. al. 1983), and it is not our intention here to update or extend existing reviews of the literature. Nor do we set out to appraise the value of, and justification for, the optimality approach in general (ALEXANDER 1982, DENNETT 1983, MAYNARD SMITH 1978, OSTER and WILSON 1978). We will use examples from recent work in our laboratory to illustrate a discussion of two issues; the choice of appropriate *constraints* and *currencies* in foraging models.

To construct an optimality model one has to choose a currency (e.g. rate of energy intake in many but not all foraging models) and identify the constraints on the animals' performance. The constraints (for example, one might posit that the animal cannot eat and search at the same time) characterise the *strategy set* available to the animal (MAYNARD SMITH 1978, OSTER and WILSON 1978). A final component of optimality models, which we shall not discuss further, is the specification of how the currency relates to fitness (reproductive success or survival) (SCHOENER 1971). Modellers generally start off with assumptions based on intuition or natural history information about both constraints and currency. The currency is chosen on the basis of assumptions about what contributes to fitness, and the constraints from knowledge about how animals behave; what they can and cannot do. Subsequently the model (assuming that it correctly defines the problem faced by the animal) may be refined either by incorporating more appropriate constraints, or by modifying the currency.

In this paper we will suggest that in order to develop optimality models in this way beyond their initial stage, a knowledge of behavioural and physiological mechanisms is essential (HEINRICH 1983). In fact part of the optimality excercise itself can be seen as an attempt to unravel these mechanisms. In addition we will make two other general points: (a) The distinction between «optimal» and «suboptimal» animals proposed by some authors (e. g. JANETOS and COLE 1981) can be reduced to a question of constraints. A suboptimal or satisficing animal is one operating under a larger set of constraints than initially assumed (KREBS et. al. 1983, MAYNARD SMITH 1983). (b) Optimality models can be used at two levels in behavioural ecology; as a tool for investigating constraints and currencies and as a method for assessing quantitatively how well animals are adapted to their environment. In both kinds of analysis it is assumed that animals are designed by natural selection to fit their environment (an assumption questioned by some, e. g. LEWONTIN 1983). The former of the two levels is concerned with understanding *how* they are designed and the latter with how *well* they are designed.

2 Mechanisms and constraints

2.1 Constraints

The classical foraging models (e. g. MACARTHUR and PIANKA 1966) were designed to be general and therefore include only the minimum number of assumptions about constraints. As we suggested in the Introduction, one way to improve upon these early models might be to include more realistic and accurate constraint assumptions. But where does the information on which to base these modifications come from? It comes from a detailed knowledge of behavioural and physiological mechanisms: once we know the cues used by the animal to detect its prey, its nutrient requirements, its capacity to remember past experiences and so on, we can accurately delineate the constraints and build a more refined optimality model. An example of this comes from work on prey selection by great tits *(Parus major)*. The classical diet model (MACARTHUR and PIANKA 1966, CHARNOV 1976 a) assumes that prey types are perfectly discriminable, while a signal detection analysis of prey selection by great tits showed that they made discrimination errors in the set-up used for testing (RECHTEN et. al. 1983, GETTY and KREBS in press). Incorporating this perceptual constraint into the diet model generated predictions (especially relating to partial perferences) which more closely matched the observed behaviour than those of the original model.

This suggests a cyclical procedure for testing, modifying, and retesting an optimality model which is summarized and extended in figure 1. The figure is meant to be in part a description of the way foraging studies have proceded to date, and in part as a prescription of what we see as one sensible way for foraging studies to proceed in the future. It is important to note, since there is still misunderstanding in the literature (MAZUR 1983),

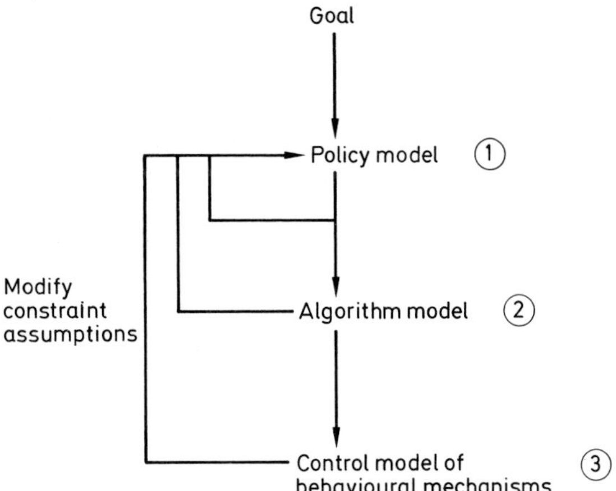

Figure 1: Levels of analysis in optimal foraging theory.

that the sequence in Figure 1, while cyclical, is not circular: modifications to the model in the form of constraint (or currency – section 4) assumptions can be independently verified. The concept of optimality itself is not under test, only the assumptions of the particular model studied (MAYNARD SMITH 1978), therefore one would not conclude, as did MAZUR (1983), that a frog starving to death while surrounded by dead flies is optimal, only that it has constraints in its visual system concerning the stimuli treated as food.

The bulk of the literature describing «tests» of optimal foraging theory has concerned itself only with the first level in Figure 1, namely tests of policy models. A policy model in our terminology is one which prescribes a general rule for maximising payoff (the goal), without identifying any specific procedure which would enable the animal to follow the rule. (A policy for maximising the goal of offspring production might, for example, be «Cross roads in a way that minimises the chance of being run over», and the procedure for doing this might be «Look in both directions and cross if clear»). A typical policy model in the literature is CHARNOV's (1976 b) marginal value theorem, which generates the intake-maximising rule for exploiting patches with resource depression (CHARNOV et. al. 1976) when the predator has perfect information. The policy, «leave a patch when the gain rate within the patch equals the average for the environment», can be used to predict a relationship between travel time and residence time within patches. This has been tested by measuring the appropriate variables by several authors (eg COWIE 1977, GIRALDEAU and KRAMER 1982, KACELNIK 1984).

The overall conclusion from studies of this sort is that provided a suitable study system is chosen, and the appropriate values measured correctly (not many supposed tests meet these requirements – KREBS et. al. 1983, KACELNIK and HOUSTON 1984), more often than not the predictions are qualitatively if not quantitatively supported (KRAMER and GIRALDEAU in prep).

Figure 1 suggests that one way to step from policy models to more specific models of a particular foraging problem is to construct *algorithms*, or procedures for solving the problem. An algorithm for patch use, for example, might be «leave patch if no prey items encountered within t seconds of arrival or since last capture». Obviously, there are a great number of possible algorithms for each foraging problem, and the criteria for choosing among them include the ability of the algorithm to describe accurately what the animal does (indeed, it may be derived inductively from the data) (PYKE 1979, YDENBERG 1982), and the performance of the algorithm in terms of payoff (i. e. how closely it approximates the optimal policy) (IWASA et. al. 1981, GREEN 1980, HOUSTON et. al. 1982, JANETOS and COLE 1981). In the context of optimality modelling, the aim would be to see if there is an algorithm which both describes what the animal does and approximates the optimal policy in terms of payoff.

Such an algorithm will almost invariably contain implicit assumptions about the animal's behavioural mechanisms. The patch use algorithm above, for example, implies the ability to measure time. As a test of these assumptions it is appropriate to refine the algorithm by investigating the actual behavioural mechanisms used in prey detection, assessment of encounter rate, and so on. Some mechanisms may be analysable at the behavioural level by looking at moment to moment choices (e. g. molecular analyses of concurrent choice experiments – STADDON 1980), others may require physiological techniques. The information about mechanisms could be used, as Figure 1 indicates, either to modify the constraint assumptions in the policy model, or to develop more realistic algorithms. As this feedback progresses, information is incorporated from one level to another and the policy model, the algorithm model, and the behavioural mechanism model (we have called it the control model in Figure 1) lose their independence.

As the policy model is modified by incorporating new constraints, the optimal policy is automatically redefined with respect to these constraints. This suggests an alternative interpretation of JANETOS' and COLE's (1981) distinction between optimal and imperfectly or suboptimal animals. An optimal animal according to these authors would be one following the predictions of a policy model, while a suboptimal one uses a rule of thumb which only approximates the optimal policy in terms of payoff (equivalent to what we have called an algorithm). Since a knowledge of the rules actually used by the animal can be incorporated as constraints in the policy model (figure 1) JANETOS' and COLE's distinction disappears: it is simply a matter of how many constraints are recognised (KREBS et. al. 1983, MAYNARD SMITH 1983). HOUSTON and MCNAMARA (in press) discuss a similar point in their distinction between local and global optima. A model of a global optimum in their terms is equivalent to our policy model, including as few constraints as possible. A local optimum is one subject to more constraints. As an example HOUSTON and MCNAMARA consider the effect of incorporating an additional constraint into the marginal value model. Animals in general appear to be unable to measure time without errors (GIBBON and CHURCH 1981), the coefficient of variation being about 0.35, and if animals use time in their patch leaving decisions (e. g. COWIE and KREBS 1979), timing errors must be treated as a constraint. With this constraint, the marginal value model predicts a different optimal residence time from that of the usual version.

2.2 Proximate and ultimate causes

BAKER (1938) was the first to clarify the distinction, now widely accepted, between accounts of biological phenomena in terms of their proximate or immediate causes, and those in terms of their ultimate causes or survival value. Where do optimality models lie in relation to this distinction?

The justification for using optimality models in biology is usually couched in terms of natural selection. Selection, it is argued, tends in the long term to maximise fitness (see GRAFEN 1984 for a discussion of fitness) and therefore it should maximise the performance of traits which contribute to fitness (DENNETT 1983, ALEXANDER 1982, KREBS and MCCLEERY 1984). One way to study the performance of these traits (loosely referred to as adaptations) is to treat them as *design features* and use optimality models to analyse the design with respect to particular performance criteria. This view places optimality analysis firmly on the ultimate side of BAKER's distinction: ultimate explanations are concerned with how traits contribute to fitness, and optimality analyses help to elucidate exactly how certain traits are designed by selection, in other words, how they contribute to fitness.

But, as we have argued in the previous sections, optimality modelling cannot proceed beyond the initial stages without a knowledge of behavioural and physiological mechanisms. The cyclical process described in figure 1 shows that optimality models incorporate both proximate and ultimate causes of behaviour: they are about both mechanisms and survival value (this contrasts with the view of HEINRICH (1983), who sees optimality models as being concerned only with ultimate causes).

2.3 Built-in obsolescence?

Figure 1 suggests that testing a policy model is only the first step. The fact that most of the optimality literature has dealt only with this step may seem in retrospect as an attempt to test whether the optimality approach in general is of use. Some would accept that the evidence is now just about convincing enough to say that it *is* of use, and figure 1 offers a possible answer to the question «what next?». In giving this answer, we join MCFARLAND (1976, 1977) and MCFARLAND and HOUSTON (1981) in stressing the intimate relationship between ultimate and proximal accounts of behaviour.

Does the prescription in Figure 1 imply that policy models will gradually die of obsolescence? Ought we to get on with the analysis of mechanisms straight away, and not bother with the window dressing of optimality (HEINRICH 1983)? We believe not, for both theoretical and practical reasons. The theoretical reason is that optimal policy models offer greater generality than do models of mechanisms (KRAMER and GIRALDEAU in prep). This follows from the notion that regardless of how they do it, many or perhaps all foraging animals are designed to do the same thing: the policy models will reveal whether this is true or not. The practical reason is that optimality models can act as guides for investigating mechanisms. To put it at its simplest, understanding how a piece of machinery such as an animal works is made easier by having some idea of what it is designed to do.

We now turn to a specific example to illustrate some of the points brought up in the preceding sections, in particular the investigation of mechanisms of foraging.

3 Mechanisms in foraging bumblebees

In this section, we discuss the work done by one of us (J.C.), on foraging bumblebees. The stimulus for this was a paper by PYKE (1979), on the intra-inflorescence ‹movement rules› of bumblebees. Pyke recorded the movements of *Bombus appositus* workers on *Aconitum columbianum,* and proposed a movement rule as a summary and explanation of his data. He then proposed a strategy set of alternative movement rules, based on assumptions about the mechanisms likely to be used by the bees. He calculated the net rate of energy intake that the bees could achieve by using each of these rules. None of them gave a higher gain rate than the observed behaviour, two ot them gave equally high rates.

Compared with other optimality studies, PYKE's paper is unusual in that the ‹policy› level is skipped; PYKE goes straight from the ‹goal› of maximising the net rate of energy intake to a set of movement rules or algorithms which do not have generality beyond the *Bombus – Aconitum* system. However, the point of interest here is the status of these algorithms with respect to the mechanisms which may be used to implement them. They are worded so as to imply the existence of specific mechanisms. For example, ‹move to the closest flower not just visited› implies that the bees can discriminate distance and have memory of which flower has just been visited. CHEVERTON (1983) discusses in detail the implications of invoking mechanisms at this stage. Here, the important point is that PYKE's paper raises questions about mechanisms which it does not answer (HEINRICH 1983), so it is a natural sequel to PYKE's study to investigate these. The same approach could be taken with a great many optimality studies, for many of these also stop at the point where interesting questions about mechanisms are raised but not answered (OSTER & WILSON 1978).

An experimental system was used in which *B. lapidarius* workers, from a captive colony within a flight room, foraged freely from artificial flowers in two-dimensional arrays. Figure 2 shows the routes taken by the bees over four different patterns. PYKE's best movement rule would predict similar or identical paths, so the present system seems to be suitable for investigating the questions of mechanism raised by his paper. Seeking the simplest explanation of the bees' movements, we attempted to model them using a procedural algorithm that took account only of the position of other flowers relative to the present position of the bee. We found that such a model could be constructed, for movements on vertical arrays, using a series of lines of equal preference (‹iseclects›) around the

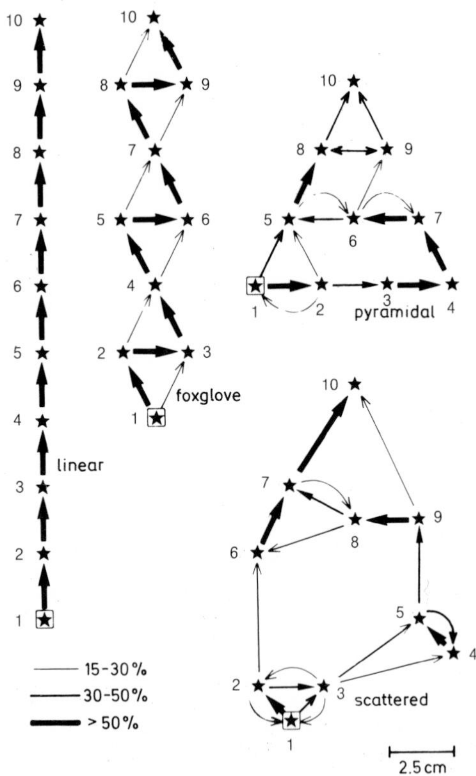

Figure 2: The movements of bumblebees over four experimental flower patterns. Filled arrow-heads represent significantly preferred choices ($p < 0.05$). Squared stars are starting points.

present position of the bee (Figure 4). The shape of these was adjusted to generate the same routes, over the four test patterns, as the preferred routes of the bees over the same patterns. Although this is no more than a description of the data, it is interesting that the same iseclects produce the preferred routes over all four patterns, suggesting that the bees use the same rule on each one. However, the success of this model depended on the inclusion of a ‹ratchet clause›, forbidding choice of the flower just visited (otherwise, for example, the model would make a revisit on 50 % of departures from flower 6 of the pyramidal array, and would alternate endlessly between flower 2 and 3 of the scattered array). This seemed to suggest that memory, at least of the most recently visited flower, was being used (HEINRICH 1983).

3.1 Memory

Several of PYKE's algorithms imply the ability to remember which flowers have already been visited on an inflorescence. Yet, PYKE expresses the opinion that:

«It is improbable that a bumblebee could have a memory and a sense of spatial geometry that would permit it to know exactly which flowers on an inflorescence it had already visited.»

This seems to contradict the implications of his strategy set. Also, it is well known that many insects have spatial memory (CARTWRIGHT and COLLETT 1982), and that bumblebees have excellent memory of the position of their nest (e.g. FREE and BUTLER, 1959), and that they can remember the position of individual inflorescences (MANNING, 1956).

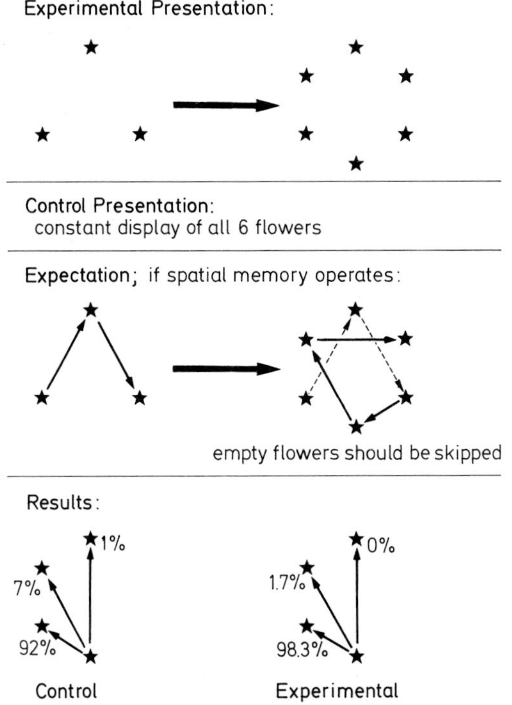

Figure 3: Experimental set-up designed to test for spatial memory, and results.

This suggests that the question of memory is worth investigating. To do so, a ‹forced choice› experiment was used, conceptually similar to those used by OLTON et. al. (1981) to demonstrate spatial memory in rats. In this experiment, a small horizontal array, 250 mm × 300 mm, was used, on which flowers could be exposed or concealed by means of a sliding shutter.

Figure 3 shows the hexagonal pattern of flowers used, three of which were exposed at first, the others being exposed in addition, when a bee was visiting the third of the original three flowers. All flowers had equal rewards, not replenished during the experiment. On the hypothesis that the bees use spatial memory to avoid revisiting previously emptied flowers, they should have skipped these, and visited only the newly exposed flowers (Figure 3, ‹expectation›). In fact, the proportion of moves which skipped adjacent flowers was lower (though not significantly) for this presentation than for the control presentation of the complete hexagon, the opposite of the prediction from spatial memory. In contrast with the papers mentioned above, the conclusion from this experiment, which was corroborated by other evidence, is either that the bees do not have spatial memory at this level, or that if they *do*, then it is not used to prompt them to skip previously-visited flowers. This is clearly a constraint which could be taken into account in any future optimality model of intra-inflorescence movements (illustrating the feedback in Figure 1). A more immediate problem was to seek an explanation for the ratchet clause of the iseclect model which did not require memory.

3.2 Posture

In conceiving the iseclet model, it was thought that the bees always sat on a flower facing directly upwards, so that their body axis, and the vertical axis of the iseclects were aligned. However, closer analysis revealed that the bees do not always orient vertically. The possible importance of posture was examined using the pyramidal array (see Figure 2). Flower ‹6› of this pattern may equally well be approached from flower ‹5› or flower ‹7›. Table 1 shows that the posture of the bees on flower ‹6› depended on which flower they had arrived from: they were tilted away from the direction of arrival.

Table 1: Contingency table demonstrating the dependence of posture on arrival direction for bees on flower 6 of the pyramidal pattern. The cells contain observed frequencies of visits.

		Posture towards:	
		Left	Right
Arrival from	Left:	4	37
	Right:	18	0

$\chi^2_1 = 43.5$ $p < 0.00001$

If the iseclects were tilted in the direction of the body axis, a ratchet clause was unnecessary to prevent choice of the previously-visited flower (Figure 4). To show that the posture was more than just an incidental variable, those few cases were examined where the posture was *not* as predicted from arrival direction (ie: where the bees were tilted towards

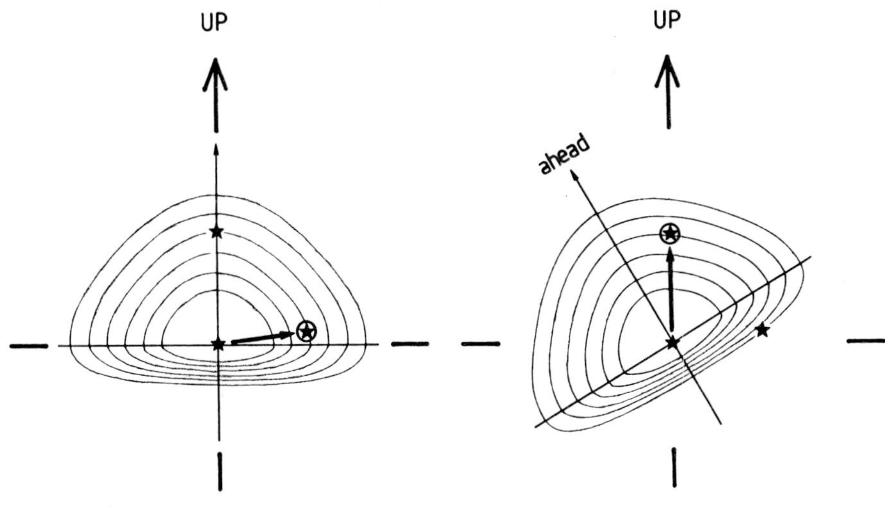

Figure 4: The ‹iseclect› model: Left: the means by which the next flower (circled star) is chosen. Preference values increase inwards. Right: the effect of aligning the iseclects with the body axis of a tilted bee.

the flower they had arrived from). In three out of four such occurences, these ‹mistakes› were carried through to the departure direction. On the hypothesis that departure mistakes were independent of posture mistakes, the probability of this occuring by chance was < 0.02 (binomial test), suggesting that departure *is* dependent on posture, rather than arrival direction *per se*. The posture variable allows for the influence of previous flowers on future choice in a simple way without the need to invoke memory, at least in the usual sense of the word.

The ‹tilting inseclect› model (Figure 4) which we have described is sub-optimal in the sense that it cannot cope with some artificial flower patterns without revisitation occurring, but once the constraints of this model are incorporated into the optimality analysis, this distinction disappears. What next? One approach (illustrating the interaction between OFT and control models shown in Figure 1) would be to pursue the question of the physiological basis of the iseclects. Cheverton (1983) has investigated the possibility that this is to be found in the distribution of ommatidia over the bees' visual field (cf. Heinrich 1983). Measurements of ommatidial density showed that the shape of the iseclects cannot be explained on the simple hypothesis that the bees fly to the flower seen by most ommatidia from their position on the present flower. However, transformation of the ommatidial density map by a typical psychophysical function relating perceived to actual intensity of a stimulus (Stevens 1961) renders much closer the fit between the behavioural and ommatidial isoclines. As an alternative approach, we could re-do the optimality analysis by accepting the iseclect model as a constraint, and trying to derive an optimum shape for the iseclects themselves.

4 Choosing the currency

Among the many examples of optimality modelling in behavioural ecology one can identify two contrasting approaches. Representing one of them, SIBLY and McFARLAND (1976) (and also McFARLAND 1976, 1977) proposed that optimality models should ideally be based on a specification of how behaviour affects states and on direct measurements in the field of the fitness costs and benefits associated with each state and type of activity. For this research program the choice of an appropriate currency ought not to be a conceptual difficulty: the currency is fitness, measured directly. Having specified both constraints and currency, optimality models can be used to test whether or not the animal performs its behavioural repertoire in a sequence that maximises fitness. These optimality models do not have the goal of identifying constraints and how behaviour contributes to fitness (although investigation of these matters may be essential for the formulation of the models) but, in SIBLY's and McFARLAND's words, their aim is to assess «... the extent to which behaviour maximises fitness.»

This route has not yet been successfully followed because of the difficulties posed by the field measurements required. Most optimality models in behavioural ecology, including OFT, have a much more limited aim. As discussed in the introduction, most models aim to gain insights into the constraints and currencies. They do this by formulating, testing and reformulating hypotheses based on particular constraint and currency assumptions, often in relation to only one or two kinds of activity. While this piecemeal approach cannot test whether observed behaviour is optimal (i. e. whether animals are optimally adapted, MAYNARD SMITH 1978) it may, as knowledge of the constraints and currencies improves, develop towards a convergence with the use of optimality as proposed by SIBLY and McFARLAND.

A major problem for the piecemeal approach is to establish the appropriate response to a discrepancy between predictions based on an optimality model and observed behaviour. How does one know whether to modify the constraint assumptions or the currency? There is no simple answer, but suitable choice of experimental systems may help to reduce this uncertainty. In general, when the primary interest lies in one of the themes, it is advisable to choose a system where the other factor offers little doubt. Thus, the study of constraints is best conducted in cases when animals have been shown to maximise a given currency (for example net energy gain) in a variety of other situations, while the study of currencies is ideally conducted when it is safe to assume that the necessary behavioural mechanisms to follow the optimal policy are available to the animal. As an example, consider the various attempts to test the choice of optimal patch residence time by animals foraging in a homogeneous environment. The usual rationale can be expressed as: if the foraging animal is subject to dimishing returns *in the relevant currency* while exploiting patches, then the optimal patch residence time is given by the marginal value rule. We will ignore here the problems arising from heterogeneous, stochastic and discontinuous environments, and refer only to cases when the instantaneous use of the marginal value rule does result in rate maximisation. If the animal can reasonably be assumed or shown to be capable of assessing marginal gain rate, failure to stay for the predicted time indicates the need for adjustements in the specification of currency. We shall discuss this in some detail in the framework of experimental work done by NUÑEZ (1982) with honey bees *(Apis mellifera)* and a theoretical extension of it elaborated by KACELNIK and HOUSTON (in prep.).

4.1 Optimal crop load in honey bees

The prediction of the marginal value rule for cases when there is no depression in the relevant currency is trivial: the animal should never abandon a patch unless it has reached its maximum loading capacity (remember that we have excluded from our analysis all cases of patch heterogeneity, etc.).

NUÑEZ (1982) performed an experiment in which nectar collecting bees had access to an artificial dispenser placed at various distances from the hive. Flow of nectar in the dispenser was not subject to depression either during or between visits. NUÑEZ correctly predicted that if overall rate of nectar collection was the maximised currency, the bees ought to follow the marginal value rule and stay for as long as necessary to fill their crops before returning to the hive. The bees failed to do so. Instead, they abandoned the patch with various degrees of crop filling depending on the distance between the patch and the hive and the rate of flow of nectar while in the patch. Crop load was reduced at close distances and low flow rates.

Facing this discrepancy, the obvious step was to correct the assumption about the currency. Because both variables affecting crop filling are correlated with time away from the hive, NUÑEZ speculated that a possible reason for the discrepancy was to be found in the social nature of food gathering in this species. The long term advantages of exchanging information with other workers would affect individual behaviour, and the maximised currency could be thought of as a compromise between individual foraging efficiency and the advantage to the hive of information exchange between workers. The appeal of NUÑEZ' suggestion derives from its testability: if information exchange is responsible for departures from individual workers' optimal behaviour, then a solitary forager like the bumblebee ought not to show those discrepancies.

While this possibility has not been tested yet, KACELNIK and HOUSTON (in prep) took a different route, improving the expression of individual workers' energetic efficiency by refining the energetic currency. They reasoned that if one chose the *net* rate of delivery of nectar to the hive instead of the rate of extraction of nectar form the patch as the maximised currency the predictions might differ.

KACELNIK's and HOUSTON's rationale can best be followed by the graphical presentation in Figure 5. The ordinate in this figure represents quantities of sugar, and the abscisa is time, with the origin at the point of arrival at the patch. Costs and gains are analysed in three periods: the outward trip, the time in the patch and the return trip. Starting at the left hand side of the graph we can follow three trajectories, corresponding to gross nectar collection (F), cumulative energy expenditure (E) and resulting net gain or load (L). Cumulative nectar collection is represented by a rising linear function while the bee is in the patch (the slope of this line is the flow rate of the feeder) and remains unmodified while the bee is travelling from or to the hive. Expenditure is represented as a decreasing function for reasons of presentation. For the purposes of the graph the effect of a varying load during the journeys is ignored, and thus cumulative expenditure during them is represented by straight lines. Rate of expenditure during the outward trip (slope of E_o) depends on the initial load, which must itself be a function of the distance between the hive and the feeder, but only one distance is considered here. Rate of expenditure during the return trip (slope of E_r) depends on the load collected during the visit ot the patch, as will be analysed later on. Expenditure during the visit itself is an accelerating function of load, as shown by the convexity of E_p. Assuming an initial load large enough to pay for the anticipated cost of the outward journey, we can follow the net load by starting at point P, and ending at point Q, when the bee reaches the hive again. Notice that the net load curve while in the patch is curvilinear during patch time (L_p), due to the increasing cost of carrying the load. The position of Q depends on time in the patch in a way that combines the course of L_p with the slope and duration of L_r. This is because both expenditure during flight and flight velocity in bees are significantly affected by the range of loads normally carried.

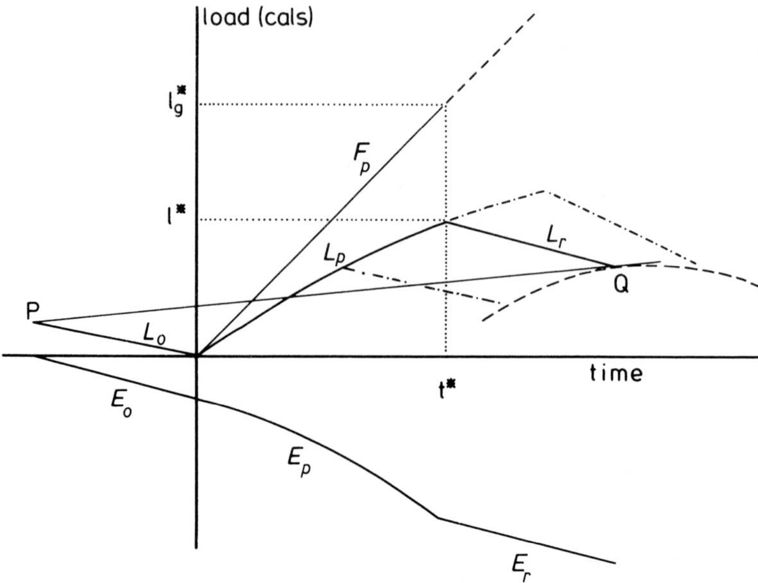

Figure 5: A graphical representation of optimal crop load.

The effect of staying for different patch times is shown by the broken lines. By leaving the patch earlier than indicated, the loss of time and energy during the return trip is reduced, but the net gain at departure from the patch is also lowered. The opposite (i. e. higher gain at departure but larger travel losses) would result from longer patch time. As result of these combined effects the position of point Q realised as a consequence of various patch times (or loads) describes an arc as shown by the broken line. Since this line represents the net load on arrival at the end of a foraging trip, the optimum patch time (t^*) can be obtained by tracing the tangent to the broken line that passes through P. The optimal time corresponds to a net load of l^*, and gross intake of lg^*. It is immediately obvious that maximum load will not always be optimum if the maximised currency is net rate of delivery. In practice, the graphical method is inappropriate and unnecesary to calculate the optimum, and KACELNIK and HOUSTON used available information of rates of expenditure combined with suitable simplifications to find an analytical version of this model. The model states that optimal patch time (t) is that one that satisfies the equation:

$$(DCFe^{-mt} + 1) [(F/m) (1 - e^{-mt}) + B/k - Be^{k\tau} ((1/k) + AD)]$$
$$=$$
$$[t + 2DA + (DCF/m) (1 - e^{-mt})] Fe^{-mt} [1 - kDC ((F/m) (1 - e^{-mt}) + B/k)]$$

where:

A : 1/ unloaded velocity
B : Flight metabolism, unloaded
C : Slope for increase in travel time with load
D : Distance hive-feeder
k : Slope of increase in flight metabolism with load
m : Slope of increase in metabolism at the feeder with load
E : Metabolic rate in feeder, unloaded
F : feeder flow – E
τ : $D(A + Cl_0)$
l_0 : Net load at departure from patch

Their predicted crop loads for NUÑEZ' conditions do fit with the qualitative observation that the bees leave the patch with larger loads at greater distances and higher flows, but the quantitative details, as shown in Figure 6, still show interesting discrepancies. One important discrepancy between model and data is that at close distance (100 m) the bees did not wait for long enough to fill their crops even when the flow was high. This points to a structural problem of the model, rather than some error in the estimate of parameters.

This analysis is still in progress, and there is no reason why the informational view taken by NUÑEZ could not coexist with the energetic analysis presented here. The relevant question is whether by following this recursive process of proposal and modification of hypotheses about currencies we learn something about bees, and our belief is (as the reader might have guessed) that we do[1].

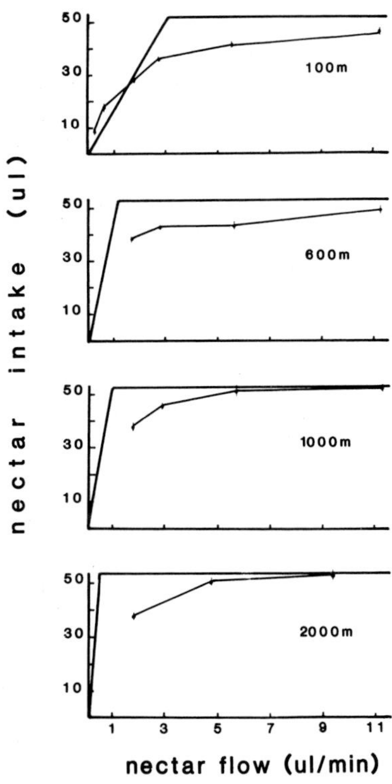

Figure 6: The relationship between nectar intake and flow-rate at four different distances from the hive. –·–·–·–: as observed by Nunez. ———: as predicted by the model.

---

[1] Further analyses and experiments (SCHMID-HEMPEL, KACELNIK, and HOUSTON in prep.) have shown that if the energetic losses while the bee is in the hive are included, a better fit is obtained by assuming that the maximised currency is the ratio of energy gained to energy spent.

5 Discussion

Our main general conclusions are as follows. (i) Optimality models include assumptions about constraints and specification of constraints requires a knowledge of behavioural and physiological mechanisms. We illustrate this point with work on foraging paths of bumblebees. (ii) The currency in an optimality model could in principle be measured directly as fitness, in which case the model could be used to test how well animals are adapted to their environment. Alternatively, optimality models can be used to evaluate different putative currencies which are thought to be related to fitness. The contrast between predictions based on gross and net energy intake as foraging currencies for honey bees illustrates this point. We now return in this concluding discussion to the question of constraints.

5.1 What are constraints?

Assumptions about constraints pose a problem: what at one time is treated as a fixed constraint may on another occassion be seen as subject to adaptive modification. How does one know, in other words, whether or not it is justifiable to treat *any* aspect of behaviour as a constraint? Is it acceptable, as would be the case for morphological or physiological properties, to infer the constraints simply from looking at the animal's behaviour?

The difficulties that may arise from inferring constraints by observing behaviour can be illustrated by two analyses of the courtship behaviour of the three-spined stickleback *(Gasterosteus aculeatus)*. When a ripe female enters the territory of a male in breeding condition, he approaches and courts her in the well-known sequence culminating in egg laying described by TER PELKWIJK and TINBERGEN (1937). Not infrequently, the normal courtship sequence is temporarily interrupted by the male, who leaves the female to visit the nest and performs a variety of nest-building or parental activities before resuming courtship. WILZ (1967, 1970) suggested that these activities are essential for the male to adjust his internal state from primarily territorial and aggressive to primarily sexually motivated. If the male is prevented from carrying out nest activities by being excluded from the nest, he subsequently tends to attack the female instead of courting her. COHEN and MCFARLAND (1979) have a different interpretation. They argue that it is «very misleading to infer function simply on the basis of causal relationships». That is, WILZ's hypothesis is based entirely on the supposed motivational constraint that the stickleback cannot switch from aggression to sex in the presence of a ripe female without doing activities at the nest. (Many parallel examples can be found in the older ethological literature). COHEN and MCFARLAND, instead of taking the motivational switch as a constraint, ask why sticklebacks are designed to have to visit the nest during courtship. They suggest that the courting male has to balance two needs: inducing the female to enter the nest and checking that the nest is in good order before the female arrives. It is the requirement to keep two balls in the water at once that has lead to the evolution of courtship interrupted by nesting activities. To the extent that visiting the nest is essential for the male to switch motivational states, this is an adaptive feature rather than a constraint. The failure of the male to continue courting after finding the nest entrance blocked is compatible with COHEN's and MCFARLAND's interpretation: since the male is unable to complete fertilisation without an accessible nest, he would not be expected to continue courtship. RIDLEY and RECHTEN (1981) offer an additional functional hypothesis for interruptions to courtship. They show that females are more likely to spawn in nests with eggs than in those without, and suggest that the male, by visiting the nest to perform parental activities during courtship, indicates the presence (perhaps deceptively) of eggs in the nest.

The stickleback example serves to illustrate the fact that there can be alternative views about what constitutes a constraint. It can also be used to suggest a possible general procedure for identifying constraints: comparison between species. If it transpired that males of species closely related to the stickleback do *not* have courtship interruptions, this would indicate that the motivational switch is not a fixed constraint (DAWKINS 1982, p 43, p 49). If, on the other hand, a wide taxonomic range of fish show interruptions, whether or not the male tends a nest, a strong case would exist for a motivational constraint. This case would be further strengthened if the assumed constraint could be shown, perhaps by cladistic methods (RIDLEY 1983), to be primitive; that is, to have persisted in independent evolutionary lines. For example, the wide taxonomic applicability of equations relating metabolic rate to body size in endotherms (ASCHOFF und POHL 1970) and the similarity of maximum sustainable work rate in several taxa (DRENT and DAAN 1980) makes it likely that these are genuine constraints. Similarly many basic features of morphology and physiology would qualify as constraints under our proposed method. Parenthetically we note that a distinction can be made between constraints which exist here and now, but could change over evolutionary time – these are the ones we are mainly concerned with – and constraints which are so universal as to be, in effect, immutable.

Although the method outlined in the previous paragraph would in principle lead to a more accurate distinction between constraints and design features, in practice the information needed to use the method is often not available. Therefore the approach adopted by most workers is to identify constraints by observing mechanisms of behaviour; in doing this one should be aware of the pitfalls.

5.2 Variability in behaviour

Variance is a universal feature of behaviour and morphology, yet most foraging models do not predict variability and many do not take it into account in their assumptions. Two contrasting ways of treating variability have been suggested reflecting the problem outlined in the previous section of distinguishing between constraints and design features. HOUSTON and MCNAMARA (in press) treat variance in a behavioural mechanism (timing ability) as a constraint subject to which an optimum can be calculated (see section 2.1). Since variance in timing has been shown in more than one taxon this seems to be a reasonable approach. In other cases the degree of variability in behaviour may differ between related taxa or may be flexible depending on costs, suggesting that it is not simply a fixed constraint. DENEUBOURG et. al. (1983), for example, show that different ant species differ in the variance about the mean recruitment directions of foragers: some species recruit accurately and others with a wide spread. These authors go on to suggest on the basis of simulation models that the different degrees of spread are adapted for exploiting different kinds of food – accurate recruitment for concentrated clumps and inaccurate recruitment for scattered food. KACELNIK (1984) noted that the distribution of deviations about the energy-maximising value of load size in Starlings *(Sturnus vulgaris)* bringing food to their young is related to the energetic costs of deviating. The loss in reward rate was asymmetric for deviations above and below the mean load and there were fewer deviations in the direction of higher loss, suggesting that variance is not entirely a fixed constraint. The question of variability would repay further study. Most tests of foraging models to date have considered only mean values and not even taken the first step of partitioning variance into components due to inter-, intra-individual differences and so on.

Finally, if variance is a universal feature of behaviour, is there any point in building models which predict deterministic behaviour? The answer depends on the purpose of the models. If the model is derived inductively as a description of behaviour (e. g. the matching law, Herrnstein 1970) then taking into account basic features of behaviour such as variability would seem to be important. On the other hand, foraging models, and optimality models in general, are justified primarily as *a priori* predictors of behaviour rather than inductive descriptors. In this case the tractability and simplicity of a deterministic model of the optimal policy outweighs the advantage of complete descriptive accuracy.

6 Summary

1. Optimality models contain assumptions about constraints and currencies.

2. Knowledge of appropriate constraints comes from understanding mechanisms of behaviour.

3. The investigation of behavioural mechanisms of foraging is illustrated by work on bumblebees collecting nectar from artificial inflorescences in a two-dimensional array. The bees avoid revisiting, but appear to do so without remembering which flowers they have visited. Their movement patterns can be explained by a model of preference isoclines and the effects of body posture.

4. If fitness is measured directly as the currency of an optimality model, the model can be used to evaluate how well an animal is adapted to its environment. More frequently a putative currency is proposed and the model used to investigate the predictive value of currency and constraint assumptions.

5. The comparison of alternative currencies is illustrated by the problem of crop load in honey bees. A model based on maximising net energy gain per unit time correctly predicted that bees return to the hive with partially loaded crops.

7 Acknowledgements

We thank the EGI foraging lunch group for valuable comments and the NERC for financial support.

References

Alexander, R. McN. (1982): Optima for animals. Arnold, London.

Aschoff, J. and Pohl, H. (1970): Rhythmic variations in energy metabolism. Federation Proceedings *29,* 1541–1552.

Baker, J. R. (1938): The relation between latitude and breeeding season in birds. Proceedings of the Zoological Society of London *106,* 557–582.

Cartwright, B. A. and Collett, T. S. (1982): How honeybees use landmarks to guide their return to a food source. Nature (London) *295,* 560–564.

Charnov, E. L. (1976 a): Optimal foraging: attack strategy of a mantid. American Naturalist *110,* 141–151.

Charnov, E. L. (1976 b): Optimal foraging: the marginal value theorem. Theoretical Population Biology *9,* 129–136.

Charnov, E. L., Orians, G. H. and Hyatt, K. (1976): The ecological implications of resource depression. American Naturalist *110,* 247–259.

Cheverton, J. (1983): Which flower next? Small scale foraging decisions in bumblebees. D. Phil. Thesis, Oxford University.

Cohen, S. and McFarland, D. J. (1979): Time-sharing as a mechanism for the control of behaviour sequences during the courtship of three-spined stickleback *Gasterosteus aculeatus*. Animal Behaviour *27*, 270–283.

Cowie, R. J. (1977): Optimal foraging in Great Tits *(Parus major)*. Nature *268*, 137–139.

Cowie, R. J., Krebs, J. R. (1978): Optimal foraging in patchy environments. Anderson, R. M., Taylor, L. R. and Turner, B. (eds): British Ecological Society Symposium on Population Dynamics. Blackwell's Scientific Publications, Oxford.

Dawkins, R. (1982): The extended phenotype. Freeman, Oxford.

Deneubourg, J. L., Pasteels, J. M. and Verhaege, J. C. (1983): Probabilistic behaviour in ants: a strategy of errors? Journal of Theoretical Biology. 105, 259–271.

Dennett, D. C. (1983): Intentional systems in cognitive ethology: «the Panglossian paradigm» defended. Brain and Behavioural Sciences 6, 343–390.

Drent, R. H. and Daan, S. (1980): The Prudent Parent: Energetic adjustements in avian breeding. Ardea *68*, 225–252.

Free, J. B. and Butler, C. G. (1959): Bumblebees. Collins New Naturalist Series, London.

Getty, T. W. and Krebs, J. R. (in press): Lagging partial preferences for cryptic prey: a signal detection analysis of great tit foraging. American Naturalist.

Gibbon, J. and Church, R. M. (1981): Time left: linear vs. logarithmic subjective time. J. exp. Psychol. *7*, 87–107.

Giraldeau, L. and Kramer, D. L. (1982): The marignal value theorem: a quantitative test using load size variation in a central place forager, the eastern chipmunk. Animal Behaviour *4*, 1036–1042.

Gould, S. J. and Lewontin, R. C. (1979): The spandrels of San Marco and the Panglossian Paradigm: a critique of the adaptationist programme. Proceedings of the Royal Society *205*, 581–598.

Green, R. F. (1980): Bayesian birds: a simple example of Oaten's stochastic model of optimal foraging. Theoretical Population Biology *18*, 244–256.

Heinrich, B. (1983): Do bumblebees forage optimally, and does it matter?. American Zoologist *23*, 273–282.

Herrnstein, R. J. (1970): On the law of effect. Journal of Experimental Analysis of Behaviour *13*, 243–266.

Houston, A. I., Kacelnik, A. and McNamara, J. (1982): Some learning rules for acquiring information. McFarland, D. J. (ed): Functional Ontogeny. Pitman Advanced Publishing Program, London.

Houston, A. I. and McNamara, J. (in press).: Variability of behaviour and constrained optimization. Journal of Theoretical Biology.

Iwasa, Y. Higashi, M. and Yamamura, N. (1981): Prey distribution as a factor determining the choice of optimal foraging strategy. American Naturalist *117*, 710–723.

Janetos, A. C. and Cole, B. J. (1981): Imperfectly optimal animals. Behavioral Ecology and Sociobiology 9, 203–210.

Kacelnik, A. (1984): Central place foraging in starlings *(Sturnus vulgaris)* I: patch residence time. Journal of Animal Ecology. 53, 283–299.

Kacelnik, A. and Houston, A. I. (1984): Some effects of energy costs on foraging strategies. Animal Behaviour.

Krebs, J. R. and McCleery, R. H. (1984): Optimization in behavioural ecology. Krebs J. R. and Davies N. B. (eds). Behavioural Ecology (2nd Edn.) Blackwell Scientific Publications, Oxford.

Krebs, J. R., Stephens, D. W. and Sutherland, W. J. (1983): Perspectives in optimal foraging. Clark G. A. and Brush, A. H. (eds) Perspectives in Ornithology. Cambridge University Press, New York.

Lewontin, R. C. (1983): Elementary errors about evolution. Brain and Behavioural Sciences 6, 367–368.

MacArthur, R. H. and Pianka, E. R. (1966): On the optimal use of a patchy environment. American Naturalist *100*, 603–609.

McFarland, D. J. (1976): Form and function in the temporal organization of behaviour. Bateson, P. P. G. and Hinde, R. A. (eds): Growing points in theology. Cambridge University Press, Cambridge.

McFarland, D. J. (1977): Decision making in animals. Nature *269*, 15–21.

McFarland, D. J. and Houston, A. I. (1981): Quantitative ethology. Pitman, London.

MacNamara, J. M. (1982): Optimal patch use in a stochastic environment. Theoretical population biology *21,* 269–288.

Manning, A. (1956): Some aspects of the foraging behaviour of bumblebees. Behaviour *9,* 164–201.

Maynard Smith, J. (1978): Optimization theory in evolution. Annual Review of Ecology and Systematics *9,* 31–56.

Maynard Smith, J. (1983): Adaptation and satisficing. Brain and Behavioral Sciences *6,* 370–371.

Mazur, J. E. (1983): Optimization: result or mechanism. Science *221,* 976–977.

Myers, J. P. (1983): Commentary on *Perspectives in Optimal Foraging* (Krebs, J. R., Stephens, D. W. and Sutherland, W. J.). Brush, A. H. and Clark, G. A. (eds): Perspectives in Ornithology. Cambridge University Press, Cambridge.

Nuñez, J. A. (1982): Honeybee foraging strategies at a food source in relation to its distance from the hive and the rate of sugar flow. Journal of Apicultural Research *21,* 139–150.

Olton, D. S., Handelman, G. E. and Walker, J. A. (1981): Spatial memory and food searching strategies. Kamil, A. C. and Sargent, T. D. (eds): Foraging behaviour. Garland STPM Press, New York.

Oster, G. P. and Wilson, E. O. (1978): Caste and ecology in the social insects. Princeton University Press, Princeton, N. J.

Pyke, G. H. (1979): Optimal foraging in bumblebees: rule of movement between flowers in inflorescences. Animal Behaviour *27,* 1167–1181.

Pyke, G. H., Pulliam, H. R. and Charnov, E. L. (1977): Optimal foraging: a selective review of theory and tests. Quarterly Review of Biology *52,* 137–154.

Rechten, C., Avery, M. I. and Stevens, T. A. (1983): Optimal prey selection: why do great tits show partial preferences? Animal Behaviour *31,* 576–584.

Ridley, M. (1983): The explanation of Organic diversity. Oxford University Press, Oxford.

Ridley, M. and Rechten, C. (1981): Female sticklebacks prefer to spawn in nests with eggs. Behaviour *76,* 152–161.

Schoener, T. W. (1971): Theory of feeding strategies. Annual review of Ecology and Systematics *2,* 369–404.

Sibly, R. M. and McFarland, D. J. (1976): On the fitness of behavior sequences. American Naturalist *110,* 601–617.

Staddon, J. E. R. (1980): Optimality analyses of operant behaviour and their relation to optimal foraging. Staddon, J. E. R. (ed): Limits to action. The allocation of individual behaviour. Academic Press, New York.

Stevens, S. S. (1961). To honor Fechner and repeat his low. Science 133, 80–86.

ter Pelkwijk, J. J. and Tinbergen, N. (1937): Eine reizbiologische Analyse einiger Verhaltensweisen von *Gasterosteus aculeatus* L. Z. Tierpsychologie *1,* 193–200.

Werner, E. E. and Mittelbach, G. G. (1981): Optimal foraging: field tests of diet choice and habitat switching. American Zoologist *21,* 813–829.

Wilz, K. (1967): The organization of courtship in sticklebacks. Unpublished D. Phil Thesis, Oxford University.

Wilz, K. (1970): Causal and functional analysis of dorsal pricking and nest activity in the courtship of the three-spined stickleback. Animal Behaviour *18,* 682–687.

Ydenberg, R. C. (1982): territorial vigilance and foraging, a study of tradeoffs. Unpublished D. Phil thesis, Oxford University.

II. The Analysis of Communication Signals

Fortschritte der Zoologie, Bd. 31 · Hölldobler/Lindauer (Hrsg.): Experimental Behavioral Ecology
G. Fischer Verlag · Stuttgart · New York · 1985

The dance language of honeybees: The history of a discovery

MARTIN LINDAUER

Zoologisches Institut der Universität Würzburg, F.R.Germany

Abstract

KARL v. FRISCH did not restrict his investigations on the beedance to its description of the symbolic code and its biological meaning. He encouraged studies on the evolution of this unique communication system. He was interested in what sensory performances were based on for each information, especially in indicating the distance and the direction towards the goal.

The world of the polarized light was detected, the complicated mechanism of sun compass orientation became understandable. His students of the 1., 2. and 3. generation are still working on unsolved problems raised from the beedance: e. g. the orientation of bees in the earth magnetic field and the information processing from one generation of bees to the next.

Preamble

The bee dance had already been discovered at least 3 times before KARL v. FRISCH published his first communication «Über die Sprache der Bienen» in 1920. ARISTOTLE observed that the forager bee after having thrown off her load returns to the crop and is followed by three or four companions – «but how they do it, has not been observed».

In 1788 Pastor ERNST SPITZNER, born in my own native country in Oberammergau, described the bee dances as «ballet of the bee»: «Full of joy they twirl in circles about those in the hive, from above downward and from below upward.

In a few minutes, after these had made it known to the others, they came in great numbers to the place!»

In 1823 UNHOCH gave an almost perfect description of the rounddance: «The dance mistresses twist and turn in something more than a semicircle, now to the right, then to the left, five or six times, and execute what is a genuine round dance. – What this dance really means I cannot yet comprehend. –»

1 The dance language of the forager bees

It was by no means a sudden and sensational discovery, when in 1919 for the first time KARL v. FRISCH saw in the Jesuit-Klosterhof of the old Zoological Institute in Munich the round dance of a returning forager bee. It was 26 years before the whole corpus of information was found. In small steps the mosaic stones were collected by himself and his students. (K. v. FRISCH 1920, 1923)

However the very first discovary already represented a breakthrough in sociobiology. In the year 1920 practically no biologist imagined that an animal

could communicate exact details about the flower-species visited and the quality and profitableness of a crop area by means of symbolic dance movements;

could deliver these pieces of information in the hive far away from the location of their origin

could successfully transmit a specific order to fellow-bees waiting near the entrance.

The philologists agree on «Semanticity», «Broadcast», «Displacement», and «Productivity» for these performances and some take them as elements of a proper language.

Twentysix years later, in 1945, an additional, almost unbelievable information code was detected:

the tailwagging dance, as the long distance dance, indicates the direction of and distance to the goal.

The distance is related to the rhythm of the wagging runs, the direction of the wagging line is related to the position of the sun, whereby the angle between sun and flight route is transposed on the vertical comb into the gravity-field. All these facts fascinating as they were in different ways, and which appear to us very clear nowadays, created some confusion for KARL v. FRISCH in the early years: precise data on the direction indication sometimes revealed deviations of up to 40° to the left or to the right; these deviations were not the same on the opposite side of the comb. Even more confusing were the dances on the horizontal surface, when the comb with the dancing bees was tilted: now the dances pointed directly towards the goal by their wagging line using the sun directly as a compass; but the dances were orientated even when the sun was covered by clouds – as long as a spot of blue sky was visible. (K. v. FRISCH 1965)

The only way to solve these riddles was to pick out single facts, isolate them, and test them again and agian. As we shall see later, pioneer work has been done in this way in the field of modern sensory physiology, opening new systems to detect parts of the environment to which our senses are blind.

This may be a useful priece of advice for the younger generation of biologists: not to throw in the sponge when a hopelessly complicated situation arises – something new and interesting may be awaiting discovery.

2 Communication in swarmbees

Of special interest was the finding that the bee dance uses its semantic code not only in foraging but also in house hunting. «Interchangeability» of semantics is another essential element in a genuine language.

Scout bees announce a suitable nesting site by the same wagging dances as successful nectar- or pollen-collectors. From the beginning quite new problems concerning sociobiology arose during our observations:

a) Scouts which are recruited from old experienced forager bees have to change their motivation radically when they move from collecting sweet nectar to inspection of dusty holes. They do this without any reward. Only social exigencies regulate this drastic psychological change in the dance mood.

b) Whereas in foraging the bees are attracted by the colour, odour, shape and pattern of flowers, now the same individual has to search for dark hollows in the ground, in a wall or even in a chimney.

c) If one follows for some time the dances of the scouting bees in the cluster and records their announcements of location, one comes to a very surprising conclusion: not just one nesting place is reported, but rather announcements are given of different directions and distances, and this means that several possible dwellings are announced at the same time.

We now have to answer the difficult question of how agreement is reached in the cluster about which of the nesting places offered should be chosen. Only one of them can be selected, for the entire swarm must move with its queen and cannot split itself up into small groups.

In fact only when the good fortune of agreement finally came about was the sign given to fly off, and the swarm moved out in the announced direction.

In any case, we arrived at the following conclusions: (1) the decision concerning the choice of a nesting place lies solely with the scouting bees; (2) the choice is always for the best nesting place offered.

The dancers in the cluster give information not only concerning the location of the new dwelling place, but also concerning its quality; this quality is indicated by the liveliness and the duration of the dance.

In this way it is not merely the interest of most novices in the swarm that is concentrated on the best nesting site offered; I have found – to my own surprise – that even the scouts which had previously announced mediocre sites later followed these lively dances, visited this new place and changed their mind, i. e. they voted for this new – better – nesting site. Only in this way can agreement in such a critical situation be reached.

d) How can the scouts judge so exactly the quality of a prospective bee home? We know that size, protection against wind, ants and rain, the distance from the mother hive etc. are decisive positive factors. We should realize, however, that they had no experience on house-hunting before; they cannot be informed in this matter by specialists, since generations can pass without any need for searching for new nesting sites.

But after swarming these old forager bees have to ensure the survival of the whole colony. Even if we suppose that the details of an optimal nesting site must be encoded as a fixed genetic information, we should admire such a radical change in motivation and in the orientation mechanism due to social demands (Lindauer 1955).

3 Evolution of the bees dance

One day in 1954 K. v. Frisch called me into his office to meet the director of the Rockefeller Foundation, Dr. Pomerat. To my great surprise I was asked: «Will you go – with a fund of the Rockefeller Foundation – to India in order to study the language of the Indian bees? Since we cannot believe», K. v. Frisch explained, «that the bee dance of the European bees has come from heaven as it is, and since the Indian honeybees and the stingless bees there live in a more primitive social organisation, we should expect some phylogenetic primitive stages of the bee dance».

My stay afterwards in Sri Lanka and in Brazil confirmed the following:

1. that in all four species of the genus *Apis* essentially the same round- and tail-wagging dances are used in food gathering and in the search for housing
2. that in indicating the distance so called «dialects», which K. v. Frisch has described for the different races in *Apis mellifera,* are even more pronounced
3. that in indicating the direction a more primitive stage in the dance of *Apis florea* is realised. This lovely dwarf honeybee is unable to transpose the direction towards the goal into the gravity field; she needs a horizontal platform on the top of her single comb, which is built on a branch right in the open air. Here the wagging line points directly towards the goal using the sun's position as the exclusive reference point (Lindauer 1956). With this evolutionary aspect established it may be of interest for

future research to find out how, on a neurological basis, the increasing complexity in multimodal information processing should be understood. The stingless bees of Southeast Asia and of Brazil – there exist more than 400 species – in their still more primitive social organization use even more primitive means of communication: Meliponini newcomers are first alerted in the hive by excited but undirected wagging runs of the forager bees; then these alerted novices, waiting in a group in front of the hive are guided by scented trail marks previously laid down by the foraging pilot bees (Lindauer 1961).

There is a link to the odour marks of different ant species and this system may be a common root of the elaborate bee dance.

4 Sensory aspects of the bees' dance

Up to now hardly any other behavioural study has been analyzed in such careful detail and so comprehensively. Apart from sociobiology and evolution, *sensory physiology* in particular was promoted in an unique way. Let me demonstrate this in a few examples.

One of the most difficult problems v. Frisch had to face for years was the sun compass-orientation. To use the sun as reference point is fine in bright weather; but it was confusing indeed, when we had to state again and agian that the bees continued dancing even when the sun was covered by clouds. The famous «Ofenrohrversuch», whereby a small spot of blue sky through a stovepipe was visible for the dancers on a horizontal comb, pointed the way:

«Ich reinigte ein Ofenrohr von Russ, steckte es durch ein Zeltdach; dann zeigte ich den tanzenden Bienen auf horizontal gelegter Wabe ein Stück blauen Himmels durch die Öffnung dieses Rohres. Die Tänze waren sofort orientiert; mit einem Spiegel, der an der Öffnung des Rohres angebracht war, ließen sich die Tänze um 180° in die Gegenrichtung ablenken!»

Is there today any experimental setup in Biology which, by its «primitive arrangement» and by its simple but clear question, has brought a similar ingenious discovery and has opened to us human beings a new sensory world – the world of polarized light? The next decisive step which led to the discovery that polarized light was involved came about in a conversation with a friend in the physics department, Prof. Benndorf in Graz. His friend Krogh in Kopenhagen then provided for us a polaroid sheet from the US; by turning this sheet above a horizontal comb K. v. Frisch could deviate the wagging line of a dancing bee; a convincing demonstration that polarized light is used for orientation. Autrum suggested a model for the «Sternfolie», that could make visible the polarized pattern on the blue sky for the human eye. It opened the way in the search for the analysator in the bee's eye. But it was a stony search for the analysator until it was localized in the microvilli of the rhabdomeres based on the dichroitic absorption (for details see K. v. Frisch 1949, 1950, 1953, 1965).

The sun compass presents still another problem: As a reference point the sun changes its position from minute to minute. Do the bees calculate in their orientation this apparent sun movement? To clear up this question we had planned to train a group of bees in an unfamiliar countryside in the afternoon to a feeding place 180 m southwards. The next morning – displaced to another unknown area – these south-trained bees were to be tested on whether they prefered the eastern feeding place – which corresponded as «angle constant» with the sun (feeding table on the left side of the sun as in the afternoon before), or the true compass direction, i. e. the southern place – compensating the sunmovement

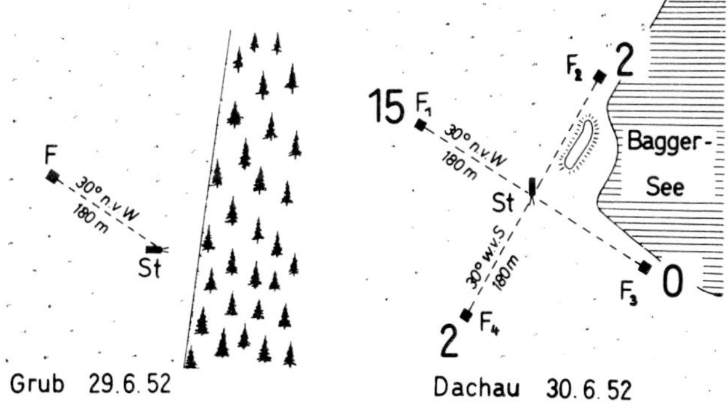

Fig. 1: In the afternoon of June 29th a group of bees was trained in an unfamiliar countryside to a feeding place 180 m NW.
The next morning 4 feeding tables in different compass directions were made available – again in an unknown area. The bees prefered the NW-place (15:2:2:0) (K. v. Frisch 1965)

(Fig. 1) «Jeder wird uns für verrückt halten, wenn wir solche Leistung von den Bienen erwarten.»

«Everyone will think we are mad to expect the bees to achieve something like that» – v. Frisch told me at that time, and I had to encourage both of us to risk the experiment; this was just a year before Kramer reported on sun compass-orientation in starlings. In the last 30 years it has been proved that many animal species can orientate themselves by means of the sun compass. It has often happened that «strange» ideas, better to say «very improbable hypotheses» have prepared the way for important discoveries – provided that challenge, risk and courage are given free rein.

5 The detour and side wind experiment

«Kleine Dinge, große Wirkung» – little things but great consequences – this holds true for many «minor»-experiments of K. v. Frisch: When the bees are forced to fly in a detour around a rock what angle will they indicate in their dance – the first or the second leg they have flown? Neither of them. Quite unexpectedly they indicate the *direct route;* only by this calculation can they avoid a mistake for the novices. Furthermore: if we move the hive and the feeding table towards the tip of the triangle so that the energy consumption in flight *over* the rock becomes lower than in the detour, the foragers choose the direct flight.

Even if we hesitate to call this calculation «intelligent» – it is an «ad hoc decision» which solves the problem – we should admire how a flying insect can have such a command of spatial coordinates. Related to this performance was the compensation performance in side wind. In such a situation the bees have to fly in an oblique body position to the goal, so that they see the sun at a wrong angle with regard to the direct route; nevertheless in the dance they keep this direct route, which is again the best solution for the novices since the wind can change from minute to minute (K. v. Frisch u. M. Lindauer 1955).

In conncetion with the «side wind problem» I should mention here again the great optimism and patience of K. v. Frisch in facing experimental difficulties. For 3 summers we had arranged our setup expecting that the famous «Brunnwind» on sunny days would blow from 10:00 up to 14:00 from the Mondsee to the Wolfgangsee in an eastward direction. On some days we were happy indeed, but in other days there was no wind whatsoever; or suddenly the Föhn (a south-wind) came up or a thunderstorm offered a strong west-wind.

«We must try again» he said and this optimism eventually resulted in some fine successes for us.

6 Problems left as a challenge for future research

The history of the bees' dance can be completed only if we now deal with the many «seedlings» K. v. FRISCH has planted during his life for his students in the 1., 2. and 3. generation – and over the whole world. (See MENZEL, WEHNER, SEELEY, GOULD, DYER, HÖLLDOBLER as referees in this symposium).

6.1 Calculation of distance

When the dancing bees indicate distance by means of their rhythm then these bees must have measured the distance themselves before. K. v. FRISCH already has found that the parameter is not the flight time but the consumption of energy. HERAN and his group have confirmed this conclusion in fine experiments. (SCHOLZE E., H. PICHLER und H. HERAN 1964) V. NEESE, my co-worker in Würzburg went a step further by asking the question: how can the bee measure so exactly the consumption of energy during flight? It is the *decreasing tension* of the honeysac-wall, which is correlated with distance flown. Cutting the «nervus recurrens» or offering an artificial dense sugar solution with agar puts the system out of order (unpublished).

6.2 Orientation in the earth's magnetic field (EMF)

It was a great surprise indeed when we – after a long and tiresome analysis – had evidence that the earth's magnetic field influences the angle orientation in the wagging dance. In cooperation with H. MARTIN we registred regular deviations which follow a typical diurnal pattern.

These diurnal curves are correlated with the dynamic variation of the EMF according to the following function:

$$Mi \triangleq [(eA_2 \pm \frac{v}{2}) \cdot (eA_2 - eA_1)^2] \cdot \lambda$$

$$Mi = (eA_2 \pm eA_1) \cdot \lambda$$

eA = effective magnetic force = $\log(\Delta f + 1) \cdot \mu \cdot \sin\alpha$
v = Rayleigh-constant ($\approx$ 0,20–0,24)
ΔF = variation of EMF
μ = permeability
α = angle of dance in the gravity field
λ = scale factor.

Fig. 2 gives an example of the dynamic change of the «Missweisung» during the day, based on the above function. The data given so far show that bees are sensitive to variations of the magnetic field in the range from 0–300 γ. However, only dynamic changes are of importance.

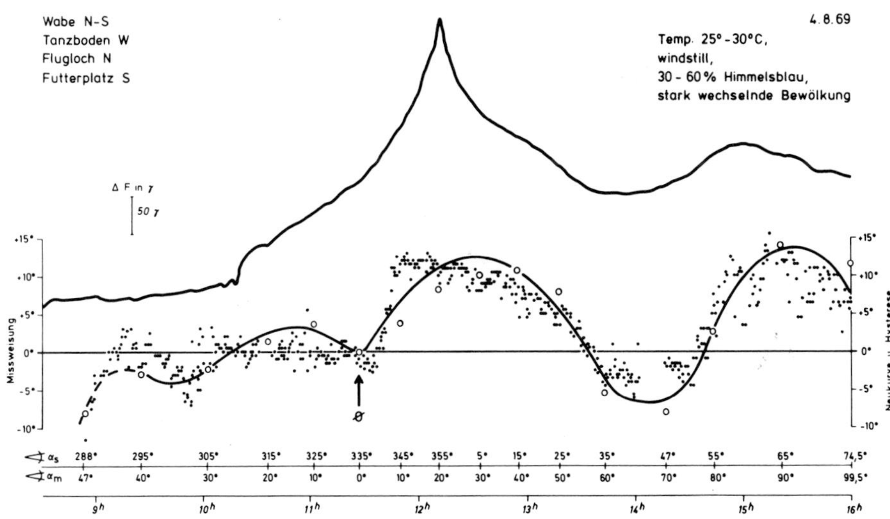

Fig. 2: The «Missweisung» correlated with the dynamic diurnal change of the EMF (see text).

When the EMF is compensated the dancers orientate exclusively with respect to gravity. Also when the wagging line corresponds with the inclination (i. e. when it is parallel to the field line of the EMF; 65° in Würzburg) the dances show no deviation. (LINDAUER and MARTIN 1968, 1972, MARTIN, LINDAUER 1973, 1977). From the beginning we called in question whether these deviations are a «Missweisung» in the sense of true mistakes for the novices. LEUCHT (1982) in an extensive analysis, has confirmed our earlier hypothesis that the EMF has a stabilizing effect against disturbance of incident light. Only during the «Nulldurchgang» i. e. when the wagging line is parallel with the inclination (where the Missweisung is zero!) does the incident light considerably deviate.

There are however two situations where the EMF becomes a proper orientation parameter for the bees:

a) Swarming bees upon entering a new site have to select by unanimous decision the compass-orientation for the new combs. The bees prefer the main compass direction; in the natural magnetic field the north-south direction is preferred. (DE JONG 1982, KIRCHNER unpublished); in certain cases the swarms select the same orientation of the combs as in the mother hive.

b) The 24-hour periodicity of the EMF – as is shown in Fig. 3. is used by the bees as a *timer,* if other «Zeitgeber» such as light dark-change are excluded. In a disturbed aperiodic magnetic field the bees no longer orientate to the feeding time they were trained to in the preceding days. They are also disoriented in time when a colony is translocated to another geographic latitude, i. e. from Würzburg to the polarcircle. Here the 24-hour periodicity in intensity and the dynamic change in direction of the inclination and declination are quite different. (MARTIN and LINDAUER 1973).

Very recently it has been realized that in extreme situations, when all exogen «Zeitgeber», known up to now – including the periodicity of the EMF – are lacking, the bees still can use some astrophysical parameter for their orientation in time:

A group of foragers was fed continuously for 24 hours under so called «constant condition» (LL, constant temperature and humidity). In the natural EMF the bees display-

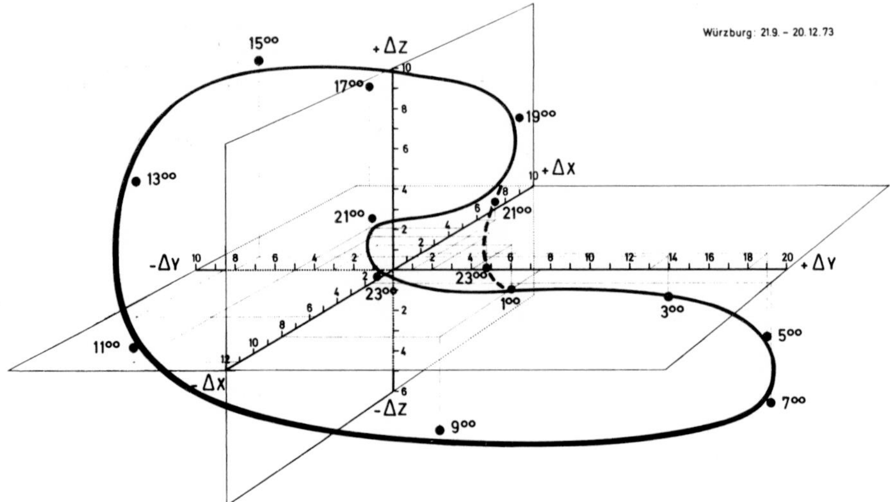

Fig. 3: The daily periodicity of the EMF in direction and intensity as a 3-Vector-Diagram.

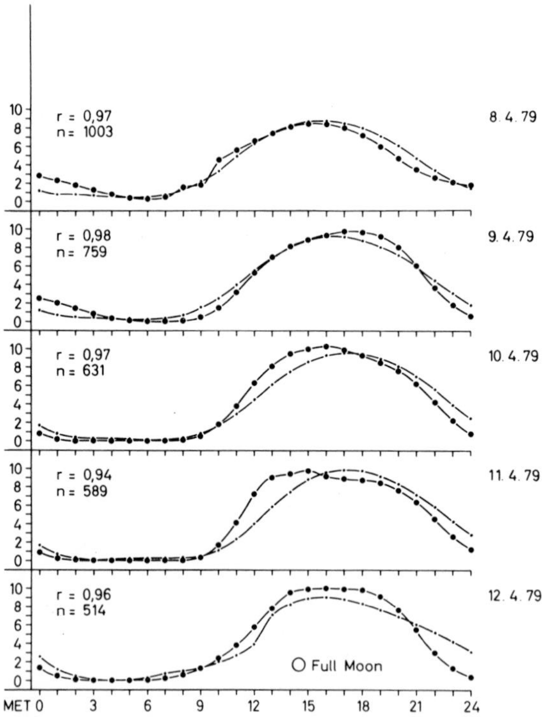

Fig. 4: In an artificial disturbed magnetic field the bees – after long adaptation – show a free-running foraging activity, correlated with the sun-moon constellation (see text) ●–●–activity of the bees; ·—·—· sun/moon constellation.

ed a preferential activity which corresponds with the dynmaic curve of the EMF. In the compensated or disturbed EMF the bees were totally disorientated in the first days, but after 2–3 weeks again the activity diagrams show a periodicity although a very smooth one. In detailed analysis MARTIN has found that besides the EMF – some additional events which originate form sun and moon affect the time sense of bees. (MARTIN, LINDAUER and MARTIN 1983) There is a nice correlation, when we take the azimuth of the sun and the moon, the altitude of both and the gravity constant of the sun.

Activity of the bees / time unit $\triangleq$ $[0{,}46 \times (\sin Az_S + \cos^2 h_S) + (\sin Az_M + \cos^2 h_M)]/\Delta t$

Az_S = azimuth of the sun
Az_M = azimuth of the moon
h_S resp. H_M = altitude of the sun resp. moon (Fig. 4).

We have to conclude from these results, that at least in honeybees – the time sense has its basis on geophysical and astrophysical parameter as exogen «Zeitgeber».

6.3 Theories concerning the perception mechanism

In the last 15 years an increasing number of data were published which demonstrate that development, metabolism, information processing in animals are affected by the EMF; in addition as we have seen (p. 135) some parameters of the EMF are used for orientation and rhythmicity for establishing rhythm of activities. Therefore the question of the perception mechanism becomes urgent. Since up to now nobody has found magnetic receptors we can only offer some theories – which may encourage further research in this field.

Besides induction within a semiconductor moving within an external magnetic field (RUSSO and CALDWELL 1971) and besides the Josephson junction concept (1969) based on the change in the permeability of cell membranes due to alternating electromagnetic field, the theory of WALCOTT et al. (1979), GOULD et al. (1978, 1980) is favored by competent researchers today: in bacteria, in the abdomen of the honeybee, in some tissues of pigeons' head and in chitons magnetic crystals ($FeO{\cdot}Fe_2O_3$) have been detected, which should be involved in the orientation in the magnetic field. The mechanism may be well understood in bacteria, where the whole organism filled with crystals of magnetite reacts «magnetotactically» as a single domain ferromagnet. In all other cases some morphological connection with the nervous system seems a prerequisite to us for special information processing coming from the magnetite crystals. Those connections have not be found.

Our own experiments, however, lead us close to the molecular niveau of perception:

As at first hand the orientation in the gravity field is affected by the EMF in bees, KORALL in our group has recorded the elementary transducer processes, i. e. the action potentials of a single gravity receptor in the neck organ of the honey bee under natural and artificial magnetic conditions (KORALL and MARTIN 1984).

The response of the receptor may be described as followed:

1. The frequency of impulses changes by tilting the vector of an artificial MF.
2. The frequency of impulses is correlated with the 24 hour periodicity of the EMF variations.
3. Early in the morning when the direction of the magnetic field vector changes very rapidly the EINSTEIN-DE HAAS-effect (gyromagnetic forces) is evident.

Some of these results and behavioural tests under distinct oscillating and static MF conditions lead us to the theory that unclear spin resonance (NMR) may be involved in the perception mechanism.

6.4 Tradition in the honeybee community?

I shall finish the history of the dance language in honeybees refering on a communication system which – as a *longterm* communication – is transmitted from one generation to the next. An essential information for the existence of a bee colony is their daily rhythm of collecting activity. I have found in the last years that besides the special collector group the whole colony is informed of such activity periods.

A group of about 300 bees was fed from April to July on a very inconventional time, from 5:00–6:00 every day. Three month later the whole group of these forager bees was removed and also all novices which had contact with the dances of this group before. Then a new group of 20 bees was trained to a feeding place for one day; however feeding time was not 5:00–6:00 but irregularly throughout the day. On the next day this group was tested – without food. Fig. 5 demonstrates clearly that these bees – as house bees – had taken over the activity period of the earlier collecting group.

This information was transmitted too the the next generation – the larvae and pupae. From the same colony I removed in July 3 combs with sealed brood. In a thermostat the young bees emerged and a new queen was added to the young colony. Two weeks later again a group of 20 bees was trained on an artificial feeding place for one day from morning to evening continuously every ½ hour for 10 minutes. Fig. 6 demonstrates that also

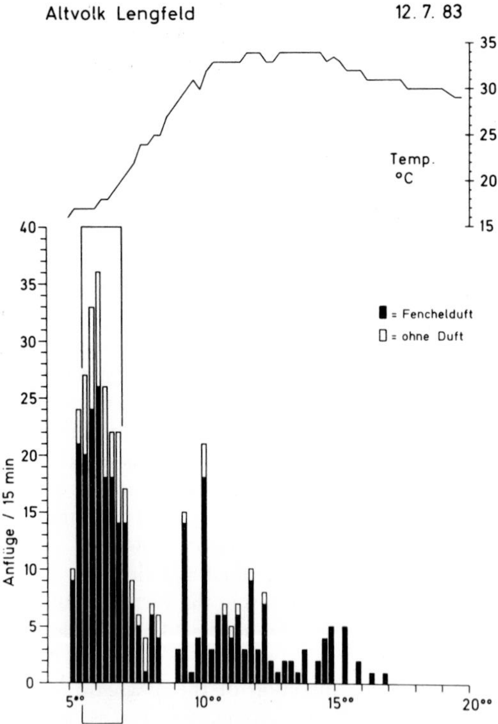

Fig. 5: From April to July ca. 300 bees could feed every morning from 5.30^h–7^h on a 2-molar sugar solution scented with Fenchel odour. After removing the whole forager group the remaining population still keeps the information concerning the odour and the activity period.

these bees – of the younger generation – were conditioned in their larval or pupal stage for the extraordinary collecting hour.

The same results we got in corresponding experiments, where the bees were trained for a feeding time 20:00–21:00.

We can only speculate how the information on the activity period is transmitted from the forager bees to all the house bees and to the larvae. No doubt in such an intense foraging activity some information is spreading out by the dances, due to food exchange and due to vibration over the combs produced by the sound in the wagging line (240 Hz). These vibrations should be perceived also by the larvae and the pupae in the brood cells. In this way the very long life period before the imago stage – 3 weeks is about ⅓ the total life span in honeybees – is used for an important communication between 2 generations (LINDAUER, unpublished).

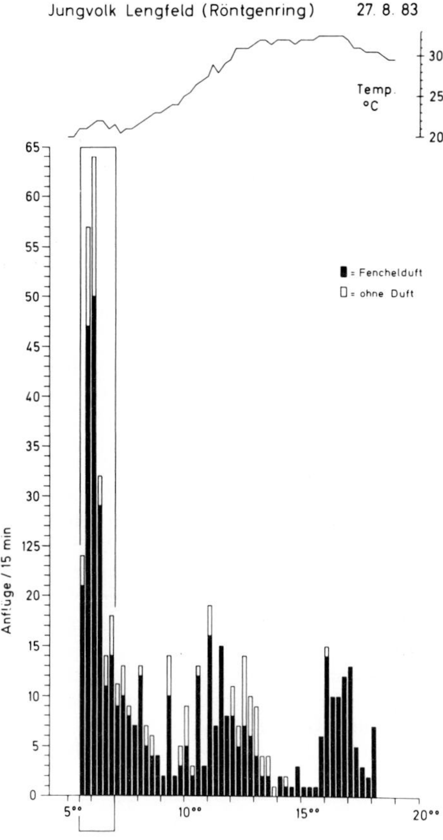

Fig. 6: After eclosion from the pupal stage the next generation (of the mother colony in Fig. 5) also preferred the same odour and showed up as «early riser».

References

De Jong, D. (1982): Orientation of comb building by honeybees, J. comp. Physiol. *147,* 495–501.

Frisch, K. von (1920): Über die «Sprache» der Bienen. Münch. med. Wschr. *566–569.*

Frisch, K. von (1923): Über die «Sprache» der Bienen, eine tierpsychologische Untersuchung. Zool. Jb. (Physiol.) *40,* 1–186.

Frisch, K. von (1949): Die Polarisation des Himmelslichtes als orientierender Faktor bei den Tänzen der Bienen. Experientia (Basel) *5,* 142–148.

Frisch, K. von (1950): Die Sonne als Kompaß im Leben der Bienen. Experientia (Basel) *6,* 210–221.

Frisch, K. von (1953): Die Richtungsorientierung der Bienen. Verhandl. d. Deutsch. Zool. Ges. in Freiburg 1952, 58–72, Leipzig.

Frisch, K. von u. M. Lindauer (1955): Über die Fluggeschwindigkeit der Bienen und ihre Richtungsweisung bei Seitenwind. Naturwissenschaften *42,* 377–385.

Frisch, K. von (1961): Über die «Mißweisung» bei den richtungsweisenden Tänzen der Bienen. Naturwissenschaften *48,* 585–594.

Frisch, K. von (1965): Tanzsprache und Orientierung der Bienen. Springern, Berlin, Heidelberg, New York.

Gould, J. L.; Kirschvink, J. L.; Deffeyes, K. S. (1978): Bees have magnetic remanence. Science *201,* 1026–1028.

Gould, J. L.; Kirschvink, J. L.; Deffeyes, K. S.; Brines, M. L. (1980): Orientation of demagnetized bees. J. expl. Biol. *86,* 1–8.

Josephson, B. D. (1969): Weakly coupled superconductors, in: Superconductivity (R. D. Parks, ed), Vol. 1, pp. 423–447, Marcel Dekker, New York.

Korall, H. and H. Martin: The influence of Geomagnetic and Astrophysical Fields on the Impulse Responses of Mechanoreceptors in Apis mellifera. J. Comp. Physiol. (in Preparation).

Leucht, T. (1982): Responses to light under varying magnetic conditions in the honeybee, Apis mellifica. J. Comp. Physiol. A (1984) *154: 865–870.*

Lindauer, M. (1955): Schwarmbienen auf Wohnungssuche. Z. vergl. Physiol. *37,* 263–324.

Lindauer, M. (1956): Über die Verständigung bei indischen Bienen. Z. vergl. Physiol. *14,* 521–557.

Lindauer, M. (1961): Communication among social bees. Cambridge Mass., Harvard University Press.

Lindauer, M.; Martin, H. (1968): Die Schwereorientierung der Bienen unter dem Einfluß des Erdmagnetfeldes. Z. vgl. Physiol. *60,* 219–243.

Lindauer, M.; Martin, H. (1972): Magnetic effect on dancing bees in: Animal orientation and navigation (Galler, S. R. et al., eds.) *559–567,* Washington D. C.

Martin, H.; Lindauer, M. (1973): Orientierung im Erdmagnetfeld. Fortschr. Zool. *21,* 211–228.

Martin, H.; Lindauer, M. (1977): Der Einfluß des Erdmagnetfeldes auf die Schwereorientierung der Honigbiene (Apis mellifica). J. comp. Physiol. *122,* 145–187.

Martin, H.; Lindauer, M.; Martin, U.: Zeitsinn und Aktivitätsrhythmus der Honigbiene – endogen oder exogen gesteuert? Sitzungsbericht: Bayer. Akademie der Wissenschaft. Mathematisch Naturwissenschaftliche Klasse. 1983. (Vorgelegt von Hansjochem Autrum in der Sitzung v. 7. Mai 1982)

Russo, F.; Caldwell, W. E. (1971): Biomagnetic Phenomena: some implications for the behavioural and neurophysiological sciences: Genetic Psychology Monographs *84,* 177–243.

Scholze, E.; H. Pichler, u. H. Heran (1964): Zur Entfernungsschätzung der Bienen nach dem Kraftaufwand. Naturwissenschaften, *51, 69–70.*

Spitzner, M. J. E.: Ausführliche Beschreibung der Korbbienenzucht im sächsischen Churkreise, ihrer Dauer und ihres Nutzens, ohne künstliche Vermehrung nach den Gründen der Naturgeschichte und nach eigener langer Erfahrung. Leipzig 1788.

Unhoch, N.: Anleitung zur wahren Kenntnis und zweckmäßigsten Behandlung der Bienen. München 1823.

Walcott, C.; Gould, J. L.; Kirschvink, J. L. (1979): Pigeons have magnets. Science *205,* 1027–1029.

Fortschritte der Zoologie, Bd. 31 · Hölldobler/Lindauer (Hrsg.): Experimental Behavioral Ecology
G. Fischer Verlag · Stuttgart · New York · 1985

Recent progress in the study of the dance language

James, L. Gould, Fred C. Dyer, and William F. Towne

Department of Biology, Princeton University, Princeton, N. J. 08544, USA

Abstract

The dance language and orientation of honey bees is probably the best understood example of a complex, highly integrated behavior. Work over the past decade has revealed this remarkable behavioural system to be even more interesting than had been thought. Progress has been made on how the language works (both for the dancers and the recruits), on how bees recognize the cues used for orienting their flights and dances, and how bees can resolve ambiguities in these cues in order to communicate. Moreover, we now know that bees navigate and dance when both the sun and blue sky are obscured entirely by clouds by falling back on a time-linked memory of the sun's course with respect to landmarks. Recent comparative studies among the different races of *Apis mellifera,* and between *mellifera* and the three Asian species of bees have revealed much about the likely evolution and adaptation of the dance system to a variety of physical settings. For instance, the dance has been modified to suit such alternatives as a vertical dance surface, or the darkness of an enclosed nest, or the ambient light of exposed comb. It has also been modified to accommodate a wide range of habitat types. The modifications have taken place in the sensory channels used to transmit the dance information, in the dance distance dialects and divergence, and several other such parameters.

1 The dance and communication

The discovery and elucidation of the dance communication system of bees by von Frisch and his colleagues (rev. in von Frisch 1967) is probably the single most impressive accomplishment in ethology. The work set the standard for deftly combining observational and experimental techniques, and joining together mechanistic and ecological perspectives. Moreover, it was critical, thorough, clear, and fascinating. von Frisch once characterized the honey bee as a «magic well» which could never go dry, and only this hypothesis adequately accounts for there being anything left to discover about these remarkable insects.

1.1 The dance-language controversy

Interest in the dance communication of bees was rekindled in 1967 by the experiments of Wenner and his colleagues purporting to show that the distance and direction correlations in the dance did not convey any information to recruits (rev. in Gould 1976).

Briefly, their argument ran that since they could create conditions under which bees were recruited to odor alone, and since bees must use either odor or a language for communication, the dance correlations were not used. The flaw in this line of reasoning, of course, is the assumption that bees can communicate in only one way. In an analogous situation VON FRISCH's work had already shown that bees can use either the sun or patterns of polarized light in the sky for navigation, depending on the circumstances.

Initial attempts to pit the dance system against odor supported VON FRISCH (eg., 28, 39), but the possibility that recruits were using site-specific odors for locating food could not be excluded. GOULD (1974, 1975 b, c) developed a technique for misdirecting recruits by painting over the ocelli of foragers, making them less sensitive to light. He then used a light as an artificial sun to reorient the dances (and interpretations) of untreated bees. Ocelli-covered bees, on the other hand, did not respond to this light, and continued to be oriented strictly to gravity. Thus the ocelli-painted dancers indicated a direction quite different from that read out by untreated dance attenders. As predicted by the dance language hypothesis, recruits did indeed fly preferentially to the distance and direction specified by the misdirected dances (Fig. 1).

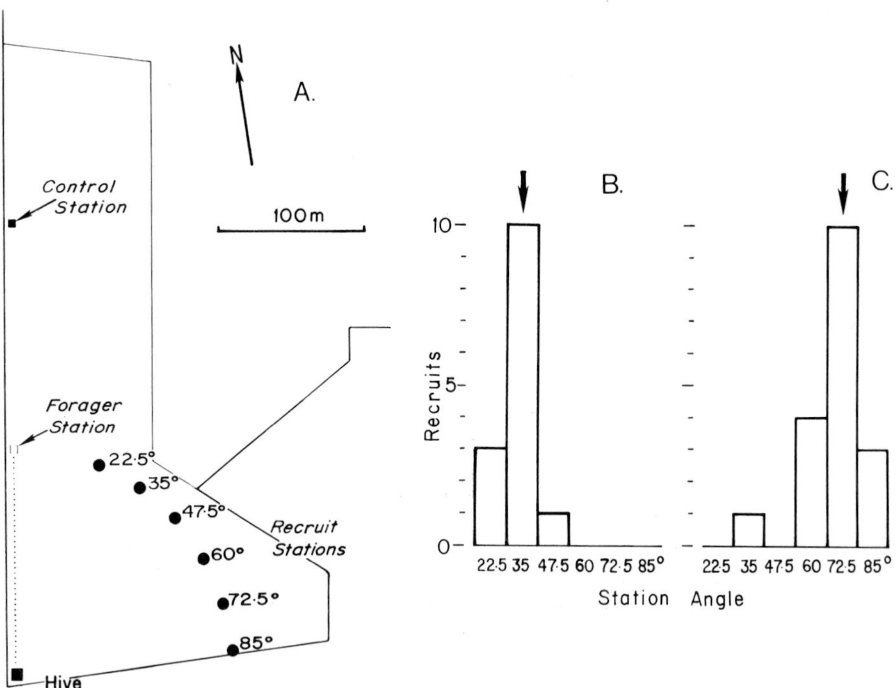

Fig. 1: Redirection experiments. a) Ocelli-painted foragers were trained to the forager station 150 m NNE of the hive. An array of recruit stations was set out offering the same food being collected at the forager station. A light was used to reorient the dance attenders (as judged by the reorientation of untreated foragers visiting the control station). In this example the redirection was first 35° clockwise and then 72.5° clockwise. b) Recruits during the first part of the experiment favored the 35° station, but c) recruitment shifted toward the 72.5° station after the light was moved. After GOULD (1975 a, 1976).

At about the same time, SCHRICKER (1974) found that poisoning foragers with sublethal doses of parathion causes them to overestimate the distance to a food source. Recruits arrived preferentially at the (incorrect) distance specified by the dance.

1.2 Dance integration

From measurements of individual waggle runs, VON FRISCH and JANDER (1957) concluded that dance attenders must average over several runs to account for the accuracy of searching recruits. WENNER and his colleagues, in stressing the possible importance of odor cues, dismissed experiments in which the forager station was near the recruit array: the recruits could be using odors generated by the foragers at the forager station to increase their accuracy. LINDAUER (1971), on the other hand, suggested that the very poor accuracy of WENNER's bees might be accounted for if recruits attended and averaged together separate dances by different bees. GOULD's experiments (1975 b, c), in which the forager station is well away from the recruit array, showed that VON FRISCH's accuracy estimates were affected only slightly by the placement of the training station, but that recruits probably do not average separate dances. Instead, the poor accuracy of WENNER's bees seems mostly attributable to his training techniques.

2 The dance and navigation

When a forager reaches a food source, even if she has flown an indirect route, she knows the distance and direction of the hive and is able to fly directly home (rev. in VON FRISCH 1967). The dance is basically a readout of the information stored in a forager's navigational computer. VON FRISCH began almost immediately to exploit the dance to learn how bees orient to environmental cues and process the information they collect.

2.1 Sun recognition

By a series of ingenious mirror experiments, VON FRISCH showed that the sun is the primary celestial landmark for bee navigation as well as the standard reference point for the dance language (VON FRISCH 1967). He obtained some indirect evidence which suggested that the sun is recognized as a UV disc in the sky, but pointed out that this was physically implausible. BRINES and GOULD (1979) investigated this question further by offering a variety of artificial suns to trained foragers dancing on a horizontal surface. The bees would either orient to the pattern as though it were the sun, a patch of sky, or something to which they could not orient. The bees interpret virtually any patch of light low in UV and smaller than about 15° as the sun, regardless of its shape, color, or polarization. Hence a green, triangular, 100 %-polarized patch 10° across will substitute for the white, circular, unpolarized 0.5° disc of the sun. The low-UV requirement probably acts as a sign stimulus for the sun.

2.2 Sun compensation

The sun is a problematical landmark because it moves. Moreover, the rate at which the sun's azimuth changes depends on the latitude, date, and time of day. VON FRISCH showed very clearly that bees compensate for this movement, even when in the hive dancing, and pointed out that bees might 1) know how the sun moves innately, or 2) learn it, or 3) use a rough approximation (VON FRISCH 1967). LINDAUER (1959) showed that they have to learn that it moves, and in which direction, but the question of exactly what they learn about its movement remained open. GOULD (1980) attempted to discover the exact mechanism by training bees at one location, trapping them in the hive out of sight of the sun for two hours, and then releasing them at a new site. (It is essential to change locations since experienced foragers frequently ignore the sun and orient their outward flights to familiar landmarks (VON FRISCH and LINDAUER 1954)).

In one set of tests, the hive was closed from 11:00 to 13:00, near the summer solstice, while in the other the closing ran from 12:00 to 14:00. Around solar noon during this part of the year the sun's rate of azimuth movement changes rapidly (Fig. 2 a). Foragers trapped at 11:00 underestimated the subsequent change in sun azimuth, while those trapped at 12:00 overestimated (Fig. 2 b, c). The results most closely fit the hypothesis that bees can extrapolate the rate of movement of the sun.

A closer look at the data suggests that the extrapolation rate used is roughly 20 min out of date. This is reminiscent of LINDAUER's (1963) discovery that foragers dancing to a moving feeder never «catch up,» but instead dance toward the location of the food source

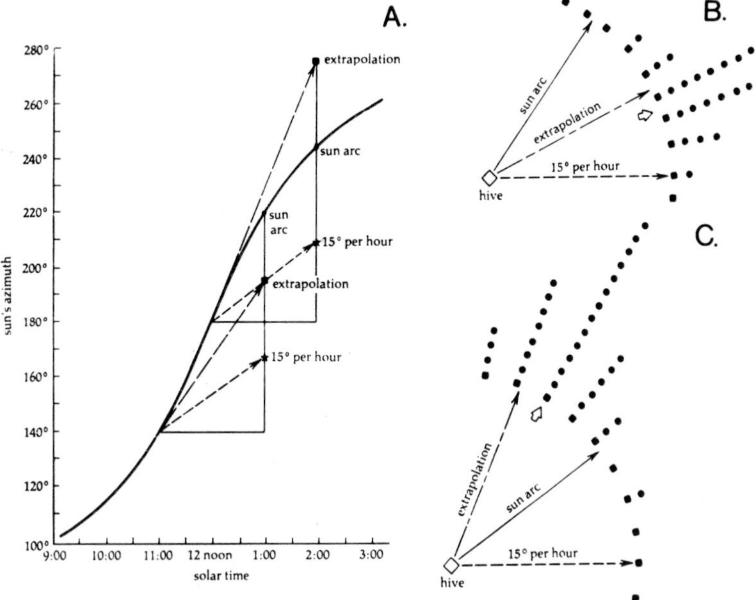

Fig. 2: Sun compensation. a) Near the solstice the sun moves very quickly in azimuth near noon, but more slowly both before and after. If foragers are prevented from seeing the sun for the two hours beginning at 11 : 00, their estimate of the azimuth of a training station after their incarceration will be in error if they have not memorized its course. For instance, simply extrapolating the sun's movement would lead to a moderate underestimate since the rate of azimuth movement increases after 11 : 00, whereas a 15°/hr approximation would be seriously in error. A bee unable to see the sun from 12 : 00 to 14 : 00, on the other hand, would overestimate the azimuth since the sun's rate of azimuth movement slows after 12 : 00. b) When the foragers are prevented from seeing the sun from 11 : 00 to 13 : 00, they tend to fly to stations near the extrapolation prediction, as though the rate used in extrapolation was measured at about 10 : 40. c) When the closing is from 12 : 00 to 14 : 00, foragers overestimate the sun's movement as though they were extrapolating the rate at about 11 : 40. After GOULD (1980).

several trips in the past. When LINDAUER moved the station direction in discrete steps to the west every 30 min, the dances first indicated the former site, and then gradually shifted toward the new location. They never fully compensated, however.

GOULD (1984) followed up this work by moving a feeder 30° east or west every hour. The dancers «caught up» with the move after about 41 min regardless of direction. Both this and the sun-compensation data are consistent with a running-average strategy of information processing. By using a sensory window of 40 min, inaccuracies in the inherently difficult task of measuring sun-azimuth movement (in which the location of an 0.5° disc must be measured with a visual system whose resolution is 2–3° at best) can be averaged out. The result is an extrapolation rate about 20 min slow and an ability to shift dances to compensate for the movement of a station over 40 min.

When bees have useful landmarks, they employ quite a different system of sun compensation: they memorize the azimuth of the sun with respect to prominent landmarks through the entire course of the day (DYER 1984 and DYER, GOULD 1981). Hence a recruit bee can correctly interpret a dance even on an overcast day (see below). It is not yet clear

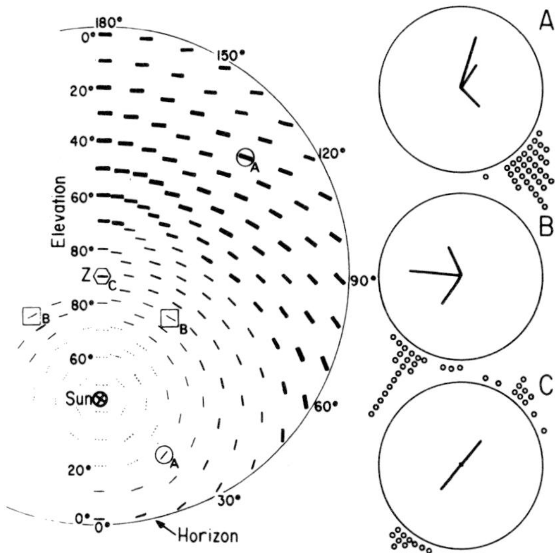

Fig. 3: Resolving polarization ambiguities (left) the pattern of polarized light is the sky is centered on the sun and symmetrical about the solar meridian (only the right half is shown here). At most elevations the same e-vector angle is found at two distances from the sun (as indicated by the two points marked «A»). Vertical polarization, which is found at all elevations above the sun's, is present at two places equidistant from the sun (as, for instance, the two points marked «B»). The polarization at the zenith (marked «C») and all along the solar meridian is horizontal. a) When bees are presented an oblique pattern such as A, they orient their dances as though the pattern were the alternative more distant from the sun. The long line in the diagram is the direction expected for the dance if the patch of light were interpreted as the sun, while the short lines represent the two polarized light alternatives. Each circle is the dance of one bee. b) If the pattern is vertical, as for the points marked B, the bees interpret it as if it were the alternative to the right of the sun. (Note that the vertical patterns appear 60° from the sun, and that the dances of these bees are clearly more nearly oriented to 60° than 90°.) c) Dances to a zenith pattern (C) are ambiguous, with both alternatives being indicated. From BRINES and GOULD (1979).

whether bees using this system accurately compensate for the sun's changing rate of azimuth movement, or instead employ a time-averaged rate.

2.3 Polarized-light orientation

VON FRISCH discovered more than three decades ago that bees will orient their horizontal dances to the patterns of polarized light in blue sky (VON FRISCH 1967), using patches as small as 15° to deduce the sun's azimuth. He left open the question of how bees manage this deduction, and how they resolve the ambiguity which arises from the frequent existence of two polarization patterns with the same e-vector angle at most elevations (Fig. 3, left).

BRINES and GOULD (1979) found that dancers ignore many potentially useful cues such as color, degree of polarization, and intensity gradients, and instead interpret the patch arbitrarily as the one existing furthest from the sun (Fig. 3 a). When the e-vector is vertical – which, as a consequence of the geometry of scattering, occurs at points equidistant from the sun – bees choose the alternative to the right of the sun (Fig. 3 b). A zenith pattern, however, is ambiguous, and bees dance in both directions (Fig. 3 c).

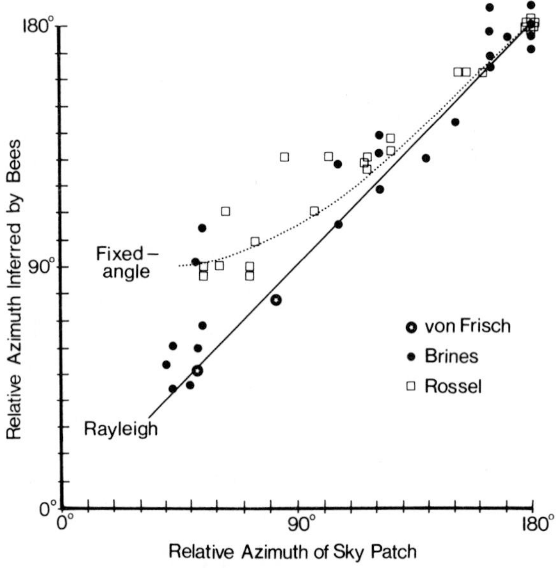

Fig. 4: Determining the sun's azimuth from polarized light. Both simple calculations based on Rayleigh scattering and a photograph-matching system would result in bees interpreting sky polarization relations accurately. In this case, dances should fall along the solid line marked «Rayleigh.» If bees use a system which ignores the elevation of the patch of sky, the dances should fall on the dotted line marked «Fixed-angle:» Data from VON FRISCH (only on two occasions does VON FRISCH report enough information to plot in this way), Brines, and Rossel are plotted, each based on a separate experiment. In general, the Brines and von Frisch data are more consistent with the Rayleigh line, while Rossel's better fit the Fixed-angle prediction. From DYER and GOULD (1983). (VON FRISCH's data comes from written reports and are of a qualitative nature – that is, he states that the dance angles correspond to the Rayleigh prediction – and so are not nearly as precise as other data; a ± 15° discrepancy would probably have gone unreported.)

ROSSEL et al (1978) performed similar experiments, and their data show a clear tendency for dancers to favour one direction, though they did not point this out. Subsequently they did report (ROSSEL and WEHNER 1982) that their bees dance for the further alternative, but still dance bimodally for vertical e-vectors. It is not obvious why the dances to vertical polarization should be unambiguous for one group's bees and not for the other.

ROSSEL and WEHNER (1982) have offered an intriguing explanation for how bees deduce the sun's position from polarization patterns. Whereas most researchers have concluded that bees take into account the elevation of the sky patch to either calculate the location of the sun or match an observed pattern against a stored one, ROSSEL and WEHNER suggest that the bees ignore elevation altogether and simply interpret all vertical patterns to be 90° away, all horizontal patterns to be 180° away (which actually is the case) and all intermediate patterns to be some defined intermediate distance away (a 45° pattern, for instance, is roughly 135° from the sun). Their data (Fig. 4) fits this hypothesis relatively well, but data from BRINES (1978) and VON FRISCH (1967) often does not fit, although VON FRISCH did sometimes report bimodal dances. Again, the source of this sometimes striking discrepancy is not clear.

BRINES favors the idea that bees remember the polarization patterns as a photograph (analagous to the mechanism of canopy orientation in ants reported by HÖLLDOBLER (1980)) which they then match with the patch observed in the sky. The basis of this suggestion is his observation that dances to artificial polarization patterns on overcast days are more scattered than on clear days, even though the ability to calculate sun azimuth ought not to be affected. By using an artificial sun, BRINES (1978) was able to show that the scatter did not arise from any uncertainty about the direction of the sun.

2.4 Cloudy-day orientation

Bees are able to orient and dance on overcast days when neither sun nor blue sky is visible (BRINES, GOULD 1982). DYER and GOULD (1981) discovered how bees know where the sun is when they cannot see it by taking advantage of VON FRISCH and LINDAUER's (1954) observation that experienced foragers can be tricked into navigating along prominent landmarks in a new location if they resemble the landmarks the bees had become familiar with during training, even if the compass direction is different. DYER and GOULD trained bees along a tree line next to an open field, moved the hive overnight to a similar site where the tree line was was rotated 130° clockwise (Fig. 5 a), and recorded the dances of foragers visiting each of two foraging stations: one located along the landmarks and to which most bees flew, the other in the field in the actual compass direction used for training. On sunny days, the dances accurately reported the directions flown (Fig. 5 b) while under overcast the dances were oriented as though the bees were still at the former site (Fig. 5 c). Clearly the foragers could not see the sun or they would have oriented correctly as they did on clear days. Instead they relied on a time-linked memory of the sun's position relative to prominent landmarks. In more recent experiments, we have shown that this memory can persist through several days of cloudy weather (DYER 1984).

Subsequently, DYER (1984) has found that *Apis cerana* (the Asian hive bee which, like *A. mellifera,* lives in enclosed cavities, rather than on open branches like Asia's *A. dorsata* and the dwarf honey bee, *A. florea*) also orients its dances by means of a time-linked sun-vs-landmark system under overcast. *A. dorsata,* which also dances on a vertical surface, may well use the same system. As we shall see below, *A. florea* needs quite a different strategy for successful dance communication during cloudy weather.

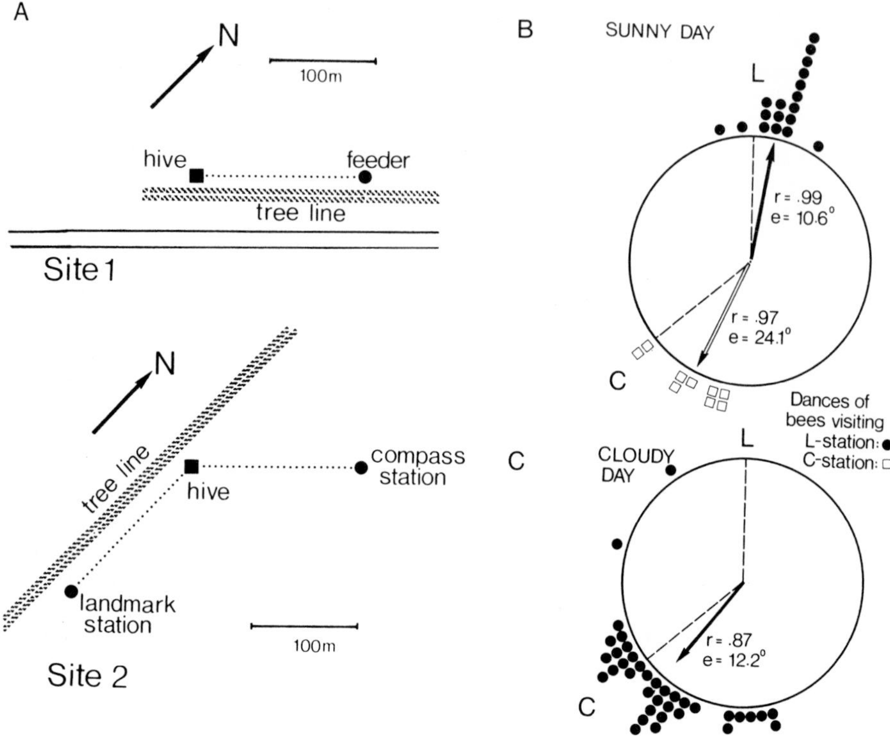

Fig. 5: Cloudy day orientation. a) Bees were trained to the feeder along a tree line at Site 1 and then moved overnight to Site 2 where the tree line was in a different direction. b) When the hive was opened under clear skies, eight of the ten foragers flew along the tree line and danced to indicate the true direction of the outward flight (marked «L» for landmark). The two foragers who found the compass station also danced relatively accurately (direction «C»). c) When the hive was opened at Site 2 under overcast, however, all foragers flew along the landmarks to find the landmark station, but danced as though they were still at Site 1, apparently relying on their memory of the sun's position relative to the direction of the tree line along which they were trained one day earlier. After Dyer and Gould (1981).

2.5 Nocturnal orientation

Apis dorsata, the large tropical honey bee, has often been reported to fly at night when the moon is sufficiently bright. Dyer (1984), working in India, has confirmed this recently and at the same time observed vigorous and apparently normal dances to natural food sources. To determine whether these dances are oriented according to the moon or to the (unseen) sun, he chose nights when the sun set almost vertically, so that dances of bees flying to the same source over a given interval and using the sun as a reference (either the actual solar azimuth or the azimuth extrapolated from the sun's movement before sunset) would change only very slowly. The moon, on the other hand, was at a high elevation during the experiments so that its azimuth shifted very rapidly. If dances were oriented to the moon, they too should shift direction very quickly over the same period.

It was relatively easy to distinguish a group of foragers feeding on one source by sampling pollen loads carried by the dancers and measuring the distance they indicated. The dances of a group identified in this way would change very little over a period when the moon's azimuth shifted by as much as 160°; this implicates the invisible sun as the reference point for the dances. Bees might use the sun despite its absence by memorizing its setting azimuth relative to landmarks (by analogy with *A. mellifera's* cloudy day system (DYER, GOULD 1981)). However, DYER's observations of pre-dawn dances (under moonlight) which did not change after the sun came up (and so were presumably oriented by the sun all along), suggest that the bees might possess a knowledge of the entire nocturnal course of the sun, or be able to use their knowledge of the sun's azimuth at sunrise to anticipate the location. Given the very scattered orientation (not to mention the rarity) of marathon dances (rev. in VON FRISCH 1967), the nocturnal dances of *A. dorsata* are probably the best way to answer this question. Observations at times of year during which the sun's azimuth changes more rapidly at sunset, or observations near midnight will be necessary for clear-cut results.

2.6 Locale maps

VON FRISCH (1967) pointed out that bees trained along an indirect route would indicate the true straight line direction in their dance even if the topography prevented them from ever flying that route. Clearly bees are able to integrate two or more legs of a journey. VON

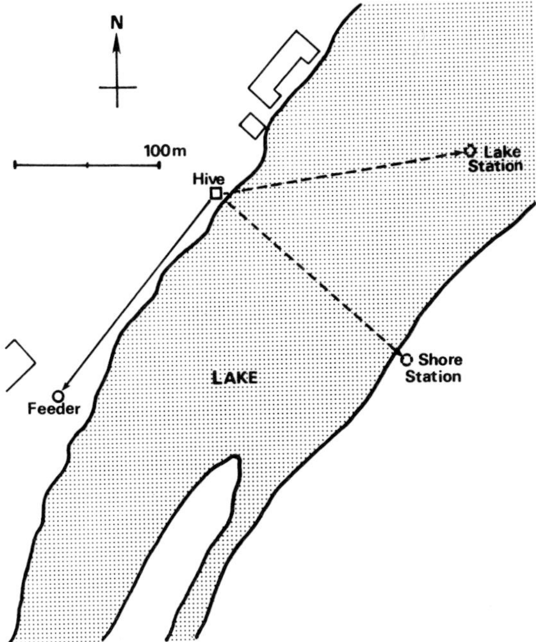

Fig. 6: Lake experiment. Dances indicating the lake station do not result in recruitment, whether the dancers are visiting that station or the one marked «feeder.» At the same time, dances to the equally distant shore station or feeder elicit heavy recruitment. One possible interpretation is that the recruits are able to «place» the location being advertized on the basis of the dance coördinates and judge its suitability without actually flying there. From GOULD (1984b).

Frisch also found that even when a direct flight is possible, many experienced foragers prefer using the training route on the outward journey, particularly if it is marked by prominent landmarks, though they often return directly. This suggests that a sequence of landmarks can be remembered. After reviewing a wide body of insect literature, Wehner (1981) concluded that bees must indeed store a series of «snapshots» of the landmarks along familiar routes, and consult these pictures as they fly.

Two recent results suggest that these snapshots or their equivalent may be integrated into the dance system (rev. in Gould 1984b). Von Frisch and his colleagues had shown that information gathered on the way to a food source is much more important to dancers than information collected on the way back. Brines (1967) attempted to see if experience from the return trip could be used in the absence of an outbound trip. He captured departing foragers at the hive entrance and transported them in darkness to feeders several hundred meters away and out of sight of the hive. Older bees (as judged by wing wear) were able to navigate directly to the hive, and some even performed accurate dances on their first return. Brines concluded that although the transported bees probably had never fed at the feeder site (frequently a large, asphalt parking lot 50 m from the nearest vegetation) they recognized their surroundings (perhaps by matching with an *en route* snapshot for some familiar but more distant natural food source) and so were able to «place» themselves with respect to home.

The second observation comes from an experiment in which Dyer (1984; rev. in Gould 1984b) trained foragers along the shore of a lake and redirected their dances to a spot in the lake (Fig. 6). Recruits did not turn up at a boat with a feeder located there, though when the redirection was set to zero, they turned up at the shore feeder in large numbers. Thinking a site odor might be involved, Dyer trained foragers to the boat itself, but still no recruits arrived. When he and the boat progressed to the far shore, however, recruits began ito arrive in substantial numbers. One (though by no means the only) interpretation is that the coordinates specified in the dance allowed recruits to consult a mental locale map or the appropriate «snapshot» in their collection, and see that the spot indicated was unsuitable. This flexibility, if it is real, may mean that the storage and organization of landmark memory is more fully integrated into the dance system than we had thought.

3 Dance divergence

Von Frisch was the first to notice that sequential waggling runs in a dance frequently alternate from left to right in direction, creating what he has called a divergence in the dance. In a divergent dance the bisector of the angle created by the two waggling directions points toward the food (Fig. 7, inset). The simplest explanation for the divergence is that it arises in some way as a result of the dancer's difficulty in determining the exact direction in which it should waggle. The divergence is greatest when the food source being indicated is nearby – that is, when the forager has only a brief flight during which to gather directional information (Fig. 7b) – and further , it is unusually large when the foragers are forced to fly a detour in reaching the food or when they are forced to dance on inclined rather than on vertical surfaces, weakening the gravity cues to which they orient (rev. in v. Frisch 1967).

Some recent evidence casts doubt on this simple explanation, however. One experiment, carried out by Towne (1984), was designed to test the hypothesis proposed by Edrich (1975) that under normal conditions the divergence depends upon the duration of the outward flight to the food. Using a manipulation invented by Heran (see von Frisch 1967, p. 115), Towne was able to hold the duration of the outward flight constant while

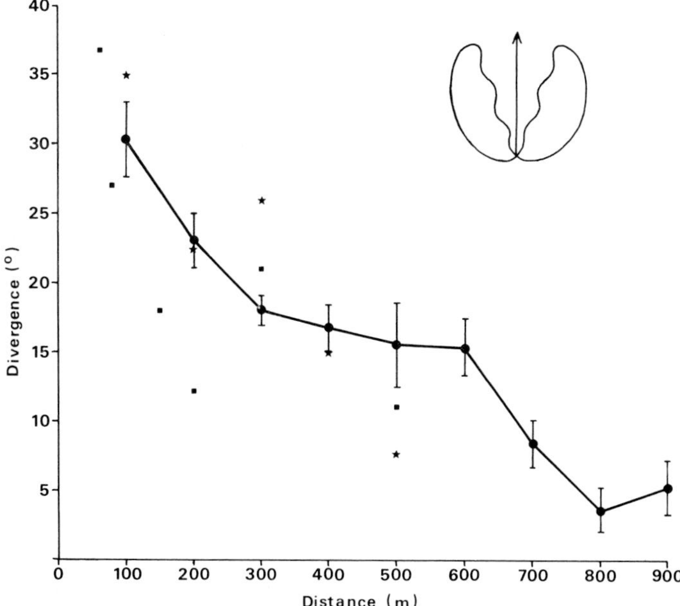

Fig. 7: Dance divergence. The inset shows the actual course followed by a dancer doing a divergent dance. The arrow indicates the true direction of the food (after VON FRISCH (1967)). The plot shows the relationship between the divergence and the distance to the food, the divergence being defined as the mean angle between consecutive waggling runs. Bars indicate standard error of the mean. N = 207 dances. Stars are data of VON FRISCH; filled squares are from EDRICH (1975). From TOWNE (48).

varying the bees' estimates of the distance flown as reflected in the durations of their waggling runs. (This confirms HERAN's original findings and supports VON FRISCH's conclusion that the bees do not use flight time to measure distance.) In addition TOWNE found that even though flight time was held constant throughout, the divergence decreased as the durations of the bees' waggling runs increased. Thus the decrease in divergence routinely observed in association with an increase in the distance to the food is most likely not a result of the increased flight time.

This result is further supported by some recent observations by TOWNE (1984a) on the three Asian honey bees. Although all four species of bees fly at roughly the same speed, they show strikingly different divergences for a given distance to the food. *A. dorsata* and *A. florea* show no noticeable divergence when the food source is 100 m from the nest, and *A. cerana* shows a divergence of only about 15° – one–half that observed in *A. mellifera.*

These observations taken together suggest that the divergence – at least under good navigating conditions (good dancing cues and no flight detours) – should not be viewed as unavoidable noise arising out of the bees' inability to learn, recall, and utilize their navigational cues with precision. Why, then, is it there? As it turns out, HALDANE and SPURWAY (1954) had suggested a possible function for the divergence almost 30 years ago. Their argument was simple: assuming that the divergence leads to imprecision in the communication and to the scattering of recruit bees around the particular flowers being visited by the dancer, it may nonetheless be adaptive if recruits that miss the mark are likely to encounter other open flowers of the species indicated. They hint further that there may be an optimal amount of scatter, the magnitude of which would presumably depend upon the spatial distribution of open flowers of the species being visited. (Since, however, the divergence is the same whether the goal is a single small feeder, a nest site, or a large patch of flowers (TOWNE 1984 a), this optimal amount of scatter would have to be a grand average which takes into account all of the resources typically used by a colony).

GOULD (1975 a) has extended this line of reasoning by pointing out that if there is some average optimal amount of scatter, there is no reason to think it should change with the distance of the food source from the nest. These arguments lead to testable predictions. First, the divergence should indeed lead to a scatter of recruit bees around the food source being visited by the dancer. More specifically, the angular error of recruits should be greatest when the food source is nearby – that is, when the divergence is greatest. Second, as suggested by GOULD, the error should be «tuned» so as to keep the absolute area over which recruit bees search roughly constant with changing distance to the food source. Finally, if the scatter of recruits has evolved in response to the type and distribution of flowers exploited by the bees, then bees which use different types of resources, say temperate vs. tropical, should show predictable differences in their divergences (GOULD 1975 c).

The third prediction is clearly upheld by TOWNE's (1984 a) observations of the tropical bees mentioned earlier. All three species – at least in Puna, India where the observations were made – show little or no divergence compared to *A. mellifera carnica,* a temperate zone bee. This makes sense since the tropical bees most commonly frequent large, insect-pollinated trees which constitute massive and highly localized resources as compared to the more diffuse resources typically visited by *A. mellifera* (although exactly what we should consider to be a typical *A. mellifera* resource is, of course, problematic).

The first prediction, that the angular error of recruits should be greater for nearby food sources, is also upheld by the evidence. Both VON FRISCH (rev. in VON FRISCH 1967) and GOULD (1976), using arc-shaped arrays of recruit stations, have performed experiments to determine recruit directional accuracy, and both found greater recruit errors for more nearby food sources. Because, however, the experiments of each involved only two distances and a relatively small number of recruit stations – VON FRISCH's widely spaced, and GOULD's closely spaced so that the entire array was relatively narrow – TOWNE (1984 a) has recently extended these observations. Capturing recruits at a wide (300 m across) array of 26 traps baited with scent, he found that 50 % of recruits arrive within 16° of the correct direction when the food source being visited is 100 m from the hive. This error decreases to about 8° when the food source is 300 m distant and to about 6° when the food source is between 500 and 700 m from the hive. The second and most quantitative prediction, that the area searched by recruits should remain constant with changing distance to the food, is supported only roughly by these data, however. The angular errors just mentioned translate into absolute median errors of 30 m for food sources at a distance of 100 m from the hive, 40 m for food sources at 300 m, 52 m for food sources at 500 m, and 73 m for food sources 700 m distant. Thus there is a slow but steady increase in the area searched with increasing distance to the food.

Taken together we feel that these observations support the hypothesis that the divergence has evolved to increase the efficiency with which a colony collects its food. The reason for the slow increase in the area searched with increased distance to the food is not clear, but the increase is certainly much less dramatic than it would be if the angular error were to remain constant.

4 Dance dialects

BOCH (1957) demonstrated that most races of honey bees have different dialects, and VON FRISCH (1967) suggested that this dialect variation probably reflected racial differences in foraging range. Colonies living in regions where foraging most often takes place far from the hive would be better served by a longer-distance dialect (that is, one with fewer waggles per km) than those whose general habitat permits a more restricted flight range. GOULD (1982) demonstrated the likely correctness of VON FRISCH's guess by taking advantage of an observation by LINDAUER (1955). LINDAUER created an artificial swarm of German honey bees *(A. m. carnica)* and permitted it to choose between a nest box at 30 m and another at 250 m. The swarm flew to the more distant box. This makes sense because the swarm and the parent colony are relatively closely related, so for a swarm to move far enough to reduce the overlap of its own average foraging range with that of the parent hive would be adaptive. Indeed, LINDAUER (1955) found that amoung 22 natural swarms of German bees, none moved less than 300 m; while SEELEY and MORSE (1978) observed only one of 13 swarms to fly less than 300 m.

Considering the costs and benefits of swarming, then, it seems likely that moving too short a distance risks competing with relatives for resources, while moving too far has other risks, particularly in terms of the lower certainty that the abundance of resources which have served to sustain the parent hive exist in regions less familiar to the bees. If the dialects reflect the relative foraging ranges of the various races, we would expect that bees should have a preferred swarming distance which should be greater for the long-dialect German bees and shorter for the medium-dialect Italian race *(A. m ligustica)*.

GOULD (1982) tested this prediction by creating artificial swarms along forest edges and setting out a array of nest boxes at roughly 100 m intervals. German swarms invariably flew further than Italian swarms (690 ± 110 m, n = 7, vs 155 ± 50 m, n = 9) and, when offered a choice, always chose a 60 l hive box, while Italian swarms chose a 30 l box on five of the six occasions they were offered a choice. This led GOULD to suggest that colony size might be a relevant variable. The more bees in a hive, the farther a forager will have to fly on average to find an unexploited source. Noting the correlation between latitude and dialect – more northerly races tend to have longer dialects than more southerly ones – GOULD speculated that perhaps climate was the crucial parameter affecting dialect. Longer, colder winters probably require larger winter stores and winter clusters.

5 Horizontal dances

VON FRISCH discovered almost four decades ago that *Apis mellifera,* though it normally dances on the vertical sheets of comb in the darkness of the hive, will dance on a horizontal surface and orient to celestial cues when necessary (VON FRISCH 1967). Later LINDAUER (1956) reported that the dwarf honey bee *(A. florea)* always dances on the top of its exposed combs, while he was unable to persuade the giant honey bee *(A. dorsata)* to do the same. Although *dorsata* also lives on exposed combs, the comb is not built up and over tree limbs in the manner of *florea,* and so there is normally no opportunity to dance on a horizontal surface.

The dances of the three species of tropical bees are interesting because *Apis* almost certainly evolved there. One of the most plausible evolutionary sequences (rev. in GOULD 1982a) sees the movement into cavities as a response to predator pressure. As argued so cogently by SEELEY et al. (1982), predation has certainly been an important force in the evolution of tropical bee societies, and the move into cavities brought with it a two-fold advantage related to predation. First, the cavity walls of *A. cerana* nests provide excellent protection against both large and small predators. Second, as the walls protect the colony, they also emancipate a large fraction of the worker force from the task of hanging in the three- to six-bee thick protective curtain which can always be seen covering colonies of both *A. florea* and *A. dorsata* (SEELEY et al 1982). Regardless of the exact details of predator avoidance, the move into cavities necessitated changes in the dance; the communication now had to take place in the dark and in the absence of celestial orientation cues. We suspect that an understanding of how the dance orientation strategies of the four species compare will help us to understand how dancing, orientation, and nesting have evolved. We will consider horizontal dancing here, and the sounds and postures in the dance in a later section.

5.1 Dances of the dwarf honey bee

KOENIGER et al (1982) have recently showed that *A. florea* dancers orient their dances to the sun when it is available, and, in the absence of the sun, are reoriented by a polarized filter interposed between the dance area and the sky. They also reported that the dances remained oriented in the absence of celestial cues, suggesting yet another orientation system. This would be adaptive for a species which always dances on horizontal surfaces, and otherwise must depend on a view of celestial cues through forest canopy and, frequently, clouds.

DYER (1984) has carried this work further and shown that *A. florea* dancers rely heavily upon landmarks near the colony and visible from the dance floor as reference points for the dance. DYER positioned a striped card on one side of the dance floor, for instance, and in the absence of celestial cues was able to rotate the dances by moving the card. This stands in sharp contrast to *A. mellifera* which appears unable to use any non-celestial cue in its horizontal dances.

A. florea might have yet a further backup system. When DYER removed the striped card, leaving a view of only the circular arena around the dance floor and the white ceiling of the hut sheltering the colony, the dancers still showed a net orientation in one direction, though scatter was considerable and the dance direction did not always correspond to the direction of the food. This orientation was unaffected by rotation of the comb; thus comb topography (and therefore gravity) could not have served as the reference. Nor could the dances be changed by rotating the arena or by observing from a different direction. (DYER watched the bees through narrow slits cut in the arena.) The only manipulation that produced an effect was to cast a shadow on the ceiling above the dancers. This resulted in a random distribution of dances, and implies that the bees could orient relative to subtle gradients of reflected light that could not be totally eliminated. Another possibility that has not been excluded is that the dancers can refer to the earth's magnetic field as a last resort.

DYER has also been able to induce *A. florea* to dance on the vertical side of the colony by denying access to the horizontal and giving the bees on the vertical a clear view of the sky. In this situation the bees continue to try to point their dances directly toward the food using celestial cues, though this means that when the plane of the dancing surface is rotated they must adopt a different angle relative to gravity. Dances do not shift as the sun moves across the sky, supporting LINDAUER's (1956) proposal that *A. florea* lacks entirely the ability – shared by the other species – to transpose the horizontal angle into a vertical one in which gravity is taken to refer to the sun. Interestingly, dances on the vertical are the same whether or not the bees can see blue sky or the sun, but it is likely that landmarks visible from the colony could provide the reference points for these dances, as is true on a horizontal surface.

5.2 Dance of the giant honey bee

Dyer (1984) has been able to induce *A. dorsata* foragers to dance on horizontal surfaces by preventing them from seeing any blue sky on the sides of the comb where they normally dance. Since his colony had been cut from its original site and suspended by a bamboo clip at ground level, the top of the colony was a large horizontal platform which the bees readily turned into a dance floor when it became the only part of the colony exposed to the sky. (Why *dorsata* should prefer a view of the sky for its normally vertical dances (Lindauer 1956) is another question.) The horizontal dances were accurately oriented and could be rotated by a polarizer, confirming that *A. dorsata*, like the other three species, can use the polarization of blue sky for orientation. The horizontal dances are also interesting because there are no known natural circumstances under which horizontal dancing would be used by *A dorsata*. This behavior may, therefore, be an evolutionary holdover.

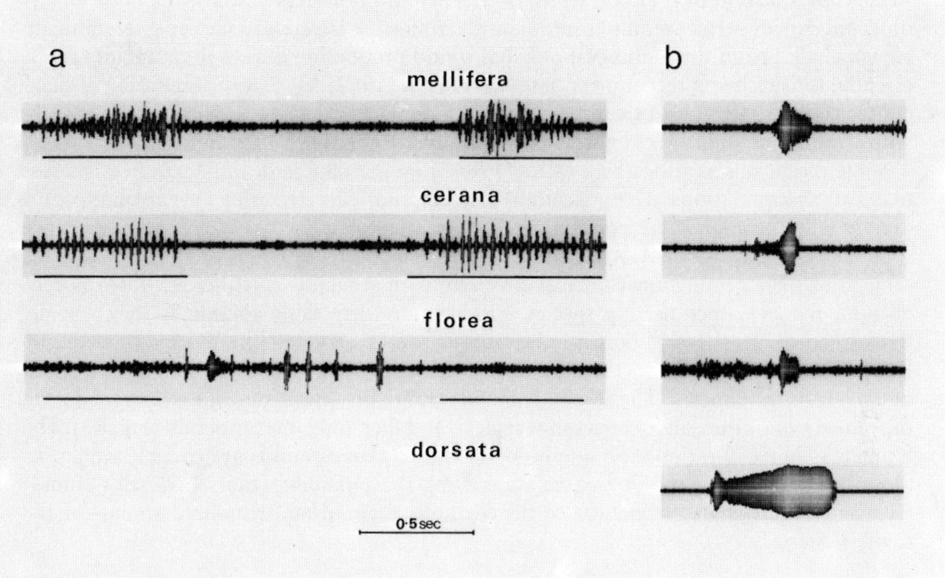

Fig. 8: Oscilloscope traces of the dance sounds (a) and «stop signals» (b) of all four species of *Apis*. In a), two waggling runs (indicated by lines under tracings) separated by a single circling return run are shown for the three sound-producing species. In the *A. mellifera* and *A. cerana* tracings, individual sound bursts produced within each waggling run can be clearly seen, while the individual sound cycles are just barely resolvable. The fundamental frequency and burst rate of the *A. mellifera* and *A. cerana* sounds are roughly the same for both species at 290 Hz and 30 bursts/sec, respectively. The fundamental frequency of the *A. florea* sound is about 250 Hz. In b), a single «stop signal» is shown for each species. The frequencies of the signals seem to vary even within species, but they are about 300 Hz in *A. mellifera,* 445 Hz in *A. cerana,* 475 Hz in *A. florea,* and 375 Hz in *A. dorsata.* The microphone was always held between about 5 and 10 mm from the bees during recording but since the amplification was varied in the face of varying levels of background noise, the relative amplitudes of the sounds as shown here should not be considered accurate. From Towne (1984b).

6 How the dance information is transmitted to recruits

Wenner (1959, 1962) and Esch (1961) independently discovered that *A. mellifera* dancers produce faint bursts of sound during the waggling run portion of their dances (see Fig. 8 a). Both have suggested that the sound carries the distance information used by recruits, but since the duration of sound production and the duration of waggling – both of which correlate extremely well with distance to the food (52, rev. in 15) – are so tightly intertwined, it has not been possible to separate them experimentally. In closing his review of the several correlates of distance in the dance, von Frisch (1967, p. 104) wrote, «... we regard the acoustically emphasized duration of waggling as the index of distance.» Although von Frisch is almost certainly correct, we still do not know the relative importance of the two cues. This applies to the transfer of directional as well as distance information.

6.1 The dance sounds

Esch (1963, 1964) observed that neither silent dances, which frequently occur when the sugar concentration of the food is low and dancing is not vigorous, nor the occasionally observed dances with continuous sound production were effective in inducing recruitment. He concluded from these observations that sound production during the waggling run is essential for successful recruitment, but von Frisch (1967) felt that the inefficacy of silent dances could be a result of their overall lack of vigor, and that the observations of dances with continuous sound were not extensive enough to permit a firm conclusion.

Some recent observations by Towne (1984b) on the dances of the Asian bees suggest that Esch's conclusions were essentially correct and, further, offer an explanation for exactly why the sounds are essential. *A. cerana,* the Asian cavity-nester, produces dance sounds just like those of *A. mellifera* (Fig. 8 a). The basic structure of the sounds is the same, and their production coincides precisely with waggling. Neither *A. florea* nor *A. dorsata,* the two open-nesting species, produce any detectable sounds as they waggle, though dance attenders of both species produce «stop signals» like those discovered by Esch (1964) in *A. mellifera* (Fig. 8b). Sounds are produced by occasional (about one out of ten) *A. florea* dancers. These sounds, however, are produced almost exclusively during the *return run* of the dance (between waggles), and then only intermittently (Fig. 8 a). The frequency, burst duration, and amplitude of the *A. florea* sounds are roughly similar to those of *A. mellifera* and *A. cerana,* suggesting the possibility that *A. florea's* sounds represent evolutionary precursors of the highly structured and ritualized sounds of the cavity nesters.

6.2 Dance postures

The association between the production of sound during the information-rich waggling run and the habit of living in cavities strongly supports the hypothesis – hinted at by both Esch (1963, 1964) and Wenner (1962) – that the sounds have evolved to aid dance attenders in their efforts to extact the dance information while following dances in the dark. A second aspect of this hypothesis is the notion that both *A. florea* and *A. dorsata* dance attenders use vision in extracting the information from their dances. Another set of observations by Towne (1984b) is consistent with this idea.

As *A. mellifera* and *A. cerana* dancers waggle, their abdomens are depressed just slightly, and the waggling motion is strictly in the horizontal plane of the dancer's body (Fig. 9, top). The waggling postures of *A. florea* and *A. dorsata,* on the other hand, are quite different. In both species the abdomen is elevated and curved upward, causing it to protrude slightly above the mass of bees milling about the dance floor as the dancer waggles (Fig. 9a). Immediately upon completion of a waggling run, dancers of both species quickly drop their abdomens to their normal resting postures. In addition, as observed by LINDAUER (1956, 1957) in *A. florea,* the abdomens of the dancers as they waggle show a dorso-ventral motion superimposed upon the normal side-to-side waggling of the rest of the body. Although LINDAUER describes the dorso-ventral excursions of the abdomens as «occasional,» TOWNE (1984b) found with the aid of video recordings that the dorso-ventral motion in both species is regular, with a frequency of several cycles per second. (The low resolution of TOWNE's video recordings does not allow precise measurements.)

The dance postures and waggling motions of both open-nesting species thus seem designed to help dance attenders to follow the dances visually. If this is true, it could explain LINDAUER's (1956) observation, confirmed by DYER (1984), that *A. dorsata* dancers prefer to dance on the most well-lit side of the comb. One additional observation made by TOWNE (1984a), is that *A. dorsata* dance attenders, unlike those of the other three species, almost never turn with the dancer as she circles around in her figure-eight pattern on the dance floor. GOULD (1975a) has described the circling behavior of *A. mellifera* dance attenders in detail. (*A. florea* and *A. cerana* attenders show basically the same pattern.) Groups of *A. dorsata* attenders instead usually form a more-or-less stationary semi-circle behind the dancer and «watch» as she performs her dance before them. In addition, although the attenders do step forward when the dancer starts to waggle, they

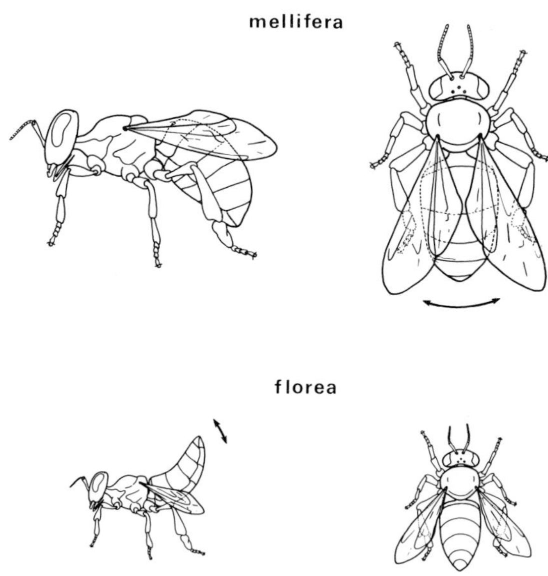

Fig. 9: The dance postures and waggling motions of *A. mellifera,* a cavity nester (top), and *A. florea,* an open nester (bottom). The drawings are to scale. The posture and motions of *A. cerana,* the Asian cavity nester, are the same as those of *A. mellifera.* Those of *A. dorsata,* the large Asian open nester, are similar to those of *A. florea* except that *A. dorsata*'s abdomen is relatively longer, and the amplitude of its dorso-ventral motions relatively greater. The upward curvatures of the abdomens of *A. florea* and *A. dorsata* are undoubtedly facilitated by their morphologies, which show a dorso-ventral flattening compared to those of the cavity nesters. From TOWNE (1984b).

rarely get close enough to make actual antennal contact with the dancer. This, too, is unlike the other species which routinely make antennal contact. Perhaps *A. dorsata* attenders are simply too big to wheel about after the dancer as the other species do.

In summary, there seem to be two different strategies for enhancing the ability of recruits to obtain the information from the waggling run. Bursts of sound are produced by the cavity nesting bees which normally dance in the dark. The open nesting species, which dance in the presence of the ambient light, instead employ exaggerated body postures which may facilitate the visual perception of waggling. The widely discussed notion that the duration of sound production is the indicator of distance in *A. mellifera* might still be correct, but we suspect that VON FRISCH's (1967) more general interpretation of the sounds as «emphasis» upon the waggling run is closer to the truth.

7 The evolution of the dance language

Much of the progress we have just dicussed can best be summarized in the form of a brief, updated review of our ideas on the evolution of the dance. Previous reviews of this sort can be found in LINDAUER (1961), VON FRISCH (1967), ESCH (1967), WILSON (1971), and MICHENER (1974).

7.1 Stingless bees

As pointed out by LINDAUER (1956) and ESCH et al. (1965) among others, the simplest and probably most primitive form of recruitment communication in the Apidae is the agitated jostling run, often accompanied by intermittent sound production, of several species of stingless bees (tribe Meliponini), such as *Trigona iridipennis* in Sri Lanka (LINDAUER 1956) and *T. silvestrii* in Brazil (KERR, ESCH 1965). In such species there is no communication regarding the location of the food source; their «dances» are merely invitations to go out searching for food of the same scent as that clinging to the «dancer's» body (31).

From this basic strategy at least two others have apparently arisen. The first, typified by *T. ruficrus* in Brazil (LINDAUER, KERR 1960), involves agitated running in the nest accompanied by the laying down of a pheromone trail by the foragers. Sometimes, perhaps always, recruits are led along the trail with the help of guiding flights by foragers (LINDAUER, KERR 1960).

The third and seemingly most advanced communication strategy among the stingless bees has been described by ESCH et al. (1965) for two species of *Melipona,* morphologically the most *Apis*-like of the Meliponines (MICHENER 1974). Dancers of these species, instead of producing intermittent and irregular sound bursts as they run, produce bursts that are evenly spaced and whose durations correlate well with the distance to the food. In addition, they seem to use only abbreviated guiding flights and no odor trails. ESCH et al. suggest that the sound burst duration symbolically communicates distance information to recruits, but, unfortunately, their observations are not conclusive on this point.

Both VON FRISCH (1967) and MICHENER (1974) give excellent and more detailed reviews of stingless bee communication.

7.2 From *Melipona* to *Apis*

ESCH et al. (1965) have argued that symbolic communication in *Apis* has arisen from the simpler form they observed in *Melipona,* but even if communication in *Melipona* is symbolic, the two languages are probably not homologous. First, the «dance» sounds of stingless bees are substrate borne (1960). The *Apis* sounds on the other hand, are probably airborne (GOULD 1975 a). Second, the lack of regular sound production in *A. florea* and *A. dorsata* argues against the homology of the sounds of the two groups of bees. In any case, the homology between communication in stingless bees and honey bees is not well established, suggesting the interesting possibility that symbolic communication has arisen twice in the Apidae.

7.3 Changes within Apis

No one has yet discovered any forms of communication intermediate between the «dances» of *Melipona* and those of *Apis,* but it seems probable that the *Apis* dance first appeared in a bee with habits like *A. florea* (LINDAUER 1956, v. FRISCH 1967). Dances were on a horizontal surface, and agitated movements gave direction with reference to the sun or blue sky and distance in the duration of waggling. Although the use of polarized skylight occurs in many insects, the rules used to eliminate ambiguity in using these cues may have arisen first at an early stage in *Apis* because of the requirement that the bees dance on a platform where large parts of the sky might be blocked from view by clouds or vegetation. As far as we know, all that would need to be added to the behavioral repertoire thus assembled to produce the dance/orientation system of *A. florea* is the secondary backup system that resorts to landmarks when celestial cues are blocked entirely. Landmarks are important to bees during the flight (VON FRISCH, LINDAUER 1954, DYER, GOULD 1981) – and probably to most mobile animals – so the early *A. florea* might have been preadapted to incorporate landmarks into the dance.

The next step, which brings us to *A. dorsata*'s strategy, most likely involved the move from horizontal to vertical dance surfaces and the transition from celestial to gravity orientation (LINDAUER 1956). We can speculate that there was an intermediate step with vertical dances oriented to celestial cues (*A. mellifera* will still use celestial cues when dancing on the vertical if the cues are strong enough [VON FRISCH 1967]), but we cannot be certain. We can also make some reasonable guesses as to why the first vertical dances were necessary. The horizontal dance surfaces of *A. florea* are extensions of the comb which wrap around the relatively thin branches on which the colonies hang. Most likely as a result of increasing colony size, the need to avoid mammalian predators (SEELEY et al 1982), or both, the bees probably began to hang their single sheets of comb from the undersides of large branches or rock outcrops, the favored nest sites of *A. dorsata* today (LINDAUER 1956, SEELEY et al 1982). Since the comb could no longer wrap around the substrate, the only surfaces available to the dancing bees were vertical ones, and so dancing on the vertical became essential. Although *A. dorsata, A. cerana,* and *A. mellifera* now seem rarely (except in experiments) to have any use for their ability to dance on the horizontal – even *A. mellifera*'s nest-site dances on the swarm (LINDAUER 1955) normally occur on the more-or-less vertical sides of the hanging ball of bees – all three species have retained it (DYER 1984).

Regardless of exactly how it arose, gravity-oriented dancing brought with it the need for a new backup orientation system; dancers had only gravity for their orientation and so needed to know the position of the sun relative to the direction of their outward flights in order to communicate. This was probably made possible in a *dorsata*-like ancestor by the evolution of the ability to memorize the sun's course with respect to landmarks, creating a robust backup system that would allow normal activity under almost any weather conditions, or even at night.

Something else that dancing on the vertical brought with it was a preadaptation for the move into enclosed cavities. This move, most likely compelled by predator pressure

(GOULD 1982a, SEELEY et al 1982) eliminated the opportunity to use vision in the communication, thereby necessitating a change in the strategy used to transmit the dance information. The ancestors of *A. mellifera* and *A. cerana* seem to have responded with the introduction of sounds into the dance, the loss of their exaggerated dance postures, and the placement of a new emphasis on the side-to-side waggling motion (TOWNE 1984b). All the while the distance dialects probably drifted in accordance with changes in the colony sizes and foraging ranges of the bees (GOULD 1982b), and the cloudy day back-up system, based on a memory of the sun's course with respect to landmarks (DYER, GOULD 1981), remained essentially unchanged.

The cavity-dwelling habit of *A. mellifera* and *A. cerana* was, in its turn, probably a preadaptation for the movement of these bees out of the tropics and into the temperate regions (of Europe and West Asia for *A. mellifera,* and East Asia for *A. cerana*). The former, because of the relatively broad and diffuse resources available in the temperate zones, and particularly intense selection for efficiency in exploiting them because of the necessity of storing sufficient honey to survive the brutal winters (SEELEY 1983), may have led to selection for a certain amount of imprecision in the dance language, ultimately giving rise to the divergence in the dances of contemporary *A. mellifera.*

It remains to be seen what other adaptations may contribute to the efficiency and flexibility of this fascinating system. At least we can say with some certainly that VON FRISCH's magic well is still a long way from going dry.

References

Boch, R. (1957): Rassenmäßige unterschiede bei den Tanzen der Honigbiene. Z. vergl. Physiol. *40,* 289–320.

Brines, M. L. (1978): Skylight polarization patterns as cues for honey bee orientation: physical measurements and behavioral experiments. Ph. D. Thesis. New York, Rockefeller University.

Brines, M. L., Gould, J. L. (1979): Bees have rules. Science *206,* 571–573.

Brines, M. L., Gould, J. L. (1982): Skylight polarization patterns and animal orientation. J. Exp. Biol. *96,* 69–91.

Dyer, F. C. (1984): Ph. D. Thesis. Princeton University.

Dyer, F. C., Gould, J. L. (1981): Honey bee orientation: a backup system for cloudy days. Science *214:* 1041–1042.

Dyer, F. C., Gould, J. L. (1983): Honey bee navigation. Amer. Sci. *71,* 587–597.

Edrich, W. (1975): The waggle dance of the honey bee; a new formulation. Fortschr. Zool. *23,* (1), 20–30.

Esch, H. (1961): Ein neuer Bewegungstyp im Schwanzeltanz der Bienen. Naturwiss. *48,* 140–141.

Esch, H. (1961 a): Über die Schallerzeugung beim Werbetanz der Honigbiene. Z. vergl. Physiol. *45,* 1–11.

Esch, H. (1963): Über die Auswirkung der Futterplatzqualität auf die Schallerzeugung im Werbetanz der Honigbiene. Verhandl. Deut. Zool. Ges. *26,* 302–309.

Esch, H. (1964): Beiträge zum Problem der Entfernungsweisung in den Schwanzeltänzen der Honigbiene. Z. vergl. Physiol. *44,* 534–546.

Esch, H. (1967): The evolution of bee language. Sci. Am. *216,* 96–104.

Esch, H., Esch, I., Kerr, W. E. (1965): Sound: an element common to communication of stingless bees and to dances of honey bees. Science *149,* 320–321.

Frisch, K. v. (1967): The Dance Language and Orientation of Bees. Cambridge, Mass., Harvard Univ. Press.

Frisch, K. v., Jander, R. (1957): Über den Schwanzeltanz der Bienen. Z. vergl. Physiol. *40,* 239–263.

Frisch, K. v., Lindauer, M. (1954): Himmel und Erde in Konkurrenz bei der Orientierung der Bienen. Naturwiss. *41,* 245–253.

Gould, J. L. (1974): Honey bee communication. Nature *252,* 300–301.

Gould, J. L. (1975 a): Honey bee communication: the dance-language controversy. Ph. D. Thesis. New York, Rockefeller University.
Gould, J. L. (1975 b): Communication of distance information by honey bees. J. Comp. Physiol. *104,* 161–173.
Gould, J. L. (1975 c): Honey bee recruitment: the dance-language controversy. Science *189,* 685–692.
Gould, J. L. (1976): The dance-language controversy. Quart. Rev. Biol. *51,* 211–244.
Gould, J. L. (1980): Sun compensation by bees. Science *207,* 545–547.
Gould, J. L. (1982 a): Ethology: The Mechanisms and Evolution of Behavior. New York, Nortons.
Gould, J. L. (1982 b) Why do honey bees have dialects? Behav. Ecol. Sociobiol. *10,* 53–56.
Gould, J. L. (1984 a): Processing of sun-azimuth information by honey bees. Anim. Behav. *32,* 149–152.
Gould, J. L. (1984 b): The natural history of honey bee learning. *In* The Biology of Learning, ed. P. Marler and H. Terrace. Berlin, Springer-Verlag (in press).
Gould, J. L., Henery, M., MacLeod. M. C. (1970): Communication of direction by the honey bee. Science *169,* 544–554.
Haldane, J. B. S., Spurway, H. (1954): A statistical analysis of communication in *Apis mellifera* and a comparison with communication in other animals. Insectes Sociaux *1,* 247–283.
Hölldobler, B. (1980): Canopy orientation: a new kind of orientation in ants. Science *210,* 86–88.
Kerr, W. E., Esch, H. (1965): Communicação entre as abelhas sociais brasileiras e sua contrabuicau para o entendimento da sua evolução. Ciencia e Cult. (Sao Paulo) *17,* 529–538.
Koeniger, N., Koeniger, G., Punchihewa, R. K. W., Fabritius, Mo., Fabritus, Mi. (1982): Observations and experiments on dance communication in *A. florea.* J. Apicult, Res. *21,* 45–52.
Lindauer, M. 1955): Schwarmbienen auf Wohnungssuche. Z. vergl. Physiol. *37,* 263–324.
Lindauer, M. (1956): Über die Verständigung bei indischen Bienen. Z. vergl. Physiol. *38,* 521–577.
Lindauer, M. (1957): Communication among the honeybees and stingless bees of India. Bee World *38,* 3–14, 34–39.
Lindauer, M. (1959): Angeborne und erlernte Komponenten in der Sonnenorientierung der Bienen. Z. vergl. Physiol. *42,* 43–62.
Lindauer, M. (1961): Communication among Social Bees. Cambridge, Mass., Harvard Univ. Press.
Lindauer, M. (1963): Kompaßorientierung. Ergebn. Biol. *26,* 158–181.
Lindauer, M. (1971): The functional significance of the honey bee waggle dance. Amer. Nat. *105,* 89–96.
Lindauer, M., Kerr, W. E. (1960): Communication between the workers of stingless bees. Bee World *41,* 29–41, 65–71.
Michener, M. C. (1974): The Social Behavior of the Bees. Cambridge, Mass., Harvard Univ. Press.
Rossel, S. R., Wehner, R. (1982): The bees' map of the e-vector pattern in the sky. P. N. A. S. *79,* 4451–4455.
Rossel, S. R., Wehner, R., Lindauer, M. (1978): E-vector orientation in bees, J. Comp. Physiol. *125,* 1–12.
Schricker, B. (1974): Der Einfluß subletaler Dosen von Parathion (E 605) auf die Entfernungsweisung bei der Honigbiene. Apidologie *5,* 149–175.
Seeley, T. D. (1983): The ecology of temperate and tropical honey bee societies. Am. Sci. *71,* (3), 264–272.
Seeley, T. D., Morse, R. A. (1978): Nest site selection by the honey bee. Insectes Sociaux *25,* 323–337.
Seeley, T. D., Seeley, R. H., Akratanakul, P. (1982): Colony defense strategies of the honeybees in Thailand. Ecol. Monogr. *52,* 43–63.
Towne, W. F. (1984 a): Ph. D. Thesis. Princeton University (in preparation).
Towne, W. F. (1984 b): Acoustic and visual cues in the dances of four honey bee species. Behav. Ecol. Sociobiol. (in press).
Wehner, R. (1981): Spatial vision in Arthropods. *In* Handbook of Sensory Physiology, ed. H. Autrum, pp. 287–616. Berlin, Springer-Verlag.
Wenner, A. M. (1959): The relationship of sound production during the waggle dance of honey bees to the distance to the food source (an abstract). Bull. Entomol. Soc. Am. *5,* 142.
Wenner, A. M. (1962): Sound production during the waggle dance of honey bees. Anim. Behav. *10,* 79–95.
Wilson, E. O. (1971): The Insect Societies. Cambridge, Mass., Harvard Univ. Press.

Fortschritte der Zoologie, Bd. 31 · Hölldobler/Lindauer (Hrsg.): Experimental Behavioral Ecology
G. Fischer Verlag · Stuttgart · New York · 1985

Manipulation, modulation, information, cognition: some of the riddles of communication

HUBERT MARKL

Fakultät für Biologie, Universität Konstanz – F.R.Germany

Abstract

Whether communicating animals exchange information for mutual benefit, or whether communication is actually a process of receiver exploitation by the manipulative powers of a sender's signals without any reliable information transfer, has recently been controversially discussed. This contribution argues that exploitative, unbalanced communicatory relationships between conspecifics should be transitory or only characteristic for special cases, while communication within social groups should be as a rule cooperative and mutually informative. In contrast, intrusion of third participants (e. g. mimics) into tuned sender-receiver relationships can more easily stay unilaterally exploitative (Tables 1,2). There are remarkable differences between signals used in manipulative relationships and those found in mutualistic communication. While the first often have the «classical» conspicuous releaser charcteristics, the latter are typically found to be unobtrusive, low-range, highly addressed, and exclusive. They frequently modulate receiver reactivity with respect to other contextual stimuli rather than triggering responses in all-or-none fashion. Ways of quantifying communication efficiency, and the meaning of different statistical information measures in a bidirectional communication loop (Fig. 1) are discussed. An analysis is made of how the ability for internal communication, which vastly expands the cognitive and therefore also behavioral powers of an organism, might derive from role-switching bidirectional social communication. Finally it is pointed out that mutualistic communication welds members of social groups into cooperative supersystems which develop all the properties of problemsolving, higher-order cognitive entities. In that respect, social communication processes can be profitably compared with functions of the central nervous system of individual organisms.

1 Introduction

It seems appropriate in a symposium dedicated to the memory of KARL VON FRISCH – a pioneer of the scientific investigation of communication in animals – to ask what animal communication is. Since «communio» means sharing, what is it that they share? What evolutionary and developmental conditions are necessary for them to participate in such a process? And if this process is the result of a history of coadaptive selection – be it through natural selection in the evolutionary history of a genepool or through selective learning by an individual – what are the advantages of this behavioral coadaptation, who receives the benefits payed out in fitness currency? Are they also equitably shared? Or do we also have to consider disadvantages and fitness costs of communicating, and how they are distributed between the communicating partners? Was KARL VON FRISCH right to summarize his astounding discovery of the «dance language» in honeybees by writing that they «können ... ein Ziel, das für *ihr Volk von Wichtigkeit ist,* den Kameraden mitteilen und seine Lage *so genau beschreiben*, daß es von den Stockgenossen in selbständigem Flug, ohne Geleit, auf kürzestem Wege gefunden wird – selbst in kilometerweiter Entfernung» (VON FRISCH 1965, p. 538, italics mine)? Or should we rather believe RICHARD DAWKINS and JOHN KREBS (1978, p. 309) who suggest quite categorically: «If information is shared at all it is likely to be false information, but it is probably better to abandon the concept of information altogether»? In the latter point they are at first glance even in full agreement with KARL VON FRISCH who seems not once to have used the word «information» in his authoritative 578 pages-magna summa of more than 40 years work on dance language and orientation in bees. This should, however, not be misread, because he readily speaks about the «Nachrichten», «Mitteilungen» and «Weisungen» contained in and transmitted by the dances – which is, of course, the very information language that DAWKINS and KREBS advise us against; even when they accept (p. 286) that in the bees' dances «the benefit is, in a sense, mutual», they «would still prefer to avoid «information terminology» and in fact even to abandon the word communication altogether» (p. 283). That this is not only an issue of semantic preferences, a disagreement about words, but relates to the very base of our understanding of animal communication, is abundantly clear from their lucid exposition and from the discussion it has aroused.

In this contribution I will further analyze this problem by first asking how we recognize communication in animals (i. e. how we define it operationally); second, why organisms communicate and who reaps the benefits and bears the costs involved; third, what determines the quality of communication processes (i. e. how we can quantify communication efficiency); finally, what is it that is communicated. I will then end with a somewhat daring comparison of communication between members of an animal society and that between neurons in a central nervous system. In doing this I will not only cast a metaphorical glimpse at the superorganism concept but argue that through the benefits of its communicatory powers a highly developed society is actually a higher level cognitive system with unpredictable (though not irreducible) emergent cognitive properties.

2 Manipulation – or the profit of communication

To communicate, it takes at least two; even if one communicates with oneself one somehow has to split into two roles: that of the *sender* and that of the *receiver* of messages. We don't know whether an animal can really communicate with itself (I will come back to that question later) therefore at least two of them have to participate as sender and receiver to make communication both possible and recognizable in animals. It is important to realize that sender and receiver of signals are temporary *roles* for most practical purposes (I will come to exceptions) and not immutable *characteristics* of individuals. Mostly, communication is bidirectional with interactants switching between these two roles (see MARKO 1983). I think it is important to emphasize this triviality because much of the argument about who is benefitting more or at all from the communicatory interaction, and who is selected to communicate, the sender or the receiver, becomes rather irrelevant in a bidirectional, role-alternating process.

There are scores of definitions of communication (see e.g. BURGHARDT 1970, TEMBROCK 1971, SEBEOK 1977, CHERRY 1978, GREEN and MARLER 1979, LEWIS and GOWER 1980, HALLIDAY and SLATER 1983 for reviews). I am inclined to follow and slightly expand a version used by WILSON (1975), who maintains that *communication occurs when the action of or cue given by one organism is perceived by and thus alters the probability pattern of behavior in another organism in a fashion adaptive to either one or both of the participants.* DAWKINS and KREBS (1978) have argued – and I think one cannot disagree with them in this point – that the sender (whom they call the actor) of a signal is selected to manipulate the behavior of the receiver (their reactor) to the adavantage of the sender. In order to avoid the impression that a sender is a sender is a sender and never anything else it might be preferable to say that a *signal* is a trait that is selected to manipulate an addressee (which, however, may not be the only receiver, see below) to the advantage of the sender of this signal. That manipulation should be for the advantage of the manipulator follows not only straight from natural selection logic, it is also very much at the core of most serious communication theories. It is also the gist of GOFFMAN's influential «Strategic Interaction» (1969), oddly not cited by DAWKINS and KREBS, though it may be even more to their point than PACKARD's Hidden Pursuaders. However, GOFFMAN never lost sight of the fact that manipulation to the benefit of the sender is only half the story of communication. Of course, addressees are not sitting around in extra-evolutionary space offering unmodifyable releasing mechanisms just waiting to be manipulated; in fact it is well known that there is hardly anything that can be more easily modified both by evolution and by individual experience – where we call it focussing of attention – than reaction thresholds and response selectivity of releasing mechanisms or sensory-neural pattern-recognition devices. Except in maladjusted transitory stages of communication, in cases of structural asymmetries of manipulative powers (e.g. between parents and offspring), and in exploitative communication triads (see below) nothing in the world can keep receivers dancing like puppets on the strings of the senders' signals – unless it is to their own advantage, too, to be manipulated: *mundus vult decipi,* successful manipulation begs the question about the sucker's interest in it.

Let me recall that sender and receiver are roles alternatingly taken over by the same individual in most processes of communication: how could we imagine the same individual being always overpowering in one role and a moment later overpowered in the next? Frequent change of sender – receiver roles between the interacting pair, the well-known intertwined communication chain of so many agonistic or sexual displays, could be *the* most important mechanism to avoid unilateral exploitation, just as exemplified by the simultaneously hermaphroditic serranid coral fish *Hypoplectus nigricans,* which takes turns in spawning and fertilizing eggs (FISCHER 1980).

Communicating animals have to be leninist: trust is good, control is better, but that holds for *both parties,* if the degeneration of manipulation into exploitation is to be avoided. If DAWKINS and KREBS dub their view of communication as cynical, I am afraid they have gone only half the way; it's like explaining the evolution of sex from the advantage for the male or, to borrow a metaphor from GREGORY BATESON, like clapping applause with one hand. We should be twice as cynical as they advise us to be: the receiver has to look at the matter just as cynically or rather realistically as the sender in order to stay in business. What is most important, half-cynicism may be unstable whereas full-cynics may open the door to a mutually satisfactory cooperative relationship even though, and even because, both partners continuously make every effort to manipulate each other. The agent behind this has been well known for over 200 years: it's ADAM SMITH's «invisible hand», and what's good for the marketplace may not be too bad for the genepool either: here we get a first glimpse of how communication produces social systems with emergent cognitive qualities: they «discover» that it can be good for others to be good to oneself. The «invisible hand» is, as it were, the hand of providence, and providence certainly is a cognitive capacity.

As AXELROD and HAMILTON (1981; AXELROD 1984) have convincingly shown, in order to arrive at a stable situation of reliable mutual profit, communication partners should preferably not only interact once in their lives but repeatedly, so that common interest can prevail; this becomes particularly effective if the communicators recognize each other individually (VAN RHIJN and VODEGEL 1980). This is typical for many important social interactions, e. g. between parents and offspring or between mates. The condition of common interest is of course no less well fulfilled in genetically closely related social groups, like insect societies, where individuals recognize each other as members of a joint family venture of investment for gaining fitness. Prolonged, addressed, role-switching partnership in communication is thus the gateway to the superior organizing powers and profitability of mutualism in complex organismal systems, be it in a lichen, a beehive, a family, or in Japan Incorporated. Whenever the profit in fitness for one partner is not decreased but rather increased by gains of the other, natural selection should drive the system up the hill of mutual social cooperation, manipulation, cynicism and selfish genes notwithstanding. It is, of course, fully appropriate to emphasize the potential for conflict and unilateral exploitation in all of these relations – as done from TRIVERS (1971, 1972, 1974) to DAWKINS and KREBS (1978, 1979). There can also be no question that such unbalanced communicational relationships are not uncommon even in highly organized social systems which are always in danger of slipping into the trap of losing in the long run from gaining in the short run. To leave it at that and see conflict-solutions with unbalanced distribution of profits as almost paradigmatic for communication in animals, seems to be tantamount to stopping talking about communication where it actually becomes most interesting: namely, where it overcomes the more primitive stage of individuals striving in isolation for fitness, and organizes higher-order social entities with common interests and cooperative action to serve them. There is nothing suspiciously altruistic about ganging up to make a higher profit together instead of going it alone. It is the amazing adaptiveness, the astounding evolutionary success of cooperation between so many organisms that impresses us when we look into Nature. We see that this cooperation depends on sophisticated mechanisms of communication between the cooperating individuals. So it seems strange to denounce emphasis on cooperation in the study of communication as idealistic and obsolete.

The main reason why DAWKINS and KREBS (1978) concentrated on non-cooperative solutions of communication problems is probably that they focussed predominantly on the example of aggressive competition for resources in developing their view. Evidently, that is not exactly the terrain to expect an outcome of mutual gain – much rather one of MAD, mutual assured destruction. As MAYNARD SMITH and others (see MAYNARD SMITH and PARKER 1976, MAYNARD SMITH 1982 a, b, CARYL 1979, 1981, 1982, VAN RHIJN 1980, VAN RHIJN and VODEGEL 1980, HINDE 1981, PARKER and RUBENSTEIN 1981, MOYNIHAN 1982) have discussed, an easy solution to decide such conflicts would be to use an asymmetry between the contenders, preferably one of Resource Holding Power (RHP PARKER 1974): the more powerful wins, the weaker loses, rationalized as «der Klügere gibt nach», the rule of the dumb – literally, of those who cannot communicate – or rather, as MAYNARD SMITH also pointed out, of those who, with brutal candour, project the message of their superior RHP. Although the loser at least avoids being damaged in such an asymmetrical fight, he certainly cannot gain resources by giving in, so to call this mutualism would really be at best a Brezhnev doctrine of cooperation.

Still, even there, second thoughts lead to second doubts: what if a subordinate can still profit somewhat by joining one whom he cannot beat, because battling alone would be even worse? We then arrive at a social system with a dominance hierarchy. If the winner in future contests can win even more with the loser in tow, the stage is set for mutual benefit, and even with intragroup competition for dominance continuing, intergroup competition might tend to favor cooperative solutions rather than disruptive competition, especially if the majority of group members invest in common family bonds. The facilitating importance of individual recognition in stable groups has in this connection been stressed by VAN RHIJN and VODEGEL (1980), VAN RHIJN (1980); see also BARNARD and BURK (1979).

With symmetry of RHP, the outcome is less transparent. First raised by game theorists, the argument goes that although the value of the contended resource and therefore the intention to fight for it may be different for both opponents, faithfully signalling by graded displays and thus accurately informing the opponent about one's motivation would not be a stable strategy since the expression of high interest could be too easily faked at little cost. «Bluff» and «deceit» would follow with the consequence that at the end of such a development all signals of intention would have to be disregarded in deciding a conflict – the symmetrically matched opponents would be back to fighting it out. This is the basic argument for rejecting the idea that any information – much less accurate information – about intentions and activity probabilities might be transmitted in animal communication.

Those who make that point confess, however, to find it a puzzle why in reality we often find gradually escalating pre-fight display in animal contests and why faked signals seem to be relatively uncommon in intraspecific communication. DAWKINS and KREBS (1978) have already suggested that if costs of fights are high – and they should be high in the RHP-symmetry case here considered – an inability to grade threat displays may be too costly for both parties to be stable. It should rather be to the advantage of the sender not to force the receiver to step up to violence immediately. Since in this match each of the opponents is playing the roles of sender and receiver simultaneously or in turn, this advantage accrues for both participants. This may not make them honest informers, but at least should lead them to make cautiously escalating threats.

The real problem with this case may be elsewhere, however, namely in the basic assumption of symmetry. If truly symmetrical matchings are sufficiently rare in reality, we might be looking at these homerian pre-fight displays from the wrong side in the first place. The problem is one of lack of sufficient information: if those who enter into a contest cannot assess at first glance the relative aggressive potential of the adversary, how should they best proceed?

Not enough emphasis may have been put on the fact that resource holding intention and resource holding power cannot be treated as completely independent variables, since intention determines how much of its available power an individual will be prepared to bring to bear in a given conflict; actual resource holding investment is thus a composite of motivation and available potential, as it were, of brains (or rather: guts) and brawns. There is a tradeoff between the two: the highly motivated weakling may prevail over the not sufficiently interested strong guy. Now, when entering into a contest, an individual may well gauge immediately the opponent's RHP and compare it to its own, but how should it gauge what is more decisive: how much of this RHP the adversary will mobilize in order to acquire the contended resource? As PARKER and RUBENSTEIN (1981), CARYL (1981) and others have emphasized, whatever happens will be strongly determined by the cost functions of different tactics and by the payoff-value of the resource: but the rating of both depends on «private» evaluation, the criteria of which are not fully accessible to the opponent. Thus we would expect each contestant to begin to find out at as low cost as possible how much the other one is prepared to invest in order to dominate. The lowest stage might be little more than indicating interest, like laying a claim. From that onwards we would expect a mutual probing of reactions at escalating costs for prober and reactor, before finally an all out fight sets in if no one withdraws from the highest level of probing. «Bluff» and «deceit» would thus be easily recognized and not pay out, because they forced the opponent to turn to an intensity of escalation which entailed the risk of serious damage to the «bluffer».

It may well be that a weaker individual – if not too badly mismatched – may be forced to step up escalation earlier and faster in a war of attrition since it can less afford to wait through, having fewer energy reserves and carrying higher opportunity costs, than the stronger, giving an observer a paradoxial impression of RHP relationships.

Rather than looking at graded escalation of displays as graded indications of a signaler's intentions, we might have to see in them a scale of (for both sides more and more costly) probes of the opponents' mobilizable aggressive potential, not only because probing can disclose «cheats» (Barnard and Burk 1979), but because it is necessary to check how much the opponent cares to become involved. Rather than informing the adversary of one's own intentions this would serve to inform oneself about the adversary's potential to resist by measuring the costs he shows himself prepared to pay. We would then also not be surprised to find displays poor indicators of what the displayer will do next (as shown by Caryl 1979), since what he will do next will have to be a function of the receiver's reaction to each move (see also Hinde 1981, Caryl 1982).

Maybe the occurrence of really symmetrical resource holding *potentials* (as defined above) are rare enough in the field to make the predictions about the advantages of «bluff» and «deceit», derived from a «war of attrition» model, only rarely applicable. Since in a scheme like this it must be dangerous to play «hawk» unless one is prepared to carry it through – forcing an opponent with unexpectedly high potential to react violently – this seems not only a way to understand why graded escalation in aggressive conflicts is common, but also a good reason to assume that «bluff» may not be a viable tactic and «honesty» in indicating the chances of one's potential paying out better. All these considerations apply with even more force if contestants know each other from previous fights and interact repeatedly (see Barnard and Burk 1979, van Rhijn 1980, Bossema and Burgler 1980) and even more so if genetic relatedness between them matters (Breed and Bekoff 1981).

This means that even in aggressive conflicts there are inherent selection pressures for arriving at arrangements which save costs for both sides. Communicative interactions may not establish perfect harmony, but they do at least create some kind of security partnership to mutual advantage, notwithstanding continuing competition for resources and for fitness. Although communication with unbalanced outcome during aggressive competition in species leading solitary lives may occur, I would argue that this is not the rule in more truly social interactions, such as between mates, between parents and offspring and in most kinds of interactions between members of a social group. In the latter cases we should rather expect strategies which yield the increased benefits of mutual coöperation for all participants. That cooperative social systems are ecologically extremely successful is abundantly clear from comparative sociobiology (see Wilson 1971, 1975) and of course from our own species. In that view, non-mutualistic interactions would be examples of more primitive stages or of transitory imbalances or of degenerative side tracks into blind alleys of the evolution of communication rather than paradigmatic for the real essence of communication.

Wiley (1983) has recently further developed a more balanced view of communication as manipulation by considering the possibilities of increase and decrease both of a signaler's and a receiver's fitness during communication. This approach can be further refined in two ways to cover animal communication as comprehensively as possible.

One should first distinguish – as has of course often been done before (see SEBEOK 1977, SMITH 1977 for review and examples) – between two types of traits of a *sender*. One type are *signals* (or *ritualized displays*) which can be assumed to have been selected to act on the perceptual mechanisms of specific addressees in order to influence their behavior. To produce such signals entails costs – e. g. of energy expenditure for signaling, of risks by making oneself conspicuous etc. – which have to be more than balanced by the fitness benefits derived from having them produced. The second type of a sender's traits are those properties of a sender which we might call *cues*. They also may make costs to acquire, but the costs are not incurred because these cues have been specifically selected to serve in communication. Examples would be unritualized intention (better: preparation) movements (e. g. getting ready for a jump), a flying bird's silhouette, the trampling of a stampeding herd, waves produced by a swimming fish, or the sheer impression of the body size and the quality of predatory weapons of a top carnivore.

On the *receiver's* side we can also discern between two kinds of conditions for reception. First, perceptual devices (releasing mechanisms or detectors) installed to allow a receiver to detect specific signals or cues. Though costs for developing these may be negligeable, they usually are combined in usage with quite important costs of search (i. e. selected appetitive behavior) for the appropriate stimuli and with costs of lost opportunities, (e. g., if a predator rejects an edible mimic) which again have to be more than compensated by the benefits accruing from being receptive. Second, a receiver can react to signals or cues without having to pay any costs for getting the opportunity to do this, e. g., because he encounters them by chance or because the sender has to pick up the bill for search risk and effort. Note that – in contrast to the approach of WILEY (1983) – the costs considered up to here – though ultimately also booked in fitness equivalents – are not the costs derived from the outcome of the communicatory interactions (e. g. from the loss of a territory or a mate), but the costs incurred in making preparations and efforts for being able to interact. These costs have then to be added to those of the outcome and to be put into balance with the benefits which only accrue from the outcome of a communicatory interaction.

The second refinement of argument should be always to make clear whether we are talking about a *dyadic* communicatory interaction, in which signals or cues act on receivers which have common interests (be it in getting the same resource or in cooperation), or a *triadic* relationship, in which a third party – as sender or as receiver – is involved that exploits such a dyadic signal-detector relationship (as first clearly expressed by WICKLER 1967). We could call the dyadic relationship bilaterally tuned, while the exploitative third participant in a triad is unilaterally tuning in to an existing dyadic relation. In dyadic mimesis a sender imitates stimuli to which the receiver is tuned *not* to react, e. g. inanimate background patterns.

Only if we consider all of these relationships, as done in Tables 1 and 2, do we see the whole universe of organismic communications and the richness of opportunities given there.

The triadic relationship is in some respects profoundly different in consequences from the dyadic one. In the dyadic system we have, as it were, a symmetry of evolutionary and developmental means. Selection acts in just the same way on the sender to manipulate the receiver more efficiently to the advantage of the sender, as on the receiver to be as discriminating as necessary to derive its own optimal profit from reacting. This makes unequal utilities (benefits – costs) for sender and receiver evolutionarily and developmentally unstable and favors balanced mutualistic relationships. Accordingly, outright receiver or sender exploitation (types 2, 3 in Table 1) should be *rare* in dyadic relationships.

The situation in the triad is quite different at least in the more common types 1 and 3 of

Table 1: *Communication dyads:* bilaterally tuned relationships between sender and receiver (or imitation by sender of stimuli which influence the receiver's behavior in a way advantageous for the sender: mimesis). The profit balance – total benefits minus total costs of interaction – of the participants is given as: gain (+) or loss (−) of fitness equivalents because of the communicative interaction of dyad members. Costs can be incurred for preparation and during interaction (sender: signaling effort, increased risks because of conspicuous signaling, search effort for receiver, lost opportunities because of signaling and searching; receiver: search effort for sender, lost opportunities from spending time on searching) and from the outcome of communication (e.g., loss of the contended resource). If the sender incurs no costs for preparation, the stimuli provided by the sender, to which the receiver reacts, are called *cues*; the receiver may incur no preparation costs if it encounters the sender by chance or if the sender carries the search costs. Benefits can be derived during interaction (receiver: avoided risk of injury by combat; sender and receiver: gain of experience) and from the outcome of the interaction (e.g. win of resource).

Type	Sender	Receiver	Examples of relationships
1	+	+	Intraspecific: mutualistic cooperation (e.g. male-female, parent-offspring, members of social group). Interspecific: symbiosis, aposematic signaling. Perfect, if gains are equal for sender and receiver; imperfect, if one gains more than the other (e.g. in agonistic communication).
2	+	−	Exploitation of receiver by sender. Intraspecific: ritualized combat (victory by threat) or social dominance hierarchy; cuckolded male investing alloparental care; immature males gaining experience by courting female who has to carry interference costs. Interspecific: mimesis, interspecific interference and dominance by threat.
3	−	+	Exploitation of sender by receiver. Intraspecific: kleptogamic interloper attracted by signals of male competitor. Interspecific: detector of palatability of mimetic prey which concentrates on hunting for the mimic.
4	−	−	Wasted efforts of sender and receiver. Intraspecific: disturbed and unsuccessful courtship; both opponents of fight – after undecided threat duel – severely damaged, none of them winning resource; fighting or courting pairs attracting predator. Interspecific: inexperienced predator suffering from killing dangerous aposematic prey.

Table 2, where the third party acts like a «parasite» or «predator» of the dyadic relationship. There the intruding participant derives benefit as a receiver from reacting to the signals of a sender I which are also under selective control in a mutual relationship with receiver I; or it derives that benefit from producing signals as sender II (mimic) that act on the detector devices of receiver I which are also under selective control in a relationship with sender I (the model). Since the mutually tuned communicators both derive fitness from their sender-receiver relationship – and only if so, which again demonstrates the importance of mutual benefits of communication – selection cannot make them completely unresponsive to the mimic's signals or stop producing their own signals in order to get rid of the parasitic or predatory exploiter, because they would at the same time jeopardize or even cut off the fitness derived from their own dyadic cooperation in system I.

Table 2: *Communication triads:* exploitation of communication dyad (I) by third participant (II) who can tune in as receiver II or as sender II. Profit balance of interactants: gain (+) or loss (−) of fitness equivalents because of communicative interaction between member of dyad and exploiter.

Type	Dyad		Participant II as		Examples of relationship
	Sender I	Receiver I	Sender II (Mimic)	Receiver II	
1	−	model		+	Sender I exploitation by predator or parasite which detect prey/host (= sender I) by their signals/cues which are selected to influence receiver I; food parasite/competitor attracted by «food calls».
2	+	model		+	Generalized cooperation, intraspecific in courtship chorus, lek displays (attraction of males by male displays); interspecific in mixed flocking or cross-specific alarm calls.
3	model	−	+		Mimicry, detrimental for receiver, e.g. Batesian mimicry (receiver loses opportunity, search effort) or aggressive mimicry.
4	model	+	+		Mimicry, advantageous for receiver, e.g., Müllerian mimicry or cichlid egg mimicry.

Clearly, the evolutionary means of the host/prey-dyad (I) and of the intruding exploiter (II) are not equal, since little can restrain the latter from becoming ever more perfectly tuned ot the dyad's signal-receiver system. Defence against exploitation by a third party can therefore not be selected as easily and as perfectly as can defence against mutual exploitation between the members of the dyadic set, unless the exploited one manages to embrace the exploiter at a point of common interest in a symbiotic relationship. This may also be the reason why intraspecific signal faking is more rare than that between species (see Dawkins and Krebs (1979) for a lucid discussion of further aspects of this problem). The two types of relationships should therefore always be clearly kept apart and not be used interchangeably as exemplifying communication in animals.

It follows that persistently effective third party exploitation can take advantage of a dyadic communication system which is highly important for the dyad's fitness. Since this is also the condition most favoring selection against intradyadic exploitation and for intradyadic cooperation, we should expect to find such exploitative third party intrusion most spectacularly combined with very cooperative and mutually beneficial social relationships. Ant mimics and social parasites yield particularly impressive examples (Hölldobler 1977,). We could also put this more simply: a cooperative system is more productive than comparable non-cooperative ones; a productive system is a valuable resource; any valuable resource will by itself select its skilled intruders and efficient exploiters, whether these are predators or parasites, whether interspecific or intraspecific ones.

Thus this section can be summed up by concluding that there are good reasons to see manipulation and counter-manipulation at work in communication of organisms. From this very fact it follows that there are also good reasons to expect the development of communication that enhances mutual cooperation between communicators both in their roles as senders and receivers. This should make «bluff» and «deceit» the exception rather than the rule in dyadic relationships at least in animals living in stable social groups. On the other hand «bluff» and «deceit» should be the rule rather than the exception in triadic communication – the exception being exemplified by types 2 and 4 of Table 2: mutual interest and common gain of members of dyad I and participant II. Finally, there are good reasons to assume that these cooperative communicators are under constant pressure from outside exploiters of their communicatory relationship (types 1 and 3 of Table 2).

In those highly developed forms of communication where mutual benefit for senders and receivers derives from communicating most efficiently in order to make the system most productive for common advantage – dance communication in honeybees being the paragon animal example – we would therefore expect a number of characteristics which enhance the efficiency of communication and make it as resistant as possible to uncooperative parasitic cheaters from within the group and against exploitation by intruders from outside. A list of these might read like the exact opposite of what DAWKINS and KREBS have emphasized as characteristic features of exploitative communication. Mutualistic communication should be deictic, i. e. directed at specific addresses (GREEN and MARLER 1979) in the most economical way (optimizing the benefit/cost ratio), informing partners as accurately as necessary to achieve the communicative purpose (e. g. by indexing external referents), controlling (e. g. by frequent role change) for fair reciprocation and against in-group exploitation, keeping the whole exchange as private as possible so that unwelcome eavesdroppers and freeriders stay excluded. We therefore would expect such communication to be narrow-cast rather than broadcast, unobtrusive, difficult to discover and recognize by outsiders and unaddressed receivers (including observers).

That means, the very reasons that cause us to understand communication as manipulation force us to conclude that the most developed forms of cooperative communication should not be characterized by spectacular showiness but rather by finely adjusted subtlety (see MARKL 1983). Cooperative communication should express itself rather by delicate directed modulation of the receiver's behavioral tendencies by the sender's signals than by the overwhelming effects of conspicuous advertisement, which leads us to consider the efficiency of communicaton.

3 Modulation – or the efficiency of communication

Efficiency of communication is a vaguely defined and variably used concept. Quite evidently it cannot be measured by the number of kilowatts put into signal production. A lion's thundering roar, a howler monkey's sonorous bellow, a blackbird's resounding aria that precipitates me out of bed at dawn – they are impressive enough for an observer, but they may actually not have very much effect on the addressed receiver and therefore need so much volume to overcome «sales resistance» (see DAWKINS 1982). This reminds us once more that it needs a responsive receiver to be an impressive signaler and very little signal energy may have to be expended if it is awaited by an acutely sensitive detector.

MARKL (1983, see also WILSON 1975, SMITH 1977, SEBEOK 1977, MARLER and VANDENBERGH 1979) has considered four approaches to quantify the effectiveness of communication:

1. The *reach* R of a signal, as measured by the distance over which it can be perceived by the addressed receiver, defines the *active space* of communication with respect to that receiver. Since signals can have extremely variable persistance – from practically indefinite (light allowing) in visual badges, over more or less slowly decaying chemical signals to extremely rapidly fading acoustic ones – active space of animal communication has always to be also described in the time dimension. Environmental conditions – by providing the signal-transmitting channel and by loading it with noise – are as decisive for signal reach and detectability as source intensity and receiver sensitivity. The active space – which of course not only determines the area of reach for the addressed receiver but also for any «parasitic» eavesdropper – can for some signal modalities be actively set by the sender by varying radiation intensity (e.g. very easily for sound, hardly at all for visual signals). It can probably always be gradedly controlled by the receiver by modifying response threshold. Receiver-control is ultimately always decisive as every troubadour must doubtlessly soon have found out. Thus we can say that active space is not only physically constrained but also motivationally controlled. Since sender and receiver motivations can also be under the influence of internal and external stimulation context, signal reach and active space are usually severely dependent on this context (see SMITH 1977 for thorough discussion).

2. Secondly, a signaling system may be rated according to how many receivers can be activated by one sender's signal: we could call this the amplification effect of communication or its *gain* G. It makes, for instance, quite a difference whether a dose of an alarm pheromone activates a dozen workers of an ant colony or whether it makes the whole nest-population pour over an intruder. Evidently this measure of effectiveness is as dependent on signal reach and active space as it is on the population structure of given receivers. This aspect has been quantitatively studied with respect to food or enemy alarm signals in social insects (see VON FRISCH 1965, WILSON 1971, HÖLLDOBLER 1977) or to predator alarm in rodents (e.g. SHERMAN 1981). It has to be emphasized that there must be no direct correlation between ultimate efficiency (see below) and communication gain (though there may be, as in evacuation alarm when social insect colonies are physically endangered).

3. Another aspect of effectiveness of communication – though closely related to the preceding measure – is the thightness of coupling between the occurrence of a signal and the receiver's reaction to it, a variable which is usually quantified by absolute or relative *transmission T* as defined in statistical communication theory (see SHANNON and WEAVER 1949, STEINBERG 1977, LOSEY 1978, MARKO 1983). Gain G quite clearly depends on T, a measure of the probability with which a sender's signal is followed by a change of behavior in the receiver or of the decrease of uncertainty about the receiver's behavior for an observer if he knows that the signal has occurred (p.181). Calculation of simple conditional probability of reaction will only do if there is only one clearly defined response to one clear-cut signal. Whenever we have probability distributions of signal and response fields, the influencing and constraining effect of the former on the latter is most easily quantified by T. At first glance this might seem like an ideal measure for signal effectiveness: what could more clearly show a signal's excellent quality than if all receivers without exception promptly react to it upon perception. I will come back to this question and show why this view can be erroneous. As pointed out for active space, transmission, too, is very much under the control of the receivers' motivation that sets reaction threshold and depends itself on internal and external context of stimulation and potentially also on previous experience (though this could be regarded as another internal context variable).

If these factors vary, transmission will hardly ever be all or none (0 or 1); it will depend on all these variables how effectively a signal can manipulate addressed receivers.

4. Though R, G and T quantify measures which can all be somehow related to real

effiency of communication there can be only one way to judge this on an ultimate scale, viz., by determining fitness increments and decrements Δ W for sender and receiver that are the result of their communicatory interactions. That means, we have to find out whether such interactions are of adaptive significance and if so, for whom: the sender, the receiver or for both. Although it is sometimes possible to measure the consequent change in fitness by severing the link between sender and receiver – e. g. if a muted cricket male fails to attract females – often the problem is not as easily solved: how does one eliminate a grin from a chimpanzee's repertoire, or chemical contact signals between bees? Nor will we as a rule be able to measure directly the resulting changes in reproductive success. The usual way to handle this is to select some intermediate indicator as fitness equivalent (e. g. amount of food collected or of weight gained) if it can be safely assumed that there is a monotonous relation between it and the real fitness effect. Once such a measure is available it can now be used to see how the other signs of effectiveness of communication (reach, gain and transmission) relate to it. To rate the efficiency of a particular mode of communication, one has to consider the profit ratio of benefits/costs. That means that communication is always under ergonomic control. Reach R, gain G and transmission T must be set by the selective forces optimizing fitness W (or its equivalent) and are not under selection to maximize R or G or T. Quite evidently there are cost functions of increasing active space (e. g. increased energy expenditure, increased risk of being detected by predators or of attracting competitors), of increasing gain (e. g. if more helpers than necessary to defend and exploit a food source are activated) and of increasing transmission (e. g. if a recruiting signal to one source of food detracts workforce engaged in exploiting another more valuable one). These cost functions have only rarely been estimated; they are specific for each kind of communication in a given species under given environmental conditions and only rarely can they be expected to be linear over their whole range; typically they should go through a minimum.

This allows me now to return to the question left at the end of the preceding section : how should different types of communication systems be most effectively organized? Clearly enough, if the effectiveness in attracting a mate, in deterring a competitor or in alarming from a predator depends on signal intensity and persuasiveness we would expect to find active space, gain and transmission maximized (at least as long as fitness effects for the receiver increase, too; beyond that limit the receiver will set resistive controls). However, even in these examples, signaling intensity will often come under the limiting constraint of exploitative eavesdroppers, as for example when frog-hunting bats or cricket-hunting parasitoids or firefly-hunting fireflies locate their prey by courtship signals (Ryan et al. 1981, Tuttle et al. 1982, Cade 1979, Lloyd 1977, 1981). Not surprisingly, we often find that broadcast courtship displays are immediately replaced by low-range types of signals as soon as a mate has been attracted at close distance; lovers guard intimacy even in the «cynical» world of animals.

This brings us to the other extreme of communication. While broadcast communication has often been thoroughly analyzed (because it is easily observed, not because it is necessarily most important: ethologists tend to look for the key where the light is, too!) we know much less about that interesting realm of «private» communication which is at the very heart of complex and highly organized animal societies. I have already given the reasons why we should expect communication in such systems to be predominantly low-intensity, narrow-range, low-gain, low-transmission, but as accurately indexical and highly addressed as possible and altogether quite intimate and exclusive to guard from exploitation by unwelcome intruders. We can find *all* of these characteristics most impressively assembled in the honeybees' dances (see von Frisch 1965 and Lindauer; Gould et al, this volume).

I want to consider here another group of examples from social insects from our own work, which make the extremely different character of this type of communication as compared to broadcast displays even more strikingly evident: communication by *modulating signals* in ants (see MARKL 1967, 1983, MARKL and FUCHS 1972, MARKL and HÖLLDOBLER 1978, FUCHS 1976 a, b, LENOIR 1982 for more details). The most typical example is communication by vibration signals which thousands of species of the subfamilies Ponerinae, Pseudomyrmecinae, Myrmicinae (and the living fossil *Nothomyrmecia:* TAYLOR 1978, MARKL and TAYLOR, unpubl. observ.) produce by abdominal stridulation whenever they are prevented from moving freely. More rarely, stridulation has also been recorded in freely mobile ants, e.g. when in contact with food, enemies, during nest-disturbance or copulation.

The stridulation signal is a unitary signal, though message and meaning (SMITH 1977) can vary exceedingly (see below). The *reach* of these signals is small because of their low intensity (10^{-8} W) and their exclusive transmission either in direct contact or as substrate vibrations over a few centimeters or maximally decimeters distance to the vibration-sensitive receivers. The *gain* depends on circumstances, in direct contact it is as low as one, usually less than 10 and only exceptionally does it exceed 100 (e.g. when acting as rescue signal of buried *Atta* queens). If *transmission* is measured by exposing workers, in a paired control situation either inside or outside the nest, to stridulation signals alone, without additional chemical signals, the effect of the signal on the workers' subsequent behavior – though somewhat dependant on the receivers' present occupation, e.g. as food collectors, diggers, scouts or brood attenders – is always exceedingly small. Normalized transmission values (corrected for bias according to FAGEN 1978) range between 1 and a few percent (MARKL and HÖLLDOBLER 1978, MARKL unpubl.). That means that maximally a few percent of the information contained in the signal is actually transmitted to the receivers, or that the uncertainty of an observer (inversely related to his probability of guessing it correctly) about what the ants will do after having noticed the stridulation signal is only decreased by these few percent.

According to these three measures, ant stridulation would therefore have to be called a most inefficient way of communication. It barely makes the statistical significance level to qualify as a communication signal at all. This would, however, be a premature conclusion. If one sets up field experiments and lets the signal act in the relevant context of spontaneous usage, it can release quite remarkable behavioral effects in the receivers and can therefore clearly be interpreted as adaptive, i.e. as favoring the colony's fitness. Two points are of interest in this connection: first, the semantic contents (message-meaning) of stridulatory communication depend to a great extent on the appropriate situational context, in fact, they are almost completely determined by it; second, in some situations the effect becomes only fully evident if the stridulation signal is acting combined with other, especially chemical, signals. To give only three examples which have been more thoroughly analyzed up to now:

1. Leafcutting ants *(Atta)* stridulate persistently when trapped under soil in a cave-in of their ground nest. If such a catastrophe – which will destroy the whole colony if the single queen is also trapped and cannot be rescued – has occurred (e.g. after heavy rain or through a predator's attack) many nest mates which have not been caught react sensitively and promptly to soil-conducted stridulatory vibrations by approaching the epicentre of the vibrations and digging for the calling nest-members. This reaction depends not only much on the general stimulus context of the disaster but also on worker occupational «caste»: while soldiers contribute very little, the task class of workers who perform inside nest digging duties are most active at rescue digging, transmission values of signal effect on them sometimes coming close to 50%. The reach of the signal hardly exceeds a few

decimeters even for the strongest source (the queen). Depending on location of the occurrence up to several hundred workers can be activated by one sender, but again a few tens is more typical. Accuracy of message and mutual benefit for sender and receiver are evident.

2. Some species of harvester ants *(Pogonomyrmex)* of southern US desert regions are exceptional for ants, in that they mate not on the wing but on the ground (HÖLLDOBLER 1976, MARKL et al 1977). While no stridulation signals seem to be involved in mate finding, courtship, and initiating copulation, stridulatory contact vibration of the female is an effective indicator of her unwillingness to accept further copulations after her spermatheca has been filled up; males who still try to lock up with her are thus discouraged and leave the no longer receptive female. Number of observations was not large enough to make a reliable estimate of transmission for this «female liberation signal», however, it is clearly much larger than if one exposes the males to the signals when they are not trying to mate a female. In this situation, the signal acts on the receivers only in direct contact (at minimal reach); the gain can range between the one and a few males that crowd around one female. It can be shown that this communication mutually benefits females and males (who are, as a rule, not genetically related) by saving the males time and thus lost mating chances, and by allowing the females to escape the rather dangerous grasp of overeager lovers. There are therefore also good reasons to assume that the female correctly informs about her lack of inclination for further copulations.

3. These two examples demonstrate that ant stridulation signals are extremely context-dependent as far as their semantic message-meaning contents are concerned. This is also true for the third example which introduces yet another aspect of a modulating signaling system. Workers of the American desert ant *Novomessor* have a remarkable ability to cooperate in retrieving prey which is too large to be dragged to the nest by a single ant. If a worker has for some time unsuccessfully tried to move an attractive piece of food, she begins to release a short-range recruiting pheromone and to stridulate while continuing with her efforts to dislodge the object.

If another nest mate is attracted traveling upwind towards the source of the chemical signal, it will upon contact with the prey and after receiving the stridulatory vibrations produced by the recruiting ant and transmitted through the prey, immediately join in chemical signaling (without being stimulated by vibrations this would take much longer), in stridulating and in trying to drag off the prey. Depending on the size of the object this process continues with rising effect until a large enough workforce is assembled to transport the bounty. Since competition for valuable food is fierce among ants of different species in the desert environment, it could be shown that the vibration-enhanced recruitment process can give *Novomessor* a decisive edge over their many competitors for prey. The reach of the vibration signal is again minimal, but through the combination with pheromones it is amplified up to several meters. The gain is small, from one to at most a dozen workers being activated at a time. Transmission, if measured without the accompanying chemical signal, is again only in the range of a few percent; however, in the relevant context of contact with a fixed prey-object, transmission rate immediately increases with a large proportion of receivers reacting fast to the signal by joining in stridulation, discharging additional recruitment pheromone and tearing wildly on the object.

The important point is here that a signal which by itself and in inappropriate contexts is almost completely unable to influence the behavior of the receivers, can strongly modulate their reaction probability to the contextual stimulation. We have therefore proposed to call such signals *modulating signals,* if they do not release a clear-cut response by themselves but rather alter (modulate) the responsiveness of receivers to other sources of stimulation, either facilitating or inhibiting reaction to them in a way which could be described as changing their state of arousal or attentiveness to other signals.

Mutuality of benefit is in the present example quite obvious. One could also conclude that the signal «informs» the receiver accurately, because signal production is stopped as soon as the prey-object is finally being moved, so that no more workers are attracted than are necessary for the task, an ergonomic measure of communicative efficiency. I have also never seen a finder calling for help and letting others do the carrying for her – as a rule she will be among the ones who drag off the prey.

From the last two examples it is also obvious that the limited reach and small gain of the signal are not signs of its inefficiency but seem rather well adapted to address exactly those receivers which ought to get the message without broadcasting further than necessary and useful.

Stridulatory vibrations are by no means the only mechanical communication signals produced by ants for modulating nestmates' reaction tendencies. In *Camponotus*, MARKL and FUCHS (1972), FUCHS (1976 a, b) have shown that drumming with mandibles and abdominal tip on the thin wooden shells of their tree-nests serves as modulating – facilitating or inhibiting – signal in a wide variety of conditions in highly context dependent ways, thus influencing different forms of alarm reactions, brood transport, colony removal (the latter function can also be taken over by stridulation signals in other species: MASCHWITZ and SCHÖNEGGE 1983, MASCHWITZ pers. comm.). Again, the limited reach, gain and transmission involved seem best suited for a signal that acts to organize limited numbers of workers in a contextually adaptive and ergonomic way. HÖLLDOBLER (1971) found that in *Camponotus socius* contact vibrations motivate workers to follow a recruiting leader to a food source. This is, of course, very similar to the more primitive forms of «dance» communication in bees (LINDAUER 1961, VON FRISCH 1965, MICHENER 1974). Modulating alarm by contact vibration is probably even more widespread in ants. In the primitive ponerine *Amblyopone* it is a prominent response to diverse kinds of disturbances (HÖLLDOBLER 1977, TRANIELLO 1982, HÖLLDOBLER and MARKL, unpubl.) and helps to organize defensive reactions in a variety of ways. In the Australian bulldog ant *(Myrmecia)* sham attacks – the most simple precursor of contact vibration signals – serve as a form of modulating alarm which sets colony members in a state of alert and distributes them over the territory to be defended in the fashion of a most economical random search pattern (MARKL unpubl.) as proven in computer models of these interactions (FREHLAND, KLEUTSCH and MARKL, in preparation), LENOIR (1982) found antennal palpations in *Myrmica* to be a modulating signal involved in controlling trophallactic food flow.

A particularly good example was found in *Oecophylla* by HÖLLDOBLER and WILSON (1978) where rectal gland secretion is combined with a variety of motor signals, depending on context. Most striking is the ritualized «attack-sequence» used as a signal in defence recruitment.

In all of these cases and a number of similar ones from other social insects (e. g. termites: HOWSE 1964) it is clear that the various forms of mechanical communication, often combined with contact- or close-range pheromonal ones, show all the characteristics listed above for an efficient, accurate and ergonomic means of coordinating behavior to the mutual benefit of senders and receivers (who need not even be closely related members of the same colony, see the *Pogonomyrmex* mating example!).

Although these examples have been picked only from social insects this should by no means suggest that modulatory communication as defined here is restricted to these groups. Quite to the contrary, even a cursory search of literature on communication both in animals and in man demonstrates that especially in species living in social groups, but also in solitary ones during more intimate social interactions (as e. g. during mating and broodcare) low-R, G, T-modulating signals for fine control of behavioral coordination are exceedingly common, though they have rarely been thoroughly investigated for obvious methodological reasons. All kinds of signal modalities can be involved with chemical, mimic-gestural, contact-mechanical and low-level acoustical ones predominating, and extreme context dependence seems always characteristic. What is most important, modulating signals are not the exception but rather the rule in communication during close interactions in exclusive social groups, and many examples leave the impression that mutuality of interests and fine-graded accuracy of information transmission may also be more typical than not. In the following, I pick only a small number of examples which can profitably be analyzed under this viewpoint (for sources see Hinde 1972, Ekman 1973, Krames et al. 1974, Weitz 1974, Argyle 1975, Wilson 1975, Argyle and Cook 1976, Sebeok 1977, Smith 1977, Brown 1979, Marler and Vandenbergh 1979, Morris et al. 1979, Pliner et al. 1979, Lewis and Gower 1979, Eibl-Eibesfeldt 1979, 1980, Peters 1980, de Waal 1982, Eisenberg and Kleiman 1983, Halliday and Slater 1983). Rather than referring to numerous specific examples, it may be more useful to try a rough and preliminary semantic classification of those message-meaning relationships in which modulating signal effects are most often to be found.

The «classical» examples are *social facilitation and social inhibition signals* (stigmergie, Stimmungsübertragung, see review by Clayton 1978), whenever the stimulating or inhibiting social influence is not only derived from perceiving other individuals perform «normal» maintenance routine (such as feeding, sleeping, walking, defecating, scratching) but comes from more specific behaviors which can be assumed to be performed or rather to have been selected for exerting a modifying influence on the behavior of receivers. Many mild alarm signals – both intra- and interspecific ones (often intention movements of escape behavior) – do not in fact release escape behavior in receivers by themselves but merely alert them and prepare them to react to additional stimuli, e. g. from an approching predator. For example according to the detailed analysis by Seyfarth et al. (1980, see also Cheney and Seyfarth 1982) the enemy markers of vervet monkeys serve primarily as such behavior modifying alerting signals notwithstanding their referential semantic significance, which makes them «specified» alerters.

Intention movements for leaving a place in group-living species are other typical examples for modulating signals, from jackdaws to baboons or hunting dogs. No less specific modulators can be found in social inhibition, e. g., if animals indicate by special sounds that they are in a socially «withdrawn state», i. e. want to be left alone and are unwilling to interact (see for example for rats Anisko et al. 1978). For a gregarious cricket, Boake and Capranica (1982) have described the function of «courtship» chirping as a comparable interference inhibitor signal.

A second group of modulating signals are *contact signals* which occur very commonly in nearly all kinds of social relationships (mates, parents-offspring, other members of social groups), sometimes in generalized form (e. g. colony odors), in vertebrates more often as individualized signals. They can be olfactory (group and individual odors), gestural («glancing,» see Chance 1967) or acoustical (see for primate examples Snowdon and Cleveland 1980, Cheney and Seyfarth 1982, and for bird duetting Wickler 1980, Wickler and Seibt 1980, 1982). The perception of familiar and regular contact signals keeps the animals in a socially connected and relaxed condition. Loss of contact, i. e., failure to perceive the familiar signals, can then effect behavioral changes, e. g. search. Evidently, there is a close similarity and partly even identity between such modulators and the signals used as familiarity cues which control kin-related behavioral preferences or aversions (see review in Markl 1980; and also Bateson 1980, 1982, McGregor and Krebs 1982, Holmes and Sherman 1983).

A third group of modulating signals is to be found in those cues and behaviors which regulate antagonistic relations in social groups by indicating *social status, dominance* and *submission.* Though we often find threat or submission displays which immediately release defined behaviors in the receivers, this could be almost regarded as the breakdown of the normal uninterrupted flow of more subtle status signals, in which mimic, gestural, proxemic, acoustic and other expressions merge into a continuous multimedia show of rank confidence and rank acceptance. Many or most of these elements have to be classified as modulating signals which are as a rule extremely context dependent (see PRUSCHA and MAURUS 1975 for one example of a detailed analysis). STAMM (1974) has even suggested that in laughing gulls there may be a specific «codon» serving similar multi-purpose pacifying functions as does «please» in human language. Olfactory scent marks, so prominent in mammals, may be regarded as sign posts exerting modulating influence during spatial competition between different individuals or groups. As such they probably alert for the detection of other more specific signals, as can introductory alerting notes of some bird songs (similar to the familiar gong introducing a news broadcast).

A fourth category of modulating signals is difficult to define but may nevertheless be of more than negligible importance in controlling behavioral tendencies: we may call them *self-communicatory signals* (see e. g. GREEN and MARLER 1979, MILLER and INOUYE 1983). «Getting oneself into the mood» by acting as sender and receiver at the same time is, of course, well known to us from introspection. There is suggestive evidence (e. g. from acoustic selftutoring) that this also occurs in animals. Evidently, this is the ultimate example of sender-receiver mutuality!

Modulating signals as discussed here have some similarities with priming signals (BOSSERT and WILSON 1963), although these have primarily developmental influences, and with signal modifiers (see e. g. JENSSEN 1979) which, however, qualify or grade displays rather than modify a receiver's motivations, as modulating signals do. What VON FRISCH (1965) described as the «vivacity» of the dances of bees which indicates nectar quality and thus influences foraging motivations of recruits has every characteristic of a modulating signal, the effects of which depend on the context of a colony's supplies and offers from other sources.

While modulating signals in animals, though widely occurrent, are certainly only insufficiently understood in syntactic, semantic and pragmatic respects, the literature on human non-language communication is full of thoroughly analyzed examples of signals that continuously knit and mend the intricate web of human social relations: mimic, gestural, postural, kinesic, proxemic, paralinguistic, prosodic and other cues permeate every human interaction in all the semantic contexts just listed. As has been often pointed out, most recently in relation to animal communication by DAWKINS and KREBS (1978) and EIBL-EIBESFELDT (in press), all kinds of expressions of human art and ritual, most convincingly manifested in music, song and dance, play the communicative roles of motivation modulators – to the extreme of hypnotic spell and trance. One need not, however, dive quite as deeply into the arcana of the human mind to find suitable examples for modulating signals in our life: traffic signs, to pick just one, demonstrate the whole gamut from high-effect releasers to the most subtle and context-dependent modifiers of alertness and attention, the communicative function of which – when studied in overt behavior and not in linguistic prescription! – is hardly different from some of the forms of alarming vibration signals in ants.

This brings most forcefully back to our mind that efficiency of communication is only rarely and under special circumstances a matter of maximizing reach, gain and transmission of signals, as so easily suggested by a purely information-technical approach to animal communication. Conspicuous is not necessarily important, unobtrusive is not negligible in animal (and human) communication. Setting R, G, and T rather low sets the threshold for communicative effects accordingly high, but this avoids the trappings of false alarm, of overreaction by exaggerated response and of parasitic intrusion.

To come back to the example of information transfer in agonistic display: a game-theoretical analysis of advantages and disadvantages of accurately signaling fighting potential and motivation or of concealing this information leads us to look at conspicuous agonistic displays from a different angle than has traditionally been the case (see Andersson 1980, Moynihan 1982, Maynard Smith 1982). However, one of the consequences of this is that what impresses us most by pompous conspicuousness in threat displays of animals may not be at all what most effectively influences the behavior of the receivers. These displays might well be examples of rather evacuated rhetoric which were formerly used to gauge a sender's potentials and intentions but which, because of this very reason, have been much devalued by imitation and bluff so that no experienced receiver cares for them much any more. At the same time the subtle, not yet ritualized, epiphenomenal indicators of behavioral intentions may be what actually counts in communication – and because this is so, will again stand a good chance of being made more conspicuous by ritualization and then devalued by imitation in due time.

Waves of rising and devaluating fashions of signals and displays would thus propagate through animal social systems just as in our social world propaganda campaigns and fashions become en vogue and fade again in public awareness and published opinion. If there is some truth to this view, theoretical and empirical ethologists alike would be well advised to pay more attention to those subtle signs – however unobtrusive they may seem – which are actually and presently employed to coordinate behavior, the informal ways of handling things, as it were, than to be too easily distracted by the empty rhetoric of highly formalized and stereotyped fossils of yesterday's communication.

4 Information – or transmission in communication

This brings us inevitably to ask once more what it is that is transmitted during communication. Is this a sharing of information as traditionally assumed, or is it the manipulative power of macchiavellian senders which lets receivers dangle on the strings of their mesmerizing signals? But if so, would this not mean too that controlling information has to be transferred to and exerted on the receiver's sensory-nervous-muscular system? Evidently, questions like these cannot be answered lest we first try to agree on how to use the word «information». Hundreds of publications have dwelt on this question. To avoid misunderstandings we have to distinguish between two different meanings of «information».

In one sense it designates, as in common usage, a semantic and pragmatic concept, describing the contents of the signal's message (as inferred from the sender's context-dependent behavior when emitting the signal) and its significance, the meaning for the receiver (as inferred from the receiver's context-dependent reactions upon perceiving the signal), and how both relate to the costs and benefits, i. e. to the fitness effects resulting for sender and receiver from their interaction (see Smith 1977 for details). In this sense information refers to the *biological importance* of the communicative process *for the communicators* under conditions in which it has evolved and developed and to which it is therefore adapted. Obviously, it is very much in the eye of the beholder how this semantic information is evaluated. Information in this sense is not «transferred» at all, it is *by definition* a property of every communicating system, because without adaptive significance for sender or receiver or for both we need not talk about communication at all. Semantic information refers to the coding and decoding rules of the communicators and why they use them. It is also evident that it does not make sense to discuss whether this interactional semantic property is true or false, because this could only be defined by interpretation from an *outside* observer or by postulating that the communicators have *subjective* expectations which can be fulfilled by true or disappointed by false informations, an assumption which not only cannot by proven for animals but mostly is quite unnecessary (see below). Finally, information in this sense must be discussed separately for each specific communication system. Specifically, there is no defined relation between how important a given signal is and how surprising, unexpected, uncertain it is for an observer and how easily he could have guessed its occurrence. If an opponent attacks after a particular highlevel threat display this may not be surprising at all, but it is nevertheless highly important and therefore certainly semantically very informative for the interactants. Just as well, a rare display of a rare animal may occur very unexpectedly and nevertheless be without much significance for a given receiver. As explained above (p. 173), the importance of a piece of information contained in a signal can also not simply be equated with the probability of reaction of a receiver to a signal: it is defined and it is only defined by the fitness consequences resulting from the communicatory interaction for the communicators.

The second sense in which information is used is that of *statistical information theory* (see SHANNON and WEAVER 1949, QUASTLER 1958, ATTNEAVE 1959, STEINBERG 1977, FAGEN 1978, LOSEY 1978, MARKO 1983). Here it is a measure of an *observer's uncertainty* about the occurrence of a particular signal event or of his probability of guessing this occurrence by chance: the more uncertain an observer is about the occurrence of the signal, the less probable it is that he can guess its occurrence, the more information he gains – about this question of non-occurrence or occurrence! – by receiving the signal. This is related to the surprise value of the signal for the observer and the evidence it provides for his decision between different hypotheses about it (PALM 1981). This gain of information in an observer, or the reduction of his uncertainty when observing the occurrence of an event x_i from a probabilistic event field X with k possible event states ($x_1, x_2 \ldots x_i \ldots x_k$), is usually quantified as Shannon's statistical information measure

$$H(X) = - \sum_{i=1}^{k} p_i \log_2 p_i,$$

with p_i the probability of the ith event of X, which is estimated by n_i/N (n_i number of observations of x_i, N total number of observations of event field X). H(X) is often, though misleadingly, called the entropy of event field X. If A and B are two interacting individuals with – at a given moment – A being regarded as the sender emitting signal a_i from a signal repertoire with s signals ($a_1, a_2 \ldots a_i \ldots a_s$) and B as the receiver, reacting with response b_j from a repertoire of r responses ($b_1, b_2 \ldots b_j \ldots b_r$), the average reduction of uncertainty in an observer when signal a_i is produced is H(A), and that when response b_j occurs is H(B). The uncertainty for joint occurrence of a_i b_j is given by H(A, B). Communication can be said to occur if $H(A,B) \neq H(A) + H(B)$ or – as usually expressed – if $T(A, B) = H(A) + H(B) - H(A,B)$ is significantly larger than zero.

For two interacting indiviudals, T measures the tightness of the connection between the probability fields of their signals and responses though it does not establish a causal link nor indicate whether A influences B or vice versa (they could both be dependent on a hidden third variable). To establish causal interdependence, the time relationship between a_i and b_j has to be observed as a first indication – if time relation is not significantly unidirectional, the connection cannot be causal. To really confirm a_i as cause of b_j, the occurrence of a_i should be given by experimental intervention and tested whether b_j follows as probabilistically expected.

The quantity t (A,B) = T (A,B)/H(B) is used to express in normalized fashion (in %) how much uncertainty about the response of B is reduced by knowing which signal A has produced. Unfortunately it has become customary to call the quantity T, which only indicates the tightness of correlation between two nominally scaled probabilistic variables, «transmission», «information transmitted» «transinformation», «shared information» or «synentropy», thus implying that T expresses the quantity of information flowing from sender to receiver, the higher T, the more informative the signal for the receiver. From what has been said it should be clear that it is misleading to thus look at this measure. What it really expresses is the reduction of uncertainty *in an observer of this interaction,* and in this sense the information thus gained by *him,* about the behavior of B upon knowing the behavior of A and viceversa in a given interaction. T says nothing about the significance of the interaction for A or B or both and therefore does not quantify the biological importance of their interaction; and it does not measure what B «gains» from receiving A's signal. It indicates how much the probability distribution of B's behavior changes in an average statistical sense in connection with A's signaling as compared to B's behavior in the absence of A's signal. The expressions «shared information» and «synentropy» if thus interpreted are somewhat less misleading. In fact it might be better to translate T as «tightness» (of correlation) rather than as «transmission».

These considerations seem at first glance to settle the question about whether information flows from sender to receiver: as far as the semantic contents are concerned it does not make much sense to say that the meaning comes from information flowing with the message from the sender to the receiver (except in the trivial sense that the signal precedes the reaction and has to travel – at best with the speed of light – from the first to the latter). As far as SHANNON's statistical measure of covariance between the interactants' behavior is concerned, it is again quite misleading to state that y bits of information «flow from» A to B or are shared between them – the flow is in fact one to an outside observer, reducing *his* uncertainty about these joint events, and what is shared is the causal link between signal production and response to it that has been programmed – by evolution or experience – into this relationship. It should be emphasized that in both respects the concept «information» cannot be used without considering both sender and receiver involvement; it is in a semantic as well as in a statistical sense *an interactional property of a relationship between systems,* not some kind of «substance» which either of them can have, share, give, fake or deny.

It would be, however, premature to leave it at that. The reason is that A and B in a bidirectional communication loop really are not only sender and receiver but can also act as observers in the sense described above, thus really drawing information, i. e. reducing their uncertainty, about important aspects of their communicative relationship. This can be best explained by referring to a diagram (Fig. 1) which analyses the different kinds of possible information relations between A and B and their respective external and internal environments.

To understand this, it is important to realize that any probabilistic relationship between two event fields X and Y can be characterized by a T-value which indicates the tightness of correlation between the event states in the two connected fields or the informational, not necessarily causal constraints exerted by one on the other; if the connection is or can be assumed ot be a causal one (X→Y), an observer can draw particularly reliable information on what will occur in field Y upon observing event x_i in X, or what had occurred (but was unobservable) in X, from knowing the resultant event in Y (always referring to probabilistic conclusions). With animal A and B communicating (both alternatingly as senders and receivers) one has to consider the informational relationships indicated in Fig. 1, each represented by a T-value, with T significantly > 0 defining the possibility for the observer to

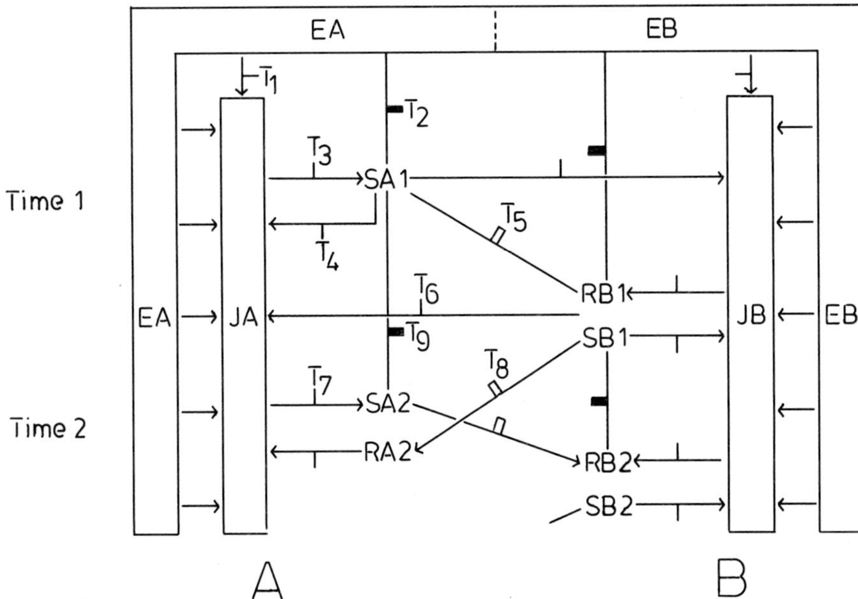

Fig. 1: Communicative relationships between animals A and B. EA, EB: overlapping external situations (environmental context) for A or B, with B's presence being part of EA and viceversa. IA, IB: Internal (motivational) states of A or B. A begins interaction by sending signal SA 1 at time 1; B responds with reaction RB 1 which is at the same time a signal SB 1 that influences response RA 2 of A at time 2; this again is the next signal action SA 2 of A etc. T: statistical information measure («transmission») for the tightness of coupling between the event states connected by lines. Lines with an arrow at one end indicate flow direction of a causal influence; lines without an arrow indicate a purely informational relationship. For the three different types of T see text.

extract useful information from T on the relationships between A, B, their external conditions and their internal constitutional and motivational states.

If A begins to act as a sender of signal SA 1 (e.g. a threat or alarm display), there is a causal relationship between his external situation EA (environment, stimulus context, which may also include the presence of B), his internal state of behavioral tendencies IA (motivation) and the signal selected for emission from an available repertoire. Therefore we can define the three informational relationships T_1, T_2, T_3. T_2 for example, being that between context and signal, i.e. what can be inferred about the external situation of A from knowing that A has produced signal SA 1, thus quantifying the reliability of connection between the signal and an external referent, e.g. the presence of a specific enemy or a mate. If A can monitor his own signal production and feed this back to his own internal state, we have another relation between SA 1 and IA, characterized by T_4, which in recurrent relationship with T_3 relates to what we might call «display management» of A or «self-communication» (see p. 186).

Assume that B perceives SA 1 and reacts by response RB 1 to it (e.g. a counter-threat display). This connection is represented by T_5, the classical «transmission» of statistical communication analysis, a measure of the constraint imposed on B's behavior by A's signaling. It need not be pointed out in detail that the relationships between B's environmental context EB (which may include A's presence again), its internal motivational state IB, the response RB 1 and its feedback on IB are all symmetrical to the conditions described for A and defined by their respective T-values (drawn but not numbered in Fig. 1 so as to avoid making the Figure even more complicated).

As a rule, response RB 1 will be monitored by A, thus closing the bidirectional communication loop («response monitoring»), influencing A's behavioral state (T_6), and allowing A to change its consecutive actions accordingly («impression or effect management», see Goffman 1959, Snyder 1979). In this respect, response RB 1 acts at the same time as a signal SB 1 which influences A's behavior. This influence adds at that stage to the combined action of EA and IA in determining the next action of A (represented by T_7) which can be regarded as a new signal SA 2 (e.g. a form of attack) or as response RA 2 to B's signal SB 1. The relationship between SB 1 and RA 2 is characterized by T_8 (which could be called «backward tranmission» if T_5 is seen as the «forward transmission»). Another important informational (not causal!) relationship is that between signals SA 1 and SA 2, given by T_9. SA 2 then can release response RB 2 (e.g. escape) in B, and so on.

Highly abstract as this may seem, all of these communicative or informational relationships are of importance in fully accounting for this simple, two-stage communication process between A and B. These 9 different relationships and T-values fall into 3 different groups according to whether and for which observer they are available from empirical evidence, reducing the observer's uncertainty.

1. T_1, T_3, T_4, T_6, T_7 all involve the «internal motivational state» of A which is not directly observable, let alone measurable, either for B or for an outside observer, who, however, may try to estimate it from presumed physiological correlates of IA. These T-values are therefore for principal reasons rather *theoretical constructs* than empirical variables. IA would, of course, be accessible for A himself: however, to treat $T_{1,3,4,6,7}$ as «real» for A, means postulating that A contains an «internal observer» as an own cognitive entity, separate from IA, that can compare the state of IA with those of other event fields to which IA is connected, and thus gain the information represented by $T_{1,3,4,6,7}$. This is an epistemologically critical issue and I will return to it in a moment.

2. T_5 and T_8 are only available to an *outside observer* – no wonder that they play a decisive role as «transmissions» in theory and practice of animal communication (see «transmission T», p. 173). They do not, to repeat it again, represent «pieces of information» flowing from A to B or back, but probabilistic constraints imposed upon each other's behavior by the communicating individuals. At first glance, T_5 and T_8 might be also regarded as being available for the interactants themselves (T_5 for B, T_8 for A). This need not imply «internal observers» in the same sense as just discussed. However, it seems fallacious for another reason: B cannot reduce any uncertainty from «observing» that it has responded with RB 1 to SA 1, because it already knows for certain that it has done so; therefore: no uncertainty reduced, no information gained, no transmission to talk about, – except in a rather schizophrenic view of animal B containing an «internal observer» which has no access to IB and therefore responds with surprise when noticing B (that is: itself!) produce RB 1 upon receiving SA 1.

3. Finally, T_2 and T_9 are most important, because they represent variables which are available from empirical observation to the *communicating animals* (at least in principle). Both represent informational, not direct causal links, since EA acts on SA 1, and SA 1 on SA 2, not directly but only through IA! From T_2 B can learn about the environmental context of A when producing SA 1, unless B itself shares this context completely with A; it is therefore representing *referential information* given by A on its external environment. With $T_2 > 0$ and B being able to extract this information from SA 1 (which is not proven by $T_2 > 0$, but has to be established *separetely* in suitable experiments, see for example the

honeybee dance language controversy about this, GOULD 1976), as has been demonstrated for honey bees, vervet monkeys (SEYFARTH et al. 1980) and chimpanzees (see review by RISTAU and ROBBINS 1982) this can be called *indexical* or even *descriptive communication*.

T_9 represents the reduction in uncertainty – for observer B if it is programmed to or has learned to extract this information, but of course also for an outside observer – about A's action SA 2 (RA 2) if its action SA 1 is known to the observer. It therefore represents *predictive information* about A for the observer. The discussion (referred to above, p. 168f) about whether signals accurately inform about the sender's intentions (IA), and therefore are reliable predictors of the sender's future actions, refers to *this* information relationship only. As can be easily seen now, the first part of this question relates to T_3 and T_7 and is unanswerable because of the principal methodological constraints mentioned; it is therefore not a particularly useful question to ask at all (nor can it be regarded as useful to answer it in the affirmative from intuitional conviction) because it mixes statements pertaining to the subjective and the objective realms of experience. The second part is a useful and empirically answerable question – a CARYL (1979) has convincingly demonstrated – but, alas, there is no general theory which could predict that T_9 has to be either 0 (no predictive value of SA 1 for SA 2 whatsoever) or 1 (totally deterministic relation between the two). Much rather we would expect $0 < T_9 < 1$ (as CARYL has found indeed) with the actual values having to be determined for each particular communicative process. What we can expect from theory is that in some (e. g. agonistic) relationships T_9 should be smaller than in others (e. g. broodcare).

Evidently, whether SA 1 is accurate or «bluff» has actually nothing at all to do with the magnitude of T_9. In fact, this question pertains to an entirely different way of looking at the system. What could it mean to say that SA 1 is «deceptive»? Leaving the question of intention for a moment aside, the only way to sensibly introduce the concept of «deception» or «honesty» into the system represented in Fig. 1, would be to call an equality between T_7 and T_9 honesty and to speak of deception if T_9 is not equal to T_7, because the «honest informer» concept assumes that signal SA 1 given at time 1 accurately predicts the conditions in A (IA) at time 2 which will make A produce SA 2. However, simply for logical reasons, it cannot be assumed that $T_7 = T_9$, for instance because T_7 is the result of influences (represented by T_6) on A which occur *after* A has produced SA 1, not to mention other possible influences (e. g. from EA) and state changes of IA (e. g. represented by T_4) between times 1 and 2.

Does that mean that signal SA 1 cannot be an «honest predictor» and therefore must be «deceitful»? This would be a rash conclusion, mainly because it amounts to an entirely inappropriate usage of words and concepts. Even if A «honestly» expressed IA at time 1 by SA 1 and IA at time 2 in SA 2, and so on, SA 1 might still be a poor predictor of SA 2. Evidently, it is wrong to use the terms «honest» or «deceptive» at all in this connection. It is wrong, first, because neither B nor an outside observer can observe IA, T_3 or T_7 directly, thus making it impossible to empirically test the hypotheses $T_7 = T_9$ or $T_7 \neq T_9$. Second, and more important, because for any acceptable way of usage of the concept «deception» it seems impossible to not postulate an «internal observer» in A which not only has to be able to «extract» T_3 and T_7, but also to intervene in the production of signals by A so that not the true state IA is expressed by the signal but some virtual state, so that B's behavior is misled. Ascribing not only the cognitive entity «internal observer» to A but also giving it the power for deceitful intervention into signal production by A is tantamount to ascribing «intention» – namely to deceive – to A (which may imply to ascribe consciousness). Thus either one has to take this position and then may talk about «honest» or «deceptive» signaling – though without being able to operationally define and empirically test the correctness of these attributions – a standpoint which will be difficult to reconcile with accepted scientific procedure; or, one cannot apply these attributions; or, if one does so, must use the words relating to them incorrectly. The disclaimer that no assumption of conscious intent is meant to be implied with the usage of these terms, can thus only be regarded as pious self-deception. I would therefore suggest to regard the question of «honesty» or «deceptiveness» of displays as terminologically and epistemologically fallacious, a red herring of animal communication research at least as long as we deal with animals for which we are not compelled by convincing evidence to postulate an «internal observer».

The fully legitimate and important quenstion of whether signals should allow the prediction of future actions of a sender or not, cannot be answered theoretically in a general way. The answer depends, as discussed above (p. 168) on the ratio of benefit to cost of different values of T_9 for the sender, on the selective pressures exerted by the sender on the receiver and viceversa (via the ways represented by T_5, T_8) in communication, and thus ultimately on the fitness effects for the interactants. All this has nothing to do with one interactant informing the other more or less honestly; it rather means that all «transmission» values considered in Fig. 1 are under selection both on A and B for optimizing their fitness. If it increases the fitness of A to increase, e. g., the «referential information» T_2, or the «predictive information» T_9, selection will favor such increase; if not, not. If it increases the fitness of B to extract these informations from A's behaviors, selection will favor development of such observers' qualifications in B; if not, resistance in B will be favored against reacting to A's signals. Selection on the sender will favor exploitation of the receiver and selection on the receiver resistance against being exploited. If both parties gain from unexploitative, equitable communicative cooperation, such mutualism will be favored by selection on both of them.

The question posed in the beginning of what is being transferred, and, specifically, whether accurate information is transmitted by a signal in animal communication, thus resolves itself on close inspection into a series of separate questions about the manifold informational relationships in a communicative process. This leads to the hardly surprising general conclusion that the fitness effects of communication on both communicators determine how «informative» (in the sense of statistical information theory) these relationships will be for different types of observers. It does not make much sense to make a general statement that information is transmitted in communication; information is a property of the communicative interaction which cannot be treated in isolation from either of both interactants. Nor does this mean that no information can be transmitted under any circumstances. It all depends on what one is talking about.

5 Cognition – or the power of communication

To be critical about the unwarranted introduction of an «internal observer» in explaining animal communication in general should not, however, lead us to disregard the evident fact that at least in one species – our own – we know that such a conscious internal observer participates most influentially in communication. As the quip goes it makes all the difference whether «the brain has been switched on before operating the mouth» or not. Referring again to Fig. 1, where we only find the most parsimonious assumption about self-communication represented by the feedback-lopp $T_3 - T_4$, we have to expand this model in order to accommodate the participant mind by adding a separate entity – the internal observer – which stands in bidirectional informational relationship with IA, that we can call *internal communication*. In addition to the feedback from a partner's response

– which serves in *impression* or *signal effect management* – and the self-monitoring feedback from A's own signals – which serves in *display management* – we thus have a third level of communicative control, *self-image management,* which coordinates the mental representations of the actor, his theory of the self with its properties, abilities, fears and desires, with the mental representation of the environment, his theory of the outside world. Evidently, this additional level of internal communicative performance supplies us with the capacity to control our outward directed communication by experience gained from internal communication within our model of self and world representations.

Animals on the level of Fig. 1 can only cope with the challenges posed by interactions with other animals by communicating according to the present state of their programs – as introduced into them by evolutionary endowment or individual experience – at best correcting and adjusting these programs step by step according to the results of behavioral performance.

In contrast, a system that is capable of internal communication within conceptual models can self-instruct its programs according to internally pretested and thus anticipated consequences of using them. The vastly expanded power of this capability for behavior preconsidered by internal communication, whether resulting in deliberate altruism or premeditated felony, need not be pointed out. Honesty or deceit are no longer non-parsimonious modes of metaphorical parlance but real attributes of systems working with internally observing, thinking minds.

It is therefore only too natural that the «cognitive ethologist» should try to use the door opened by communication to an animal's inner world in order to find out whether the machine is inhabited by such an internal agent or only equipped to act out its programs (see Griffin 1981 for the discussion of the aspirations and problems raised by this endeavour). Since the «perfected» state of mental development of our species is the result of a long evolutionary history, efforts have been made to trace the development of the way from animal mind to human mind (see Crook 1980, Griffin 1982). A most important step for giving a system the superior power provided by internal communication, is to enable it to put itself in the place of the external communication partner (empathy), or more correctly, to let it conceive an internal model of the partner's properties, intentions and behaviors and to interact cognitively with this model (see Stent 1978 for further discussions fo this view).

Now, even the most primitive forms of animal communication contain what may be the most decisive prerequisite for this development; the ability, even the necessity for each animal to function simultaneously and alternatingly as sender *and* as receiver of signals. This entails that – if the other with whom one communicates is not entirely different from oneself and it cannot be so, if at least some kind of reciprocal communication is to come about – the partner is in some respect already represented in oneself in the form of one's own alternate state of communicating. The animal that emits a specific signal contains at the same time at least some information on how to react to a similar signal which must overlap with that present in the receiving partner, and vice versa. The evolution of the «internal observer» could thus be viewed as the operational concentration of all the information that can represent the other interactant and the establishment of this representational unit in the central nervous system so that an *internal dialogue* can begin that precedes the overt external signal exchange. If there is some truth to this view this would mean that the development of the conscious mind is inseparable from the development of an internal operational representation of the communication partner in order to control more efficiently one's interaction with one's social environment. Thus the self could only define, i. e. demarcate itself by conceptually incorporating the interactant as partner of one's internal communication.

The development of advanced and intensive social cooperation would thus seem a prerequisite of the development of a conscious self. Just as – in the view of Ch. S. Peirce and C. W. Morris (see Morris 1946) nothing can be a sign without an object to which it refers and without an interpreter that responds to it, making the information contained in the sign a triadic relational quality, it could be that the power of the internal information processing that we call thinking could not develop without the internal representation of both communicators with the vis-à-vis being modeled from the ambivalent self-experience of a janus-headed sender-receiver-communicator. The ability to deceive the outside partner by a lie would then depend on the ability to tell its internal representation the truth, because only by internally rating the partner's presumed response to a signal can it become possible to change the display until the anticipated response suits one's own purposes.

This is, however, only one way of looking at the relationship between communication and cognition.

What is cognition? We may call it the ability to relate different unconnected informations in new ways and to apply the results of this in an adaptive manner, i. e. to solving one's problems. We know that (though hardly how) we do this by thinking in our minds. We are, however, also aware that there is a specific kind of thinking that is fueled by communicating with others. J. von Liebig is said to have confessed to rarely having had a good idea, except if somebody else suggested anything: then he immediately had a better idea. This wellknown experience deserves further elaboration in the present context.

Communication in a social environment can not only sharpen an individual's acuity of cognition. The social organization can, by communicatory processes, itself develop the properties of a problem solving, cognitive system that transcends the cognitive capacities of individual participants even though it is entirely based on their contributions to the communicative interactions. To regard societies as supraorganismic systems with emergent properties is of course not new. We can find this view – in a sometimes more, sometimes less metaphorical fashion – in leading students of social insects (like W. M. Wheeler or A. E. Emerson) as well as in many eminent observers of human sociology from Adam Smith to F. von Hayek, and even applied to the entire biosphere (see Lovelock 1979). What deserves emphasis in the present connection is that among the emergent properties of such supraindividual systems there is one that seems most intimately related to the communicative mechanisms that coordinate the complex workings of these systems: the invention of social adaptations which have all the characteristics of cognitive problem solving, of relating the information available to the single individuals in such a way as to produce novel solutions for the challenges and exigencies of life, not at the individual, but at the group level. The «invisible hand» that seems to guide the communal activities for the benefit of the whole system could then be seen as directed by an «invisible brain», the functional elements of which are the individual communicating members which are connected by a web of informational relationships, the most primitive core of which would be represented by the transmission-network in Fig. 1.

Ensemble performances are, of course, particularly evident in the amazing constructions of many social animals as Karl von Frisch so vividly described late in his life (1974). Although it could be argued that a honeybees' comb gives expression of the recognition of basic laws of material science, fluid and structural mechanics, and of solving the problems of storing large quantities of highly concentrated and perishable food over many months, we would ordinarily not be inclined to see more than metaphor in expressing these facts in a language inplying «cognition». However, there is a gradual, not categorical transition from these crystallized relics of behavior to behavioral processes, where we have to concede that the interaction group performs through communication and cooperation in a way that is not principally different from the outward behavioral expression of problem solving of a cognitive system (the assumption of conscious thinking being neither necessary in the one nor in the other case).

This idea can be elaborated by a few examples from social insects in an ascending scale of complexity:

– A most simple case is represented by the workers of the ant *Novomessor* which cooperatively transport large prey (see p. 176). Although the single worker is neither able to estimate the weight of a grasshopper, too large to be carried alone, nor to judge how many helpers would be needed to move the object, a complex communication system makes sure that the necessary workforce – and not much more – assembles fast and thus performs the transport at low cost and with high efficiency (Hölldobler et al. 1978, Markl and Hölldobler 1978).

– Colonies of the bulldog ant *Myrmecia* have to guard an area of dozens of square meters around their nest against intruders with a workforce of only a few scores of individuals. They achieve this by a contact alarm system in combination with a random search technique that allows them to cover the area thoroughly with minimum personnel (Markl, Frehland, Kleutsch unpubl.).

– Workers of different colonies of the honeypot ant *Myrmecocystus* engage in fighting tournaments in which they rate the strength of the enemy force in ritualized mass displays, thus gaining information on relative fighting strength (RHP) and deciding accordingly to attack, withstand or withdraw, an ensemble decision with all characteristics of a cognitive evaluation of the situation (Hölldobler 1983, Lumsden and Hölldobler 1983).

– One of the most complex achievements of an «ensemble cognitive process» is the decision finding about moving into a new home by a honeybee swarm: scouts that have discovered a suitable dwelling and investigated its quality according to well-defined bee public housing norms, return to the swarm and advertise their discovery in competition with other bidders, indicating the quality of their find by the vivacity and obstinacy of their dance. In an extended and sometimes even swarm-rending procedure, the receivers of these messages decide in ensemble fashion which offer to prefer (Lindauer 1955, 1961, Seeley 1977). It is hardly possible to describe what occurs in such a swarm without using vocabulary that implies that a higher-order cognitive performance is achieved during this process.

This brings us back once more to the importance of modulating signals in these communication processes. Many of the signal effects observed during these performances are not of the high-transmission, releaser kind, but rather ones that subtly modify the behavioral properties of the receivers, alerting them and focussing their attention to the situational context. It seems inevitable to compare the coordinating mechanisms active in such a «cognitive ensemble» with the interactive neuronal processes that presumably endow a central nervous system with its acknowledged cognitive capacities. It is remarkable that in the brain as in the social system we find mechanisms that set the overall level of arousal.

Tonic sensory input form a variety of sensory organs and spontaneous activity of neural arousal systems perform in the nervous system the functions that unspecific modulating alarm signals serve in social organizations.

In both forms of organization we further find more specific, as it were, regionalized or addressed mechanisms of focussing the attention to a specified subset of stimuli in a given context. Furthermore we find that within the larger systems mechanisms exist which modulate in graded fashion the activity probability of small dedicated subpopulations of neurons or individuals which are thus recruited to performing specified tasks. It is probably more than chance that the neurophysiologist arrives at describing these mechanisms as local modulating interactions between neurons just as the student of social communication in animals independently finds it appropriate to qualify basic processes of social organization in this way. It seems more convincing to accept that at the level of the highly organized, complex society we find a functional systemic correlate of what also occurs at the multicellular level of an individual organism (for a review of arousal and modulating mechanisms in nervous systems of invertebrates and vertebrates, see Bullock and Horridge 1965, Florey 1967, Schmitt et al. 1976, Evans 1978, Evans and O'Shea 1978, Hoyle and Dagan 1978, Dismukes 1979, Laming 1981, Frederickson and Geary 1982, Jones 1982, Bloom 1983, Iversen 1983).

Lumsden (1982) has applied a very similar view to an explanation of the division of labor by dividing up the workforce of an insect society into distinct physical castes, and has shown that these supraorganismal achievements can be treated by rigorous theory (see also Franks and Wilson in this volume). This opens the even wider view of regarding one of the most basic elements of cooperation between members of an animal species – that between the sexes – as an expression of supraorganismal problem solving.

To view social systems as higher-level «organisms» characterized by emergent qualities notwithstanding the fact that each of their members strives to maximize its own (though inclusive) fitness, is – as was said before – not at all new.

The study of communication goes beyond the metaphor, however, by analysing the mechanisms that enable a society to achieve these ensemble performances and to unravel the mutualistic informational networks that are so tightly knit as to assume the properties of cognitive, problem-solving entities that process information in highly coordinated fashion and adapt to changes in the environment by discovering new solutions, by learning from experience and by establishing stores of knowledge in social traditions.

There is a long way to this stage of development from two solitary animals opposing each other over a desired resource, trying to arrive at their goal at minimal cost by employing browbeating signals rather than beating each other by violent force. There is, of course, nothing wrong about looking at animal communication from below, where it starts. Only one should not stop there. It would seem that one cannot get a comprehensive impression of what animal communication can be unless one also has looked at it from the top of achievement, where it is at its best in welding a cooperative whole out of numerous independently acting parts. It seems only appropriate to call this the Karl von Frisch-view of animal communication.

Acknowledgement

Bert Hölldobler and John Krebs have read and commented on the manuscript. Their suggestions were highly appreciated.

References

Andersson, M. (1980): Why are there so many threat displays. J. theor. Biol. *86,* 773–781.

Anisko, J. J., Suer, Sh. F., McClintock, M. K., Adler, N. T. (1978): Relation between 22-kHz ultrasonic signals and sociosexual behavior in rats. J. Comp. Physiol. Psychol. *92,* 821–829.

Argyle, M. (1975): Bodily communication. Methuen, London.

Argyle, M., Cook, M. (1976): Gaze and mutual gaze. Cambridge Univ. Press, Cambridge.

Attneave, F. (1959): Applications of information theory to psychology. Holt, Rinehart, Winston, New York.

Axelrod, R., Hamilton, W. D. (1981): The evolution of cooperation. Science *211,* 1390–1396.

Axelrod, R. (1984): The evolution of cooperation. Basic Books, New York.

Barnard, C. J., Burk, T. (1979): Dominance hierarchies and the evolution of «individual recognition». J. Theor. Biol. *81,* 65–73.

Barnard, C. J., Burk, T. (1981): Individuals as assessment units. J. theor. Biol. *88:* 595–598.
Bateson, P. P. (1980): Optimal outbreeding and the development of sexual preferences in Japanese quail. Z. Tierpsychol. *53,* 231–244.
Bateson, P. P. (1982): Preferences for cousins in Japanese quail. Nature *295,* 236–237.
Bloom, F. E. (1983): The endorphins: a growing family of pharmacologically pertinent peptides. Ann. Rev. Pharmakol. Toxicol. *23,* 151–170.
Boake, Ch., Capranica, R. R. (1982): Aggressive signal in «courtship» chirps of a gregarious cricket. Science *218,* 580–582.
Bossema, I., Burgler, R. R. (1980): Communication during monocular and binocular looking in European jays *(Garrulus g. glandarius).* Behaviour *74,* 274–283.
Bossert, W. H., Wilson, E. O. (1963): The analysis of olfactory communication among animals. J. theor. Biol. *5,* 443–469.
Breed, M. D., Bekoff, M. (1981): Individual recognition and social relationships. J. theor. Biol. *88,* 589–594.
Brown, R. E. (1979): Mammalian social odors: a critical review. Adv. Study Behav. *10,* 103–162.
Bullock, Th. H., Horridge, G. A. (1965): Structure and function in the nervous systems of invertebrates. 2 vol. Freeman, San Francisco.
Burghardt, G. M. (1970): Defining communication, p. 5–18, in: J. W. Johnston, D. G. Moulton, A. Turk (eds.) Communication by chemical signals. Appleton-Century-Crofts, New York.
Cade, W. (1979): The evolution of alternative male reproductive strategies in field crickets. p. 343–380 in: M. S. Blum, N. A. Blum (eds.) Sexual selection and reproductive competition in insects. Academic Press, New York.
Caryl, P. G. (1979): Communication by agonistic displays: what can games theory contribute to ethology? Behaviour *68,* 136–169.
Caryl, P. G. (1981): Escalated fighting and the war of nerves: games theory and animal combat. p. 199–224 in: P. P. G. Bateson, P. H. Klopfer (eds.): Perspectives in Ethology *4,* Plenum, New York.
Caryl, P. G. (1982): Animal signals: a reply to Hinde. Anim. Behav. *30,* 240–244.
Chance, M. R. (1967): Attention structure as the basis of primate rank orders. Man *2,* 503–518.
Cheney, D. L., Seyfarth, R. M. (1982 a): Recognition of individuals within and between groups of free ranging vervet monkeys. Amer. Zool. *22,* 519–529.
Cheney, D. L., Seyfarth, R. M. (1982 b): How vervet monkeys perceive their grunts: field playback experiments. Anim. Behav. *30,* 739–751.
Cherry, C. (1978): On human communication. MIT Press, Cambridge.
Clayton, D. A. (1978): Socially facilitated behavior. Quart. Rev. Biol. *53,* 373–393.
Crook, J. H. (1980): The evolution of human consciousness. Clarendon Press, Oxford.
Dawkins, R. (1982): The extended phenotype. Freeman, San Francisco.
Dawkins, R., Krebs, J. R. (1978): Animal Signals: Information or manipulation? p. 282–309 in: J. R. Krebs, N. B. Davies (eds.) Behavioural Ecology. Blackwell, Oxford*.
Dawkins, R., Krebs, J. R. (1979): Arms races between and within species: Proc. R. Soc. Lond. B *205,* 489–511.
de Waal, F. (1982): Chimpanzee Politics. Cape, London.
Dismukes, R. K. (1979): New concepts of molecular communication among neurons. Behav. Brain Sciences *2,* 409–448.
Eibl-Eibesfeldt, I. (1979): Human ethology: concepts and implications for the sciences of man. Behav. Brain Sciences *2,* 1–57.
Eibl-Eibesfeldt, I. (1980): Strategies of social interaction p. 57–80 in: Emotion. Theory, Research and Experience. Vol. *1,* Theories of Emotion.
Eibl-Eibesfeldt, I.: The biological foundations of aesthetics (in press.).
Eisenberg, J. F., Kleiman, D. G. (eds.) (1983): Advances in the study of mammalian behavior. Amer. Soc. Mammal. Special Publ. No 7.
Ekman, P. (ed.) (1973): Darwin and facial expression. Acad. Press, New York.
Evans, P. D. (1978): Octopamine: from metabolic mistake to modulator. TINS *1,* 154–157.

---

* Note added in proof: the revised 2nd edition (1984) of this book has not been available before this article was finished.

Evans, P. D., O'Shea, M. (1978): The identification of an octopaminergic neurone and the modulation of a myogenic rhythm in the locust. J. exp. Biol. *73*, 235–260.

Fagen, R. M. (1978): Information measures: statistical confidence limits and inference. J. theor. Biol. *73*, 61–79.

Fischer, E. A. (1980): The relationship between mating system and simultaneous hermaphroditism in the coral reef fish, *Hypoplectus nigricans* (Serranidae). Anim. Behav. *28*, 620–633.

Florey, E. (1967): Neurotransmitters and modulators in the animal kingdom. Fed. Proc. *26*, 1164–1178.

Frederickson, R. C., Geary, L. E. (1982): Endogenous opioid peptides: review of physiological, pharmacological and clinical aspects. Progr. Neurobiol. *19*, 19–69.

Fuchs, S. (1976 a): The response to vibrations of the substrate and reactions to the specific drumming in colonies of carpenter ants (*Camponotus*, Formicidae, Hymenoptera) Behav. Ecol. Sociobiol. *1*, 155–184.

Fuchs, S. (1976 b): An informational analysis of the alarm communication by drumming behavior in nests of carpenter ants (*Camponotus*, Formicidae, Hymenoptera). Behav. Ecol. Sociobiol. *1*, 315–336.

Goffman, E. (1959): The presentation of self in everyday life. Doubleday Anchor, Garden City.

Goffman, E. (1969): Strategic Interaction. Basil Blackwell, Oxford.

Gould, J. L. (1976): The dance-language controversy. Quart. Rev. Biol. *51*, 211–244.

Green, S., Marler, P. (1979): The analysis of animal communication, p. 73–158 in: P. Marler, J. G. Vandenbergh (eds.): Handbook of Behavioral Neurobiology 3. Social Behavior and Communication. Plenum, New York.

Griffin, D. R. (1981): Animal Mind-Human Mind. Dahlem Konferenzen 21 Springer, Berlin.

Griffin, D. R. (1982): The Question of Animal Awareness. Rockefeller Univ. Press, New York.

Halliday, T. R., Slater, P. J. B. (eds.) (1983): Animal Behavior. 2. Communication. Blackwell, Oxford.

Hinde, R. A. (ed.) (1972): Non-verbal Communication. Cambridge Univ. Press, Cambridge.

Hinde, R. A. (1981): Animal signals: ethological and games-theory approaches are not incompatible. Anim. Behav. *29*, 535–542.

Hölldobler, B. (1971): Recruitment behavior in *Camponotus socius* (Hym. Formicidae). Z. vergl. Physiol. *75*, 123–142.

Hölldobler, B. (1976): The behavioral ecology of mating in harvester ants (Hymenoptera: Formicidae: Pogonomyrmex). Behav. Ecol. Sociobiol. *1*, 405–423.

Hölldobler, B. (1977): Communication in social hymenoptera p. 418–471 in: Th. A. Sebeok (ed.): How animals communicate. Indiana Univ. Press, Bloomington.

Hölldobler, B. (1983): Chemical manipulation, enemy specification and intercolony communication in ant communities. p. 354–365 in: F. Huber, H. Markl (eds.): Neuroethology and Behavioral Physiology. Springer, Berlin.

Hölldobler, B., Wilson, E. O. (1978): The multiple recruitment systems of the African weaver ant *Oecophylla longinoda* (Latreille) (Hymenoptera: Formicidae): Behav. Ecol. Sociobiol. *3*, 19–60.

Holmes, W. G., Sherman, P. W. (1983): Kin recognition in animals. Amer. Sci. *71*, 46–55.

Howse, P. E. (1964): The significance of the sound produced by the termite *Zootermopsis angusticollis* (Hagen). Anim. Behav. *12*, 284–300.

Hoyle, G., Dagan, D. (1978): Physiological characteristics and reflex activation of DUM (octopaminergic) neurons of locust metathoracic ganglion. J. Neurobiol. *9*, 59–79.

Jenssen, T. A. (1979): Display modifiers of *Anolis opalinus* (Lacertilia: Iguanidae) Herpetologica *35*, 21–30.

Jones, R. S. G. (1982): Tryptamine: a neuromodulator or neurotransmitter in mammalian brain? Progr. Neurobiol. *19*, 117–139.

Iversen, L. L. (1983): Nonopioid neuropeptides in mammalian CNS. Ann. Rev. Pharmacol. Toxicol. *23*, 1–27.

Krames, L., Pliner, P., Alloway, Th. (eds.) (1974): Nonverbal communication. Plenum, New York.

Laming, P. R. (1981): The physiological basis of alert behavior in fish. p. 202–222 in: P. R. Laming (ed.) Brain mechanisms of behaviour in lower vertebrates. Soc. Exp. Biol. Sem. Ser. 9 Cambridge Univ. Press, Cambridge.

Lenoir, A. (1982): An informational analysis of antennal communication during trophallaxis in the ant *Myrmica rubra* L. Behav. Processes *7,* 27–35.

Lewis, D. B., Gower, D. M. (1980): Biology of Communication. Blackie, Glasgow.

Lindauer, M. (1955): Schwarmbienen auf Wohnungssuche. Z. vergl. Physiol. *37,* 263–324.

Lindauer, M. (1961): Communication among social bees. Harvard Univ. Press, Cambridge.

Lloyd, J. E. (1977): Bioluminescence and communication. p. 164–183 in: The. A. Sebeok: How Animals Communicate. Indiana Univ. Press, Bloomington.

Lloyd, J. E. (1981): Mimicry in the sexual signals of fireflies. Sci. Amer. *245,* 111–117.

Losey, G. S. (1978): Information theory and communication. p. 43–78 in: P. W. Colgan (ed.): Quantitative Ethology. Wiley, New York.

Lovelock, J. E. (1979): Gaia. Oxford Univ. Press, Oxford.

Lumsden, Ch. J. (1982): The social regulation of physical caste: the superorganism revived. J. theor. Biol. *95,* 749–781.

Lumsden, Ch. J., Hölldobler, B. (1983): Ritualized combat and intercolony communication in ants. J. theor. Biol. *100,* 81–98.

Markl, H. (1967): Die Verständigung durch Stridulationssignale bei Blattschneiderameisen. I. Die biologische Bedeutung der Stridulation. Z. vergl. Physiol. *57,* 299–330.

Markl, H. (1983): Vibrational Communication. p. 332–353 in: F. Huber, H. Markl (eds.): Neuroethology and Behavioral Physiology. Springer, Berlin.

Markl, H. Fuchs, S. (1972): Klopfsignale mit Alarmfunktion bei Roßameisen (*Camponotus,* Formicidae, Hymenoptera). Z. vergl. Physiol. *76,* 204–225.

Markl, H., Hölldobler, B. (1978): Recruitment and food-retrieving behavior in *Novomessor* (Formicidae, Hymenoptera). II. Vibration signals. Behav. Ecol. Sociobiol. *4,*183–216.

Markl, H., Hölldobler, B., Hölldobler, T. (1977): Mating behavior and sound production in harvester ants (*Pogonomyrmex,* Formicidae): Insectes Sociaux *24,* 191–212.

Marko, H. (1983): Information theory and communication theory. p. 788–794 in: W. Hoppe, W. Lohmann, H. Markl, H. Ziegler (eds.): Biophysics. Springer, Berlin.

Marler, P., Vandenberg, J. G. (1979): Handbook of Behavioral Neurobiology 3. Social Behavior and Communication. Plenum, New York.

Maschwitz, U., Schönegge, P. (1983): Forage communication, nest moving recruitment, and prey specialization in the oriental ponerine *Leptogenys chinensis.* Oecologia (Berlin) *57,* 175–182.

Maynard Smith, J. (1982 a): Evolution and the theory of games. Cambridge Univ. Press, Cambridge.

Maynard Smith, J. (1982 b): Do animals convey information about their intentions? J. theor. Biol. *97,* 1–5.

Maynard Smith, J., Parker, G. A. (1976): The logic of asymmetric contests. Anim. Behav. *24,* 159–175.

McGregor, P. K., Krebs, J. R. (1982): Mating and song types in the great tit. Nature *297,* 60–61.

Michener, Ch. D. (1974): The social behavior of the bees. Harvard Univ. Press, Cambridge.

Miller, S. J., Inouye, D. W. (1983): Roles of the wing whistle in the territorial behavior of the male broad-tailed humming-bird *(Selasphorus platycerus).* Anim. Behav. *31,* 689–700.

Morris, C. W. (1946): Signs, language and behavior. Prentice-Hall, New York.

Morris, D., Collett, P. Marsh, P., O'Shaughnessy, M. (1979): Gestures. Their origins and distribution. Cape, London.

Moynihan, M. (1982): Why is lying about intentions rare during some kinds of contests? J. theor. Biol. *97,* 7–12.

Palm, G. (1981): Evidence, information, and surprise. Biol. Cybern. *42,* 57–68.

Parker, G. A. (1974): Assessment strategy and the evolution of animal conflicts. J. theor. Biol. *47,* 223–243.

Parker, G. A., Rubenstein, D. J. (1981): Role assessment, reserve strategy, and acquisition of information in asymmetric animal conflicts. Anim. Behav. *29,* 221–240.

Peters, R. (1980): Mammalian Communication. Brooks/Cole, Monterey.

Pliner, P., Blankstein, K. R., Spigel, I. M. (1979): Perception of emotion in self and others. Adv. Study of Communication and Affect. Vol. 5. Plenum, New York.

Pruscha, H., Maurus, M. (1976): The communicative function of some agonistic behavior patterns in squirrel monkeys: the relevance of social context. Behav. Ecol. Sociobiol. *1,* 185–214.

Quastler, H. (1958): A primer in information theory. In: Symp. information theory in biology. Yockey, H. P., Platzman, R. L., Quastler, H. (eds.) p. 3–49. Pergamon, New York.

Ristau, C. A., Robbins, D. (1982): Language in the great apes: a critical review. Adv. Study Behav. *12,* 141–255.

Ryan, M. J., Tuttle, M. D., Taft, L. K. (1981): The costs and benefits of frog chorusing behavior. Behav. Ecol. Sociobiol. *8,* 273–278.

Schmitt, F. O., Dev, P., Smith, B. H. (1976): Electrotonic processing of information by brain cells. Science *193,* 114–120.

Sebeok, Th. A. (1977) (ed.): How animals communicate. Indiana Univ. Press, Bloomington.

Seeley, Th. (1977): Measurement of nest cavity volume by the honey bee *(Apis mellifera).* Behav. Ecol. Sociobiol. *2,* 201–227.

Shannon, C. E., Weaver, W. (1949): The mathematical theory of communication. Univ. of Illinois Press, Urbana.

Seyfarth, R. M., Cheney, D. L., Marler, P. (1980): Vervet monkey alarm calls: semantic communication in a free-ranging primate. Anim. Behav. *28,* 1070–1094.

Sherman, P. W. (1981): Kinship, demography, and Belding's ground squirrel nepotism. Behav. Ecol. Sociobiol. *8,* 251–259.

Smith, W. J. (1977): The behavior of communicating. Harvard Univ. Press, Cambridge.

Snowdon, Ch. T., Cleveland, J. (1980): Individual recognition of contact calls by pygmy marmosets. Anim. Behav. *28,* 717–727.

Snyder, M. (1979): Cognitive, behavioral, and interpersonal consequences of self-monitoring, p. 181–201. In: Perception of emotion in self and others. Pliner, P., Blankstein, K. R., Spigel, I. M. (eds.) Plenum, New York.

Stamm, R. A. (1974): Ein Codon for «Bitte!» im Verständigungssystem der Lachmöve (*Larus ridibundus* Linnaeus; Aves, Laridae). Revue Suisse Zool. *81,* 722–728.

Steinberg, J. B. (1977): Information theory as an ethological tool. p. 47–74. In: Quantitative methods in the study of animal behavior. Hazlett, B. A. (ed.) Academic Press, New York.

Stent, G. (ed.) (1978): Morality as a biological phenomenon. Dahlem Konferenz 9 Verlag Chemie, Weinheim.

Taylor, R. W. (1978): *Nothomyrmecia macrops:* a living-fossil ant rediscovered. Science *201,* 979–985.

Tembrock, G. (1971): Biokommunikation I/II. Akademie-Verlag, Berlin.

Traniello, J. F. A. (1982): Population structure and social organization in the primitive ant *Amblyopone pallipes* (Hymenoptera: Formicidae). Psyche *89,* 65–80.

Trivers, R. W. (1971): The evolution of reciprocal altruism. Quart. Rev. Biol. *46,* 35–57.

Trivers, R. W. (1972): Parental investment and sexual selection. p. 136–179. In: Sexual selection and the descent of man. Campbell, B. (ed.) Heinemann, London.

Trivers, R. W. (1974): Parent-offspring conflict. Amer. Zool. *14,* 249–264.

Tuttle, M. D., Taft, L. K., Ryan, M. J. (1982): Evasive behaviour of a frog in response to bat predation. Anim. Behav. *30,* 393–397.

van Rhijn, J. G. (1980): Communication by agonistic displays: a discussion. Behaviour *74,* 284–293.

van Rhijn, J. G., Vodegel, R. (1980): Being honest about one's intentions: an evolutionary stable strategy for animal conflicts. J. theor. Biol. *85,* 623–641.

von Frisch, K. (1965): Tanzsprache und Orientierung der Bienen. Springer, Berlin.

von Frisch, K. (1974): Tiere als Baumeister. Ullstein, Frankfurt.

Weitz, Sh. (1974): Nonverbal communication, Oxford Univ. Press, New York.

Wickler, W. (1968): Mimikry. Kindler, München.

Wickler, W. (1980): Vocal dueting and the pair-bond: I. Coyness and partner commitment, a hypothesis. Z. Tierpsychol. *52,* 201–209.

Wickler, W., Seibt, U. (1982): Song splitting in the evolution of dueting. Z. Tierpsychol. *59,* 127–140.

Wiley, R. H. (1983): The evolution of communication: information and manipulation, p. 156–189. In: Animal Behaviour. 2. Communication. Halliday, T. R., Slater, P. J. B. (eds.) Blackwell, Oxford.

Wilson, E. O. (1971): The insect societies. Belknap Press of Harvard Univ. Press, Cambridge.

Wilson, E. O. (1975): Sociobiology. The new synthesis. Belknap Press of Harvard Univ. Press, Cambridge.

Fortschritte der Zoologie, Bd. 31 · Hölldobler/Lindauer (Hrsg.): Experimental Behavioral Ecology
G. Fischer Verlag · Stuttgart · New York · 1985

Auditory brain organization of birds and its constraints for the design of vocal repertoires

H. Scheich

Zoological Institute, Technical University Darmstadt, F. R. Germany

Abstract

Vocal repertoires of birds and mammals frequently contain a large number of calls with overlapping spectral components.In addition such calls may show meaningful variability. It is demonstrated in the Guinea fowl using the 2-deoxyglucose method how species-specific calls are represented in the tonotopically organized field L, the primary auditory projection area of the telencephalon. The main findings are 1. that calls are represented in restricted parts of the tonotopic map. The representation of calls with a wide-band spectrum is not according to the frequency bands which contain the largest energy but according to spectral bands with the largest intensity jump. This means that call analysis in field L chiefly relies on spectral contrasts. 2. the representation of spectral contrast was spatially separated in field L for the calls tested. 3. for frequency modulated calls the representation of the spectral contrast is enhanced by a special mechanism of lateral suppression. These results are at variance with previous postulates of environmental acoustics, namely that the strongest frequency components of calls are chosen according to optimal transmission in the atmosphere. Obviously atmospheric transmission of those parameters of complex calls which are most relevant for auditory analysis and recognition in the brain, namely spectral contrast, still awaits testing.

1 Introduction

Acoustic communication in birds and mammals commonly involves recognition of multiple sounds which together constitute the repertoire of a species. Some of these vocalizations may show a high degree of acoustic complexity and also meaningful variability comparing individuals or different situations (Green et al. 1977, Maier 1982). It is *multiplicity* and *variability* of signals in a repertoire which pose a formidable problem for the organization of neuronal networks in the auditory system that has to recognize them. While recognition of a *unique* and *stereotyped signal,* even if acoustically complex, is a relatively trivial task for a specialized network, multiplicity and variability are not, especially if there is partial overlap of parameters among different sounds. This lesson has been learnt by all investigators trying to perform acoustic, or other signal recognition by auto-

Supported by the Deutsche Forschungsgemeinschaft, SFB 45.

mata (Schroeder 1977). The solution to this problem that the brain has found during evolution is the representation and analysis of a multiplicity of variable signals in *neuronal maps* (Mountcastle 1957, Hubel and Wiesel 1962). These maps are organized in such a way that neurons representing a continuum of important stimulus variables are laid out along spatial dimensions within the anatomical structure. I shall follow this hypothesis while describing some of the experiments carried out in this laboratory in order to elucidate the neuronal mechanisms of sound recognition in birds.

All available data from birds and mammals favor the view that a basic organizational feature of auditory maps is *tonotopy* which is projected from the cochlea onto most of the central auditory structures in the brain (Aitkin 1976, Scheich 1979). This tonotopy is a simple means of systematically monitoring and comparing frequencies. Implicitly tonotopy has consequences for the strategy of processing complex auditory signals. It has emerged from our studies of auditory maps in birds that there are important constraints on the analysis of vocal sounds of species once these auditory maps were formed during evolution, and consequently that there may be constraints on the design of these sounds.

2 Coding of vocal sounds and evolution

In communication messages can be encoded by various types of physical signals. In the domain of acoustic signals the three physical dimensions which together encode messages are: frequency, intensity of individual frequency components, and temporal structures of sounds. Even though all acoustic signals can be described by these general dimensions, more complex entities are useful to characterize the majority of communication sounds. This is because one of the dimensions or a combination of them may show the highest complexity and thus contain most of the information. Harmonic tones, noisy sounds, frequency or amplitude modulations and pulse sequences are some examples. It is obvious that such complex entities are most easily envisaged as *patterns* in a three-dimensional space of frequency, intensity and time. Displays like sonagrams or running power spectra (Fig. 1) facilitate visualizing these patterns.

Communication sounds of species were shaped as a result of evolutionary pressures. In other words, the rules of encoding particular messages may not be considered arbitrary. In following this idea it should be possible, in principle, to trace the external and internal factors which in each species have favored the use of particular sounds for certain messages. In reality there is no direct evidence of this evolution, since vocal records are lacking. Consequently, this approach relies entirely on comparative studies of extant species. Key factors for the use of particular sounds may then be derived from physical properties of biotopes and from the life style of species (Wiley and Richards 1978). There are also some attempts to relate the sounds in use to basic mechanisms of vocal organs (Dürrwang 1974).

In terms of one of the above acoustic dimensions, namely frequency, there is probably one concept which truly relates to the evolution of sound production (Scheich et al. 1983). The original forms of vocalization early in evolution were most likely sounds with a *wide-band spectrum*. In other words, it was only in the case of later specializations that vocal apparatus was capable of producing pure tones and controlled modulations of tones. Instead, the construction of primitive vocal organs may have permitted the generation of noisy sounds and harmonic tones. Similarly information bearing sounds from inanimate sources are never pure tones. The wide-band acoustic environment was most probably the reason for the evolution of a cochlea in higher vertebrates which permitted a systematic resolution of simultaneously occurring frequency components of sounds. The central auditory system in turn started to be capable of recognizing those components as a pattern in a three dimensional space of frequency, intensity and time. The possibility has not attracted considerable attention that the basic need for recognition of wide-band sounds may have favored particular mechanisms in the central auditory system which relate to tonotopy. These central mechanism of sound analysis, once they were established in phylogeny may subsequently have determined the design of the vocal repertoire in each species as much as physical and ecological constraints. It is the aim of the present paper to provide some evidence for this idea.

3 Physical and ecological constraints on acoustic communication

The strongest evidence that the choice of vocal sounds is critically dependent upon acoustic properties of the environment comes from studies in bats. Even though echolocation superficially has little bearing on communication, the physical constraints working on the choice of signals may be comparable.

In bats the sounds in use, i. e. tonal or harmonic ultrasounds with or without frequency modulations and short noise bursts, are probably adaptations for echo-location in various ecological niches. In studying bats, the concept of optimizing signals has been successfully exploited by showing that species with hunting grounds in open air, close to the surface or in dense foliage, all use different signals. In each case arguments can be found that echos from certain sounds contain relevant information about nature, speed or trajectories of targets or reduce sound scatter coming from the environment, i. e. improve signal to noise ratio (Simmons and Stein 1980, Neuweiler 1983, Schnitzler and Ostwald 1983).

Following the same line of reasoning there have been attempts to explain the frequency content of long distance communication sounds of birds and mammals in various terrestrial habitats. Acoustic transmission within a few meters off the ground in open spaces or in forests favors the band between 1 and 4 kHz for distance communication. In spite of somewhat different mechanisms in both habitats this band is a compromise between *ground attenuation,* which effectively reduces frequencies lower than 1 kHz and *scattering* by plants and other obstacles, which strongly attenuates frequencies above 4 kHz (Embleton 1963, Aylor 1971, Linsken et al. 1976, Marten and Marler 1977, Marten et al. 1977, Wiley and Richards 1978). This finding is in agreement with the general observation that the highest energy in bird calls is in this medium frequency range and that hearing in most birds (and mammals) has lowest thresholds in this band (Konishi 1970, Dooling et al. 1971, Sachs et al. 1978, Kuhn et al. 1982, Theurich et al. 1984).

More interesting in our context than this general match are reports on differences between birds of tropical forests and of open habitats. In accordance with the fact that forest scattering effectively attenuates high frequencies, birds in forests appear to use somewhat lower pitched calls than species in open spaces of the same area (Chappius 1971, Morton 1975).

Similar observations on preference for lower pitched sounds stem from a comparison of African monkeys which keep different typical distances between social groups (Waser and Waser 1977). The lowest frequencies are used by the species with the largest typical distance. Note however that the frequencies involved are way below those which are used by the birds discussed above (around 500 Hz).

In summary there are several examples where physical constraints seem to have influenced the choice of frequencies in calls. In view of the fact that these examples cover only a small proportion of the repertoire of the species and that only an extremely small proportion of species has been studied, the evidence appears to be weak. It is also unfortunate that nobody has looked for counter examples where an apparent acoustic constraint has not led to an adaptation.

The use of complex parameters in long distance communication is even less understood than the choice of peak frequencies described above. There is good evidence that amplitude modulations become heavily distorted by so-called nonstationary heterogeneities such as air turbulence (WILEY and RICHARDS 1978). Nevertheless, amplitude modulations are characteristic of many bird and mammalian vocalizations (see GREENEWALT 1968, SEBEOK 1977). Among frequency modulations, slow modulations of a tone and fast repetitive modulations (trills) should be distinguished. Slow modulations between 1 and 4 kHz are not easily masked by any one of the discussed physical mechanisms. Therefore they are quite useful for long distance communication and are heard frequently from birds. Similarly trills are abundant in bird and mammalian repertoires. In theory, however, their usefulness is limited because they are masked by reverberations of the sound from dense arrays of obstacles such as leafs. It was claimed that birds of the *understory* in tropical forests avoid trills (MORTON 1975).

From these considerations it becomes apparent that so far there is only limited congruence between acoustic rules and preference or avoidance of certain parameters in vocalizations. One of the reasons may be that it is difficult to make a clear-cut distinction between long- and short distance communications. With all mechanisms discussed, the masking effect increases with distances and may be negligiable over a short range. In social animals the majority of sounds are short distance signals anyway.

There is a more fundamental objection to directly correlating the avoidance of some acoustic parameters with their vulnerability by physical influence in the environment. This has to do with the fact that there is little information on which parameters are actually used by the auditory system. As we shall see below, it is even doubtful that for the auditory system a prominent feature such as the peak frequency content (frequency with the highest energy) is an important parameter in all sounds. Similarly for more elaborate parameters such as amplitude or frequency modulations it is not intuitive in which way the system makes use of them and which kind of perturbations of these parameters the system is resistent to.

4 Functional organization of the auditory field L in the bird telencephalon

I shall describe here results which have been obtained in field L of the Guinea fowl. This field in the caudal neostriatum of the bird telencephalon was cytoarchitectonically distinguished from the sourrounding neostriatum by ROSE (1914) and later identified as the primary projection area of the thalamic auditory n. ovoidalis by KARTEN (1968) using degeneration technique. Trancing with labeled aminoacids in our laboratory has revealed that the extent of field L as shown by the earlier studies corresponds to only one layer, i. e. the chief input layer L_2 of terminals from n. ovoidalis (BONKE et al. 1979a). Two other layers, L_1 dorsal and L_3 ventral to L_2 were determined to be autitory as well, but to have little direct input. Microelectrode recordings showed that all three layers have a prominent and corresponding tonotopic organization with isofrequency contours oriented approximately perpendicular to the layering (BONKE et al. 1979b).

The 2-deoxyglucose method (SOKOLOFF et al. 1977) has enabled us to confirm this result and to demonstrate other features of field L (Fig. 1). The glucose analogue 2-deoxyglucose is incorporated into active neurons together with glucose proportional to its concentration in the blood. After phosphorylation it is not further metabolized and consequently accumulates. When animals injected with 14-C-2-deoxyglucose are stimulated, autoradiographs of brain sections show the spatial extent of neuronal ensembles which were activated by the stimulus. The three-dimensional organization of the field is visualized by the method. Each isofrequency contour extends into depth across the layers and with the second dimension rostro-caudally (Scheich et al. 1979a, Scheich and Bonke 1981). The crystalline structure of field L thus contains an orderly frequency representation, a map of isofrequency planes which is primarily excitatory.

Three questions arise from this type of organization. First, which mechanisms of processing or recognizing sounds benefit from the orderly representation of frequencies? Second, which is the significance of representing each frequency in a two-dimensional neuronal space. In other words, why is one locus along the one-dimensional cochlea represented in a plane? Third, what is the role of the different layers?

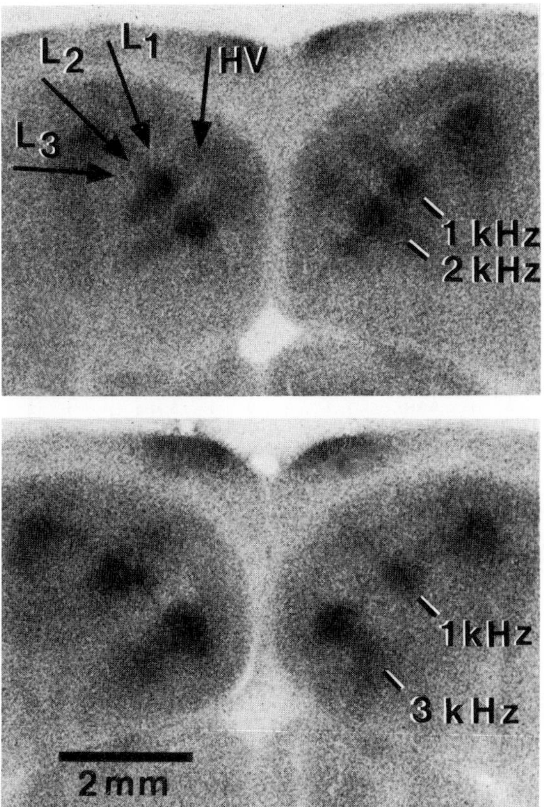

Fig. 1: 2-deoxyglucose autoradiographs of field L of Guinea fowls stimulated with pure tones. The sections were cut in the transverse plane and are taken from an intermediate rostrocaudal level of field L. The bird in the top figure was stimulated with 1 and 2 kHz alternatingly at the rate of 3/sec. Two tone-activated stripes of labeling are shown which cut perpendicularly across the three layers L_1, L_2 and L_3 and produce a focus of high intensity labeling in L_2. Note that the stripes reach into the auditory hyperstriatum ventrale (HV). The bottom figure was obtained with 1 and 3 kHz alternating tones. The 1 kHz stripe is at the same position along the L_2 layer as in the top experiment while the 3 kHz stripe is shifted towards the ventromedial end of L_2.

4.1 Representation of calls in field L

Experiments with complex stimuli including species specific calls have provided some preliminary answers to these questions. With respect to two-dimensional representation of frequencies in planes 2-deoxyglucose results with frequency modulated tones and species specific calls in comparison to simple tone burst have given the first clue for the second dimension. The representation of tone bursts diminishes in strenght along a gradient from caudal to rostral through field L. In the caudal and intermediate parts of L, tone bursts produce a stripe-like pattern across all three layers with a focus of higher optical density in the input layer L_2 (Fig. 1). Towards the rostral end of L the labeling in L_1 and L_3 becomes weaker. This demonstrates that the influence of the dominant frequency on neurons in the rostral part of isofrequency planes is small. In contrast more complex sounds which contain this dominant frequency label an isofrequency plane throughout (Scheich and Bonke 1981). These sounds may be frequency or amplitude modulated tones with small bandwidth or narrow band noise.

Similar 2DG results are obtained with calls of the Guinea fowl which contain such complex dimensions. Fig. 2 shows three calls of the Guinea fowl which were selected from the repertoire of the species (for a comprehensive review of the vocal ethology of the Guinea fowl see Maier 1982). One of the presented calls is the kecker, which is a harsh

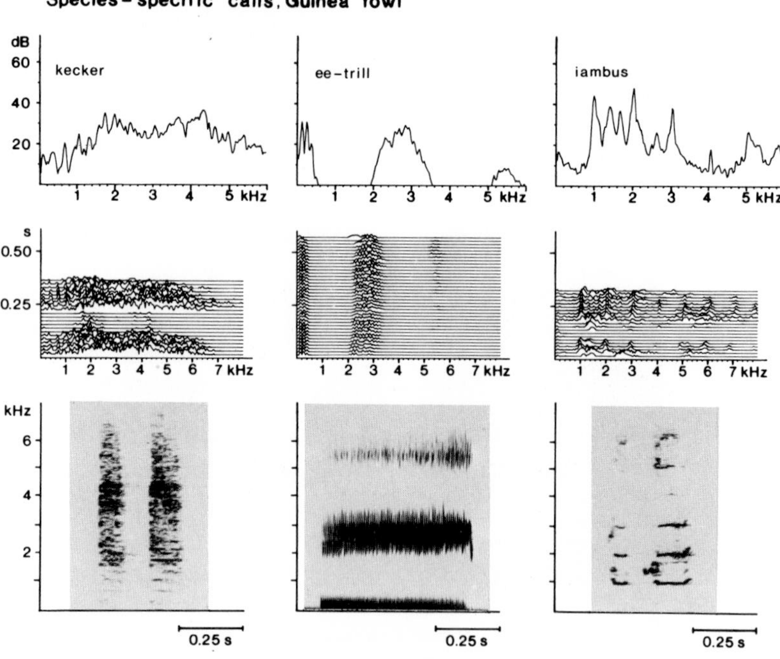

Fig. 2: Analysis of three calls of the Guinea fowl in terms of average spectra, running spectra, and sonagrams. The sonagrams at the bottom show most clearly the large frequency overlap of the three wide-band calls (for explanation of the calls see text). The average spectra on top represent the average intensity in each frequency band for the full duration of each call shown in the sonagrams. The running spectra in the middle allow an appraisal of intensity fluctuations in all frequency bands with a time window of 30 ms for each profile. The time on the ordinate runs from bottom to top.

sounding utterance. It is used by males and females in a number of agonistic situations. The kecker has a noisy wide-band spectrum with quite variable frequency limits extending from less that 1 to more than 6 kHz. The spectrum is relatively flat but has energy peaks usually between 1.5 and 5 kHz. The group of trills covers calls which direct attention of group members towards food, strange objects and animals or strange conspecifics. The ee-trill used in the present experiment has a strong alarming connotation. Its frequency limits are between less that 100 Hz and 6 kHz. The chief bands of energy (formants) are below 400 Hz and between 2 and 3.5 kHz. The iambus is a hen call which serves to keep contact between a hen and her mate or the hen and her chicks. The spectral composition of the double-note iambus is quite complicated. It consists of a harmonic carrier with 1 kHz fundamental frequency and sidebands due to amplitude modulation at a rate of about 300 Hz. The spectrum is wide-band and overlaps with both that of the kecker and of the ee-trill. In contrast to the other calls, the iambus has several energy peaks of slightly different height, namely at 1, 2 and 3 kHz.

The exposure of Guinea fowls to calls of their repertoire produced several interesting effects in the 2-deoxyglucose autoradiographs of field L (Bonke 1982, Bonke and Scheich, in prep.). Figs. 3 and 4 provide an overview of the main results. 1. Stimulation with species-specific calls leads to labeling throughout the rostrocaudal extent of the field. There was also strong labeling in the rostral third where tones in previous experiments caused very little 2 DG incorporation (Scheich and Bonke 1981). 2. In spite of the large band-width of the calls, particularly of kecker and iambus, all calls produced stripe-like activation along isofrequency contours, similar to the effect of pure tones. 3. Some calls, namely the frequency modulated ee-trill, in addition to stripe-like zones of activation, produced adjacent zones of suppression of activity. 4. The calls not only produced stripes across all three layers of field L but also markedly enhanced the activity of the overlaying hyperstriatum ventrale (HV, see Fig. 3, B3). Note that this part of the HV was determined by physiological and anatomical methods (Bonke et al. 1979a, b) to be a higher order auditory field.

4.2 Representation of spectral contrast in field L

From these four main results several rules may be derived concerning how species-specific calls are processed in telencephalic auditory maps. Most important, it is evident that such calls activate only restricted frequency bands of the tonotopic map. This is astonishing in view of the fact that some of the calls have energy over a frequency range much wider than the tonotopic area which is activated by them. A detailed comparison of call spectra (Fig. 2) with the tonotopic organization known from microelectrode recordings (Bonke et al. 1979b) and 2 DG labeling with pure tones (Fig. 1) provides a frame for understanding this effect.

For instance, the spectrum of the iambus shows several sharp peaks of energy, namely at 1, 2 and 3 kHz. There are also several smaller peaks which are either harmonics of 1 kHz or sidebands due to the 300 Hz amplitude modulation of the harmonic carrier. The largest peak is at 2 kHz with the 1 kHz peak following next. The profiles of 2 DG labeling in Fig. 4 show two sharp peaks of labeling which on the tonotopic scale of the L_2 layer correspond to frequencies around 1 and 2 kHz. Similar to the iambus frequency spectrum the 2 kHz peak is the largest (Fig. 4, C1–5). Thus there is a very distinct representation of the band of highest energy in the call spectrum by the metabolic activity of neurons. Other bands of energy, even if only 10 dB waker lead to much smaller activation. For instance, the prominent band of energy at 3 kHz (Fig. 2) obtained very little representation in the profiles. Obviously the main issue in frequency representation of this call in field L is to

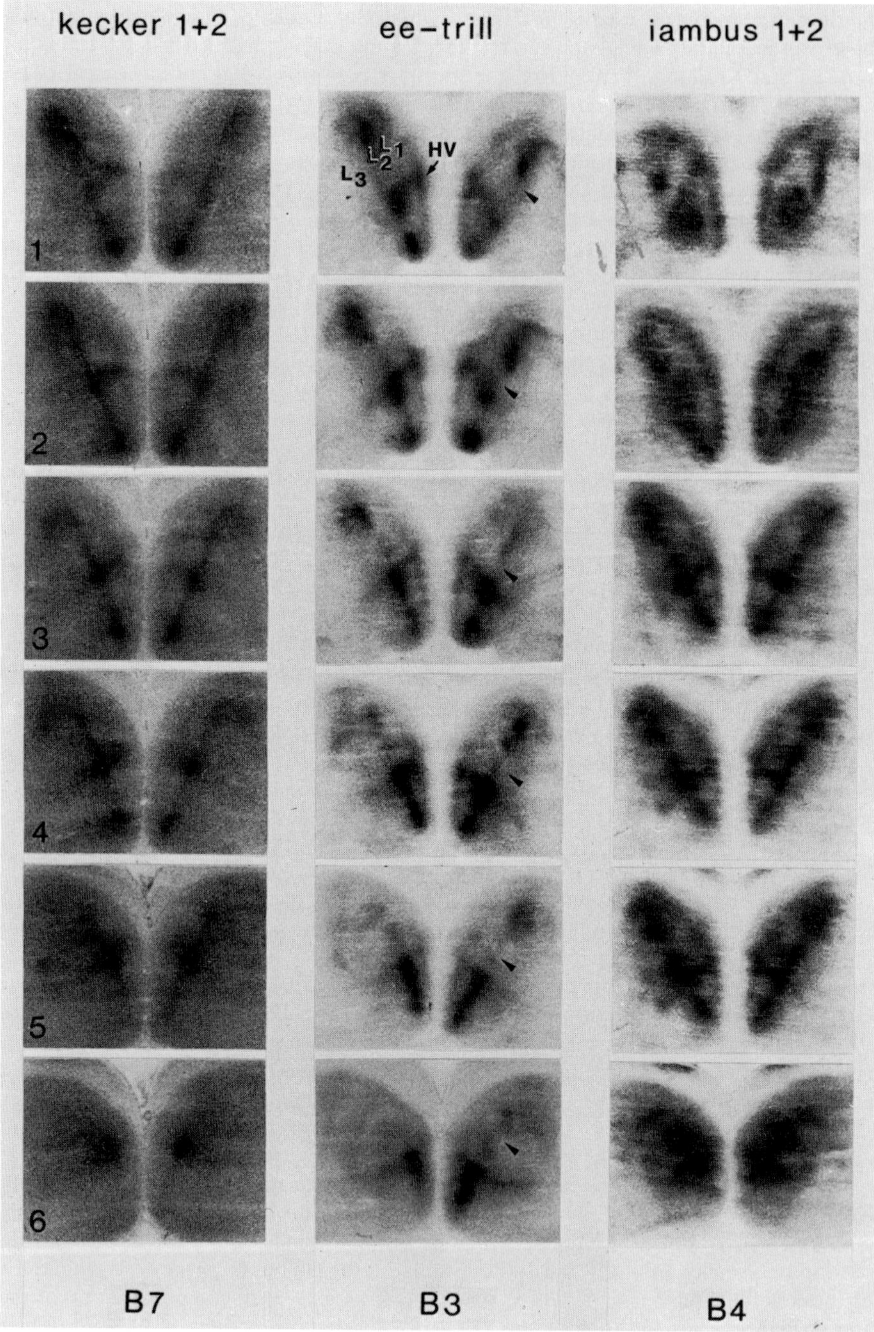
kecker 1+2
ee-trill
iambus 1+2
1
2
3
4
5
6
L1
L2
L3
HV
B7
B3
B4

emphasize the spectral peak. These 2 DG results closely correspond to the results of systematic microelectrode penetrations through field L while stimulating with a large number of natural variations of iambus calls. Most of these variations had spectral peaks between 1.7 and 2 kHz. We found that the tonotopic area around 1.8 kHz contained the largest number of units which responded to iambus calls or preferred them over calls (BONKE et al. 1979a, SCHEICH et al. 1979b, BONKE 1982). In view of the poor spatial resolution of microelectrode penetrations in a grid of 0.2 × 0.2 mm this is a very good match.

The iambus is a call which has a wide-band spectrum composed of relatively discrete lines with little energy in between. Other calls like the kecker or the ee-trill have a different type of spectrum. There the energy is more densely spaced. In the case of kecker, noise is the chief component while in the case of the ee-trill the frequency is modulated at a rate of about 100/s through the full bandwidth. Neither call has sharp peaks of energy.

For instance the spectrum of the kecker extends up to 6 kHz with a rather flat plateau of energy roughly between 2 and 4 kHz. In spite of this the profiles of 2 DG labeling (Fig. 4, B1–5) show one sharp peak of labeling which on the tonotopic scale of the L_2 layer corresponds to frequencies between 1 and 1.5 kHz. This is an extremely narrow band of tonotopic labeling with respect to the band-width of the kecker spectrum (see Fig. 2). In fact, the stripe of labeling is not wider than those obtained for pure tones (compare Fig. 1 with Fig. 3, B7/3). Obviously there is a special cue in the call spectrum which neither corresponds to the band-width nor to the spectral peak and yet determines the metabolic response of field L. Judging from Fig. 2 there is a 20 dB rise of average intensity in the kecker spectrum between 1 and 2 kHz. This is the largest rise of intensity in the whole spectrum. The preliminary conclusion which can be drawn from this result is that it is some kind of a spectral contrast which is represented in the case of wideband spectra lacking a clear spectral peak. The results from ee-trill stimulation further substantiate this idea.

The ee-trill has at least two bands of prominent energy. The main band is between 2 and 3.5 kHz and corresponds to the frequency modulated carrier. The other band is below 400 Hz and shows a first peak at 110 and a second at 220 Hz. These peaks correspond to that frequency (and its second harmonic) which modulates the carrier in the 2–3.5 kHz band. The 2 DG profiles in Fig. 4B show one prominent band of metabolic excitation with a peak at 2 kHz. Again, as in the case of kecker, it is not a prominent spectral frequency which is represented. Instead the steep rise of intensity at the low frequency end of the main spectral component elicits the largest metabolic response. In other words the system optimally reflects what I would like to call the *spectral contrast* of the call.

How can we now relate the results from the iambus experiment to those of ee-trill and kecker stimulation? In the first case the spectral peaks were represented, in the second a strong spectral contrast led to maximum labeling. There is no fundamental contradiction between these cases. As explained above the iambus has a line spectrum with strong tone-like frequency components separated by gaps with little or no energy. Consequently the

---

Fig. 3: 2-deoxyglucose autoradiographs of right and left field L of Guinea fowls stimulated with species-specific calls. Insets 1 through 6 are equidistant transverse sections from rostral to caudal through the field. Note that all three calls chiefly produce stripe-like activation which cuts across the three layers of field L similar to pure tones. This effect is most pronounced in the intermediate sections 2, 3 and 4. Autoradiographs from the frequency modulated ee-trill show marked suppression of metabolic activity in L_2 adjacent to the stripe of activation (arrow heads). The activation of the hyperstriatum ventrale (HV) in the rostral sections is most pronounced with this call.

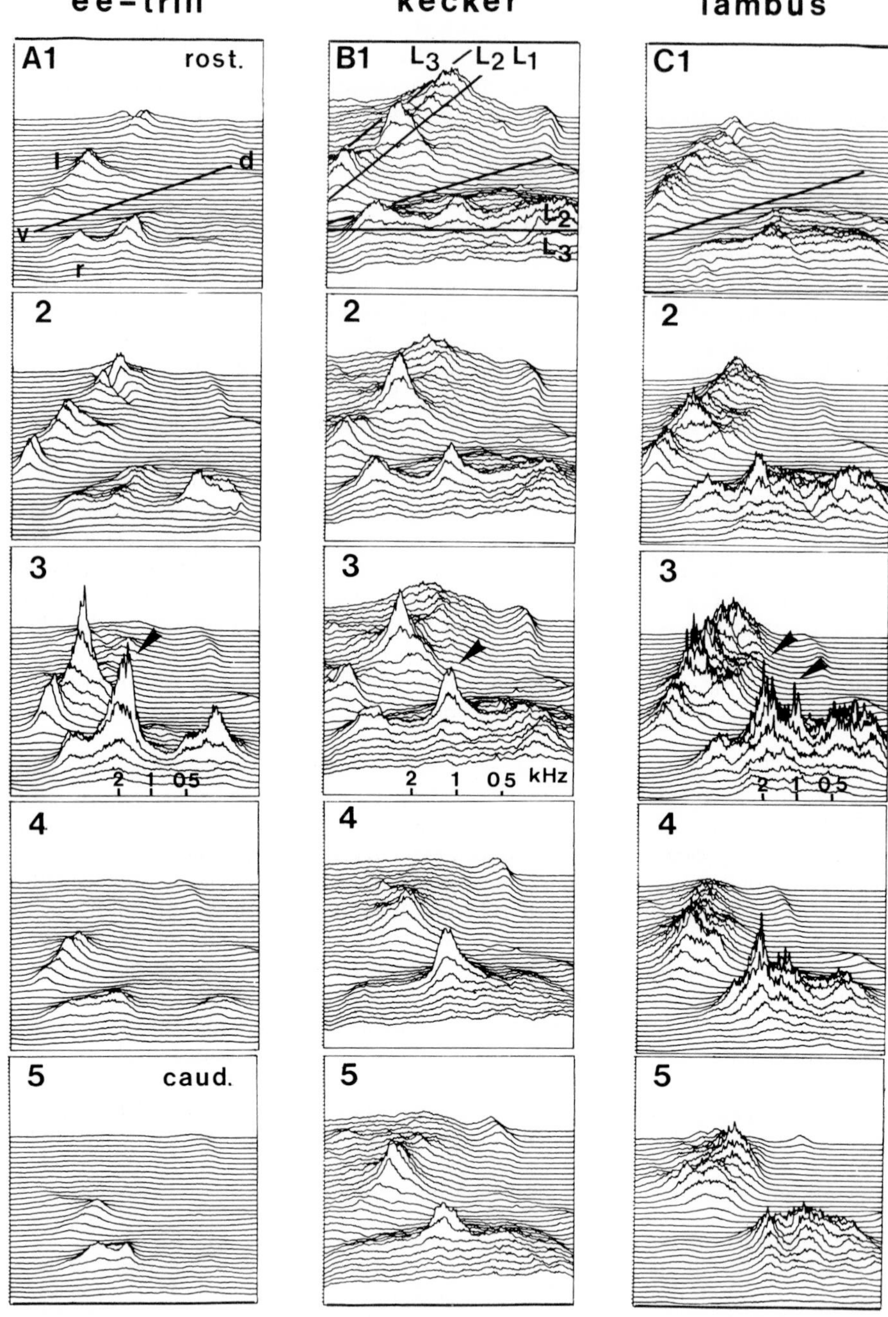
ee-trill
kecker
iambus
A1
rost.
l
d
v
r
B1
L3
L2
L1
L2
L3
C1
2
3
2
1
0.5
kHz
4
5
caud.

largest spectral constrasts correspond to the largest spectral peaks. This is similar to pure tones where the spectral peaks and the maximum contrasts are identical. The fact that not the spectral peak but some frequency at the slope of this peak is chiefly represented in field L has been reported already by Bonke (1982) from both 2 DG results and microelectrode recordings. In summary it may be said that in terms of the largest metabolic response, field L appears to be interested chiefly in spectral contrasts of calls. This should be taken into account as an important point when speculating about the design of bird calls.

If this property of field L, of representing chiefly spectral contrasts, were a general property of telencephalic auditory maps which recognize auditory patterns, there are several basic implications for the optimal design of vocal patterns. First, if an animal uses a variety of different calls they should have their largest spectral contrast in different regions of the frequency scale. This would provide for an optimal distinction in the representational maps of the auditory system. It is implicit that the wide-band nature of calls with a considerable spectral overlap among different calls creates little complication as long as the contrasts are spaced out. It may also follow from this need to space out spectral contrasts of calls in a large repertoire that the species is not completely free to choose any suitable frequency for long distance calls. At least there may be pressures from physical constraints and perceptual constraints interacting to make this choice.

Second, if a call type is variable and thereby gradually shifts its meaning, the representation in the tonotopic map is affected only if the maximum contrast changes. In the case of the iambus which codes individual voices of females a detailed study of variability has shown that, in spite of considerable differences in terms of strength of harmonics and sideband, the energy peak was always between 1.5 and 2 kHz (Scheich et al. 1979b). With respect to the total band-width of the iambus this is a small range of variability. We shall see below that this variability may not be in conflict with the representation of the other calls.

4.3 Enhancement of spectral contrast in field L

Distribution of spectral constrasts of calls along the frequency scale and consequently separation in the neuronal space of field L is one basic strategy for the distinction of calls. There is another neuronal mechanism which enhances this effect. In Figs. 3 and 4 the 2 DG profiles of the ee-trill show an area of metabolic suppression in the L_2 layer adjacent to the large peak of labeling. This suppression is below the level of spontaneous labeling of L_2 which is obtained in animals kept in silence. It almost reaches down to the background labeling in the remainder of the brain sections. This type of suppression is typical for frequency modulated sounds (Scheich and Bonke 1981) and is related to inhibitory

Fig. 4: Densitometric profiles of 2-deoxyglucose labeling from the cases of Fig. 3. The autoradiographs of Fig. 3 were analyzed on X-rax film by a TV camera and transformed into a matrix of 256 × 256 picture points, each point with a relative intensity resolution of 256 grades. The average intensity of 8 rows of points respectively was integrated and transformed into an intensity profile. In that way 32 profiles represent the labeling in a given area. Note that the orientation of field L was rotated 45° with respect to Fig. 3. In A1 the midline between right and left field L is indicated by a line pointing dorsally (d) and ventrally (v). In that way the right field L (r) has a horizontal orientation and the perspective view of the profiles of labeling is from lateral. This allows a good overview of the layers L_3 and L_2. As indicated in B1 the left field L is seen from a medial aspect which unconvers mostly the layers L_1 and L_2. In A3, B3 and C3 a frequency scale of the tonotopic organization of the L_2 layer as obtained from tone experiments is fitted to the profiles. In that way the large peaks of labeling in L_2 (arrowheads) can be referred to frequencies.

sidebands of units in L_2 (Scheich et al. 1983). It is not seen in the other calls of Figs. 3 and 4. Only occasional weak suppression has been obtained with pure tone stimulation.

The net effect of suppression together with a large adjacent peak of activity corresponds to an enhancement of contrast. It may also be considered as an improvement of signal to noise ratio for the representation of a sound. There may be a fundamental implication of this effect for the design of vocal sounds in the repertoire of the Guinea fowl and of other animals. It is probably not a pure coincidence that all calls in the Guinea fowl repertoire which have an alerting connotation are frequency modulated sounds (see Maier 1982). Frequency modulations may be particularly suitable for this purpose because they increase signal to noise ratio in the auditory map and thereby facilitate recognition against a background of noise.

Frequency modulated sounds are widespread in the animal kingdom and their systematic study may subsequently disclose a common use as alerting signals. At least one such case of a fast modulated call has been describes for the dwarf mongoose (Maier et al. 1983). The well known cases of so called hawk alarm calls in European songbirds (Marler 1959), which are slow downward modulations of a tone, may be in the same category. Scheich and Bonke (1981) have shown that slowly modulated tones (3 Hz modulation frequency) also lead to sharp zones of suppression of metabolic activity. Finally the widespread use of frequency modulated sounds as alarm signals in our human civilization should be noted.

5 The mosaic pattern of call representation

We have seen that representation of spectral contrasts and enhancement of contrast by sideband-suppression are basic rules of operation of field L. For the three calls presented here and for a few others which have been tested these mechanism provide some rough distinction in the spatial map of the field. However, there are about 20 different call types in the repertoire of the Guinea fowl (Maier 1982), which all have to be distinguished. Some of these call types similar to the iambus show meaningful variations which among other cues may also involve small shifts of spectral contrast. Consequently there must be additional mechanisms which provide for a distinction of more subtle differences.

Close inspection of the 2 DG profiles in Fig. 4 shows that there are small secondary peaks of labeling beside the main peaks. The distribution of secondary peaks is uneven from rostral to caudal in the L_2 layer. In the intermediate third, i. e. particularly in A2, A3 the ee-trill produces small peaks below 500 Hz. They most probably correspond to a spectral contrast of the modulation frequencies in Fig. 1. In C3, C4 the already mentioned 1 kHz fundamental of the iambus or its spectral contrast produces a secondary peak of labeling. For the kecker the secondary peak is less pronounced but may be identified at the ventral high frequency end of L_2 in B2, B3. It probably reflects another intensity jump, i. e. the spectral contrast between 4 and 6 kHz (see Fig. 1). The representation of at least two separate spectral components in the tonotopic map may be the way to overcome the difficulty of identifying a great number of calls. In that way the combination of the strongest or primary representational component with a secondary one could easily lead to an unmistakable representational pattern. In other words the recognition of calls may reside in a mosaic of spatial excitation of neuronal ensembles in the map. It is not clear yet why the secondary component is most prominently represented in the intermediate third of the field.

The reader should be reminded here of the introductory statements about the three-dimensional organization of field L. So far I have focussed only on the tonotopy of the L_2

layer in the transverse plane and not mentioned differences of call representation along the rostrocaudal extent of this layer nor differences of call representation in L_1 and L_3. The description of these differences would be beyond the scope of this paper. I shall only mention that there are indeed such differences in the responsiveness and selectivity of neurons to calls along those spatial dimensions, some of which have been described in electrophysiological studies (SCHEICH et al. 1979b, BONKE et al. 1979b, SCHEICH et al. 1983). Their significance in terms of complex dimension, i.e. combination of frequency, intensity and time was clarified only in part.

6 Environmental acoustic versus neuronal call representation

The ultimate goals for the design of all communication sounds are a largely undisturbed transmission to and an optimal perception by the receiver. For long distance communication and to a lesser extent for short distance communication environmental acoustics impose certain constraints on transmission in the atmosphere and therefore may have influenced the choice of call frequency in species. This was claimed by a number of authors as discussed above and is best summarized in the article by RICHARDS and WILEY (1978). The results presented here on call representation in a forebrain auditory map of a bird are at variance with some of these claims or at least should make one cautions about the *consequences* of some results in environmental acoustics for auditory perception. Before drawing any conclusions from vocalizations it is important to determine first which parameters of a call the auditory system of the receiver is chiefly interested in. As we have seen, except for narrow band tonal stimuli, the bird system is interested in spectral contrasts. This means that in wideband calls the acoustic transmission of the strongest spectral components is not always the relevant parameter. I would like to propose therefore that future measurements of complex sound transmission in various biotopes should focus on spectral contrasts in order to elucidate physical constraints on this parameter.

Moreover, the discovery that frequency modulations produce extremely high contrast representations in the tonotopic map is important in the light of sound transmission. It suggests that in spite of a certain vulnerability of frequency modulated patterns by sound reverberation (RICHARDS and WILEY 1978), the representation of such a call may be preserved in the auditory map by the special mechanism of contrast enhancement. Again as in the case of other wide-band calls it is a spectral contrast and not the spectral peak which is represented. This should also be borne in mind when interpreting effects of environmental acoustics on frequency modulated calls.

References

Aitkin, L. M. (1976): Tonotopic organization at higher levels of the auditory pathway. Intern. Rev. Physiol. (Neurophysiology II) *10,* 249–279.

Aylor, D. (1971): Noise reduction by vegetation and ground. J. Acoust. Soc. Amer. *51,* 197–205.

Bonke, B. A., Bonke, D., Scheich, H. (1979): Connectivity of the auditory forebrain nuclei in the Guinea fowl (Numida meleagris). Cell Tissue Res. *200,* 101–121.

Bonke, D. (1982): Processing of species-specific calls in the field L in the Guinea fowl (Numida meleagris) demonstrated by the 2-deoxyglucose technique and y systematic electrophysiological recordings. Verh. Dtsch. Zool. Ges., Stuttgart, Gustav-Fischer-Verlag, *302.*

Bonke, D., Scheich, H., Langner, G. (1979): Responsiveness of units in the auditory neostriatum of the Guinea fowl (Numida meleagris) to species-specific calls and synthetic stimuli. I. Tonotopy and functional zones of field L. J. Comp. Physiol. *132,* 243–255.

Chappuis, C. (1971): Un exemple de l'influence du milieu sur les émissions vocales des oiseaux: l'évolution des chants en forêt équatoriale. Terre et Vie *118,* 183–202.

Dooling, R. J., Mulligan, Y. A., Miller, J. D. (1971): Auditory sensitivity and song spectrum of the common canary. J. Acoust. Soc. Am. *50,* 700–709.

Dürrwang, R. (1974): Funktionelle Biologie, Anatomie und Physiologie der Vogelstimme. Ph. D.-Thesis, University of Basel, Switzerland.

Embleton, T. F. W. (1963): Sound propagation in homogeneous deciduous evergreeen woods. J. Acoust. Soc. Amer. *35,* 1119–1125.

Green, S. M., Darwin, C. J., Evans, E. F., Fant, G. C. M., Fourcin, A. J., Fujimura, O., Fujisaki, H., Liberman, A. M., Markl, H. S., Marler, P. R., Miller, J. D., Milner, B. A., Nottebohm, F., Pisoni, D. B., Ploog, D., Scheich, H., Stevens, K. N., Studdert-Kennedy, M. G., Tallal, P. A. (1977): Comparative Aspects of Vocal Signals Including Speech (Group Report). In: Dahlem Workshop on Recognition of Complex Acoustic Signals (T. H. Bullock, Ed.), Berlin-Dahlem Konferenzen, pp. 209–237.

Greenewalt, C. H. (1968): Bird Song. Acoustics and Physiology. Washington D. C., Smithsonian Institution Press.

Hubel, D. H., Wiesel, T. N. (1962): Receptive fields, binocular interaction and functional architecture in the cat's visual cortex. J. Physiol. *160,* 106–154.

Karten, H. J. (1968): The ascending auditory pathway in the pigeon (Columba livia). II: Telencephalic projections of the nucleus ovoidalis thalami. Brain Res. *11,* 134–153.

Konishi, M. (1970): Comparative neurophysiological studies of hearing and vocalizations in songbirds. Z. Vergl. Physiol. *66,* 257–272.

Kuhn, A., Müller, Ch., Leppelsack, H.-J., Schwartzkopff, J. (1982): Heart-rate conditioning used for determination of auditory threshold in the starling. Naturwiss. *69,* 245.

Linskens, H. F., Martens, M. J. M., Hendricksen, H. J. G. M., Roestenberg-Sinnige, A. M., Brouwers, W. A. J. M., van der Staak, A. L. H. C., Strik-Jansen, A. M. J. (1976): The acoustic climate of plant communities. Oecologia (Berl.) *23,* 165–177.

Maier, V. (1982): Acoustic communication in the Guinea fowl (Numida meleagris): Structure and use of vocalizations and the principle of message coding. Z. Tierpsychol. *59,* 29–83.

Maier, V., Rasa, O. A. E., Scheich, H. (1983): Call-system similarity in a ground-living social bird and a mammal in the bush habitat. Behav. Ecol. Sociobiol. *12,* 5–9.

Marler, P. (1959): Developments in the study of animal communication. In: Darwin's Biological Work: Some Aspect Reconsidered (P. R. Bell, Ed.), Cambridge, Cambridge University Press.

Marten, K., Marler, P. (1977): Sound transmission and its significance for animal vocalization. I. Temperature habitats. Behav. Ecol. Sociobiol. *2,* 271–290.

Marten, K., Quine, D., Marler, P. (1977): Sound transmission and its significance for animal vocalization. II. Tropical forest habitats. Behav. Ecol. Sociobiol. *2,* 291–302.

Morton, E. S. (1975): Ecological sources of selection on avian sounds. Amer. Nat. *109,* 17–34.

Mountcastle, V. B. (1957): Modality and topographic properties of single neurons of cat's somatic sensory cortex. J. Neurphysiol. *20,* 408–434.

Neuweiler, G. (1983): Echolocation and adaptivity to ecological constraints. In: Neuroethology and Behavioral Physiology (F. Huber, H. Markl, Eds.), Springer, pp. 280–302.

Rose, M. (1914): Über die cytoarchitektonische Gliederung des Vorderhirns der Vögel. J. Physiol. Neurol. (Leipzig) *21,* 278–352.

Sachs, M. B., Sinott, J. M., Hienz, R. D. (1978): Behavioral and physiological studies of hearing in birds. Fed. Proc. *37,* 2329–2335.

Scheich, H. (1979): Common principles of organization in the central auditory pathway of vertebrates. Verh. Dtsch. Zool. Ges., Stuttgart, Gustav Fischer Verlag, pp. 155–166.

Scheich, H., Bock, W., Bonke, D., Langner, G., Maier, V. (1983): Acoustic communication in the Guinea fowl (Numida meleagris). In: Advances invertebrate neuroethology (J.-P. Ewert, R. R. Capranica, D. J. Igle, Eds.), Plenum Publishing Corp., 731–782.

Scheich, H., Bonke, B. A. (1981): Tone-versus FM-induced patterns of excitation and suppression in the 14-C-deoxyglucose labeled auditory «cortex» of the Guinea fowl. Exp. Brain Res. *44,* 445–449.

Scheich, H., Bonke, B. A., Bonke, D., Langner, G. (1979): Functional organization of some auditory nuclei in the Guinea fowl demonstrated by the 2-deoxyglucose technique, Cell Tissue Res. *204,* 17–27.

Schnitzler, H.-U., Ostwald, J. (1983): Adaptations for the detection of fluttering insects by echolocation in horseshoe bats. In: Advances in vertebrate neuroethology (J.-P. Ewert, R. R. Capranica, D. J. Ingle, Eds.), New York, London, Plenum Press, pp. 801–827.

Schroeder, M. R. (1977): Machine processing of acoustic signals: What mechanisms can do better than organisms (and vice versa). In: Dahlem Workshop on Recognition of Complex Acoustic Signals (T. H. Bullock, Ed.), Berlin-Dahlem-Konferenzen, pp. 183–207.

Sebeok, T. A. (Ed.) (1977): How animals communicate. Bloomington, London, Indiana University Press.

Simmons, J. A., Stein, R. A. (1980): Acoustic imaging in bat sonar: Echolocation signals and the evolution of echolocation. J. Comp. Physiol. *135,* 61–84.

Sokoloff, L. (1975): The coupling of function, metabolism and blood flow in the brain. In: Brain Work (H. D. Ingvar, N. A. Lassen, Eds.), Copenhagen-Munksgaard, pp. 385–388.

Waser, P., Waser, M. S. (1977): Experimental studies of primate vocalization: specializations for long-distance propagation. Z. Tierpsychol. *43,* 239–263.

Wiley, R. H., Richards, D. G. (1978): Physical constraints on acoustic communication in the atmosphere: Implications for the evolution of animal vocalizations. Behav. Ecol. Sociobiol. *3,* 69–94.

Fortschritte der Zoologie, Bd. 31 · Hölldobler/Lindauer (Hrsg.): Experimental Behavioral Ecology
G. Fischer Verlag · Stuttgart · New York · 1985

Representational vocal signals of primates

PETER MARLER

The Rockefeller University, Center for Field Research, Millbrook, N. Y., USA

Abstract

The prevailing view of the vocal signals of non-human primates is that they are signs of emotion and nothing more. A distinction is made between affective communication and symbolic communication, and the latter is thought to be a special human prerogative. Recent work on calls used by monkeys to signal alarm in the presence of predators, and to recruit aid from companions during agonistic encounters with other group members suggests that this view is incorrect. Vervet monkeys appear to use alarm calls as though they actually represent or symbolize particular predators. Rhesus macaques give screams while fighting that are usually thought of as addressed primarily to the opponent. In fact they are calls for help. The signals are discretely organized, and do not constitute a graded continuum as previously thought. Field experiments on responses to tape recorded vocalizations confirm the conclusions drawn from descriptive studies that each call appears to represent to other listening monkeys not only the severity of the fight but also the class of opponent present, as defined by rank and kin relationships to the caller. We conclude that the calls of non-human primates reflect both affective and cognitive processes. The signs of these two kinds of process are often intermingled and conveyed by different acoustic vehicles, as in human speech.

1 Introduction

The modern era of research on vocal signals and communication in non-human primates began in 1962, with the investigations by ROWELL and HINDE of rhesus monkeys in captivity (ROWELL 1962; ROWELL and HINDE 1962), followed soon afterwards by ANDREW's studies of a variety of captive primate species (ANDREW 1962, 1963). Prior to that time virtually nothing was known about non-human primate vocal behavior. The situation now is quite different. An impressive amount of descriptive information has been gathered on the physical structure of the vocalizations of monkeys and apes, and we are beginning to understand the relationships of vocal behavior to phylogeny, localizability and contexts of use (reviewed in STRUHSAKER 1970; GREEN 1975; GAUTIER and GAUTIER-HION 1977; MARLER and TENAZA 1977; SNOWDON, BROWN and PETERSEN 1982). The stage is now set for a more analytical approach, except for two serious gaps. We still know almost nothing about the ontogeny of primate vocal behavior (NEWMAN and SYMMES 1982) and perhaps most surprising of all, remarkably little is known about what is

Table 1: Interpretations of Monkey Signals as Manifestations Only of Affect, Emotion or Motivation

A. «Nonhuman primates can send complex messages about their motivational states; they communicate almost nothing about the state of their physical environments.» (LANCASTER 1965, p. 64)

B. «... all signals appear to be clearly related to the immediate emotional states of the signaling individuals and their levels of arousal.» (BASTIAN 1965, p. 598)

C. «The use of both the face and the voice by rhesus monkeys in their natural habitat seems to be restricted to circumstances that connote emotion.» (MYERS 1976, pp. 747–748)

D. «... the nonhuman primate does not use the auditory medium to communicate whatever conceptual knowledge it possesses. The vocal repertoire appears to relate to affective rather than cognitive dimensions, the nature of the signal reflecting the emotional disposition of the caller.» (MARIN, et al. 1979, p. 184)

E. «... the signal emitted by an animal is an expression of its affective condition, and the reception of the signal indicates the infection of others by the same condition – nothing more.» (LURIA 1982, p. 29)

F. «Man has both affective and symbolic communication. – All other species, except when tutored by man, have only the affective form.» (PREMACK 1975, p. 593)

actually being expressed in the calls of monkeys and apes and what they mean to other animals (MARLER 1977, 1978a, b).

ROWELL and HINDE (1962) made it clear that they view rhesus macaque calls as signs of affect or emotion. Their paper opens with the statement, «The noises made by monkeys express their mood, and are effective in communicating it to others». In their judgement the calls express not thoughts but affective moods. Until very recently this had been the prevailing view of animal vocal signals in general. To demonstrate how widespread and firmly rooted this interpretation has been over the past 20 years, I have assembled a series of quotations selected from different disciplines (Table 1). All take the position that there is nothing symbolic about primate vocalizations, and that the essential substrates for the calling behavior of non-human primates are to be found in the emotions. The present paper seeks to challenge that position by demonstrating that there is much more to primate vocal behavior than just affect.

The first quotation is from LANCASTER (1965), a physical anthropologist. BASTIAN (1965) is a linguist, one of the few who has worked actively on animal communication. MYERS (1976) is a neuroanatomist, representing another widely held view. MARIN et al. (1979) presents the viewpoint of a clinician, and LURIA (1982) that of a cognitive psychologist. Finally there is an expression of the same position from PREMACK (1975). The latter's viewpoint is especially important because of his demonstrations of the cognitive prowess of primates in the laboratory (PREMACK and PREMACK 1983). If we assume that nothing more than affect is involved in their vocal behavior, this raises the apparent paradox that there is a reserve of cognitive abilities that is simply not manifest in their normal communicative behavior.

2 Environmental representation: the alarm calls of monkeys

Some six years ago I decided to take a fresh look at this problem. As a starting point, I had already been preoccupied with the fascinating repertoire of alarm calls that STRUHSAKER (1967) had described in the vervet monkeys of Amboseli in Kenya while working as a student in my laboratory. Within a rather complicated vocal system there were three calls in particular that, if his data were valid, could provide the basis for a challenge to the «emotion» or «affect» interpretation of primate calls. According to Struhsaker, distinct alarm calls were employed for leopards, for snakes, and for eagles, used in ways that seemed to imply more than was compatible with a simple affect interpretation (MARLER 1977). Soon after, I invited DOROTHY CHENEY and ROBERT SEYFARTH to join forces with me to reexamine the natural patterns of use of vervet alarm calls, and then to conduct experiments under field conditions to study responses of the monkeys to playbacks of tape recordings of their alarm calls in the absence of predators.

The basic question was whether the alarm calling behavior of the vervets was indeed compatible with an affect interpretation. In this case different alarm calls would presumably reveal different degrees of fearfulness in the caller and evoke equivalent fearful states in other animals. Is this a sufficient explanation for what takes place, or is something more involved? As it turned out, the «something more» loomed larger and larger in our minds until we found ourselves entertaining a quite different interpretation.

The problem to be faced is the nature of the chain of events that intervenes between an eagle coming into view, a vervet giving a particular signal, and eliciting in another monkey the pattern of escape behavior appropriate for evading an eagle attack. In the case of a martial eagle the appropriate response would be to rush out of open areas and especially to drop rapidly from the treetops, into dense cover, which is the place where a monkey is safest from a stooping raptor. Field experiments in which alarm calls of animals of known identity were played back from concealed loudspeakers showed unequivocally that this behavior is triggered by the call when there is no predator present (SEYFARTH, CHENEY and MARLER 1980a, b). Note moreover that if the predator is a leopard, for which another alarm call is given, then a different escape reaction is appropriate, almost the opposite of that for an eagle. The safe response to a leopard is to rush out into the open or up into the tops of trees, away from dense cover which is where a leopard, which hunts by ambush, is most likely to be. If you play a recording of a leopard alarm call this is the behavior the signal evokes.

Note the difficulty that immediately confronts us if we try to explain these antithetical responses to predators on the basis of a simple affect interpretation of alarm calling behavior. In both cases the animals are clearly fearful, but in one situation they rush towards dense cover, and in the other they rush away from it. This contrast becomes comprehensible, however, if we propose that, in addition to the involvement of emotion, the alarm calls in some sense *represent* particular predators to other animals. What I am presuming here is that, after examination of the visual stimulus of an eagle while it is still only a silhouette in the sky and consulting schemata stored in the brain from past encounters with different species of flying birds, the calling monkey decides that this is indeed a martial eagle, serious enough as a threat, especially to infants in the group, to merit a signal. As I interpret the situation, the alarm call that is uttered elicits in monkeys hearing the call an equivalent mental concept to that experienced by the caller when it first saw the eagle. The appropriate sequel, on the basis of past experience, is to run for cover.

Fear will be of course a component in the response, galvanizing the animal into action, no doubt with all of the appropriate autonomic accompaniments. But although fearfulness plays a role, the essential key to this behavior must, I believe, be found in cognition rather than affect. The experimental result that convinced us more than any other that representation must be involved was to see the vervets, confronted with an alarm call played from a hidden speaker, but with no predator in sight, suddenly going bipedal when a snake alarm call was played and searching anxiously in the grass around them for a snake that the call told them must be there, or looking up into the sky for the non-existent eagle announced by playback of an eagle alarm call (SEYFARTH, CHENEY and MARLER loc cit.).

We are dealing of course with difficult problems that philosophers have argued about for centuries (GRIFFIN 1981). It is difficult to prove unequivocally that cognition is involved. At some point, the decision to adopt a cognitive interpretation hinges on an appeal to Occam's razor. You could, as some have suggested, invoke triggered reflexes of looking up or looking down. You could fractionate the categories of affect more and more finely so that distinct affective states would be evoked by different classes of predators, a solution that reduces the power of the concept of affect to a point where it loses all explanatory value. The fact is that the monkeys behave as though the alarm calls make them think that a predator is indeed present. I am using the term «think» advisedly, in the sense of evoking a mental concept. As a related accompaniment, there must be an expectation that a visual search will lead to detection of a member of a particular class of stimuli with specified properties, such as those of a soaring martial eagle.

The specificity of alarm calling behavior as manifest in adult animals is dramatized by the behavior of younger animals, who are prone to make mistakes. Inspection of the array of objects to which an infant vervet will give eagle alarm calls (Figure 1) reveals that, although martial eagles are occasionally responsible, they are no more common as a stimulus than a great variety of other birds, some predatory, others not (SEYFARTH, CHENEY and MARLER 1980a; SEYFARTH and CHENEY 1980). Similarly with leopard alarm calls, an infant will give this call to a lion or a hyena, or even to an impala or a warthog. As the Amboseli vervets mature, their calling behavior gradually becomes more and more focussed on very particular stimuli. Thus in adulthood the class of referents for the eagle

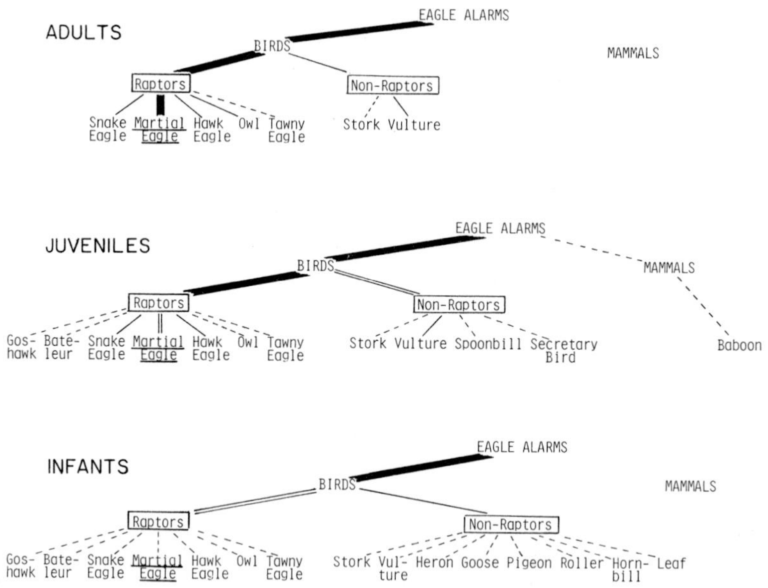

Fig. 1: A diagrammatic representation of data from SEYFARTH, CHENEY and MARLER (1980b) on stimulus objects evoking «eagle alarm» calls in wild vervet monkeys. Results for adult, juvenile and infant monkeys are given separately so that a developmental sequence can be discerned. Only the species connected by lines were effective. The width of the lines provides a rough index of the relative frequency of elicitation of alarm calling. In adult vervets, martial eagles were the major referent for eagle alarm calling, but a minor one for infants, along with many other non-raptorial birds as well.

alarm call consists not just of eagles in general, of which there are about 20 at Amboseli, but of one in particular, the martial eagle. The referents for other vervet alarm calls range from the general to the highly specific depending on the functional connotations of the situation.

Once it is mature, an adult vervet will display extraordinary visual acuity in identifying a soaring eagle long before a human observer is even aware that a raptor is present let alone able to identify the species. Thus in adulthood both the referents for calls and the responses that they evoke are specific. As arbitrary, non-iconic signals, the calls satisfy basic criteria for symbolic behavior. The calls seem to represent the predators to other animals that hear them for purposes of social communication. According to this view affect is indeed involved in this complicated predator situation that confronts vervets at Amboseli, but it plays a secondary role, more generalized than has been presumed in previous interpretations. By its very nature a purely affective system is unable to specify the particulars that are so crucial to vervets in making the right behavioral decisions both in selecting a signal for production and in choosing an appropriate response when a signal is received.

So we find ourselves in diametric opposition to the overwhelmingly prevalent view that affect is the crucial component of monkey vocal signals – even the only component – underlying their production and their interpretation by others. The question must be faced, are we dealing with an obscure special situation that arises because of peculiar selection pressures that confront the Amboseli vervets (SMITH 1977) or is representational signaling widespread, and perhaps undetected until now only because we have failed to investigate other cases in the right way? The case of the vervet monkey alarm calls was chosen because it was felt to be a relatively simple one. The notion of symbolic representation is difficult at best to submit to empirical investigation, but it is more manageable if that which is represented is a stable object in the environment, identifiable unequivocally by an observer, and subject to withholding or presentation by an experimenter. It is not inconceivable that representation of environmental referents, likely to be biologically salient objects such as predators, or perhaps in some cases food items, will prove to recur in non-human primates, and perhaps in other taxa as well. What of the possibility that animals might use their vocalizations to represent something more abstract than classes of objects, such as relationships? We can visualize this possibility most easily in social contexts, in the form of dominance relationships or kinship, engaging phenomena that are of interest in their own right as a challenge to the cognitive capacities of animals, and likely a special focus of their natural intelligence (CHENEY and SEYFARTH 1982a; HUMPHREY 1976).

3 Vocal representation of social relationships: agonistic recruitment calls of macaques

In a search for animal communication systems that are amenable to investigation of the possibilities of representational signaling about social relationships, my colleagues Harold and Sarah Gouzoules and I conducted an analysis of a system of vocal signals used by rhesus macaques to recruit help from relatives when they became embroiled in an agonistic encounter with a member of the social group. This pattern of social interaction proves to be peculiarly suitable for representational analysis. Macaques and baboons commonly display recruitment behavior during agonistic interactions. One animal gets into a fight and seeks help from another group member. If recruitment is successful, the two combine forces, so neutralizing or countering the aggressive threat (e. g. KAPLAN 1978).

Other investigators had already noted that there is commonly a lot of vocalization during recruitment, especially screaming. We set out to study the role of screams in the recruitment process, and to assess the adequacy of the classic «affect» view of primate calls in the interpretation of our findings. We knew already from the work of others that recruitment is by no means a random process. Recruits are much more likely to be drawn in if the opponent of the victim is a non-relative than if the altercation is between two members of the same family, for example. Thus one abstract dimension that is relevant in recruitment is kinship. It is also likely that dominance relationships between caller and opponent provide another dimension by which the potential severity of the threat could be assessed by other animals, and perhaps represented to them by a process of vocal encoding. An extended program of research was launched to explore the role, if any, of the calls in the recruitment process, and to see whether, if they are involved, they reflect any of these abstract aspects of social relationships. We also sought to establish what part affect plays, and whether something like the threat of the victim being hurt is sufficient to explain what takes place.

The subjects chosen were the rhesus macaques living in semi-natural conditions in Puerto Rico on the island of Cayo Santiago. The choice was also influenced by the fact that this is the species with which ROWELL and HINDE (1962) originally worked. They concluded that rhesus vocalizations do not form discrete categories, but rather are highly intergraded. This has since been widely cited as a classic case of graded vocal signals (e. g. MARLER 1965).

Our first undertaking was to assemble a catalogue of calls and the circumstances of use (GOUZOULES, GOUZOULES and MARLER 1984, in press). Contrary to expectations, it was found that the screams used by rhesus in recruitment are largely discrete. Five distinct classes of screams are used (Figure 2). More than 90 percent of the nearly 600 scream bouts that were analyzed contained only calls from one of these five types. Moreover, most bouts consisted only of a single type and only 9 percent included more than one of these five call types. Our data thus appear to indicate a discrete signaling system, contradicting earlier conclusions (ROWELL 1962). There is nevertheless a great deal of call variation. Individual differences occur and there are variations within a call class, but we found little or no intergradation between classes. This was reassuring for our purpose because the strategy of playback experimentation is more complicated with graded signals than with discrete ones.

Next we addressed the question of whether the different classes of calls were regularly associated with different probabilities of actual agonistic contact. This proved to be the case, as indicated in Figure 2. With one class, the noisy screams, there was a high probability that the caller would actually be contacted by the aggressor. With another call, the pulsed screams, the probability was also high, though significantly lower than for the first. Two scream types were associated with an intermediate probability of aggressive contact, tonal screams and undulated screams, and one with a low probability, the arched screams.

It is relatively straightforward to then rank the call types in an order that, on the basis of an «affect» interpretation of calling behavior, should predict the likelihood of an ally being recruited. If the primary issue is the likelihood of harm to the caller, translated into the degree to which the caller is aroused or fearful, and then manifest in one of the five scream types, we have a firm basis for predicting how likely each call type is to elicit recruitment. On the other hand, more abstract aspects of agonistic interaction, such as kinship and dominance, that are less bound up with the immediate arousal state of the caller, as will be discussed in a moment, also have potential significance to allies. If they proved to be involved, this might lend support to a representational interpretation of what is taking place. Our investigations in fact confirmed that different classes of calls are significantly associated with different kin and dominance rank relationships between caller and opponent, as well as with the probability of contact (Figure 2).

The next stage was to conduct playbacks of recorded calls to animals of known identity and descent. In early pilot studies we found these to be quite effective. If we recorded a sequence of screams from the victim of an aggressive encounter and subsequently played them back to a resting group of monkeys that included the caller's mother, she responded quickly and vigorously, looking towards the speaker, getting up and approaching, and even investigating the speaker box. However it also became clear that the response to a long string of calls was difficult to interpret, partly because of variation in call timing and quality, and partly because other animals sensed the disturbance of the mother and joined in. For this and other reasons, we decided to use the briefest samples of each call type possible, and to limit the responses studied to the viewing reaction of the mother. Re-

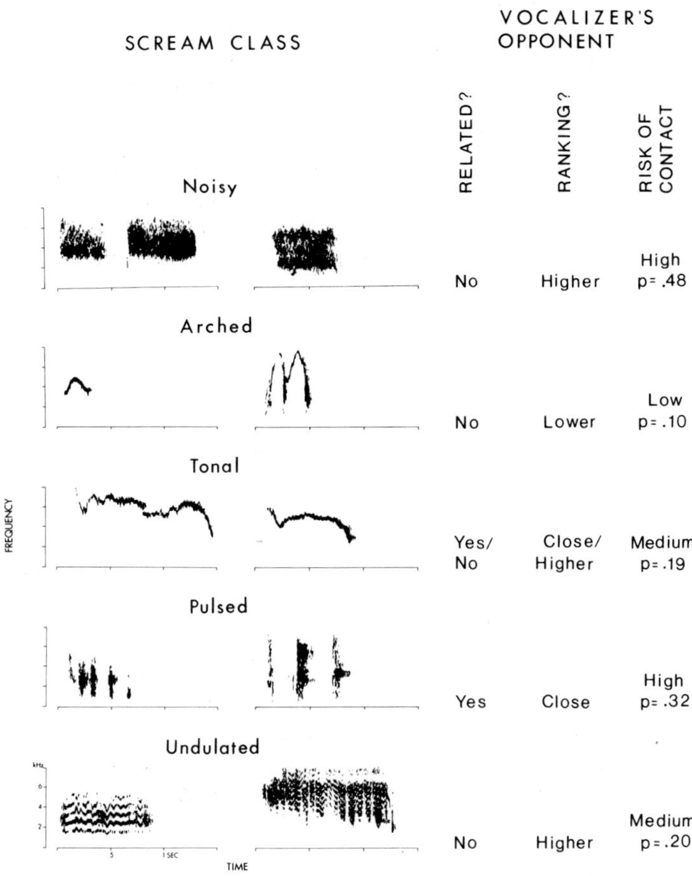

Fig. 2: Spectrograms of the five rhesus monkey scream classes used in recruiting the help of others in agonistic encounters. Degrees of relatedness and relative dominance rankings of typical opponents, and risks of contact are indicated for each scream type. (After Gouzoules, Gouzoules and Marler, in preparation.)

sponse measures were looking at the speaker, the latency of the response, and the duration of looking.

As test sounds, good examples were recorded of several examples of four of the five scream classes, as given by the juvenile offspring of 11 adult female rhesus, all members of the same troop. We were unable to get adequate samples of undulated screams from juveniles for playback purposes. Effects of the playback were quite dramatic, with a strong response evoked from the mothers, as detected and measured by filming the animals before and after playback.

Analyses of the data revealed significant differences in the probability of responses to the different classes of call. When these were ranked, they did not fall out in the order predicted by the affect hypothesis. While the call that elicits the strongest response from mothers, the noisy scream, is the one with the highest probability of agonistic contact the call that ranked next, also evoking strong responses from mothers, was the one associated with the lowest probability of agonistic contact.

At this point more must be said about the organization of aggressive behavior in a rhesus group. While in the short term the likelihood of physical harm in aggressive encounters is obviously a serious issue, in the long term, what are called rank challenges are of equal if not more importance to other members of the group. If a lower ranking non-related animal, a member of some other matriline, challenges and intimidates a young animal, this defeat may have reverberations not only for the victim, but for the status of the entire matriline to which it belongs. Thus the security of a mother's rank might be endangered if such a challenge was not countered, even if the immediate risk of harm to the juvenile was minimal. Interestingly, this is the typical situation in which arched screams occur. With this call contact is rare, and the opponent is a lower ranking non-relative (Figure 2). According to the affect hypothesis, at least in simple form, the response of mothers should be weak. If on the other hand we are ready to admit that a mother can infer the nature of this challenge from the occurrence of arched screams, then her response should be strong. This is in fact what the playbacks show, requiring something more complex than a simple affect interpretation.

At this stage of the investigation we began to entertain the possibility that these calls actually encode abstract information about the nature of the victim's opponent, providing a secure basis on which the mother can judge whether she should respond strongly, or weakly, or not at all. It has already been mentioned that recruitment is less frequent if opponents are relatives, and the results of the playbacks conform to this finding. The two calls given when opponents are likely to be relatives, the likelihood being strong in one case and weaker in the other, are the same calls, tonal and pulsed screams, that evoked the weakest responses in the playbacks. This was the case despite the fact that one, the pulsed scream, is associated with a high probability of agonistic contact. Thus the kinship relationship between the victim and its opponent is reflected in the patterns of call use and in the response to the calls of potential recruits, in this case the mother.

As the experiments progressed, our skepticism about the adequacy of a simple affect interpretation of this recruitment signaling system increased. Thus we are more and more inclined to favor the notion that these calls do in some sense represent to a potential recruit information both about kinship and about dominance relationships between caller and attacker. There are things we need to know before we can be sure, such as how the system develops. It is conceivable, for example, that mothers teach young whom they should especially fear. On the other hand, a sequelae analysis failed to yield clear evidence of variations in fearfulness of the young across call types. The behaviors of callers were tallied while they were producing the different call classes, or immediately afterwards. If different calls were based on different levels of emotional arousal, we would expect the likelihood that call types would be accompanied by withdrawal, holding ground, or advancing on the opponent, to vary accordingly. In fact the sequelae to production of different scream types were remarkably similar, a further indication of the inadequacy of a simple affect interpretation of call production. The fact that calls tend to occur in bouts of one type even though the emotional state of a caller must be varying radically as a fight develops also argues against an affect interpretation, and is at least consistent with the representational hypothesis, as is the discreteness of call structure.

For all these reasons we favor the view that this is a system of representational signals that conveys information not just about classes of objects but about relationships between individuals. These are issues that we can be sure are very much on their minds, likely to be strong foci for their intelligence, and of great potential relevance for survival and reproduction in a species in which the social factors have a strong impact on the prospects of the individual.

4 Conclusions

The inadequacies of previous interpretations of primate vocal behavior are evident. The explanatory value of a simple affect interpretation as applied to rhesus recruitment calls, to vervet alarm calls and apparently to vervet signals with social connotations as well (Cheney and Seyfarth 1982a) is strictly limited. I am increasingly convinced that it will prove inadequate with other primate communication systems, though probably not with all. Monkeys surely do experience affect, and affect is surely sometimes manifest in certain aspects of their signaling behavior. We are forced to acknowledge, however, that in some non-human primate signaling behavior, as in our own speech, signs of affect and signs of cognitive activities must be intermingled, functioning as a synergistic blend in which different acoustic vehicles convey different kinds of information. Loudness, pitch, and rates of delivery are all features of monkey calls that are likely to reflect variations of affective arousal in the caller, with the potential to convey information of value to others (Green 1975). But the quality of the information, serving to shape the *nature* of the response, rather than its *intensity*, seems to be conveyed by other structural features of the vocalizations. As arbitrary, non-iconic signals these appear to function as symbols, in a representational fashion, involving some of the same powerful cognitive machinery that hitherto has been thought to be largely latent in the natural behavior of monkeys, unless revealed by the techniques of laboratory experimentation. We thus see the prospect of closure of the gap between the intelligence that monkeys display in the laboratory and the apparently simple, emotion-based automaticity of natural communicative behavior as it has previously been interpreted. These two studies on vocal communication in the vervet monkeys, and rhesus macaques invite us to contemplate a much richer view of the vocal behavior of primates, with profound implications both for cognitive ethology (Griffin 1981) and for behavioral ecologists concerned with the functional implications of communicative behavior in animals. The occurrence of representational signaling in animals also bears on speculations about the origins of human language. It is self evident that these animals are not using language. But if they are indeed displaying symbolic behavior as our data indicate, this implies possession and use in their natural behavior of some of the cognitive underpinnings and cross-modal capabilities that are necessary prerequisites for the emergence of human language.

Acknowledgements

This paper is based on the research of four colleagues, Drs. Dorothy Cheney and Robert Seyfarth, of the University of California, Los Angeles, and Drs. Harold and Sarah Gouzoules of the Rockefeller University, New York. I am deeply indebted to them all for permission to quote this work, some of which is still unpublished. Research was supported by NIMH postdoctoral fellowship F32 MH08473 to S. G., NIMH postdoctoral fellowship F32 MH08533 to H. G., NSF grant BNS 8023423T to P. M., and Biomedical Research Support Grant PHS RR07065–15 to the Rockefeller University.

References

Andrew, R. J. (1962): The origin and evolution of the calls and facial expressions of the primates. Behaviour *20,* 1–109.

Andrew, R. J. (1963): Trends apparent in the evolution of vocalization in the old world monkeys and apes. Symp. Zool. Soc. Lond. *10,* 89–101.

Bastian, J. (1965): Primate signalling systems and human languages. In: Primate Behavior: Field Studies of Monkeys and Apes, I. DeVore (ed.). New York, Holt, Rinehart & Winston.

Cheney, D. L. and R. M. Seyfarth (1982a): How vervet monkeys perceive their grunts: field playback experiments. Anim. Behav. *30,* 739–751.

Cheney, D. L. and R. D. Seyfarth (1982b): How monkeys see the world: a review of recent research on East African vervet monkeys. In: Primate Communication, C. T. Snowdon, C. H. Brown & M. R. Petersen (eds.). Cambridge, Cambridge University Press.

Gautier, J.-P. and A. Gautier-Hion (1977): Communication in old world monkeys. In: How Animals Communicate, T. A. Sebeok (ed.). Bloomington, Indiana, Indiana University Press.

Gouzoules, G., S. Gouzoules and P. Marler (in press): External reference in mammalian vocal communication. In: The Development of Expressive Behavior: Biology-Environment Interactions, G. Ziven (ed.). New York, Academic Press.

Gouzoules, S., H. Gouzoules and P. Marler (1984): Rhesus monkey (*Macaca mulatta*) screams: representational signalling in the recruitment of agonistic aid. Anim. Behav. *32:* 182–193.

Green, S. (1975): Communication by a graded vocal system in Japanese monkeys. In: Primate Behavior, Vol. 4, L. A. Rosenblum (ed.). New York, Academic Press.

Griffin, D. R. (1981): The Question of Animal Awareness. New York, The Rockefeller University Press.

Humphrey, N. K. (1976): The social function of intellect. In: Growing Points in Ethology, P. P. G. Bateson and R. A. Hinde (eds.). Cambridge, Cambridge University Press.

Kaplan, J. R. (1978): Fight interference and altruism in rhesus monkeys. Am. J. Phys. Anthropol. *49,* 241–250.

Lancaster, J. (1965): Primate Behavior and the Emergence of Human Culture. New York, Holt, Rinehart & Winston.

Luria, A. (1982): Language and Cognition. Cambridge, Mass., MIT Press.

Marin, O., M. F. Schwarz and E. M. Saffran (1979): Origins and distribution of language. In: Handbook of Behavioral Biology II. Neuropsychology, M. S. Gazzaniga (ed.). New York, Plenum Publ. Corp.

Marler, P. (1965): Communication in monkeys and apes. In: Primate Behavior, I. DeVore (ed.). New York, Holt, Rinehart & Winston.

Marler, P. (1977): Primate vocalization: Affective or symbolic? In: Progress in Ape Research, G. Bourne (ed.). New York, Academic Press.

Marler, P. (1978a): Affective and symbolic meaning: some zoosemiotic speculations. In: Sight, Sound and Sense, T. A. Sebeok (ed.). Bloomington, Indiana, Indiana University Press.

Marler, P. (1978b): Vocal ethology of primates: Implications for psychophysics and psychophysiology. In: Recent Advances in Primatology, Vol. 1, D. J. Chivers & J. Herbert (eds.). New York, Academic Press.

Marler, P. and R. Tenaza (1977): Signaling behavior of apes with special reference to vocalization. In: How Animals Communicate, T. A. Sebeok (ed.). Bloomington, Indiana, Indiana University Press.

Myers, R. E. (1976): Comparative neurology of vocalization and speech: proof of a dichotomy. In: Origins and Evolution of Language and Speech, S. R. Harnad, H. D. Steklis and J. Lancaster (eds.). New York, New York Academy of Sciences.

Newman, J. D. and D. Symmes (1982): Inheritance and experience in the acquisition of primate acoustic behavior. In: Primate Communication, C. T. Snowdon, C. H. Brown and M. R. Peterson (eds.). Cambridge, Cambridge University Press.

Premack, D. (1975): On the origins of language. In: Handbook of Psychobiology, M. S. Gazzaniga and C. B. Blakemore (eds.). New York, Academic Press.

Premack, D. and A. J. Premack (1983): The Mind of An Ape. New York, W. W. Norton & Co.

Rowell, T. E. (1962): Agonistic noises of the rhesus monkey *(Macaca mulatta)*. Symp. Zool. Soc. Lond. *8,* 91–96.

Rowell, T. E. and R. A. Hinde (1962): Vocal communication by the rhesus monkey, *(Macaca mulatta)*. Proc. Zool. Soc. Lond. *138,* 279–294.

Seyfarth, R. M. and D. L. Cheney (1980): The ontogeny of vervet monkey alarm calling behavior: A preliminary report. Z. Tierpsychol. *54,* 37–56.

Seyfarth, R. M., D. L. Cheney and P. Marler (1980a): Vervet monkey alarm calls: semantic communication in a free-ranging primate. Anim. Behav. *28,* 1070–1094.

Seyfarth, R. M., D. L. Cheney and P. Marler (1980b): Monkey responses to three different alarm calls: evidence of predator classification and semantic communication. Science *210,* 801–803.

Smith, W. J. (1977): The Behavior of Communicating. Cambridge, Mass., Harvard University Press.

Snowdon, C. T., C. H. Brown and M. R. Peterson (eds.) (1982): Primate Communication. Cambridge, Cambridge University Press.

Struhsaker, T. T. (1967): Auditory communication among vervet monkeys *(Cercopithecus aethiops)*. In: Social Communication Among Primates, S. A. Altmann (ed.). Chicago, University of Chicago Press.

Struhsaker, T. T. (1970): Phylogenetic implications of some vocalizations of Cercopithecus monkeys. In: Old World Monkeys, J. R. Napier and P. H. Napier (eds.). New York, Academic Press.

III. Communication and Reproductive Behaviour

Fortschritte der Zoologie, Bd. 31 · Hölldobler/Lindauer (Hrsg.): Experimental Behavioral Ecology
G. Fischer Verlag · Stuttgart · New York · 1985

Reproductive behaviour in honeybees

Friedrich Ruttner

Institut für Bienenkunde, Oberursel, F. R. Germany

Abstract

The reproductive behaviour of honey bees is very complex. Three points of major significance should be emphasized:

1. The course of reproduction is a concern of the whole colony with environmental factors weighing heavily. Thus reproduction is the immediate subject of selection, as easily shown by variation of the swarming impetus.

2. A sophisticated device is required to transfer semen in sterile conditions and to store it for several years.

3. Specific characteristics of mating behaviour have the single goal of maintaining high genetic variety for the purpose of the long term survival of the population – in spite of the high individual risks for the queen and for the whole colony.

1 Introduction

Division of functions between worker and sexual castes in reproductive behaviour is a definitive characteristic in eusocial insects.

In honeybees this division is developed to the highest level. No member of the honeybee society can exist in isolation, even queen and drones during the mating period – except for the short time of the mating flight itself. Thus reproduction in honeybees is strictly linked with the process of fission of the colony, that is in most cases with swarming. The rate of reproduction does not depend on the number of queens produced, but on the number of swarms. The whole colony – primarily the worker bees – decide on swarming and production (and maintaining) of young queens and of drones. The interaction between queens and workers is to be observed in various stages of the process of reproduction.

Another characteristic in honeybees which has to be considered in order to understand the pecularities of reproductive behaviour is one of a genetic kind: The proneness of honeybees to inbreeding. The rapid appearance of inbreeding effects in honeybees, well known through apicultural experience, may be explained by the accumulation of two different systems:

1. General inbreeding effects by increasing homozygosity, as found in other animals and plants. It is assumed that a decrease in fitness will occur as soon as the factor of inbreeding is 0,25 or more (R. Moritz 1982). Decreased fitness of a bee colony is observed at two levels: in queens (and even in her haploid sons due to the insufficient equipment of the egg with maternal enzymes (Moritz 1979), and in workers.

2. Loss of brood of queens after mating with related drones. This is due to the peculiar mechanism of «complementary sex determination» first described by WHITING (1943) in *Habrobracon hebetor*, and later on also in the honey bee by MACKENSEN (1951) and WOYKE (1963): The female sex is determined by heterozygosity in a certain locus («sex locus», «X»), the male sex by homozygosity or hemizygosity in the same locus. Homozygosity occurs after inbreeding (50 % after brother-sister mating with one drone, 25 % on average after mating with several drones).

Considering this double danger of inbreeding damages to fitness it is not surprising to find a number of behavioural characteristics of honeybees which seem to have only one single goal: the prevention of inbreeding as far as possible (e. g. obligatory mating during flight, great mating distance, multiple matings at drone congregation areas etc.).

All in all, mating behaviour is a complicated sequence of exactly determined elements. The peculiar complexity in this field is best illustrated by one unique morphological characteristic, the male genitalia. The endophallus of the drone is a complicated organ, not found elsewhere. The function of its different sections can only be partly explained. The structure is even more bizarre in the two species *A. dorsata* and *A. florea*.

A third very specific aspect of this field seems to be primarily inherent in human nature. That is the obvious irresistible stimulation of human fantasy and romantic imagination by certain phases of reproduction in honey bees, e. g. by the mating flight. This event was first correctly observed more than two hundred years ago by ANTON JANSCHA and FRANCOIS HUBER. In the following period of developing experimental research in biology scientists did not further the analysis of this phenomenon, However, poets such as MAETERLINCK and others gave a very romantic story and created the dogma of the honeybee queens monogamy, which was also accepted in the scientific world. We experienced ourselves 30 years ago how much evidence was needed to break this dogma.

This preference for hypothetizing before collecting experimental data in the field of honeybee reproduction still seems to be widespread nowadays.

It is true that experiments with honeybee sexuals are much more difficult to perform than those with worker bees: the sexuals can only be obtained in a restricted part of the season and in restricted numbers. All attempts to train them have so far been unsuccessful, and many experiments depend on the weather conditions. However, this alone cannot explain fully the scarcity of experimental data in certain sections of this field. As this symposium is being held in memory of KARL V. FRISCH, I am inclined to say we need more of his spirit, which can best be expressed by his own words: «The correctly planned experiment is the magic key that forces an animal to answer a given question and this answer will never be wrong». The following problems have to be borne in mind. First of all we have to consider the complexity of the subject. Secondly the level of research in this field. In my opinion it is inadequate compared to what has been accomplished concerning the behaviour of worker bees in orientation and communication. Thirdly the sensitiveness of honeybees to inbreeding.

The sequence in the reproductive behaviour of honeybees is as follows:

(1). Swarming
   1.1. Preparation for swarming and exodus of the prime swarm with the old queen.
   1.2. Hatching of one or several young queens in the residual part of the mother colony, eventual departure of one or several after-swarms with some virgin queens.

(2.) Premating phase of queens and drones

(3.) Mating

(4.) Behaviour and physiology in the postmating phase of the queen.

Emphasis will be put on the interactions of the sexuals with the whole bee colony, and with the environmental conditions.

2 Swarming

The process of swarming is the most spectacular event in the annual life cycle of a bee colony, and many studies have been made to analyse this behaviour (Simpson 1973). Here we will deal with this phenomenon as far as it is relevant to basic aspects of reproduction.

Many factors are involved in the swarming behaviour of a bee colony including overcrowding of the hive, especially of the brood nest, amount and quality of pollen collected, the age of the queen, microclimate and genetic disposition.

It is evident that these factors – except for genetic disposition – act only as releasers, not as causal agents. Swarming is an innate behavioural characteristic, obligatorily linked with reproduction: a colony can produce as many daughter colonies with mated young queens as it produces swarms (without considering supersedure of old queens).

The frequency of swarms varies from year to year, from region to region and from strain to strain. Seeley (1978) found in feral colonies a minimum swarming frequency of 0.8–0.9 in the region of Ithaca, N. Y. A minimum average swarming frequency has to occur to maintain a given population size and to cover inevitable losses. In the temperate zones of Europe and America the average losses are as follows: during mating flight 20 %, during winter 10–20 % (in domestic hives). In feral colonies losses are probably sometimes much higher. Young feral colonies have a much higher loss rate than established ones (Seeley 1978). The amount of loss due to enemies and diseases of a largely varying kind is not predictable.

2.1 Preparation to swarm

The first step in reproduction is the raising of sexuals. Drone brood appears prior to queen brood, exactly corresponding to the time of development from egg to the mature adult (40 versus 22 days). The production of drones and queens depends on a certain stage of development of the colony – a sufficient number of brood cells and nurse bees to raise and feed a large number of drones and – later on – to give viable divisions of the colony.

As soon as the first queen cell is capped, the prime swarm with the old queen leaves the nest. The earlier the swarm in the season, the better its chance of surviving.

2.2 Young queens in the mother colony after the departure of the prime swarm

Seven days after the first swarm, the decision is made on the further course of events. One virgin queen is hatched by this time, and it is again up to the worker bees whether she is allowd to destroy all other queens and queen cells. If so, the result of reproduction is one colony with a young queen and with old combs, and a colony with the old queen (which may be superseded later on) on a new site. If not, a second swarm with one or several virgin queens will issue. This incident may recur, depending on the number of bees left and the genetic disposition.

The chance of a secondary swarm to survive is worse than that of a prime swarm since mating of the virgin queen with its risks and loss of time to develop a wintering colony is involved. Thus a high selective pressure will act on the bee colony towards a swarming behaviour in a given climate which guarantees the necessary rate of reproduction and, at the same time, viable colonies.

Generally the tendency to swarm will be rather low in the temperate zones. However, a considerable genetic variability exists, corresponding to the variable conditions from year to year – swarming being either advantageous or disadvantageous to the population. Planned selection is feasible in both directions. There exist strains with a very low propensity to swarming (selected for modern apiculture). Other strains regularly give 2–3 swarms per season, a consequence of selection for centuries in a special environment (late honeyflow).

In warm climates with no wintering problems for the bees, though sometimes with heavy losses from drought, fire or enemies, extreme tendencies to swarming are observed during favorable conditions. This is well documented for the «Africanized bee» in South America. In an oasis of Algeria, BROTHER ADAM (1968) observed a colony of *A. mellifera intermissa* which yielded seven swarms in one single season. Finally only a cup of bees was left, which developed again to a full size colony. There is a clear correlation between environmental conditions and the degree of swarming tendency in various races of bees.

The principle of monogyny is abandoned for a short period during the act of swarming in all races of the temperate zone; at the moment of exodus of the secondary swarm several virgin queens hatch simultaneously from their cells and either join the swarm or stay in the mother colony. Monogyny, however, is restored within a few hours, since in each site one virgin kills all the other queens.

In several races of the southern Mediterranean *(A. mellifera lamarckii, intermissa, sicula)* the behaviour is quite different. Virgin queens are not aggressive towards each other. In one case we found more than 50 virgin queens in one swarm of *A. m. lamarckii*, and we were able to keep a dozen virgins together for several days in one cage. As soon as the first queen was mated, however, all the others were eliminated.

This behaviour is highly interesting. It seems excellently suited to reducing losses during mating to zero – a considerable advantage in selection. It is hard to understand why this characteristic has not been adopted by other races. This observation possibly indicates that the principle of monogyny is very important in the bee colony's life, even during the short period of mating.

3 Premating phase

3.1 Drones

In newly hatched drones the mucus gland consists of solid tissue; the semen is still in the testis. Neither mucus nor semen can be ejaculated. At least 12 days are needed for full sexual maturity. Six days after hatching, the mucus glands are transformed into thin walled sacs filled with a white liquid, which rapidly coagulates if exposed to air. The migration of the spermatozoa to the vesicula seminalis is nearly completed. At full sexual maturity semen mixed with the secretion of the vesicula is ejaculated after an adequate stimulus followed by the liquid mucus (RUTTNER and TRYASKO 1975).

In the first days of adult life drones depend to a large extent on feeding by worker bees. This explains the heavy influence of the colony's condition on the maintaining of drones. Flight activity of drones for orientation starts at the age of 8 days.

3.2 Queens

The freshly hatched queen is almost completely ignored by the worker bees, corresponding to the low quantity of 9-ODA produced by the still undeveloped mandibular glands (7 μg versus 135 μg in queens 5–10 days old, BUTLER and PATON 1962). Later on the first signs of a court around the queen are observed, but also an antagonistic behaviour of the workers towards the queen: the queen is attacked and chased through the hive. These attacks are obviously a part of the process of maturation (HAMANN 1956). The queen is only attacked by older bees, not by young bees. If the queen is kept permanently among young bees she will never show a tendeny to leave the hive for mating.

Generally the queen will not start earlier than the 6th day of adult life for the first short flight (orientation flight), having been suffering increasing attacks by worker bees. Longer flights, possibly with matings, start on the 7th day.

During the premating phase important physiologcial changes occur in the young queen: simultaneously with the mandibular glands, the tergite glands increase in size (RENNER and BAUMANN 1964). Both gland systems produce the sexual odour of the queen. High activity of the neurosecretory cells in the pars intercerebralis as well as a storage of neurosecretion in the corpora cardiaca and corpora allata is observed during sexual maturation (BIEDERMANN 1964, HERRMANN 1969). At the same time the yolk protein titre in the hemolymph of the queen and of workers is increasing (ENGELS 1980).

4 Mating

4.1 Time and number of mating flights and matings

It is a well known fact that *A. mellifera* queens never mate in the hive – all the later experiments confirmed the early observations by F. HUBER in 1814 –, and no really exact observation of a mating in confinement (in flight cages) is available.

Mature queens and drones can stay on the same comb in a hive without paying any attention to each other. However, it can be shown experimentally that the olfactory sensory cells of the drone respond to the queen in any situation, even when the drone is tethered (RUTTNER and KAISSLING 1968). On the other hand, the same drone will be strongly attracted by the same queen if she is tethered to a pole outside the hive at a height of 10 meters.

With *A. mellifera*, mating flights are restricted to the hours of early afternoon (13–16h local time) – the hours of highest average temperatures. The flight of the queen takes place only under the conditions of a warm, quiet, bright summer day, i. e. temperature not lower than 20°, bright sky (not necessarily sun), little air movement (under 4 on the meteorological scale).

Queens are inflexible in maintaining the mating flight time. This is important for behavioural isolation of sympatric species. N. KOENIGER (1976) found three different flight periods of drones of three species in Sri Lanka.

In regard to the queen's flight duration two groups are found:

(1.) Shorter than 10 min. The queen returns regularly without mating sign – orientation flights.

(2.) Longer than 10 min. (up to 45 min. and more). The queen returns with mating sign – effective mating flights.

The longer the duration of flight, the greater the number of spermatozoa in the queen (WOYKE 1956).

Half of the queens start egg laying after one single effective mating flight. The other half only after two or more (up to 5) effective mating flights, mostly on successive days (ROBERTS 1944, RUTTNER 1954, 1957). From data on the quantity of sperm in the oviducts of mated queens and on the analysis of the progeny after mating with genetically marked drones, it has to be concluded that the queen mates on average with 6–8 drones in one single flight (maximum observed 16 drones, WOYKE 1956; see also TABER 1954, TRYASKO 1956). According to RUTTNER et al. 1973 and WOYKE 1975, in *A. cerana* this figure is much higher – up to 30 drones per flight.

These relatively recent observations are of great significance for the genetics and the social structure of the bee colony. Assuming mating with 8 drones occurs, the average parentage of the progeny is reduced from 75 % (mating with one drone) to 53 % – that is approximately to the level of diploid organism. This diminishes the danger of inbreeding. Recently the consequences of multiple mating in regard to fitness were analysed by Page (1980) and Page and Metcalf (1982).

However, there is another, very simple reason for multiple mating. To produce a large number of fertilized eggs during three or more years, the honeybee queen needs a) – a large number of spermatozoa in the spermatheca, b) a working device which regulates as exactly as possible the dosage of spermatozoa for fertilization of the individual egg (we calculated that on average only 15 spermatozoa are at the disposal of one egg).

Such an exact dosage is to be achieved only with a very narrow valve – as can be found in the ductus spermaticus of the queen (∅ 20–25 μ). This narrow passage is a severe obstacle to the process of filling the spermatheca. Only 10 % of the spermatozoa deposited in the oviducts reach the spermatheca. To fill the spermatheca with 5 millions, 50 mill. spermatozoa have to be injected in the oviducts. Queens inseminated with a lower quantity of sperm will soon lay unfertilized eggs (Woyke and Ruttner 1976).

5 Drone congregation areas

The first approach to a closer study of the events during the mating flight was found by the observation of drone congregation areas in 1958 (P. Jean-Prost). This term does not describe casual concentrations of drones in any location. A drone congregation area is an accumulation of flying drones in one and the same place during the whole flight season, year after year, independent of the presence of a queen (F. and H. Ruttner 1963, 1965, 1966, 1968, 1972; Gerig 1975). Such places have been recorded during the last decades in many countries in different continents and climates. Some of the congregations were observed for more than 20 years without noticeable change. The height of maximum density of drones varies (in different regions and different seasons) from 2–4 m to as much as 20–40 m.

The evidence of a congregation area can be detected by ear (loud humming), by throwing a stone in the air (the stone will be followed by a couple of drones) or – best – by attracting drones to a tethered queen, to queen extract or synthetic 9-ODA.

It is strange to observe that evidently the drones are stronger attracted to a specific location than to the odour of a queen. If a tethered queen with a swarm of drones circling round her is moved beyond the borders of the congregation area, the drones will follow her only for a short distance, and then return to the area, leaving the queen alone. This behaviour allows the queen to terminate the mating process at any time.

It was observed that queens arrive at the congregation area about one hour later than the drones, and that they mate there (Jean-Prost 1958). The assumption that queens mate mainly in drone areas is supported by the well known observation that the result of a mating flight follows the rule of «All or Nothing»: the queen returns either with the semen of many drones or without any (however very rarely with semen of one or two drones only).

6 Orientation

Direction and range of drone flights was studied by marking different drone populations in a region of the Austrian Alps and another in the Taunus Hills near Frankfurt.

More than 120000 drones were marked during several years, and genetic markers were also used. It was shown that certain flight directions were clearly preferred to others, e. g. at the congregation area «Seekopf». This proved to be an excellent experimentation area, located on a saddle at the crossing of two valleys (Fig. 1A). At this heavily visited place drones were found from all but one bee yards of the region, in a radius of 5 km, corresponding to an area of 78 km^2 with several hundred colonies. A congregation area is the meeting place of all drones of the region – resulting in an effective prevention of inbreeding.

During these experiments it became evident that drones of certain apiaries visited this area more frequently than expected, in other cases less frequently. Apiaries located in the SW, NE and ENE from where a sharp mark on the mountain silhouette was clearly visible were overrepresented, those in the N with a less clear mark were underrepresented. From one apiary in NW with no visible horizon mark visible in this direction (Fig. 1B) not a single drone was found in two consecutive years in spite of the relatively short distance of 1700 m.

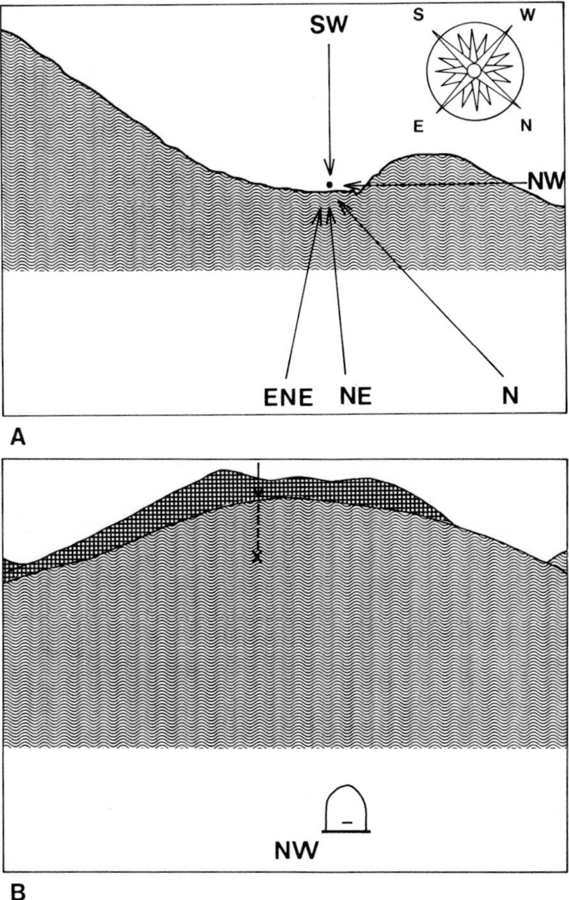

Fig. 1: *A* Drone congregation area «Seekopf». Lines indicate directions to apiaries with populations of marked drones. *B* View from NW – no horizon mark visible.

This orientation behaviour is not learned, it is present as an innate characteristic as the following experiment showed. We knew the frequency of drones arriving from one apiary ENE of the congregation area Seekopf. We transported 4 colonies with differently marked drones to a place 50 m away from this apiary. If the orientation behaviour were learned an increasing number of drones would have been expected during the following days. This was not the case. On the very first day we caught newcomer drones at just the frequency expected, without an increase on the following days (F. and H. Ruttner 1972).

In less mountainous regions other marks, e. g. trees may be used for orientation.

In the flight area of a bee colony several congregation places are found. Using three dominant genetic markers the approximate mating distance was determined. Maximum drone flight distance of more than 7 km, traversing a mountain ridge of 800 m was found in this case (H. Ruttner 1976) and a medium mating distance of queens of 2,5 km. The figures of flight distances of queens and drones added together give a maximum mating distance of 10–12 km. This corresponds exactly to the data of Peer (1957) from Canada, determined by a quite different method.

Thus good evidence exists regarding far distance orientation of drones and queens. However, the question remains unanswered concerning the orientation at the area itself. What makes drones circle at a certain location, within invisible boundaries, as if attracted to the area by magic powers? Several hypotheses have been discussed – condition of the ground, thermal upwind, distribution of radiation on the horizon – but not a single one can be applied to all cases observed. Thus an important question of orientation still remains to be solved.

7 Copula

From a greater distance drones are attracted by a moving object, irrespective of shape or size or odour. We have seen drones following other drones, butterflies, birds, leaves blown in the wind, or stones thrown upwards. The drone approaches the object upwind and at a closer distance it is attracted by the odour of the queen or synthetic 9-ODA, again irrespective of shape, size and movement.

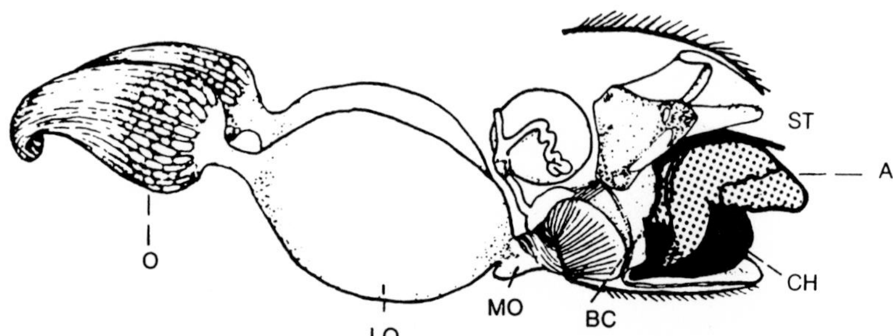

Fig. 2: Genitalia of a freshly mated queen (G. Koeniger 1983) CH Chitinuous plates of the «mating sign», pressed into the bursa copulatrix (BC) of the queen; its sticky mucus (A) will adhere to the hairy part of the endophallus of the next drone during eversion, and thus the whole plug will be removed.

LO	lateral oviduct, filled with semen
MO	median oviduct
O	ovary
ST	sting

The copula itself will be described for the most frequent case, which is also the most complicated one, namely the mating of a queen with a mating sign, i. e. all the matings of a series but the very first one. Earlier it had been an enigma how a queen can mate again as long as the «mating sign» – a part of the male endophallus with mucus – is fixed in the sting chamber (Fig. 2; THORNHILL and ALCOCK 1983). This was fully understood only recently by a study of G. KOENIGER (1983). The drone clasps the queen from behind and below and brings the tip of the abdomen to the sting chamber of the queen. In all four *Apis* species the drones show brushes of sophisticated hairs on the metatarsus (RUTTNER 1976). The last stimulus – releasing eversion and ejaculation – is the contact to the opened sting chamber of the flying queen, that is to a tube 3,5 mm wide (GARY and MARSTON 1971). Tufts of sensory hairs with a peripheric ganglion on the parmeres serve as receptors (RUTTNER 1957, SCHLEGEL 1967).

In the first stages of eversion a field of soft brown hairs appears on the ventral side of the endophallus. These hairs glue to the mucus of the mating sign and extract it from the queen while the process of eversion proceeds. Thus the bulb of the endophallus can be inserted (Fig. 3) and the semen ejaculated into both oviducts. By detaching one part of the bulb the drone separates from the queen and falls dead to the ground.

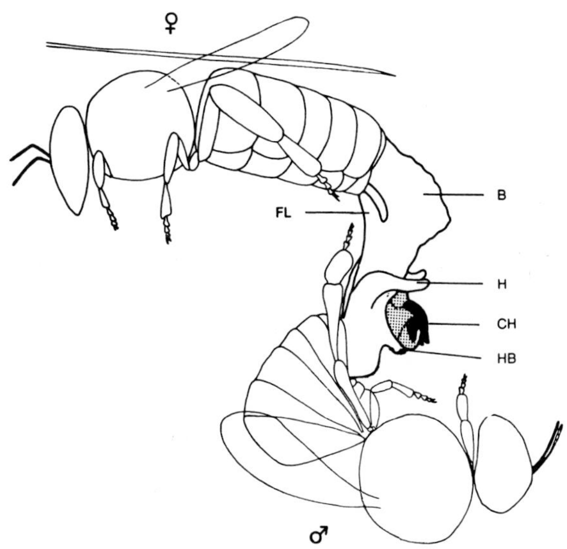

Fig. 3: Queen and drone during copula, as observed in experiments with tethered queens (G. KOENIGER 1983)

B bulb of endophallus

CH chitinuous plates of the mating sign (dotted) originating from the precedent drone, glued to a prominent hairy field (HB).

H cornua

FL fimbriate lobe of endophallus.

8 Postmating phase

The semen ejaculated into the lateral oviducts of the queen is transferred to the spermatheca within 24 hours. As shown by experiments at the laboratory at Oberursel, this is a combined process of pressure by the queen and undulating locomotion of the spermatozoa (Ruttner and G. Koeniger 1972, Gessner and Ruttner 1976). The completely filled spermatheca of a queen of a European race contains 5–7 mill. spermatozoa surviving for years in an isotonic and alkaline (p_H 8,5) liquid. A net of tracheae round the receptaculum, a one-layer epithelium on its membrane and the secretion of the spermathecal gland are essential for their survival (G. Koeniger 1970).

Oviposition starts as soon as 1½ days after the last mating. During the development of the ovaries the neurosecretion is emitted from the neurosecretory cells and the corpora allata (Herrmann 1969). The external stimuli for the start of oviposition are certainly linked to events during the mating flight. The perception mainly occurs in the sting chamber and bursa copulatrix during the copula, but semen and mating sign do not have a significant effect. Mere external contacts show a slight effect (G. Koeniger 1985).

References

Adam, Brother (1968): In search of the best strains of bees. Walmar Verlag, Zell (BRD).

Alber, M., Jordan, R., Ruttner, F. and H. (1955): Von der Paarung der Honigbiene. Z. Bienenforschung *3*, 1–28.

Biedermann, M. (1964): Neurosekretion bei Arbeiterinnen und Königinnen von *Apis mellifera L.* unter natürlichen und experimentellen Bedingungen. Z. wiss. Zool. *170*, 256–308.

Butler, C. G., Paton, P. N. (1962): Inhibition of queen rearing by queen honeybees *(Apis mellifera L.)* of different ages. Proc. R. Ent. Soc. London A *37*, 114–116.

Engels, W. (1980): Control of fertility in bees. Int. Conf. on Regulation of Insect Development and Behavior. Karpasz, Poland.

Gary, N., Marston, J. (1971): Mating behavior of drone honeybees with queen models *(Apis mellifera L.)*. Anim. Behav. *19*, 229–304.

Gerig, L. (1971): Wie Drohnen auf Königinnen-Attrappen reagieren. Schweiz. Bienenz. *94*, 558–562.

Gessner, B., Ruttner, F. (1977): Transfer der Spermatozoen in die Spermatheka der Bienenkönigin. Apidologie *8*, 1–18.

Hammann, E. (1956): Wer hat die Initiative bei den Ausflügen der Königin, die Königin oder die Arbeitsbienen? Diss. Zool. Inst. der Freien Universität Berlin.

Herrmann, H. (1969): Die neurohormonale Kontrolle der Paarungsflüge und der Eilegetätigkeit bei der Bienenkönigin. Z. f. Bienenforsch. *9*, 509–544.

Jean-Prost, P. (1957): Observation sur le vol nuptial des reines d'abeilles. C. R. Acad. Sci. *245*, 2107–2110.

Koeniger, G. (1985): Reproduction and mating behavior. In Rinderer, Genetics of honeybees. Academic Press (in press).

Koeniger, G. (1970): Bedeutung der Tracheenhülle und der Anhangdrüse der Spermatheka für die Befruchtungsfähigkeit der Spermatozoen in der Bienenkönigin *(Apis mellifica L.)*. Apidologie *1*, 55–71.

Koeniger, G. (1983): Die Entfernung des Begattungszeichens bei der Mehrfachpaarung der Bienenkönigin. Allg. Dtsch. Bienenztg. *17*, 244–245.

Koeniger, N., Wijayagunasekera H. (1976): Time of drone flight in the three honeybee species *(A. cerana, A. florea, A. dorsata)* of Sri Lanka. J. apic. Res. *15*, 67–71.

Mackensen, O. (1951): Viability and sex determination in the honey bee *(Apis mellifera L.)*. Genetics *36*, 500–509.

Moritz, R. F. A. (1978): Einfluß der Inzuchtdepression auf die Fitness der Drohnen *(Apis mellifera capensis)*. Apidologie *9*, 156–158.

Moritz, R. F.A. (1982): Maternale Effekte bei der Honigbiene *(Apis mellifera L.)* Zeitschr. Tierzücht. Züchtungsbiol. *99*, 139–148.

Page, R. E. (1980): The evolution of multiple mating behavior by honey bee queens *(Apis mellifera L.)*. Genetics *96*, 263–273.

Page, R. E., Metcalf, R. A. (1982): Multiple mating, sperm utilization, and social evolution. Amer. Naturalist *119*, 263–281.

Peer, D. F. (1957): Further studies on the mating range of the honeybee, Apis mellifera. Can. Entom. *89*, 108–110.

Pirchner, R., Ruttner, F. and H. (1961): Erbliche Unterschiede zwischen Ertragseigenschaften der Bienen. Proc. XIth Int. Congr. Entomol. II, 510–516.

Renner, M., Baumann, M. (1964): Über Komplexe von subdermalen Drüsenzellen (Duftdrüsen?) der Bienenkönigin. Naturwiss. *51*, 68–69.

Roberts, W. C. (1944): Multiple mating of queen bees proved by progeny and flight tests. Glean. Bee Cult. *72*, 281–283.

Ruttner, F. (1954): Mehrfache Begattung der Bienenkönigin. Zool. Anzeiger *153*, 99–105.

Ruttner, F. (1957): Die Sexualfunktionen der Honigbienen im Dienste ihrer sozialen Gemeinschaft. Z. vergleich. Physiol. *39*, 577–600.

Ruttner, F. (1962): Bau und Funktion des peripheren Nervensystems an den Fortpflanzungsorganen der Drohnen. Ann. Abeille *5*, 5–58.

Ruttner, F. (1975): Ein metatarsaler Haftapparat bei den Drohnen der Gattung Apis *(Hym.: Apidae)*. Ent. Germ. *2*, 22–29.

Ruttner, F. (1979): Events during natural replacement of the queen in a bee colony. In «Queen Rearing», p. 5–18. Apimondia, Bucharest.

Ruttner, F., Kaissling, K.-E. (1968): Über die interspezifische Wirkung des Sexuallockstoffes von *Apis mellifica* und *Apis cerana*. Z. vergl. Phys. *59*, 362–370.

Ruttner, F., Koeniger, G. (1971): Die Füllung der Spermatheka der Bienenkönigin. Aktive Wanderung oder passiver Transport der Spermatozoen? Z. vergl. Physiol. *72*, 411–422.

Ruttner, F., Ruttner, H. (1963–1972): Untersuchungen über die Flugaktivität und das Paarungsverhalten der Drohnen. – I., 19. Int. Beekeeping Congr. Praha, 604–605. – II. Beobachtungen an Drohnensammelplätzen. Z. Bienenforsch. *8*, 1–8. – III. Flugweite und Flugrichtung der Drohnen. Z. Bienenforsch. *8*, 332–354. – IV. Zur Fernorientierung und Ortsstetigkeit der Drohnen auf ihren Paarungsflügen. Z. Bienenforsch. *9*, 259–265. – V. Drohnensammelplätze und Paarungsdistanz. Apidologie *3*, 1972, 203–232.

Ruttner, F., Woyke, J., Koeniger, N. (1972): Reproduction in *Apis cerana*. 1. Mating behaviour. J. apic. Res. *11*, 141–146.

Ruttner, F., Woyke, J., Koeniger, N. (1973): Reproduction in *Apis cerana*. 2. Reproductive organs and natural insemination. J. apic. Res. *12*, 21–34.

Ruttner, H. (1976): Untersuchungen über die Flugaktivität und das Paarungsverhalten der Drohnen. – VI. Flug auf und über Höhenrücken. Apidologie *7*, 331–341.

Schlegel, P. (1967): Elektrophysiologische Untersuchungen an den Borstenfeld-Sensillen des äußeren Geschlechtsapparates der Drohnen *(Apis mellifica)*. Naturwiss. *54*, 26.

Seeley, T. (1978): Life history strategy of the honey bee, *Apis mellifera*. Oecologia *32*, 109–118.

Simpson, J. (1973): Influence of hive space restriction on the tendency of honeybee colonies to rear queens. J. apic. Res. *12*, 183–186.

Taber, S. III (1954): The frequency of multiple mating of queen honeybees. J. Econ. Entomol. *47*, 995–998.

Thornhill, R., Alcock, J. (1983): The evolution of insect mating systems. Havard Univ. Press, New York.

Trjasko, V. V. (1956): Multiple matings of queen bees (in Russian). Pchelovodstvo *12*, 29–31.

Whiting, P. (1943): Multiple alleles in complementary sex determination of Habrobracon. Genetics *28*, 365–382.

Woyke, J. (1955): Multiple mating of the honeybee queen (Apis mellifica L.) in one nuptial flight. Bull. Acad. Polon. Sci II *3*, 175–180.

Woyke, J. (1960): Natural and artifical insemination of queen honeybees. Pszczelinicze Zeszytyn Naukowe IV, 183–275.

Woyke, J. (1963): Drone larvae from fertilized eggs of the honeybee. J. apic. Res. *2,* 19–24.

Woyke, J. (1975): Natural and instrumental insemination of *Apis cerana* indica in India. Journ. apic. Res. *14,* 153–159.

Woyke, J., Ruttner, F. (1958): An anatomical study of the mating process in the honeybee. Bee World *39,* 3–18.

Woyke, J., Ruttner, F. (1976): Results. In F. Ruttner, Instrumental Insemination of the Honeybee Queen. Apimondia Publ., Bucharest.

Fortschritte der Zoologie, Bd. 31 · Hölldobler/Lindauer (Hrsg.): Experimental Behavioral Ecology
G. Fischer Verlag · Stuttgart · New York · 1985

Sociobiology of reproduction in ants

Bert Hölldobler and Stephen H. Bartz

Department of Organismic and Evolutionary Biology, MCZ-Laboratories, Harvard University, Cambridge, Mass. 02138, USA

Abstract

Two major components of the sociobiology of reproduction in ants are mating behavior and the division of reproductive labor in mature colonies. In this paper we review the facts and some of the theoretical issues associated with these major aspects of ant reproduction.

Mating behavior among ant species is quite diverse, but we recognize two major patterns which we have termed the «female calling syndrome» and the «male aggregation syndrome». In the female calling syndrome, new queens typically do not disperse widely, but tend to remain close to the natal nest, and secrete pheromones which attract males for mating. This syndrome has been found in several phylogenetically primitive species and in some socially parasitic, dulotic, and myrmecophylous species. Mature colonies are generally small and produce relatively few new reproductives each year. The male aggregation syndrome occurs among more phylogenetically advanced species that form large colonies and produce hundreds or thousands of alates in each reproductive season. Among these species, males congregate at mating sites and produce sex-attractant pheromones in concert. Females often come from considerable distances to these aggregations for mating. Differences between these syndroms in such features as the extent of synchronization of mating flights, the incidence of multiple insemination, and the potential for sexual selection are discussed both empirically and theoretically.

Conflict of interest between the queen and the workers in ant societies is expected to occur over the ratio of investment in male versus female reproductives and over the source of the reproductive males. The theoretical arguments concerning queen or worker control of colony reproduction are reviewed along with data suggesting that the queen has a powerful controlling influence. We argue that it is possible for workers to lose control of the colony's reproductive activities in the course of evolution.

1 Introduction

It has been estimated that there are 12,000–14,000 species of ants living on earth (Wilson 1971). With the exception of a few, clearly derived parasitic species, all form eusocial societies in which one or several individuals are aided in their reproductive effort by a much larger number of sterile or reproductively inactive individuals. The evolution and maintenance of these societies has been the subject of considerable theoretical interest (see Starr 1979, Crozier 1979, 1980 for reviews). One issue that is still debated is whether colony reproduction is controlled, in a proximate as well as evolutionary sense, by workers or by the queen. We will review the facts and theoretical arguments concerning queen or worker control of reproduction in the second part of this paper.

Another important aspect in the reproductive biology of ants is the substantial diversity in mating behavior. This diversity is only now beginning to be explored in greater detail, and in the first part of this paper we will discuss some of the most distinct mating patterns in ants in the context of modern evolutionary theory.

2 Mating behavior

2.1 Two major mating syndromes

Although mating behavior among ant species is very diverse we can recognize two major syndromes. In one of these the females remain relatively stationary, they are often even wingless or have reduced wings and they attract males from distant nests by exhibiting a special calling behavior during which they release sex pheromones (Fig. 1). This

Fig. 1: Mating sequence in *Rhytidoponera metallica.* (1) Ergatoid female (black) in calling posture, during which the female releases a sex pheromone from the pygidial gland located between the 6th and 7th tergites. (2) A male (white) approaches a calling female, touching her with his antennae. (3) The male grasps the female at the thorax and mounts her. Simultaneously he extends his copulatory organ in search for the female's genitals. (4) This finally leads to copulation (from Hölldobler and Haskins 1977).

Fig. 2: *a Pogonomyrmex maricopa* females, emerging from the nest entrance in order to take flight.
b Mating aggregation of *Pogonomyrmex desertorum* at an acacia tree. Note the upwind flight of the approaching males and females.
c Mating cluster of *P. desertorum*. A female is surrounded by two conspecific males (from HÖLLDOBLER 1976).

«female calling syndrome» has been found among several phylogenetically primitive species (HASKINS 1978, HÖLLDOBLER and HASKINS 1977, JANZEN 1967), including perhaps the most primitive living ant species *Nothomyrmecia macrops* (TAYLOR 1978, HÖLLDOBLER and TAYLOR 1983) as well as in some phylogenetically more advanced species, especially in a number of socially parasitic, dulotic and myrmecophilic ants, including species of the genera *Leptothorax, Doronomyrmex, Harpagoxenus* and *Formicoxenus* (BUSCHINGER 1968, 1971a, b, 1974a, 1975). As far as we know these females usually mate only once. Mature colonies of these species tend to be relatively small (on the order of 20–1000 workers) and produce relatively few new reproductives per year.

On the other hand, among species that form very large colonies (several thousands to more than a half million workers), another striking pattern of mating is observed. We call this the *«male-aggregation syndrome»*[1]. In these species males from many colonies gather at specific mating sites, usually marked out by conspicuous features in the landscape, like trees atop ridges, sunflecks, or tall buildings. Females fly in, also from considerable distances, to mate (Fig. 2). Multiple insemination is common among these species, and after mating females disperse widely before starting to build their founding nest chambers. Mature colonies of these species usually produce hundreds to thousands of reproductives every year.

2.2 Synchronization of mating

Nuptial flight activities in species with the *«female calling syndrome»* are inconspicuous and do not appear to be well coordinated or synchronized on the colony and population level (BUSCHINGER 1975, HASKINS 1978).

Male aggregation, however, requires precise synchronization of the mating flights within the colony as well as among conspecific colonies within a population. Indeed, most of these ant species exhibit strong seasonal periodicity in the production and release of alates, and the daily synchronicity of flights is often striking. Within minutes of one another, colonies in species like *Pogonomyrmex, Camponotus* or *Lasius* will release thousands of alates, literally filling the air with flying ants.

Experimental and observational evidence suggest that mating flight timing is, at least in part, controlled by endogenous rhythms (KANNOWSKI 1963, 1969; MCCLUSKEY 1965). For example, MCCLUSKEY (1965) showed that male ants kept in the laboratory under constant illumination showed peaks in restlessness that corresponded in time to the hours at which mating flights occurred each day in the field. Virgin alate females of *Pogonomyrmex californicus* also displayed similar daily increases in activity corresponding to flight times in nature, but stopped exhibiting such behavior once mated (MCCLUSKEY 1967). In addition, environmental cues, or «Zeitgeber», and social synchronization signals can modulate and fine tune mating flight activities within colonies and across populations (HÖLLDOBLER and MASCHWITZ 1965). For example, for the carpenter ant *Camponotus herculeanus* it has been demonstrated that the mass take off of both sexes is controlled by the workers. Males and females attempting to depart prematurely are hindered by workers, either by warning signals or by physical restraint (Fig. 3). But once the workers stop interfering, it is usually the males that fly first, each discharging a pheromone from his mandibular gland. At the peak of male activity, the concentration of this pheromone is highest, and it appears that

---

[1] We do not include here species which form polygynous unicolonial systems, or the special case of army ants, which reproduce by fission, but rather focus on species with monogynous-oligogynous societies which conduct massive nuptial flights.

this triggers the mass take-off of the females (Fig. 3) (Hölldobler and Maschwitz 1965).

It is interesting to note that in many ant species the males have well-developed mandibular glands which produce distinct mixtures of secretions (Law et al 1965; Pasteels et al 1980; for review see Blum 1981), and in many species nuptial flights are well synchronized with the males leaving first and females following shortly afterwards. It is possible that this male synchronization pheromone is a more general phenomenon, but no experimental data exist as yet.

Such synchronization of mating flights may be the result of several important selective forces. In temperate and boreal species the annual periodicity of colony reproduction follows the distinct annual cycle of climatic changes (for reviews see Brian 1979; Schmidt 1974; Wilson 1971). With a few exceptions, reproductives are produced once a year and usually they depart from their natal colonies a few days or weeks after eclosion. But for some of the species that inhabit boreal regions with severe seasonality, the period available for raising large numbers of new reproductives, releasing them for the mating flight, and enabling the young queens to found incipient colonies might be too short within one year. In such a situation selection may have favored colonies that extended the reproductive cycle over two years. This is the case in some *Camponotus* species in which alates are produced in late summer, and only after having overwintered in the mother nest, do they become ready to take off for the nuptial flight, the temperature permitting (Hölldobler 1966). In the laboratory it has been demonstrated that the alates can stay through a second winter period in the mother colony provided the temperature remains below 18°C. This

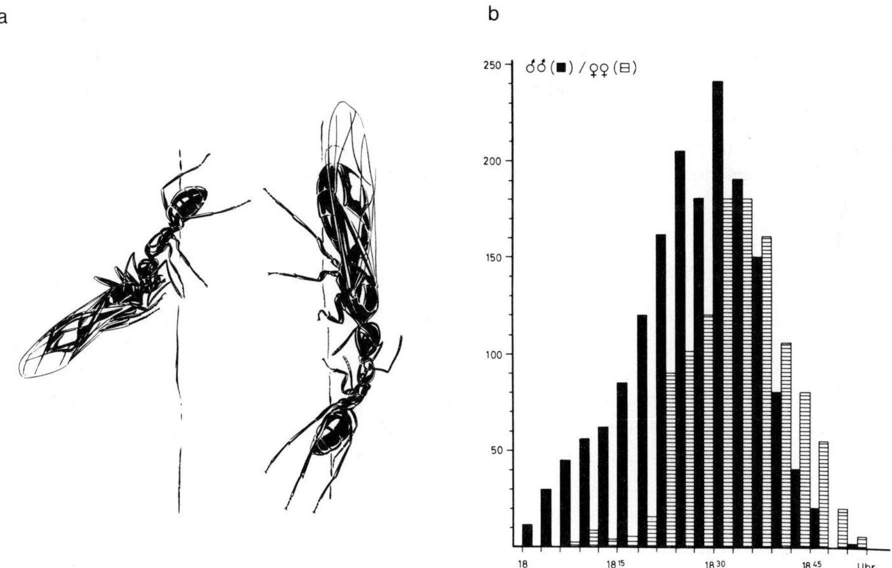

Fig. 3: *a* Males (left) and females (right) of *Camponotus herculeanus* attempting to depart prematurely are hindered by workers.
b Take off of males and females during the nuptial flight of *C. herculeanus*. At the peak of male activity (black columns), mass take off of the females (hatched columns) commences (from Hölldobler and Maschwitz 1965).

«storage-system» of young reproductives makes *Camponotus* colonies very flexible in having young reproductives always ready for nuptial flights once the climatic conditions are suitable. Although the extended pressence of young male and female reproductives in the colony puts a considerable ergonomic burden on the society, this is alleviated by the unusually high degree of social behavior demonstrated in young *Camponotus* males (Hölldobler 1966).

The annual temperature cycle is clearly the dominant environmental timing factor in temperate and boreal regions. In more arid geographic areas the reproductive annual periodicity of ants appears to be synchronized with the seasonality of precipitation. But even in the tropics with relatively little annual climatic fluctuations, species specific peaks in the production of alates during the year have been reported (Baldridge et al 1980).

Ecological factors may explain seasonal periodicity in nuptial flights, but many species also exhibit a remarkable daily periodicity in their mating activities. Among some closely related sympatric species this phenomenon is particularly evident, indicating that it may play a role in reproductive isolation. For example, four sympatric species of the harvester ant *Pogonomyrmex* in the Arizona desert tend to have nuptial flights on the same days, however the timing of their flights differ strongly. During a three years study *Pogonomyrmex maricopa* swarmed in the morning between 10:00 and 11:30 followed by *P. desertorum* (11:00–13:00) while the other two species flew in the afternoon (*P. barbatus* between 15:30 and 17:00, and *P. rugosus* from 16:30 to 18:00). Such timing was consistent from day to day (Hölldobler 1976). Another example is reported by Kannowski (1959) for three sympatric species of *Myrmica* in which alates are released daily over a period of about six weeks from mid-July to September. *Myrmica emeryana* invariably flew between 6:00 and 8:00, *M. americana* between 12:30 and 16:30, and *M. fracticornis* between 18:00 and 19:30.

Finally, to the extent that selection favors outbreeding, it will also favor synchronization of mating flights and mating aggregations. The sparse data that has been collected so far on patterns of allozyme variation in ants (Craig and Crozier 1979; Pamilo and Varvio-Aho 1979; Pearson 1983; Ward 1983a) indicates that inbreeding does not occur to any appreciable degree. Alates appear to mate preferentially with individuals from other colonies. This can only happen if the alates from several colonies are on their nuptial flights at the same time.

2.3 Mating aggregations and sexual selection

Mating aggregations and communal signalling occur in many insect groups. Among the more striking examples are communal chorusing in cicadas, crickets, and katydids (Alexander 1975), synchronous flashing of males in some firefly species (Lloyd 1983), communal chemical calling in bark beetles (reviewed by Alcock 1981), and the assemblies of honeybee drones at traditional congregation places (see Ruttner in this volume).

Many mating assemblies may be explained as the result of males congregating in places where females are likely to occur such as emergence or hibernation sites, feeding areas, or oviposition sites. This does not apply to the mating aggregations of most ant species, however. As in *Pogonomyrmex* which form large species-specific mating aggregations on bushes, small trees or on the ground, there is no competition among males for control of resources valuable to females, such as nesting sites or foraging areas, and males do not aggregate near nest entrances where females are likely to emerge. In fact, in *Pogonomyrmex* females fly off after they have mated, and only upon landing a second time, often hundreds of meters from the mating aggregation, do they break off their wings and begin to dig nest chambers in the soil (Hölldobler 1976).

Ant mating aggregations seem to be very similar to the situation described by ALEXANDER (1975) for some acoustical insects such as the periodical cicadas. He imagines that aggregations of males are the result of female choice. Females are selected to choose among males, and are thus attracted to places where males are dense. Once such a process is in operation, it is to the advantage of each male to participate in activities like synchronous calling because this increases the number of females attracted to his particular aggregation. Males attempting to «go it alone» simply fail to attract females. The evidence that *Pogonomyrmex* males release mandibular gland secretions in concert at their mating aggregations seems to suggest that this scenario describes the situation in these species.

As in other species, males far outnumber females in the mating aggregations of ants. When females arrive at the arena, there is often intense competition among the males for access to the females. In *Myrmecia,* for example, WHEELER (1916) describes the «balling behavior» at mating sites during which masses of males surround each female, often creating a ball of insects 15–20 cm in diameter. At some point, presumably when the female has been fully inseminated, the ball breaks up, the female leaves to establish a nesting chamber, and the males return to the swarm. In *Pogonomyrmex,* the males grapple

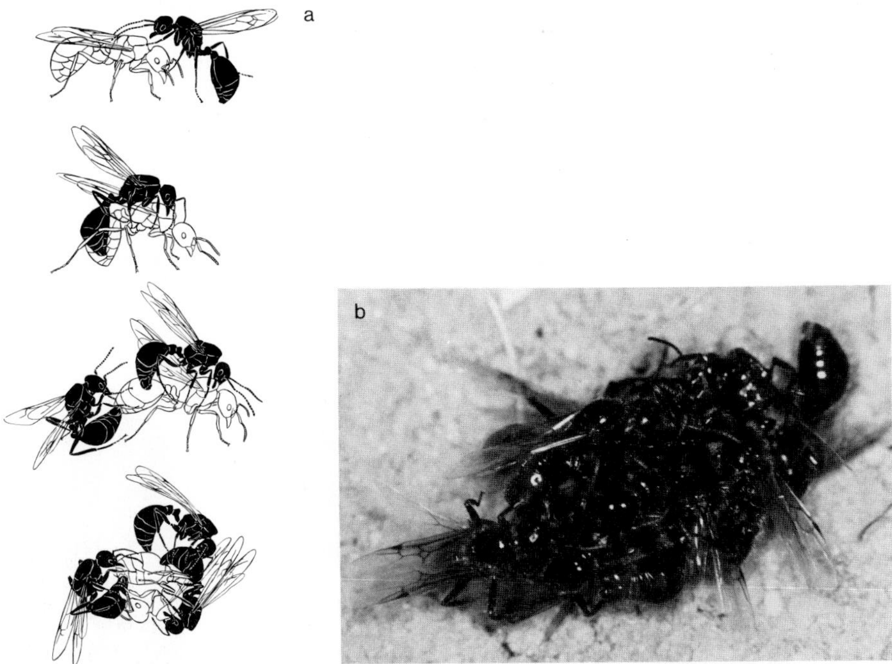

Fig. 4: *a* Behavioral sequences during mating in *Pogonomyrmex:* (1) A male (black) approaches a female (white) and touches her with his antennae. (2) The male grasps the female's thorax and attempts to insert copulatory organs into female's cloaca. (3) After successfully inserting copulatory organs the male releases the mandibular grip of the female's thorax. (4) The copulating male massages the female's gaster with his mandibles and forlegs; the female begins to gnaw at the male's gaster. A third male has arrived and grasps the second male's gaster.
b Mating cluster of *P. rugosus.* A female is surrounded by approximately ten males (from HÖLLDOBLER 1976).

with the female and each other (Fig. 4). When one male has secured a female, he holds her tightly about the thorax with his large, heavily-muscled mandibles. Other males attempt to separate the pair and vie among themselves for position in a waiting line – ready to mount the female as soon as the first male is finished. Hölldobler (1976) attributes the large mandibles of males in these species to selection for holding onto females, rather than to selection favoring aptitude in male-male fighting. Watching females being surrounded by a frenzied cluster of powerful males, one wonders how the mated female can ever free herself in order to depart from the mating site. Markl et al (1977) discovered that females, which are apparently fully charged with sperm and no longer motivated to mate, produce stridulation signals whenever they are assaulted by additional sexually excited males. The observational data indicate that males more readily release stridulating females than silent ones.

Many of these observations suggest that the mating aggregations, in *Pogonomyrmex* at least, may be similar to the mating aggregations in some other species of organisms like *Drosophila*, Sage Grouse, and Hammerheaded Bats, called leks (but see Bradbury in this volume). Ant leks differ from other such mating aggregations, however, in at least one important aspect. Ant males are constrained in a way that Sage Grouse or fruit flies are not, because each male ant ecloses with all of the sperm that he is ever going to have. Dissections of sexually mature males from such phylogenetically diverse groups as *Formica, Camponotus, Lasius, Myrmica, Pogonomyrmex* and *Nothomyrmecia* (Hölldobler 1966, and unpublished data) reveal that the males' testes are degenerated and all the sperm is stored in the expanded vas deferens (Fig. 5). Thus, male reproductive success in ants does not necessarily increase with increasing number of copulations, as it can in other species in which males continually replenish their sperm supply. In addition, it does not appear, from the few cases that have been studied so far, that males typically have enough sperm to inseminate more than one female. Indeed, in *Atta sexdens* each newly eclosed male was found to contain between 40 and 80 million spermatozoans, while newly mated females contained between 200 and 310 million spermatozoans in their spermathecae

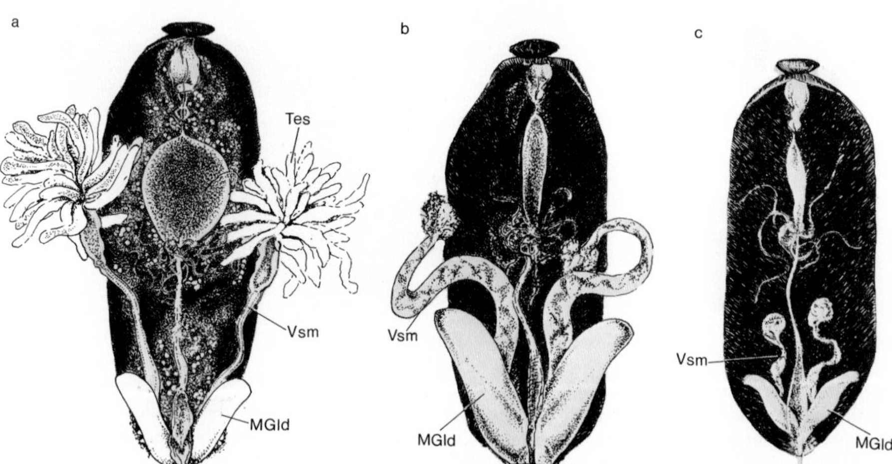

Fig. 5: Gasters of males of *Formica polyctena* were cut open to expose the gonads. (a) Freshly eclosed male; (b) eight days after eclosion; (c) after copulation. MGld = mucus gland; Tes = testis; Vsm = vesicula seminalis.

(KERR 1962). Male ants, therefore, should be under strong selection to be very particular about how they husband their sperm.

One obvious issue of concern for selection is whether it is better for a male, in terms of ultimate reproductive success, to invest all of his sperm in a single female or whether it is better to copulate with several females. Important to this issue is the fact that in ants, unlike other non-social species, a male's sperm does not all go towards effective reproduction. That is, in order for an ant colony to begin to produce any reproductive forms, it first must produce many workers. In most advanced ant societies it appears (claims to the contrary notwithstanding) that workers are rarely reproductive, and because workers are female, derived from fertilized eggs, a substantial portion of a male's sperm is used in colony growth and maintenance rather than in the direct production of new queens. So, the trade-off for a male is obvious. If he inseminates only a single female, and if she mates with no other male, then the male is certain to father any reproductives that she eventually produces. However, mortality of colony founding females is extremely high. Of the tens of thousands of female alates mated in a season's mating flight, at most only a few survive the colony founding phase and only a small portion of those will ever reach colony maturity. So, a male that inseminates only a single female puts all of his sperm in one basket. If he were to inseminate several females he increases the chance that his sperm will end up in a successful colony foundress. In this case, however, he might decrease the chance that his sperm is used by the female to make alates.

On the other hand, if males do inseminate several females, there may be selection favoring males whose sperm mixes with other males' sperm in the females' spermathecae. Mixing sperm increases the chance that each male will have at least some offspring among the new crop of alate queens. It appears indeed, from electrophoretic studies of allozyme variation in multiply-mating ant species (PAMILO 1982, PEARSON 1983, WARD 1983a), that workers in colonies have been fathered by several males.

Multiple insemination occurs relatively frequently among ant species. Of the 32 species for which COLE (1983) could find reliable accounts of the number of matings, in 22 species (69 %) females typically mated more than once. As already pointed out, multiple mating is more common among species that form large colonies, and COLE attributes multiple mating to the founding queen's requirement for more than a single male's worth of sperm. It is obvious that females should be selected to acquire sufficient sperm to last the life of the colony. It is not obvious why, as colonies increased in size, this has led to the practice of multiple insemination rather than the evolution of males with more sperm. It could be, as DAVIDSON (1982) suggests, that there is some trade-off associated with the size of males such that larger males would not represent proportionally larger reproductive returns to the colonies producing them. This issue, and the other questions surrounding the sexual size dimorphism in ants (males are usually smaller, sometimes a tenth the size of females), are far from resolved.

In species that make large colonies where females apparently must mate with several males in order to acquire sufficient sperm, it has been noted (COLE 1983) that males seldom attempt to monopolize females. In many other insects, and other organisms as well, sperm competition is an important selective force, and males are often favored to ensure that no other male copulates with his mate (PARKER 1970). In multiply-mating ant species, however, a male that prevents his mate from mating again may well be preventing her from acquiring sufficient sperm to make a colony large enough to ever produce reproductives. Males in these species therefore should be selected to mate with females that are already mated. The active vying for position in the waiting lines behind copulating pairs in *Pogonomyrmex* species (HÖLLDOBLER 1976, DAVIDSON 1982) indicates at least that males do not discriminate against females that have mated before.

Some kind of discrimination seems to be practiced, however, by *Pogonomyrmex* females. Hölldobler (1976) suggested that the intense competition among males for females arriving at the mating arena in *Pogonomyrmex* may be the result of a form of female choice – only those aggressive or persistent males would achieve copulations. Davidson (1982) confirmed Hölldobler's suspicions by showing that larger females tended to mate disproportionately often with larger males. In fact, Davidson reports that there is significant variation in the copulatory success of males, and that variation is associated with male size. As support for this hypothesis of female choice, she points out that there is a weak but positive correlation between colony productivity (number of alates produced) and male size. In the absence of data on the heritability of male size, however, this only suggests that colonies in good reproductive condition produce bigger males.

A confounding bit of data reported by Davidson (1983) is that not only do larger females tend to mate with larger males, but smaller females tend to mate with *smaller* males! If sexual selection is operating such that females choose larger males, why do small females not also choose to mate with larger males? The answer to this question may be that males are selected to be choosy as well. Recall that unlike other lek-mating species, male ants have only so much sperm at their disposal. They cannot afford to be profligate, and selection may favor males who choose to mate with larger females because there is a better chance that large females will survive to produce a mature colony. It is not clear, among ant species, which sex represents a limiting resource to the other. This is especially true in multiply-mating species, even though males may outnumber females.

Mating is only the first phase in the complex process of reproduction in ants. After this phase the males die, but their sperm lives on in the females' spermathecae; it will be used to fertilize eggs in many years to come. For the freshly mated females colony founding follows the mating phase. Only after the colony has grown to maturity, that is, a large enough worker force has been produced to secure the resources needed to raise reproductive offspring, will the second phase of reproduction begin.

3 Reproduction division of labor

The reproductive interests of queens and workers in ant societies are not always identical. Specifically, there is expected to be conflict between the queen and the workers regarding two major aspects of reproduction: the relative investment in the sexes, and the source of reproductive males (Trivers and Hare 1976).

Because of the haplodiploid genetics of the Hymenoptera, a worker in a monogynous colony is three times as closely related to her sisters than to her brothers. The queen, however, is equally related to her sons and daughters. Consequently, workers will be selected to invest more of the colony's reproductive effort in the production of female sexuals than in male production, while the queen is selected to enforce equal investment in the sexes (Trivers and Hare 1976, Benford 1978, Oster et al. 1977).

With regard to the production of males, workers are more closely related to their sons and nephews than they are to their brothers, and selection acting on workers would favor the substitution of sons or nephews for brothers in the reproductive brood. The queen is more closely related to her sons than to her daughters' sons, and selection on the queen should favor any traits that ensure that her sons, rather than her grandsons, comprise the reproductive brood (Trivers and Hare 1976).

With several exceptions (see below) the data that exist point to a powerful and important role of the queen in controlling the reproductive activities in ant societies. In the following sections we will review these data, and discuss how these facts relate to current hypotheses regarding the evolution of eusociality in ants.

3.1 Evidence for the controlling effect of the queen

There are several facts that seem to indicate that the queens control the workers' reproductive activities. Pheromones, emanating from the queen, are usually highly attractive to workers (Stumper 1956, Watkins and Cole 1966, Jouvenaz et al. 1974, VanderMeer et al. 1980, Coglitore and Cammaerts 1981, Fowler and Roberts 1982, Hölldobler and Wilson 1983). These «queen tending pheromones» appear to play important roles in ensuring that the queens are always surrounded and protected by the workers (Fig. 6), and in directing the flow of high quality food to the queens (Lange 1958, 1967). Indeed, queens seem to be the most powerful competitors for the best nutrients in the society. In some species it appears that this queen-ward direction of nutrients suppresses the rearing of sexuals by the workers. This was first demonstrated in *Formica polyctena* by Gösswald and Bier (1954a, b).

In early spring, queens of *F. polyctena* cluster in the sun on top of the nest mounds. During this «sunning period» the queens' ovaries become activated and after a few days

Fig. 6: *a* The queen of *Oecophylla longinoda* during nest emigration is totally covered by a retinue of major workers. *b* Inside the nest chamber the queen is continuously licked and fed by workers (from Hölldobler and Wilson 1983).

thousands of eggs are deposited in the upper regions of the nest. These eggs appear to be blastogenically predisposed to become sexuals (Bier 1954), but the larvae must still be fed with high quality food in order to develop into reproductives. Gösswald and Bier discovered that this is permitted because the queens move to the lower regions of the nest after egg-laying, where they rest until the first brood has been reared. This separation of queens and brood, the so-called «physiologische Weisellosigkeit» (physiological queenlessness) enables the workers to channel the high quality food to the developing reproductives. In the laboratory, if the larvae are forced to compete with the queen for these nutrients, they do not develop into reproductives. The females become workers, and the males are usually eaten in the larval stage (Schmidt 1972).

Similar inhibitory trophic effects caused by the queens have been reported for other ant species. Petersen-Braun (1977) discovered that in *Monomorium pharaonis* only queens in the period of their peak fecundity inhibit the rearing of reproductives. Very young queens and senile queens do not have such inhibitory effects. Earlier findings suggested that unmated as well as mated queens have an inhibitory effect (Buschinger and Kloft 1973), but more recent studies have shown that such effects emanate only from mated queens (Petersen-Braun 1977, Berndt and Nitschmann 1979). Buschinger and Kloft (1973) and Petersen-Braun (1977, 1982) argue that this inhibition is due to the queen's high attractiveness to workers, which has the effect that workers preferentially feed the queens and not the brood. On the other hand, Berndt (1975) and Berndt and Nitschmann (1979) propose that the inhibition is not caused primarily by trophic effects, but rather by inhibitory pheromones emanating from the queens. The experimental results are ambiguous, however, and controlled feeding experiments (Buschinger and Kloft 1973) support the hypothesis of trophic inhibition. In other ant species, however, such as *Myrmica rubra* (for a review see Brian 1979) or *Pheidole pygmaea* (Passera 1980), strong circumstantial evidence indicates that queen pheromones can indeed directly inhibit the workers rearing of reproductives.

The fact that the queens in these ant species seem to regulate the reproductive activities of colonies is not necessarily an indication that it would be in the workers best interests to behave otherwise. Queens and workers, for example, should not be in conflict over such matters as the timing of the production of sexuals or the fraction of colony resources devoted to reproduction versus growth. Workers would clearly be selected to respond to influences emanating from the queen that regulate such matters.

The regulating power in the queen is especially striking in monogynous societies. Her presence seems to maintain a colony cohesiveness, and the behavioral changes among the workers after the queen has been removed are often remarkable. In some species the workers become indifferent, and after some time will readily accept a foreign queen in the old queen's place (K. Hölldobler 1928, Goetsch 1953, Hölldobler 1962, Schneirla 1971, Fletcher and Blum 1983).

The fact that in many of these species workers will begin to lay viable, male-destined eggs after the removal of the queen is an indication that queen regulation may have become queen oppression. As mentioned earlier, workers ought to be selected to substitute sons or nephews for brothers in the reproductive brood. In species like *Novomessor cockerelli*, it appears that the presence of the queen prevents workers from reproducing. When the queen is removed, colony cohesion weakens, and workers sometimes become antagonistic towards one another. Correlated with the onset of this aggression is the production of viable, male-destined eggs (Hölldobler et al., unpublished data). In the presence of the queen, workers produce only trophic eggs, so it appears that the queen clearly affects workers' reproductive physiology.

Apparent reproductive dominance by the queen over other reproductive females or workers in functionally monogynous colonies (for definition see BUSCHINGER 1974) has been demonstrated for several ant species. FLETCHER and BLUM (1983) recently discovered that in *Solenopsis invicta* the presence of queens prevents alate virgin females from shedding their wings. Like in *Novomessor,* in *Plagiolepsis pygmaea* workers produce only trophic eggs in queenright colonies, but after removal of the queen they lay both trophic and viable eggs (PASSERA 1978). Some of the experimental results obtained by PASSERA suggest that the queen's effect in *Plagiolepis* is mediated pheromonally.

HÖLLDOBLER and WILSON (1983) recently obtained indirect evidence indicating that in the weaver ants *Oecophylla longinoda* and *O. smaragdina* the queen's reproductive dominance is based on pheromones. Weaver ant colonies are unusual among ants in combining very large size, extensive territories with multiple nests, and monogyny[2]. Therefore, if a queen control extends to the entire colony it must be of an especially potent form. Queens of *Oecophylla* have been demonstrated to exert an inhibitory influence on the workers' reproductive physiology. Like in the species mentioned above, in *Oecophylla* this effect is very specific. In the presence of a nest queen, the ovaries of young workers remain active and they produce trophic eggs which form an important part of the queen's diet. When the queen is removed, however, these same workers begin to produce viable, male-destined eggs in profusion.

Because the queen cannot be physically examined, seen, or heard by more than a small fraction of the worker population for substantial periods of time, it is reasonable to expect that the influence she exerts is mediated by pheromones rather than by signals in other sensory modalities. This assumption is strongly supported by the fact that the queen's suppressing effect on the workers' ovaries persists in the corpse of a dead queen for several month in the laboratory (HÖLLDOBLER and WILSON 1983). The presence of large complexes of exocrine glands in the head, mesosoma and gaster further suggests the existence of special queen pheromones. There are also exocrine glandular complexes associated with the queen's reproductive organs and it is possible that queen laid eggs are contaminated with queen pheromones. Workers usually carry the eggs to other leaf nests and in this way queen pheromones might also be spread in the colony.

The results of these laboratory experiments are at variance with previous findings reported by LEDOUX (1950, 1974). According to LEDOUX, groups of workers of *O. longinoda* depart from the territory of the mother colony and establish new colonies in the neighborhood. There, they produce workers and reproductive females by parthenogenesis. However, in our experimental series of 10 queenless colony fragments maintained in the laboratory, the broods produced by the workers were composed exclusively of males (HÖLLDOBLER and WILSON 1983). Thus, it seems unlikely that workers are producing viable diploid eggs by parthenogenesis.

Although there is ample circumstantial evidence that ant queens produce pheromones which affect the workers' behavior and reproductive activities, the anatomical source of the queen pheromones is largely unknown and no pheromone has been chemically identified. STUMPER (1956) was able to extract from queens of *Lasius alienus* and *Pheidole pallidula* secretions to which workers were attracted. VANDERMEER et al. (1980) found that workers of *Solenopsis invicta* are attracted to their queen's poison gland secretions, and BRIAN and HIBBLE (1963) found that in *Myrmica rubra* ethanol extracts of the queen heads, when applied to larvae inhibit their growth. Similar results were obtained with

---

2 Monogyny can never be claimed with absolute certainty, as in some species so-called monogynous colonies can become oligogynous (HÖLLDOBLER 1962). But in all our field investigations during which we opened many leaf nests, we found only one queen per colony.

Pheidole pygmaea, where circumstantial evidence suggests that an inhibitory effect on the workers' sexual brood rearing behavior emanates from the queen's head (PASSERA 1980).

It is surprising that little attention has been given to the study of the ant queens' exocrine glandular repertory, although a previous histological investigation of *Eciton* by WHELDEN (1963) has shown that queens are much more richly endowed with exocrine glands than workers. Recently we found similar results in *Oecophylla* (HÖLLDOBLER and WILSON 1983) and in *Onychomyrmex* (HÖLLDOBLER and TAYLOR in prep.). In this context it is interesting to note that in some ecitonine army ants, where males have to enter a colony in order to get access to the wingless virgin females, the males are as richly endowed with abdominal tergal and sternal glands as reproductive females (HÖLLDOBLER and ENGEL-SIEGEL 1982).

As already indicated, there appears to be a fundamental difference in the mechanism of regulation of fertility in monogynous and truly polygynous ant societies. If a single queen is in «control», passing secretions around the colony, the reproductive behavior of workers can be regulated in a direct and relatively simple manner. In truly polygynous colonies, however, a very different circumstance pervails. Large numbers of queens are often found clustered together, and so far as known, all lay eggs and are not organized into dominance orders. BIER (1958) hypothesized that in these cases the suppression of fertility in the workers is not regulated by primer pheromones, but rather by a directed flow of «profertile» substances away from the workers toward the queens. Although queens of polygynous colonies produce attractants that mediate the directed social flow of profertile substances (LANGE 1958, 1967), these queen pheromones apparently do not function as primer pheromones in suppressing worker fertility.

Finally, it has been hypothesized that if a single queen is in «control», passing secretions around in the colony, the queen can also readily serve as the sole or primary source of the colony odor (HÖLLDOBLER and MICHENER 1980). It was reasoned that if genetic differences rather than environmental odors play the dominant role in the determination of colony odors in the large ant societies, the simplest procedure would appear to be for the queen to provide the essential ingredients. Monogyny would make such a system easily operable; polygyny would tend to break it down (HÖLLDOBLER and WILSON 1977).

Experimental evidence showing that ants recognize their own queen by odor supports these considerations. For example WATKINS and COLE (1966) discovered that workers of several army ant species of the genera *Labidus* and *Neivamyrmex* were attracted to spots of absorbent paper on which the queen had been previously confined. Spots that had been occupied by the mother queen were more attractive than spots occupied by foreign conspecific queens. Similar results were obtained by JOUVENAZ et al. (1974) with *Solenopsis invicta* and *S. geminata.*

Recently, CARLIN and HÖLLDOBLER (1983) have obtained the first experimental evidence that transferred queen discriminators serve as recognition cues in monogynous colonies of the ant genus *Camponotus.* These acquired odor labels are sufficiently powerful to cause non-preferential acceptance among unrelated workers of different species, raised in artificially mixed colonies, and rejection of genetic sisters reared by heterospecific queens.

3.2 Evidence for worker control of reproduction

Support for worker control of colony reproductive activities comes from data on the sex-ratios of alates in various ant species. TRIVERS and HARE (1976) pointed out that selection on workers would favor female-biassed investment ratios, while selection on queens would favor equal investment in the sexes. They produced data on investment ratios (dry weight was used as the measure of investment for several species of ants), and showed that, as they predicted, the sex-ratios exhibited by monogynous ants were normally female biassed. Such female biassed sex-ratios seem to indicate that, in the words of IWASA (1981), «the workers held hegemony in the nest». That is, that the workers are controlling the investment in the sexes. If the queen were controlling her workers in her own interests, one would expect sex-ratios nearer 50 : 50, but it appears that it is only in polygynous ant societies or in slave-making ant species that equal investment in the sexes is seen.

These observations, especially those indicating that some ant species that often serve as slaves for slave-making species produce female-biassed sex-ratios in their own nests, but 50 : 50 ratios in the slave makers' nests where it is not possible for selection on workers to counteract the evolution of control by heterospecific queens (TRIVERS and HARE 1976), suggest that eusocial colonies are maintained because of selection on workers, and not because of selection on queens.

However, the use of sex-ratio data to discriminate between worker and queen control is difficult because of several confounding effects. First, investment is extremely difficult to measure, and dry weights can only be viewed as rough approximations of total investment. For example, the males of some species like *Camponotus* (see HÖLLDOBLER 1966), contribute to colony maintenance by aiding in the distribution of food. If males' activities aid in the production of females, or in the production of workers, and if female alates do not engage in such activities (or at least not to the same extent as the males do), then males would represent a smaller «cost» to the colony than their dry weight would indicate.

Another major problem with using sex ratio data as an indication of queen or worker control of investment is that the relatednesses of the various actors represent only one of several powerful selective forces affecting the sex ratio. HAMILTON (1967) has shown that when family members expect to be competing with one another for access to the same mates, selection will favor decreasing the investment in the competing sex. If, for example, brothers often compete among themselves for access to the same females, then selection will favor a female-biassed sex-ratio. Small, viscous populations which, incidentally, are often inbreeding, will generally result in this kind of selection occurring. ALEXANDER and SHERMAN (1977) suggest that such local mate competition may be common for ant species and that the female biassed ratios reported by TRIVERS and HARE (1976) reflect only this, and say nothing about queen or worker control of investment. There is, however, no data suggesting that local mate competition is an important feature in ant populations. In fact, data on enzyme frequencies suggest that most populations of ants are relatively outbreeding, and that local mate competition is not likely to be occurring, even in polydomous ants like *Myrmecia pilosula* (CRAIG and CROZIER 1979). Data on *Rhytidoponera* (WARD 1983a, b) also suggest that inbreeding, and thus local mate competition, is not occurring. Finally, the species in which local mate competition would be most likely to be occurring are those polygynous species in which females mate in the near vicinity of their natal nests. In these species, however, the sex ratios are often nearly 50 : 50 (TRIVERS and HARE 1976).

Another factor affecting selection for a particular sex ratio strategy is the occurrence in many ant species of colonies that specialize in the production of one sex or the other. Among some species, colonies persist after the queen dies and continue to produce broods of reproductive males. In a population at equilibrium, the consistent proportion of such senescent nests producing males would favor production of excess females among queenright colonies. In addition, extreme specialization in the production of the sexes seems to occur in polydomous species of ants. Data on alate production in *Harpagoxenus* (WESSON 1939) show the common polydomous pattern. Some nests produce almost exclusively males and some nests produce almost exclusively female alates. *Formica opaciventris*

(SCHERBA 1984) and *Formica exsectoides* (TALBOT 1953) display similar patterns, and HERBERS (1984) has shown that in the tiny acorn-nesting *Leptothorax* species, some nests produce excess females and others produce excess males. While it is not clear exactly whether this production data reflects the production of alates by one or several colonies, it does indicate that more than relatedness is important in determining particular sex-ratio strategies.

Finally, another confounding feature of sex ratio data collected to this point is that it describes alate production over only a very small portion of the life span of colonies. Relative investment at any one time is really irrelevant, because it is the total reproductive investment of a colony that must be measured in order to assess sex investment ratios. For example, many ant species seem to be protandrous (WILSON 1971, BULMER 1983), that is, males are produced early in a colony's life and only later are alate females produced. Sampling that included only one or two seasons of production of alates would give a faulty impression of colonies' investment ratios.

Either of the factors, local mate competition or specialist male-producing colonies, could easily provide selection pressure sufficiently strong to neutralize any effect of queen-worker conflict, and at the very least, serve to complicate the interpretation of sex-ratio data.

Acknowledging these complications, there is one study of the sex-ratio in ants that indicates that Trivers and Hare's prediction about control of investment and the sex-ratio is correct. WARD (1983a, b) has studied populations in the *Rhytidoponera impressa* group in eastern Australia. He finds populations with mixtures of two kinds of colonies: «A» colonies which have only one queen, and «B» colonies which each have several queens. This situation was confirmed using elctrophoretic techniques to assess the number of parents responsible for samples of workers taken from various colonies.

Using dry weights as estimates of investment, Ward has shown that type A colonies and type B colonies differ in the production of the sexes. Type A colonies tend to produce female biassed investment ratios while type B colonies tend to produce the sexes in about equal proportions. Remarkably, in populations with both A and B type colonies, the sex ratio produced by type A colonies depends on the frequencies of A and B type colonies in the population. As the frequency of type B colonies increases, the proportion of females in A colonies' broods increases. This is precisely the response that would be expected if the workers in the A type colonies were somehow assessing the population frequency of B colonies or the population sex ratio, and controlling the production of the sexes in their own interest. Incidentally, when the population consists of only type A colonies, the regression analysis predicts that there should be about three-to-one investment in favor of females. There seems to be no other way to account for A colonies changing their investment patterns with changing proportions of B colonies except on the view that workers are manipulating the sex ratio in their own interests.

3.3 Did the queen win in highly advanced species?

The data that has just been reviewed has been used to support one or the other of the two major hypotheses concerning the evolutionary origin of eusociality in the Hymenoptera. TRIVERS and HARE (1976) have argued that the female-biassed investment ratios in ant species support the hypothesis that eusociality evolved by kin selection increasing the inclusive fitness of workers. This «offspring consent» hypothesis (CHARNOV 1978) imagines that selection acts more effectively on offspring than on parents, and that when there is conflict between parents and offspring, it will be resolved in favor of the offspring.

If not completely resolved, at least the effects of selection on offspring will be manifest in those issues, like relative investment in the sexes, over which there is conflict (OSTER et al. 1977). On the other hand, the data on the controlling effects of the queen have been used by ALEXANDER (1974) and others (MICHENER and BROTHERS 1974) to support the hypothesis that eusociality arose by parental manipulation. On this view, parent-offspring conflict is always resolved in favor of the parents. This leads to the conclusion that offspring are essentially somatic extensions of the parents, and like any other organ of the parent's body, they can be modified by natural selection to function to increase parents' reproductive success, even if doing so decreases their own reproductive success. Thus, much as some vertebrate parents sacrifice one offspring of a brood so that the others may live, so the ancestors of the modern eusocial species could have forced offspring to remain on the natal nest, give up their own reproduction, and raise brothers and sisters instead. Important to this view is that there is no possibility for conflict between queens and workers, or among workers in eusocial colonies. The parent's reproductive interests are expected always to be realized.

It may, however, be inappropriate to use data on existing ant species to test hypotheses about the origins of eusociality. The oldest known ant, *Sphecomyrma freyi,* some 100 million years old, was undoubtably eusocial (WILSON et al. 1967a, b), and there are no solitary, subsocial, or presocial ant species that might represent transitional stages between solitary and eusocial states. Thus, all ant societies have a history of at least 100 million years of social evolution, and any traces of conditions affecting the origin of social behavior may be long lost. Furthermore, it is at least possible that the locus of control of reproductive activities has changed in the course of evolution, workers having lost, or perhaps gained control of various aspects of colony reproduction at different times.

A case appropriate to this point may be seen among species of ants that form especially large colonies, such as *Camponotus, Atta, Oecophylla, Acromyrmex, Lasius,* and *Pogonomyrmex*. It appears that the evolution of large colony size has been accompanied by, among other things, the evolution of multiple insemination (COLE 1983). Multiple insemination causes workers to be less closely related to the reproductives that they are working to raise, and selection might be expected to cause workers to treat more closely related reproductives preferentially. At least, there would be strong selection for workers in such colonies to produce sons rather than the more distantly related male issue of the queen or other workers. Yet, these colonies appear to be paragons of obedience to the queen. Reproducing workers appear to be rare in these species (at least they lay no viable eggs in the presence of the queen – see above), and colony investment ratios do not appear to be extremely female-biassed. The queen's presence is vital to some of these species like *Atta* in which, when the queen dies, the colony perishes as well. If not in any other species, among the phylogenetically highly advanced species that form large monogynous colonies the queen seems to be in control.

It is possible that, even if eusociality arose because of óffspring consent, the workers in these species could have lost control in the course of their evolution. If workers had been selected to be responsive to organizing and regulating influences emanating from the queen, they may have lost the ability to evolve counteractive measures to the development of queen oppression. For example, in *Camponotus*, CARLIN and HÖLLDOBLER (1983) have shown that workers derive colony-specific recognition cues from the queen. In effect, she labels the workers, and unrelated individuals reared (in the lab) in single broods treat each other like full sister colony mates, and show aggression towards noncolony members regardless of closeness of relatedness. Once workers have «permitted» such a practice to evolve, they have lost substantial control within the colony.

Acknowledgement

The authors' personal research reproted in this review has been supported by a series of grants from the National Science Foundation, the latest of which is NSF BNS–82–19060 (to B. H.).

References

Alcock, J. (1981): Natural selection and communication among bark beetles. Florida Entomologist *65*, 17–32.

Alexander, R. D. (1974): The evolution of social behavior. Ann. Rev. Ecol. Sys. *5*, 325–383.

Alexander, R. D. (1975): Natural selection and specialized chorusing behavior. In: Insects, Science, and Society (D. Pimentel, ed.) pp. 35–77, Academic Press, London.

Alexander, R. D. and Sherman, P. W. (1977): Local mate competition and parental investment in social insects. Science *196*, 494–500.

Baldridge, R., Rettenmeyer, C. and Watkins, J. (1980): Seasonal, nocturnal and diurnal flight periodicities of nearctic army ant males (Hymenoptera: Formicidae). J. Kansas Ent. Soc. *53*, 189–204.

Benford, F. A. (1978): Fischer's theory of the sex ratio applied to the social Hymenoptera. J. Theor. Biol. *72*, 701–727.

Berndt, K. P. (1975): Physiology of reproduction in the Pharaoh's ant (*Monomorium pharaonis* L.). 1. Pheromone mediated cyclic production of sexuals. Proc. 3rd Symp. med. vet. Acaroentomology, Gdansk, 25–28.

Berndt, K. P. and Nitschmann, J. (1979): The physiology of reproduction in the Pharao's ant (*Monomorium pharaonis* L.). 2. The unmated queens. Insect. Soc. *26*, 137–145.

Bier, K. (1954): Über den Saisondimorphismus der Oogenese von *Formica rufa rufo-pratensis* minor Gössw. und dessen Bedeutung für die Kastendetermination. Biol. Zentralblatt *73*, 170–190.

Bier, K. (1958): Die Regulation der Sexualität in den Insektenstaaten. Ergebn. Biol. *20*, 97–126.

Blum, M. S. (1981): Sex pheromones in social insects: chemotaxonomic potential. In: Biosystematics of Social Insects (P. E. Howse and J. L. Clement, eds.), pp. 163–174, Academic Press, London.

Brian, M. V. (1979): Caste differentiation and division of labor. In: Social Insects Vol. I (H. R. Hermann ed.), pp. 121–222, Academic Press, London.

Brian, M. V. and Hibble, J. V. (1963): 9-Oxodec-trans-z-enoic acid and *Myrmica* queen extracts tested for influence on brood in *Myrmica*. J. Ins. Physiol. *9*, 25–34.

Bulmer, M. G. (1983): The significance of protandry in social Hymenoptera. Am. Nat. *121*, 540–551.

Buschinger, A. (1968): «Locksterzeln» begattungsbereiter ergatoider Weibchen von *Harpagoxenus sublaevis* Nyl. (Hymenoptera: Formicidae). Experientia *24*, 297.

Buschinger, A. (1971a): «Locksterzeln» und Kopula der sozialparasitischen Ameise *Leptothorax kutteri* Buschinger (Hym. Form.) Zool. Anz. *186*, 242–248.

Buschinger, A. (1971b): Weitere Untersuchungen zum Begattungsverhalten sozialparasitischer Ameisen (*Harpagoxenus sublaevis* Nyl. and *Doronomyrmex pacis* Kutter, Hym. Formicidae). Zool. Anz. *187*, 184–198.

Buschinger, A. (1974a): Zur Biologie der sozialparasitischen Ameise *Leptothorax goesswaldi* Kutter (Hym. Formicidae). Insect. Soc. *21*, 133–143.

Buschinger, A. (1974b): Monogynie und Polygynie in Insektensozietäten. In: Sozialpolymorphismus bei Insekten (G. H. Schmidt, ed.) pp. 862–896, Wissenschaftliche Verlagsgesellschaft Stuttgart.

Buschinger, A. (1975): Sexual pheromones in ants. International Union for the Study of Social Insects, Symposium on Pheromones and Defensive Secretions in Social Insects. pp. 225–233, Dijon.

Buschinger, A. and Kloft, W. (1973): Zur Funktion der Königin im sozialen Nahrungshaushalt der Pharaoameise *Monomorium pharaonis* (L). Forschungsber. Landes Nordrhein-Westfalen, No. 2306, 34 pp.

Carlin, N. F. and Hölldobler, B. (1983): Nestmate and kin recognition in interspecific mixed colonies of ants. Science, *222*, 1027–1029.

Charnov, E. L. (1978): Sex-ratio selection in eusocial Hymenoptera. Am. Nat. *112*, 317–326.

Coglitore, C. and Cammaerts, M. C. (1981): Étude du pouvoir agrégatif des reines de *Myrmica rubra* L. Insect. Soc. *4*, 353–370.

Cole, B. J. (1983): Multiple mating and the evolution of social behavior in the Hymenoptera. Behav. Ecol. Sociobiol. *12*, 191–201.

Craig, R. and Crozier, R. H. (1979): Relatedness in the polygynous ant *Myrmecia pilosula*. Evolution *33*, 335–341.

Crozier, R. H. (1979): Genetics of sociality. In: Social Insects Vol I (H. R. Hermann ed.). pp. 223–286, Academic Press, London.

Crozier, R. H.: Genetical structure of social insect populations. In: Evolution of Social Behavior: Hypotheses and Empirical Tests (H. Markl ed.). pp. 129–146. Dahlem Konferenzen 1980. Verlag Chemie, Weinheim.

Davidson, D. W. (1982): Sexual selection in harvester ants (Hymenoptera: Formicidae: *Pogonomyrmex*). Behav. Ecol. Sociobiol. *10*, 245–250.

Fletcher, D. J. C. and Blum, M. S. (1983): Regulation of queen number by workers in colonies of social insects. Science *219*, 312–314.

Fowler, H. G. and Roberts, R. B. (1982): Entourage pheromone in carpenter ant *(Camponotus pennsylvanicus)* (Hymenoptera: Formicidae) Queens. J. Kansas Ent. Soc. *55*, 568–570.

Goetsch, W. (1953): Vergleichende Biologie der Insektenstaaten. Geest & Portig, K. G. Leipzig, 482 pp.

Gösswald, K. and Bier, K. (1954a): Untersuchungen zur Kastendetermination in der Gattung *Formica*. 3. Die Kastendetermination von *Formica rufa rufopratensis* Gössw. Insect. Soc. *1*, 229–246.

Gösswald, K. and Bier, K. (1954b): Untersuchungen zur Kastendetermination in der Gattung *Formica*. 4. Physiologische Weisellosigkeit als Voraussetzung der Aufzucht von Geschlechtstieren im polygynen Volk. Insect. Soc. *1*, 305–318.

Hamilton, W. D. (1967): Extraordinary sex ratios. Science *156*, 477–488.

Haskins, C. P. (1978): Sexual calling bahavior in highly primitive ants. Psyche *85*, 407–415.

Herbers, J. (1984): Queen-worker conflict and eusocial evolution in a polygynous ant species. Evolution. *38*, 631–643.

Hölldobler, B. (1962): Zur Frage der Oligogynie bei *Camponotus ligniperda* Latr. und *Camponotus herculeanus* L. (Hym. Formicidae). Z. angew. Entomol. *49*, 337–352.

Hölldobler, B. (1966): Futterverteilung durch Männchen im Ameisenstaat. Z. vergl. Physiol. *52*, 430–455.

Hölldobler, B. (1976): The behavioral ecology of mating in harvester ants (Hymenoptera: Formicidade: *Pogonomyrmex*). Behav. Ecol. Sociobiol. *1*, 405–423.

Hölldobler, B. and Engel-Siegel, H. (1982): Tergal and sternal glands in male ants. Psyche *89*, 113–132.

Hölldobler, B. and Haskins, C. P. (1977): Sexual calling in a primitive ant. Science *195*, 793–794.

Hölldobler, B. and Maschwitz, U. (1965): Der Hochzeitsschwarm der Rossameise *Camponotus herculeanus* L. (Hym. Formicidae). Z. vergl. Physiol. *50*, 551–568.

Hölldobler, B. and Michener, C. D.: Mechanisms of identification and discrimination in social Hymenoptera. In: Evolution of Social Behavior: Hypotheses and Empirical Tests (H. Mark, ed.) pp. 35–58, Dahlem Konferenzen 1980, Verlag Chemie, Weinheim.

Hölldobler, B. and Wilson, E. O. (1977): The number of queens: an important trait in ant evolution. Naturwissenschaften *64*, 8–15.

Hölldobler, B. and Taylor, R. W. (1983): A behavioral study of the primitive ant *Nothomyrmecia macrops* CLARK. Insect. Soc. *30*, 384–401.

Hölldobler, B. and Wilson, E. O. (1983): Queen control in colonies of weaver ants (Hymenoptera: Formicidae). Ann. Ent. Soc. Amer. *76*, 235–238.

Hölldobler, K. (1928): Zur Biologie der diebischen Zwergameise *(Solenopsis fugax)* und ihrer Gäste. Biol. Zentralbl. *48*, 129–142.

Iwasa, Y. (1981): Role of the sex ratio in the evolution of eusociality in the haplodiploid social insects. J. Theor. Biol. *93*, 125–142.

Janzen, D. H. (1967): Interaction of the bull's horn acacia (*Acadia cornigera* L.) with an ant inhabitant (*Pseudomyrmex ferruginea* F. Smith) in Eastern Mexico. Kansas University Science Bulletin *47,* 315–558.

Jouvenaz, D. P., Banks, W. A. and Lofgren, C. S. (1974): Fire ants: attraction of workers to queen secretions. Ann. Entomol. Soc. Amer. *67,* 442–444.

Kannowski, P. B. (1959): The flight activities and colony founding behavior in bog ants in Southeastern Michigan. Insect. Soc. *6,* 115–162.

Kannowski, P. B. (1963): The flight activities of formicine ants. Symposia Genetica et Biologica Italica, *12,* 74–102.

Kannowski, P. B. (1969): Daily and seasonal periodicities in the nuptial flights of neotropical ants. I. Dorylinae. Int. Union Study of Social Insectes, 6. Congress, Bern, pp. 77–83.

Kerr, W. E. (1962): Tendências evolutivas na reproducão dos himenópteros sociais. Arquivos do Musen Nacional, Rio de Janeiro, *52,* 115–116.

Lange, R. (1958): Der Einfluß der Königin auf die Futterverteilung im Ameisenstaat. Naturwissenschaften *45,* 196.

Lange, R. (1967): Die Nahrungsverteilung unter den Arbeiterinnen des Waldameisenstaates. Z. Tierpsychol. *24,* 513–545.

Law, J. H., Wilson, E. O. and McCloskey, J. A. (1965): Biochemical polymorphism in ants. Science *149,* 544–546.

Ledoux, A. (1950): Recherche sur la biologie de la fourmi fileuse (*Oecophylla longinoda* Latr.). Ann. Sci. Nat. Zool. *12,* 313–461.

Ledoux, A. (1974): Polymorphismus und Kastendetermination bei den Weberameisen. In: Sozialpolymorphismus bei Insekten (G. H. Schmidt, ed.) pp. 533–541, Wissenschaftl. Verlagsgesellschaft, Stuttgart.

Lloyd, J. E. (1983): Bioluminescence and communication in insects. Ann. Rev. Entomol. *28,* 131–160.

Markl, H., Hölldobler, B. and Hölldobler, T. (1977): Mating behavior and sound production in harvester ants (*Pogonomyrmex,* Formicidae). Insect. Soc. *24,* 191–212.

McCluskey, E. S. (1965): Circadian rhythms in male ants of five diverse species. Science 150, 1037–1039.

McCluskey, E. S. (1967): Circadian rhythms in female ants and loss after mating flight. Comp. Biochem. Physiol. *23,* 665–677.

Michener, C. D. and Brothers, D. J. (1974): Were workers of eusocial Hymenoptera initially altruistic or oppressed? Proc. Nat. Acad. Sci. USA *71,* 671–674.

Oster, G., Eshel, I. and Cohen, D. (1977): Worker-queen conflict and the evolution of social insects. Theor. Pop. Biol. *12,* 49–85.

Pamilo, P. (1982): Genetic evolution of sex ratios in eusocial Hymenoptera: allele frequency simulations. Am. Nat. *119,* 638–696.

Pamilo, P. and Varvio-Aho, S. L. (1979): Genetic structure of nests of the ant *Formica sanguinea.* Behav. Ecol. Sociobiol. *6,* 91–98.

Parker, G. A. (1970): Sperm competition and its evolutionary consequences in the insects. Biological Reviews *45,* 525–568.

Passera, L. (1978): Demonstration of a pheromone inhibiting oviposition by workers in the ant *Plagiolepis pygmaea* Latr. (Hym. Formicidae). C. R. Acad. Sci. *D 286,* 1507–1509.

Passera, L. (1980): La fonction inhibitrice des reines de la fourmi *Plagiolepis pygmaea* Latr.: Rôle des pheromones. Insect. Soc. *27,* 212–225.

Pasteels, J. M., Verhaeghe, J. C., Braekman, J. C., Daloze, D. and Tursch, B. (1980): Caste-dependent pheromones in the head of the ant *Tetramorium caespitum.* J. Chem. Ecol. *6,* 467–472.

Pearson, B. (1983): Intra-colonial relatedness amongst workers in a population of nests of the polygynous ant *Myrmica rubra* Latreille. Behav. Ecol. Sociobiol. *12,* 1–4.

Petersen-Braun, M. (1977): Untersuchungen zur sozialen Organisation der Pharaoameise *Monomorium pharaonis* (L.) (Hymenoptera, Formicidae). II. Die Kastendeterminierung. Insect. Soc. *24,* 303–318.

Petersen-Braun, M. (1982): Intraspezifisches Aggressionsverhalten bei der Pharaoameise *Monomorium pharaonis* L. (Hymenoptera, Formicidae) Insect. Soc. *29,* 25–33.

Scherba, G. (1961): Nest structure and reproduction in the mound-building ant *Formica opaciventris* emery in Wyoming. J. N. Y. Ent. Soc. *69,* 71–87.

Schmidt, G. H. (1972): Männchendetermination im polygynen Waldameisenstaat. Zool. Anz. *189,* 159–169.

Schmidt, G. H. (ed.) (1974): Sozialmorphismus bei Insekten, Wissenschaftl. Verlagsgesellschaft, Stuttgart.

Schneirla, T. C. (1971): Army Ants. W. H. Treeman & Comp., San Francisco.

Starr, C. K. (1979): Origin and evolution of insect sociality: a review of modern theory. In: Social Insects (H. R. Hermann, ed.) pp. 35–80. Academic Press, London.

Stumper, R. (1956): Sur les sécrétions des fourmis femelles. C. R. Acad. Sci. Paris *242,* 2487–2489.

Talbot, M. (1953): Ants of an old field community on the Edwin S. George Reserve, Livingston County, Michigan. Contrib. Lab. Vert. Biol. Univ. Mich. *63,* 1–13.

Taylor, R. W. (1978): *Nothomyrmecia macrops:* a living-fossil ant rediscovered. Science *201,* 979–985.

Trivers, R. L. and Hare, H. (1976): Haplodiploidy and the evolution of social insects. Science *191,* 249–263.

Vander Meer, R. K., Glancey, B. M., Lofgren, C. S., Glover, A., Tumlinson, J. H. and Rocca, J. (1980): The poison sac of red imported fire ant queens: source of a pheromone attractant. Ann. Entomol. Soc. Amer. *73,* 609–612.

Ward, P. S. (1983a): Genetic relatedness and colony organization in a species complex of ponerine ants. I. Phenotypic and genotypic composition of colonies. Behav. Ecol. Sociobiol. *12,* 285–299.

Ward, P. S. (1983b): Genetic relatedness and colony organization in a species complex of ponerine ants. II. Patterns of sex ratio investment. Behav. Ecol. Sociobiol. *12,* 301–307.

Watkins, J. F. and Cole, T. W. (1966): The attraction of army ant workers to secretions of their queens. Texas Journal of Science *18,* 254–265.

Wesson, L. G. (1939): Contribution to the natural history of *Harpagoxenus americanus* (Hymenoptera, Formicidae). Trans. Ann. Ent. Soc. *65,* 97–122.

Wheeler, W. M. (1916): The marriage-flight of a bull-dog ant (*Myrmecia sanguinea* F. Smith) J. Anim. Behav. *6,* 70–73.

Whelden, R. M. (1963): The anatomy of the adult queen and workers of the army ants *Eciton burchelli* Westwood and *Eciton hamatum* Fabricius. J. N. Y. Entomol. Soc. *71,* 246–261.

Wilson, E. O. (1971): The Insect Societies. Belknap Press of Harvard University Press, Cambridge.

Wilson, E. O., Carpenter, F. M. and Brown, W. L. (1967a): The first Mesocoic ants. Science *157,* 1038–1040.

Wilson, E. O., Carpenter, F. M. and Brown, W. L. (1967b): The first mesocoic ants, with the description of a new subfamily. Psyche (Cambridge) *74,* 1–19.

Fortschritte der Zoologie, Bd. 31 · Hölldobler/Lindauer (Hrsg.): Experimental Behavioral Ecology
G. Fischer Verlag · Stuttgart · New York · 1985

Endocrine basis of dominance and reproduction in polistine paper wasps

PETER-FRANK RÖSELER

Zoologisches Institut der Universität Würzburg, F.R.Germany

Abstract

In several *Polistes* wasps some overwintered females can cooperate in founding a colony. The females form a linear hierarchy: the dominant α – female becomes the principal egg-layer, whereas the others act as subordinate helpers and do not reproduce. The formation of multiple foundress associations results in a greater success of that colony and in an increased number of offspring per colony. Since reproduction is increased, the dominant egg-layer gains most by the association. Subordinates benefit by joining a group only if the egg-layer is a full sister. Correspondingly, in some species mechanisms are evolved to discriminate former nestmates.

After hibernation, differences in the endocrine activity among foundresses are used to rank them in dominance. Hormonal treatment gives an advantage in dominant behaviour and increases the probability of becoming dominant. After a dominance hierarchy is established the endocrine activity in subordinates is inhibited by the dominant. That mechanism for sorting foundresses ensures that only those females become dominant egg-layers which have a high endocrine activity and thereby a high reproductive capacity.

1 Introduction

Paper wasps *(Vespidae)* are social wasps. The family includes the *Vespinae* (hornets and yellow jackets) and the *Polistinae*. Formerly, the *Stenogastrinae* (social species found in Asian tropics) were also believed to be a subfamily of *Vespidae,* but now they are considered as a subfamily of the *Eumenidae.*

The biology of paper wasps is well illustrated in some books (KEMPER and DÖHRING, 1967; EVANS and WEST EBERHARD, 1973; SPRADBERY, 1973) and reviews (WEST EBERHARD, 1969; PARDI, 1980). The species are distributed in the tropics as well as in palaearctic and nearctic regions. Nests of *Vespinae* are characterized by an envelope built around the combs. In temperate zones, colonies are founded in the spring by a solitary queen. She is the only egg-layer and inhibits egg-formation in workers by a pheromone (ISHAY et al., 1965; IKAN et al., 1969; AKRE and REED, 1983). Only toward the end of the season or when the queen disappears can workers lay eggs. In temperate zones, the colonies are annual and the societies disintegrate during autumn. In the tropics, however, nests can persist. Young queens are recruited so that nests become polygynic. Such perennial colonies can have in excess of 100.000 individuals (Spradbery, 1973a).

The majority of the *Polistinae* genera are confined to the tropics, most of them found in Central- and South-America. *Polistes* is the only cosmopolitan genus (about 150 species) distributed in the tropics as well as in temperate zones. The nests of *Polistes* consist of a single comb without an envelope. Since a colony seldom contains more than one or two hundred individuals, observations on behaviour and experiments can be easily performed. «This visibility, and the fact that the range of temperate-zone *Polistes* happens to overlap that of university biologists, helps to explain why the biology of *Polistes* is relatively well known.» (Evans and West Eberhard, 1973). That situation has not yet changed. Thus, this review on endocrine activity and reproduction in wasps must be restricted to results obtained on some *Polistes* species.

2 Biology of *Polistes*

In the tropics, the colony cycle of *Polistes* species is asynchronous, so that all stages of colony development can be found around the year. In temperate zones the colonies have a seasonal cycle. Nests are started in the spring by overwintered females. Only the dominant female, the queen, lays eggs as long as she retains her dominant position. During colony development workers are produced at first, thereafter mainly future foundresses and males emerge. In autumn the colonies die out, only future foundresses are able to hibernate. They initiate new nests in the following spring.

Interestingly, there are certain *Polistes* species in which some overwintered females often cooperate in initiating a nest. The frequency of such multiple foundress associations is different in different species and also different in various localities for the same species (Table 1). The number of foundresses that cooperate is also different. An association mostly consists of between two and five females. Very large associations, some with more than twenty foundresses, were found in *P. annularis* (Strassmann, 1981).

An association is formed when foundresses meet at nesting sites or when a foundress joins an existing colony. Foundresses of a group are arranged in a linear dominance

Table 1: Frequency of multiple foundress colonies observed in different *Polistes* species.

Species	Percentage	Authors
P. annularis	100%	Strassmann, 1981
P. canadensis	98%	West Eberhard, 1969
P. exclamans	43%	Strassmann, 1981a
P. fuscatus	43%	Gibo, 1978
	50%	Noonan, 1981
	60%	West Eberhard, 1981
P. gallicus	30%	Turillazzi et al., 1982
	40%	Pardi, 1942
P. metricus	9%	Bohm and Stockhammer, 1977
	10–75%	Dropkin and Gamboa, 1981

hierarchy ($\alpha > \beta > \gamma > \ldots$) (PARDI, 1948). The dominant foundress subordinates the associated foundresses by a specific behaviour. In *Polistes,* pheromones are not known to be involved in domination. When two foundresses of *P. gallicus* meet for the first time after hibernation, they inspect each other by intensively antennating. Then they rise high on their hind-legs, try to climb the other with the fore-legs and to make the other akinetic by biting and mock-stinging it. But even severe fights seldom result in injuries. Finally, one female becomes the dominant, chewing on the other which itself behaves akinetically, that is, the body is pressed to the ground, and antennae and legs drawn nearest the body. In all further meetings the lower ranked female will immediately act subordinately in this way. After a hierarchy has been established, a change in the dominance order seldom occurs.

In the field, the associations are not confined to the first members. After a fight a subordinated female can escape and try to become a dominant in an other group; correspondingly, a dominant foundress can be subordinated by a usurper (YOSHIKAWA, 1955). Thus, in some species, foundresses can shift from nest to nest for the first time (GAMBOA, 1978).

The dominant foundress spends most of the time on the comb and seldom leaves the nest. The dangerous tasks are performed by the subordinated foundresses which behave as workers; they collect food, water and wood pulp for cell construction. In order to distinguish them from real workers (the first daughters), they are called co-foundresses or auxiliaries. A characteristic attribute of dominance is the trophic advantage of the dominant foundress by trophallaxis.

The dominant foundress lays all the eggs, or at least the majority of them; so the α-female is called the queen. Egg-formation in co-foundresses is inhibited. But at early stages of nest development co-foundresses of *P. gallicus* can occasionally lay some eggs which are immediately eaten by the dominant queen (differential oophagy, GERVET, 1964). Also in early nests of *P. annularis* with only eggs or larvae co-foundresses sometimes lay eggs, but it is not clear whether some eggs can develop (STRASSMANN, 1981). In *P. fuscatus* only a third of the eggs laid by co-foundresses are eaten by the queen, so that approximately 33 % of the offspring derives from subordinates (NOONAN, 1981). The situation in *P. metricus* is not clear. GAMBOA et al. (1978) reported that all eggs from co-foundresses are eaten by the queen, whereas METCALF and WHITT (1977) using electrophoretic analysis of isozymes estimated a 10–20 % contribution of subordinates to the offspring. In general, the dominance hierarchy of *Polistes* finally results in a reproductive dominance of the queen which inhibits reproduction of subordinates in two ways: by inhibition of oogenesis and by egg-eating.

In *Polistes* wasps there are several overwintered foundresses which renounce their own direct reproduction, join another foundress and help to raise its progeny, though they would be able to found their own nest. The percentage of non-reproductive females within a population can be very high. In *P. fuscatus* approximately 50 % (GIBO, 1978; NOONAN, 1981), in *P. canadensis* even 79 % (WEST EBERHARD, 1967a) of the females do not reproduce.

This system of multiple foundress colonies raises certain questions: are multiple foundress nests more successful than single foundress nests? What is the advantage for the dominant female? Why do other females join a dominant foundress instead of founding their own colony? Do different types of overwintered females exist, some of them probably predetermined to become subordinates? What factors control dominant behaviour and egg-formation and are possibly involved in ranking foundresses in dominance? Some of the problems have been extensively discussed in other papers and will be only briefly reviewed here (WEST, 1967; WEST EBERHARD, 1975; GIBO, 1978; PRATTE and GERVET 1980; PARDI, 1980; NOONAN, 1981; STRASSMANN, 1981, 1981b; et al.)

3 The advantage of multiple foundress colonies

3.1 The ecological advantage

The advantage of multiple foundress associations is commonly seen in the better protection of nests against predators, parasites, and usurpation by conspecific females. Multiple foundress nests are always attended by at least one female, whereas single foundresses must leave their nests unattended during the foraging trips. In *P. metricus,* for instance, three foundress nests are left unattended for only 0.7 % of the time, whereas single foundress nests are left unattended 29 % of the time (GAMBOA, 1978).

Predators of *Polistes* wasps are mainly birds, which destroy nests or hunt females during their foraging trips. But several observations show that birds are not prevented from destroying nests by several foundresses on the comb. In *P. exclamans* many nests are regularly destroyed by birds which, on the other hand, do not attack *P. annularis* nests in the same region (STRASSMANN, 1981a). Probably, the aggressiveness of foundresses is different in different species. In general, foundresses were found to be less aggressive than workers. When workers are present, nests are not further attacked by birds. In P. *apachus* 60 % of nests (GIBO and METCALF, 1978) and in *P. fuscatus* more than 80 % of nests (GIBO, 1978) were destroyed prior to the emergence of workers. Nest usurpation also exclusively occurs during that early period. Therefore, a good strategy to reduce predation is for a late foundress to join established nests which will produce workers early in the season. In *P. metricus,* nests with multiple foundresses are established earlier and, consequently, workers emerge earlier than in single foundress nests (GAMBOA, 1980).

When nests are destroyed, they can be reestablished by the wasps. The probability for reestablishing was found to be greater in multiple foundress associations (40 % of the nests) than in single foundress nests (7 % of the nests) of *P. fuscatus* (GIBO, 1978). Also in *P. gallicus* the reestablishing activity tested in captivity is prolonged in multiple foundress associations (PRATTE and GERVET, 1980).

The best strategy for survivorship, however, is to build the combs at protected sites. Only one of 32 protected nests of *P. fuscatus* was destroyed, in contrast to 59 of 68 non-protected nests (GIBO, 1978). In two nesting-sites of *P. gallicus* we observed that all nests were destroyed which could be reached by birds. Thus, the competition for suitable nesting sites can result in high densities of foundresses at that places. GAMBOA (1978) found that in *P. metricus* multiple foundress associations are mainly established when densities are high. Single foundress nests were started at localities with low densities of foundresses.

Other strategies to reduce predation by birds or parasitism are the construction of satellite nests found in *P. exclamans* (STRASSMANN, 1981a, 1981b) and nest-sharing observed in *P. metricus* (GAMBOA, 1981).

In temperate zones, weather conditions can frequently change in late spring. The nourishment of larvae seems to be better provided by multiple foundresses which can collect more food in warm periods, than by a single foundress. But this problem has never been investigated.

The ecological advantage of multiple foundress associations results in a higher success of those nests compared with single foundress nests. The reproductive success of a colony is the number of offspring which will again be reproductive, the future foundresses and the males. Workers do not pass their genes to the next generation, except when they occasionally produce males at the end of colony development. It is, therefore, necessary to estimate the production of future foundresses and males, and moreover to know whether the probability is the same for all future foundresses to survive the winter independent of their emergence early or late in the season. Since this is difficult to estimate in the field, a random sample is normally used. For instance, the number of cells before the emergence of the first brood is compared between nests (West Eberhard, 1969; Turillazzi et al., 1982), or the number of offspring which emerge during the 24 days after the nests have been collected (Metcalf and Whitt, 1977a), or the sizes of nests at the end of the season and the number of wasps counted biweekly on the nest during the season (Gibo, 1978), or the number of offspring emerged during August and September when only reproductives are produced (Noonan, 1981).

Table 2: Production of cells or pupae at the late preemergence period in different size multiple foundress colonies. (1) West Eeberhard, 1969; (2) Gibo, 1978; (3) Turillazzi et al., 1982.

	P. canadensis[1]		*P. fuscatus*[1]		*P. fuscatus*[2]		*P. gallicus*[3]	
Number of foundresses	Number of cells	nests	Number of cells	nests	Number of pupae	nests	Number of cells	nests
1	10	1	24	40	6	48	20	8
2	18	2	42	25	11	24	27	5
3	26	4	49	10	20	17	37	3
4	37	6	65	8	21	6	40	5
5	36	6	61	4	22	2	–	–

All the different methods used in the field show that multiple foundress nests produce more offspring than single foundress nests (table 2). Noonan (1981) found in contrast to Gibo (1978) only a difference between single- and multiple foundress nests of *P. fuscatus*, but no correlation between mean reproductive success and a number in excess of two foundresses, probably because of variations among localities of nest sites. Larger associations with more than five foundresses did not generally result in larger nests, probably because of the upper limit of the queen's reproductive capacity. Gervet (1964) observed that queens of *P. gallicus* laid 5 eggs/day in the mean and 9 eggs/day as a maximum. We found a mean of 3 eggs/day in early nests of the same species in captivity. Oviposition can be increased in later stages to 9 eggs/day, a rate maintained for up to ten days. 12 eggs/day were observed as a maximum.

In captivity, colonies of *P. fuscatus* started with one, two or three foundresses showed no differences in nest size and reproduction (Gibo, 1974), but multiple foundress colonies were more likely to succeed than single foundress nest. In *P. gallicus* we obtained clear differences in the number of offspring between different number of foundresses up to three individuales (table 3).

Table 3: Number of offspring emerged during the first 4 months after nest foundation in different size multiple foundress colonies of *P. gallicus* in captivity.

Number of foundresses	Number of offspring	Number of nests
1	60	9
2	145	29
3	163	4
> 3	125	4

The studies on different *Polistes* species show two clear advantages of multiple foundress nests: a greater probability of success and an increased number of offspring. But, of course, reproduction by foundress decreases the larger an association becomes. So an advantage for the whole population is difficult to see. The total number of reproductives produced within a population might be the same, whether only single foundress colonies are established of which several are destroyed, or whether few, but more successful multiple foundress colonies exist.

3.2 The advantage for foundresses

A dominant female of a multiple foundress association gains most by the presence of subordinates. Since the queen is mostly the only egg-layer, her reproduction is increased. We observed in *P. gallicus* that in early nests a single foundress laid 2,5 eggs/day in the mean and queens of two foundress colonies 3,2 eggs/day, respectively. Two causes are thought to be responsible for the increase in reproductive capacity. Co-foundresses perform most of the dangerous and energetic expensive tasks, so that the dominant seldom leaves the nest. In addition the risk is lowered of the queen being lost during foraging trips. In *P. gallicus*, the dominant females perform only 6–22 % of the foraging flights (Pardi, 1942). Strassmann (1981) observed no hunting by *P. annularis* queens. Secondly, the queen has a trophic advantage by trophallaxis and by eating the eggs occasionally laid by subordinates.

What does a foundress gain by joining a dominant foundress as a subordinate? Females can pass their genes to the next generation either by direct reproduction (egg-laying) or by helping a genetically related female with which they share a proportion of genes to rear its offspring. Since reproduction in multiple foundress colonies is increased, a subordinated female can theoretically pass more genes to the next generation in this indirect way than by founding its own colony as a single foundress. The inclusive fitness of a foundress becomes higher than its direct fitness. But this is true only if a subordinate joins a full sister, both of them having the same father. In order to ensure this, foundresses should be able to discriminate between nestmates and non-nestmates.

Observations on several species show that associations are formed in the next spring by former nestmates. West Eberhard (1969) marked future foundresses of *P. fuscatus* on their parental nests in the autumn and recognized in the following spring 18 foundresses; 14 of them started a colony with sisters. Noonan (1981) found in the same species that 52 % of the foundresses initiated a colony exclusively with sisters. Only 7 % formed associations with non-sisters. In *P. annularis,* foundresses from the same parental nest started a colony together (Strassmann, 1981). Some observations on *P. gallicus* show that sisters can occasionally found a colony together (Heldmann, 1936; Röseler, unpubl. observations).

Females of some species have the ability to discriminate nestmates, and they try to exclude non-nestmates from an association (Noonan, 1981; Strassmann, 1981; Ross and Gamboa, 1981; Post and Jeanne, 1982). The cues involved in discrimination are nest-specific, probably genetically determined. In *P. gallicus,* in contrast, Pratte (1982) has shown that foundresses in captivity form associations with nestmates as frequently as with non-nestmates. This is supported by our experiments in which we have never observed that a foundress was not allowed to join an association.

In all the investigations it was not possible to decide whether nestmates are really full sisters. Metcalf and Whitt (1977) have shown in their genetic studies using electrophoretic differences of esterase isozymes, that females of *P. metricus* mate twice, and that sperms are not equally used, but in a 9 : 1 ratio. Polyandry means that co-foundresses emerged in the same nest can be more distantly related than full sisters. In that case subordinates should gain more if they found their own nests.

A possible explanation for the high percentage of multiple foundress associations might be the competition for suitable, protected nesting-sites. Most of the species studied tend to start their nests in the vicinity of the natal nest. In this way, nestmates come together at the same places, and the probability is increased that nestmates will form an association. Thus, primary home site recognition seems to increase the density of foundresses which itself then induces the formation of multiple foundress associations (KLAHN, 1979). Nests established away from their natal sites are more likely to be single foundress nests (GAMBOA, 1978).

Subordinated foundresses seldom have the chance to take over a nest and to start direct reproduction, because the probability of survivorship for a dominant foundress is increased in multiple foundress associations. Mortality is normally raised in subordinates which spend a lot of time away fom the nest. STRASSMANN (1981) observed only one of 20 nests of *P. annularis* in which a subordinate became the dominant. In *P. exclamans*, however, the mortality of foundresses was found to be very high, so that co-foundresses have the chance to lay eggs (STRASSMANN, 1981a). Subordinates of other species *(P. gallicus, P. metricus)* are often driven away from the nest by unknown reason at the time, when the first brood emerges (PARDI, 1948; GAMBOA et al., 1978). We have observed the same behaviour on *P. gallicus* nests in captivity. In 12 of 46 multiple foundress nests, subordinates were found dead shortly after the emergence of the first brood; sometimes we have seen that co-foundresses were chased by the queen. In 13 other nests, however, subordinated foundresses survived the queen, but only 4 of them became dominant egg-layers (in < 10 % of all the nests); the other co-foundresses seemed to be too much exhausted by the previous tasks to take over the nests. At present, we do not know what might cause the different development of multiple foundress colonies. In general, there is a tendency for increased longevity of queens in multiple foundress nests. We have found the mean longevity to be as follows: two foundress nests: α-foundress 122 days, β-foundress 106 days; four foundress nests: α-foundress 107 days, β-foundress 105 days, γ- and δ-foundress 77 days.

It has been postulated since GERVET (1964) and WEST EBERHARD (1969) that females with an inferior reproductive capacity will join a strong and more aggressive female with a high reproductive capacity. This conclusion was drawn from the observation that subordinated females have less developed ovaries than dominant females. The consequence should be that different females exist. That hypothesis was at first supported by GIBO (1974). In his studies on *P. fuscatus* in captivity he found that 40 % of single foundresses did not reproduce, and he concluded that in the field these females may have joined a dominant foundress as subordinated helpers. But in later studies on the same species using the technique of colony division, subordinate and dominant females were found to have the same reproductivity (BENDEGEM et al., 1981). DROPKIN and GAMBOA (1981) also concluded that fecundity varies among *P. metricus* foundresses. They observed that ovaries of single foundresses were less developed than in queens of multiple foundress nests, but similar to those of subordinated foundresses. That finding, however, can be interpreted in another way (TURILLAZZI et al., 1982; RÖSELER et al., 1984; see below).

We have investigated the reproductivity of dominant and subordinated foundresses of *P. gallicus* in captivity. After a dominance hierarchy has been established, the females were separated and each female together with a subordinate helper was able to found its own colony (table 4). There was no difference in total reproductivity between previously α-β-foundresses and γ-foundresses. Only the production of females was found to be significantly smaller in colonies headed by a formerly γ-foundress.

Table 4: Number of offspring during the first 4 months after nest foundation of α-, β-, and γ-foundresses of *P. gallicus* which started separately a colony each female together with a subordinate helper. Data with standard deviation; n.s. = non significant.

	α-foundresses		β-foundresses		γ-foundresses
Number of colonies	6		6		6
Number of total offspring	164 ± 40		175 ± 64		158 ± 74
P		n.s.		n.s.	
Number of female offspring	133 ± 40		142 ± 75		102 ± 35
P		n.s.		< 0.1%	

Nevertheless, there are some hints that different future foundresses may exist. Deleurance (1952) reported that in *P. gallicus* some of them performed colony tasks and showed egg-formation prior to hibernation. Strambi et al. (1981) found that two populations of overwintering *P. gallicus* females exist, the one with a high, the other with a low protein level in haemolymph. Bohm (1972) obtained a different response to daylength according to whether future foundresses of *P. metricus* had been emerged in June or in July–August. But it is not clear whether the differences observed reflect different fecundities of foundresses in the next spring. Thus, there is to date no evidence for foundresses with an inferior reproductive capacity which determines them to become a subordinated foundress.

4 Endocrine control of dominance and reproduction

Since all hibernated females have a comparable reproductive capacity, the question arises: Have all foundresses which meet at a nesting-site the same chance to become a dominant? Are there certain factors influencing dominant behaviour and favouring a female to achieve the dominant position? What are the mechanisms controlling the sorting of females into dominance hierarchy and, finally, what is responsible for the differentiation in reproductive queens and workerlike co-foundresses?

Hibernated foundresses of *Polistes* are of different sizes. Observations show that dominant foundresses are mostly, but not always, larger than their co-foundresses (Turillazzi and Pardi, 1977; Noonan, 1981; Dropkin and Gamboa, 1981). Body size, however, is a relative value, differences are valid only within a group. Therefore, the dominant foundress of an association can be smaller than subordinated foundresses of other groups. Studies on *P. gallicus* in captivitiy show that body size does not contribute to dominance. Since large females, however, were found to have an increased endocrine activity and ovary development, large females have a greater probability to become a dominant than small females (Röseler et al., 1984).

The queen lays all eggs, oogenesis in subordinated foundresses is inhibited. Dominant behaviour is connected to egg-formation. Many observations on different species have shown that the dominant foundress has the most developed ovaries of a colony, and that the social rank of co-foundresses is related to the degree of ovary development (Pardi, 1946, 1948; Gervet, 1964; Röseler, et al., 1980; Dropkin and Gamboa, 1981; Turil-

LAZZI et al., 1982). It seems very likely that the achieving of a dominant position is mediated by factors connected to oogenesis. Egg-formation is generally controlled by hormones, above all by juvenile hormone. Correspondingly, dominant foundresses of *P. gallicus* have larger corpora allata than subordinates, and the activity of ovaries is related to the activity of the endocrine glands. Differences in the volume of corpora allata were found to reflect different synthetic activities, so that the size of the glands can be used as an index of activity (STRAMBI, 1969; RÖSELER et al., 1980; TURILLAZZI et al., 1982). First investigations on endocrine activity and ovary development were performed some days or weeks after the hierarchies had been established. By this, it remained unclear whether differences in the endocrine activity and oogenesis were the consequence of the hierarchy or had caused the ranking of foundresses. We have studied that problem on *P. gallicus* in captivity (RÖSELER et al., 1981; 1984).

Foundresses of *P. gallicus* exhibit clear dominante – subordinate behaviour one day after hibernation, when they meet other foundresses. In the test, immediately after the dominance between two socially naive foundresses was established, the wasps were sacrificed and investigated. 89 % of the females which became dominant one day after hibernation had larger corpora allata than subordinated females, and 85 % of the dominants had better developed oocytes. The activity of both organs is correlated even at that early stage of development ($r = 0.49$; $n = 109$).

When we put several females together into a cage, they form a linear dominance hierarchy according to the development of corpora allata and oocytes. Aggressiveness we found to be correlated to the activity of the glands. Females with a corpora allata volume smaller than a certain limit (approx. 220×10^4 μm^3) showed no dominant behaviour, acted immediately as subordinates, and were arranged at the end of the hierarchy. VAWTER and STRASSMANN (1982) reported that high juvenile hormone production by corpora allata in vitro was correlated with aggressive behaviour and high social rank.

During oogenesis, the ovaries of some insect species produce ecdysteroids which are mainly incorporated into eggs. In some species, however, the hormone is also released into the haemolymph. This is true for *P. gallicus*, in which the ecdysteroid titre in ovaries as well as in haemolymph sharply rises during oogenesis (STRAMBI et al., 1977). But, interestingly, even during the early stages of oogenesis there is a clear correlation between size of oocytes and ecdysteroid titre in haemolymph (RÖSELER et al., 1984). The function of ecdysteroids in the haemolymph of adults is still unclear (HOFFMANN et al., 1980), except in mosquitos *(Aedes)*, in which the hormone regulates the synthesis of vitellogenins (HANAOKA and HAGEDORN, 1980), and in crickets, in which a metabolic role of ecdysteroids is postulated (HOFFMANN and BEHRENS, 1982, 1983).

After hibernation the *Polistes* foundress which has more active corpora allata and with it better developed oocytes becomes a dominant, which means that dominant foundresses have a higher titre of juvenile hormone and ecdysteroids in the haemolymph than subordinated foundresses have. We have experimentally tested whether juvenile hormone and 20-hydroxyecdysone can really influence the dominant behaviour. The hormones were injected into foundresses at the end of hibernation and the dominant behaviour was tested one day later. The injection of juvenile hormone as well as of 20-hydroxyecdysone increased the probability that such a foundress became dominant, which had smaller corpora allata and smaller oocytes than the subordinated foundress. The percentage of those foundresses was increased by the hormones to about 50 %, whereas in controls we had obtained only 11 % (foundresses with smaller corpora allata) and 15 % (foundresses with smalles oocytes), respectively (see above). Juvenile hormone and 20-hydroxyecdysone evoked a similar response, but neither hormones acted synergistically. A combined injection of both of them had the same effect as an injection of a single hormone each.

The influence of hormones on ranking foundresses can be clearly seen in foundresses which are of smaller size and have smaller corpora allata and oocytes than the competitor. In untreated controls, only 4 % of foundresses with that threefold disadvantage became dominant. By the injection of hormones the percentage of those dominant foundresses could be increased up to 43 %.

After a dominance hierarchy has been established, an injection of juvenile hormone has no effect on dominance hierarchy in *Polistes*. A subordinated foundress treated with hormone remains subordinated and does not change its behaviour. Similar results were obtained by Barth and Strassmann (1980) in *P. annularis*. Thus, the hierarchy later on is probably not only controlled by hormones, but also by social experience and specific behaviour; pheromones seem not to be involved. This is supported by the observation that workers of *P. biglumis bimaculatus* retained their social position even after they have been ovariectomized (Deleurance, 1948). The behaviour of workers, on the other hand, can be influenced by juvenile hormone treatment. Barth et al. (1975) showed that repeated topical application of juvenile hormone increased the aggressive behaviour of *P. annularis* workers and induced egg-formation, finally resulting in a disruption of the social structures of colonies.

At the end of hibernation, foundresses of *P. gallicus* have a different endocrine activity responsible for ranking foundresses. How can such differences develop among foundresses? In the following experiments we have tried to increase the endocrine activity during overwintering and to increase the chance by that treatment to become a dominant. Overwintering foundresses were divided into several groups: Some wasps remained in the darkness of the cold-room (= controls), others were daily taken out of the room and exposed to sunlight and room temperature for 15 min and 60 min respectively. After 10 days the dominance reaction was tested against controls, and the endocrine activity and oocyte growth were estimated (table 5). After treatment with one hour light and elevated temperature daily corpora allata and oocytes of foundresses were enlarged. The probability of becoming a dominant was increased to 86 %. The 15 min treatment, in contrast, had no effect, corpora allata and oocytes were not enlarged and the probability of winning an encounter was found to be 50 %. In the field, such differences in endocrine activity can be due to differences in the microclimate of hibernation sites, or due to varying amounts of time elapsing after the end of hibernation. Foundresses investigated immediately after being taken out of their natural hibernation sites showed large differences in their corpora allata volume among the foundresses of one site as well as between the two sites studied (Mean corpora allata volume: site 1 = $280 \pm 87 \times 10^4 \mu m^3$; site 2 = $356 \pm 72 \times 10^4 \mu m^3$).

Table 5: Influence of activation by light and temperature on volume of corpora allata, length of oocytes, and dominant behaviour in overwintering foundresses of *P. gallicus* in captivity.

	Activation 15 min/day	control	Activation 60 min/day	control
Number of foundresses	18	17	16	18
C. allata-volume (x 10^4 μm^3)	329	332	343	267
Length of occytes (mm)	0.54	0.55	0.56	0.48
Percentage of dominants	50%	50%	86%	14%

The dominant foundress tries to monopolize reproduction. The mechanisms so far known to be involved in the control were reviewed by WEST EBERHARD (1977). In *Polistes* wasps, queen control is performed by a specific dominant behaviour and by differential egg-eating. So far there is no evidence that pheromones could be involved. When a dominance hierarchy is established, the endocrine activity and oogenesis of subordinates become more and more inhibited, whereas the endocrine and reproductive activity of the dominant is markedly increased. During that divergent development a very interesting phenomenon occurs. The corpora allata of the dominant *P. gallicus* foundress are larger and oogenesis is stimulated more, the more subordinated co-foundresses are arranged beneath them (TURILLAZZI et al., 1982; RÖSELER et al., 1984; see also DROPKIN and GAMBOA, 1981). Thus, a β-female of a multiple foundress association can have more active corpora allata than a single foundress or than a β-female of a two foundress association.

The regulation of endocrine activity in Polistes wasps is probably not only a simple switch on – off mechanism, but the activity can be adjusted to different levels. This result is difficult to explain. It seems that subordination by a dominant can be in part compensated by dominating other subordinates. TURILLAZZI et al. (1982) discussed the importance of trophic advantage. The queen seldom leaves the nest, so that energetically expensive tasks like hunting are reduced; moreover, the dominant experiences the trophic advantage by trophallaxis and, occasionally, egg-eating. In this way, the β-female of a multiple foundress association with more than two foundresses can also exploit subordinates. Behavioural aspects, however, should also be considered (RÖSELER et al., 1984). A β-female in a two foundress colony experiences all the domination by the queen, whereas in larger associations the domination is shared among all foundresses. In general, domination not only subordinates co-foundresses, but also has a positive effect on the dominant itself.

5 Discussion

Polistes species which form multiple foundress colonies seem to have developed different strategies to ensure the reproductive success of nests and of foundresses. At the end of hibernation a female is faced with three options: founding a nest as a single foundress, becoming a dominant foundress in an association, or joining a dominant foundress as a subordinate. Since subordinate foundresses in most species have no chance for direct reproduction, the reproductive success of subordinates is maximized only if the foundress joins a full sister and helps to raise its offspring, and if the total reproductive output is increased. This system is highly evoluted in species in which foundresses are able to discriminate between nestmates and to exclude non-sisters. In *P. gallicus*, in which nestmate discrimination seems not to exist, the probability of foundress associations formed by nestmates is increased by the behaviour that foundresses tend to return to the natal nest site.

Since some assumptions underlying the hypothesized evolution of foundress associations are not completely fulfilled, the theories cannot explain the origin of multiple foundress colonies, and the validity of the theories still remains to be investigated. A hypothesis for the evolution of foundress associations based on the co-evolution of dominant-subordinate behaviour was presented by PRATTE and GERVET (1980a). An association can only be formed if both behaviours are present. The behavioural prerequisite (dominant – subordination) was observed to exist not only in species with multiple foundress colonies, but also in species which very seldom form multiple foundress nests (PERNA et al., 1978). With this prerequisite the evolution of multiple foundress colonies requires no sister associations. Multiple foundress colonies probably evolved due to densities of foundresses competing for suitable nesting-sites and using the dominant – subordinate behaviour.

For comparison, bumblebee queens of *Bombus hypnorum* often return to their natal nest sites in the spring and compete for the limited number of suitable holes. All the queens behave aggressively and do not show subordination, so that frequently one of the competitors is killed. In some nests we have observed up to four different queens before the emergence of the first workers. A foundress association cannot be formed, in contrast to *Polistes,* because of the lack of subordinate behaviour in nestfounding queens.

An important prerequisite for the evolution of multiple foundress colonies not considered up to now is the sensitivity of the endocrine system in responding to behavioural signals. Only by that mechanism can the queen control reproduction within the colony.

After hibernation, all foundresses have a comparable reproductive capacity. There is no clear evidence for the existence of inferior females. When foundresses meet at nesting-sites, the dominance hierarchy is established. Small differences among foundresses are used for sorting them into dominance. Foundresses with the highest endocrine activity become the dominants. The correlation between endocrine activity, ovary development and dominant behaviour ensures that only those females which indicate that they have at present the highest reproductive capacity become reproductive foundresses.

The connection between reproductivity and dominance behaviour is widespread in social insects. The results on *Polistes* show that dominant behaviour might be mediated by juvenile hormone and ecdysteroids. It is, however, at present not possible to decide whether both hormones directly influence the behaviour in *Polistes.* More investigations are required before precise conclusions on the effect of both hormones on dominant behaviour can be drawn.

Acknowledgement

The studies on *P. gallicus* cited here were performed in collaboration with the research group of Alain Strambi, CNRS, Marseille, France. I thank all members of both groups in Würzburg and Marseille for their unfailingly generous help, their skilful assistance, and their valuable discussions. The work was supported by Deutsche Forschungsgemeinschaft and by Centre National de la Recherche Scientifique.

References

Akre, R. D., Reed, H. C. (1983): Evidence for a queen pheromone in *Vespula* (Hymenoptera: Vespidae). Can. Ent. *115,* 371–377.

Barth, R. H., Strassmann, J. (1980): Hormones and dominance in paper wasps. Amer. Zool. *20,* 1193 (abstr.).

Barth, R. H., Lester, L. J., Sroka, P., Kessler, T., Hearn, R. (1975): Juvenile hormone promotes dominance behavior and ovarian development in social wasps *(Polistes annularis).* Experientia *31,* 691–692.

Bendegem, J. P. van, Gibo, D. L., Alloway, T. M. (1981): Effects of colony division on foundress associations in *Polistes fuscatus* (Hymenoptera: Vespidae). Can. Ent. *113,* 551–556.

Bohm, M. K. (1972): Effects of environment and juvenile hormone on ovaries of the wasp, *Polistes metricus.* J. Insect Physiol. *18,* 1875–1883.

Bohm, M. K., Stockhammer, K. A. (1977): The nesting cycle of a paper wasp, *Polistes metricus* (Hymenoptera: Vespidae). J. Kans. Ent. Soc. *50,* 275–286.

Deleurance, E. P. (1948): Le comportement reproducteur est indépendant de la présence des ovaires chez *Polistes* (Hyménoptères Véspides). C. R. Acad. Sci., Paris, *227,* 866–867.

Deleurance, E. P. (1952): Le polymorphisme social et son déterminisme chez les guêpes. «Structure et Physiologie des sociètés animales», Colloque Int. du CNRS, Paris, 141–155.

Dropkin, J. A., Gamboa, G. J. (1981): Physical comparisons of foundresses of the paper wasp, *Polistes metricus* (Hymenoptera: Vespidae). Can. Ent. *113,* 457–461.

Evans, H. E., West Eberhard, M. J. (1973): The wasps. David & Charles, Newton Abbot, pp. 265.

Gamboa, G. J. (1978): Intraspecific defence: Advantage of social cooperation among paper wasp foundresses. Science *199,* 1463–1465.

Gamboa, G. J. (1980): Comparative timing of brood development between multiple- and single-foundress colonies of the paper wasp, *Polistes metricus.* Ecol. Ent. *5,* 221–225.

Gamboa, G. J. (1981): Nest sharing and maintenance of multiple nests by the paper wasp, *Polistes metricus* (Hymenoptera: Vespidae). J. Kans. Ent. Soc. *54,* 153–155.

Gamboa, G. J., Heacock, B. D., Wiltjer, S. L. (1978): Division of labor and subordinate longevity in foundress associations of the paper wasp, *Polistes metricus* (Hymenoptera: Vespidae). J. Kans. Ent. Soc. *51,* 343–352.

Gervet, J. (1964): La ponte et sa régulation dans la société polygyne de *Polistes gallicus* L. (Hyménoptère Vespidé). Ann. Sci. Naturelles, Zool., 12e Sér., T6, 601–778.

Gervet, J. (1964a): Le comportement d'oophagie différentielle chez *Polistes gallicus* L. (Hymen. Vesp.). Ins. sociaux *11,* 343–382.

Gibo, D. L. (1974): A laboratory study on the selective advantage of foundress associations in *Polistes fuscatus* (Hymenoptera: Vespidae). Can. Ent. *106,* 101–106.

Gibo, D. L. (1978): The selective advantage of foundress associations in *Polistes fuscatus* (Hymenoptera: Vespidae): A field study of the effects of predation on productivity. Can. Ent. *110,* 519–540.

Gibo, D. L., Metcalf, R. A. (1978): Early survival of *Polistes apachus* (Hymenoptera: Vespidae) colonies in California: a field study of an introduced species. Can. Ent. *110,* 1339–1343.

Hanaoka, K., Hagedorn, H. H. (1980): Brain hormone control of ecdysone secretion by the ovary in a mosquito. In «Progress in ecdysone research» (ed. by Hoffmann, J. A.), 467–480.

Heldmann, G. (1936): Über das Leben auf Waben mit mehreren überwinterten Weibchen von *Polistes gallica* L. Biol. Ztbl. *56,* 389–400.

Hoffmann, J. A., Lagueux, M., Hetru, C., Charlet, M., Goltzene, F. (1980): Ecdysone in reproductively competent female adults and in embryos of insects. In «Progress in ecdysone research» (ed. by Hoffmann, J. A.), 431–465.

Hoffmann, K. H., Behrens, W. (1982): Free ecdysteroids in adult male crickets, *Gryllus bimaculatus.* Physiol. Ent. *7,* 269–279.

Hoffmann, K. H., Behrens, W. (1983): Ecdysteroids – a class of metabolic hormones in adult crickets? In «Biosynthése, metabolisme, mode d'action des hormones d'invertebres», Colloque Int. du CNRS, Strasbourg, *117.*

Ikan, R., Gottlieb, R., Bergmann, E. D., Ishay, J. (1969): The pheromone of the queen of the oriental hornet, *Vespa orientalis* F. J. Insect Physiol. *15,* 1709–1712.

Ishay, J., Ikan, R., Bergmann, E. D. (1965): The presence of pheromones in the oriental hornet, *Vespa orientalis* F. J. Insect Physiol. *11,* 1307–1309.

Kemper, H., Döhring, E. (1967): Die sozialen Faltenwespen Mitteleuropas. pp. 180, Parey-Verlag.

Klahn, J. E. (1979): Philopatric and nonphilopatric foundress associations in the social wasp, *Polistes fuscatus.* Behav. Ecol. Sociobiol. *5,* 417–424.

Metcalf, R. A., Whitt, G. S. (1977): Intra-nest relatedness in the social wasp *Polistes metricus.* Behav. Ecol. Sociobiol. *2,* 339–351.

Metcalf, R. A., Whitt, G. S. (1977a): Relative inclusive fitness in the social wasp *Polistes metricus.* Behav. Ecol. Sociobiol. *2,* 353–360.

Noonan, K. M. (1981): Individual strategies of inclusive-fitness-maximizing in *Polistes fuscatus* foundresses. In «Natural selection and social behavior» (ed. by Alexander, R. D. and Tinkle, W. D.), 18–44.

Pardi, L. (1942): Ricerche sui Polistini. V. La poliginia iniziale in *Polistes gallicus* L. Boll. Ist. Ent. Univ. Bologna *14,* 1–106.

Pardi, L. (1946): Ricerche sui Polistini. VII. La «dominiazione» e il ciclo ovarico annuale in *Polistes gallicus* (L.). Boll. Ist. Ent. Univ. Bologna *15,* 25–84.

Pardi, L. (1948): Dominance order in Polistes wasps. Physiol. Zool. *21,* 1–13.
Pardi, L. (1980): Le vespe sociali: biologia ed evoluzione del comportamento. Accad. Naz. Lincei: Contrib. Centro Linc. Interdiscipl. Sc. Mat. e appl. N. *51,* 161–221.
Perna, B., Marino Piccioli, M. T., Turillazzi, S. (1978): Osservazioni sulla poliginia di *Polistes foederatus* (Kohl) (Hym. Vesp.) in dotta in cattivita. Boll. Ist. Ent. Univ. Bologna *34,* 55–63.
Post, D. C., Jeanne, R. L. (1982): Recognition of former nestmates during colony founding by the social wasp *Polistes fuscatus* (Hymenoptera: Vespidae). Behav. Ecol. Sociobiol. *11,* 283–285.
Pratte, M. (1982): Relations antérieures et association de fondation chez *Polistes gallicus* L. Ins. sociaux *29,* 352–357.
Pratte, M., Gervet, J. (1980): Influence des stimulations sociales sur la persistence du comportement de fondation chez la guêpe Poliste, *Polistes gallicus* L. (Hymen. Vesp.). Ins. sociaux *27,* 108–126.
Pratte, M., Gervet, J. (1980a): Le modèle sociobiologique, ses conditions de validité dans le cas des sociétés d'hyménoptères. Ann. Biol. *19,* 163–201.
Röseler, P.-F., Röseler, I., Strambi, A. (1980): The activity of corpora allata in dominant and subordinated females of the wasp *Polistes gallicus.* Ins. sociaux *27,* 97–107.
Röseler, P.-F., Röseler, I., Strambi, A. (1981): Rôle des corpora allata dans l'établissement de la hierarchie sociale chez *Polistes gallicus.* Bull. Int. Section Française UEIS, Toulouse, 112–113.
Röseler, P.-F., Röseler, I., Strambi, A., Augier, R. (1984): Influence of insect hormones on the establishment of dominance hierarchies among foundresses of the paper wasp, *Polistes gallicus.* Behav. Ecol. Sociobiol. *15,* 133–142.
Ross, N. M., Gamboa, G. J. (1981): Nestmate discrimination in social wasps (*Polistes metricus,* Hymenoptera: Vespidae). Behav. Ecol. Sociobiol. *9,* 163–165.
Spradbery, J. P. (1973): Wasps. An account of the biology and natural history of social and solitary wasps. pp. 408, Sidgwick & Jackson, London.
Spradbery, J. P. (1973a): The European social wasp, *Paravespula germanica* (F.) (Hymenoptera: Vespidae) in Tasmania, Australia. Proc. VIIth Int. Congr. IUSSI, London, 375–380.
Strambi, A. (1969): La fonction gonadotrope des organes neuroendocrines des guêpes femelles du genre *Polistes* (Hyménoptères). Influence du parasite *Xenos vesparum* Rossi (Strepsiptères). Thèse Doct. Sci. Nat., Paris, pp. 159.
Strambi, A., Strambi, C., Reggi, M. de (1977): Ecdysones and ovarian physiology in the adult wasp *Polistes gallicus.* Proc. VIIIth Int. Congr. IUSSI, Wageningen, 19–20.
Strambi, C., Strambi, A., Augier, R. (1982): Protein level in the haemolymph of the wasp *Polistes gallicus* L. at the beginning of imaginal life and during overwintering. Action of the strepsipterian parasite *Xenos vesparum* Rossi. Experientia *38,* 1189–1191.
Strassmann, J. E. (1981): Wasp reproduction and kin selection: reproductive competition and dominance hierarchies among *Polistes annularis* foundresses. Florida Ent. *64,* 74–88.
Strassmann, J. E. (1981a): Evolutionary implications of early male and satellite nest production in *Polistes exclamans* colony cycles. Behav. Ecol. Sociobiol. *8,* 55–64.
Strassmann, J. E. (1981b): Kin selection and satellite nests in *Polistes exclamans.* In «Natural selection and social behavior» (ed. by Alexander, R. D. and Tinkle, W. D.), 45–58.
Turillazzi, S., Pardi, L. (1977): Body size and hierarchy in polygynic nests of *Polistes gallicus* (L.) (Hymenoptera Vespidae). Monitore Zool. Ital. (N. S.) *11,* 101–112.
Turillazzi, S., Marino Piccioli, M. T., Hervatin, L., Pardi, L. (1982): Reproductive capacity of single foundress and associated foundress females of *Polistes gallicus* (L.) (Hymenoptera Vespidae). Monitore Zool. Ital. (N. S.) *16,* 75–88.
Vawter, L., Strassmann, J. (1982): Juvenile hormone production in Polistine wasps. Proc. IXth Int. Congr. IUSSI, Boulder, 22 (abstr.).
West, M. J. (1967): Foundress associations in Polistine wasps: dominance hierarchies and the evolution of social behavior. Science *157,* 1584–1585.
West Eberhard, M. J. (1969): The social biology of Polistine wasps. Misc. Publ. Mus. Zool., Univ. Michigan, Nr. *140,* 1–101.
West Eberhard, M. J. (1975): The evolution of social behavior by kin selection. Quart. Rev. Biol. *50,* 1–34.
West Eberhard, M. J. (1977): The establishment of reproductive dominance in social wasp colonies. Proc. VIIIth Int. Congr. IUSSI, Wageningen, 223–227.
Yoshikawa, K. (1955): A Polistine colony usurped by a foreign queen. Ecological studies of Polistes wasps II. Ins. sociaux *2,* 255–260.

Fortschritte der Zoologie, Bd. 31 · Hölldobler/Lindauer (Hrsg.): Experimental Behavioral Ecology
G. Fischer Verlag · Stuttgart · New York · 1985

Contrasts between insects and vertebrates in the evolution of male display, female choice, and lek mating

Jack W. Bradbury

Dept. of Biology, University of California, San Diego, La Jolla, CA 92039, USA

Abstract

Researchers with similar interests but studying different taxa often construct theoretical paradigms which reflect the particular adaptations shown in the taxon they study. It is often illuminating to apply definitions and constructs developed for one taxon to another very different taxonomic group. Current research on mating system evolution of insects and vertebrates respectively is a case in point. A number of criteria have been proposed to distinguish lek mating in vertebrates from alternative systems. When applied to insects, the most common mating patterns appear to fit some of the criteria but not others. In this chapter, it is suggested that the best way to accomodate both taxa in a common paradigm is to treat the proposed criteria as continuous variables instead of discrete alternatives and to accept that they may vary independently of each other. This in fact solves some definitional problems within each taxon as well: recent studies of exploded leks have shown that some of the original discrete criteria were met whereas others were not. Replacing each discrete criterion with an independent variable acknowledges this type of variation. While this type of approach allows for consideration of an infinite number of variable combinations, clearly not all combinations are found. A brief review of vertebrates and insects indicates both similarities and differences in the combinations of variable values found. A number of testable propositions to explain these convergences are suggested. Most focus on the different life history economics and mobilities of the two taxa. In short, it is probably not useful to argue about whether a given taxon has «leks» or not. Instead, one should consider a vertebrate lek as a special case of a suite of strategic adaptations, any one of which might take different values in other taxa facing different ecological constraints. Viewed this way, the differences between the taxa are not a hindrance, but on the contrary a necessary source of variation to characterize each adaptation and identify any correlations between adaptations.

1 Introduction

Most sociobiological research presumes that there are general rules for social evolution which apply across all taxonomic groups. Despite this, most of us tend to work with a limited number of taxa and our own perspectives are often limited by the alternatives which exist in our favored systems. Periodically, bodies of research which have evolved around separate taxa are brought together and the result is usually a major advance in our understanding of those presumed general rules. Hamilton's classic papers (1964) on kin selection were motivated by patterns observed in social insects, and although their generality was not initially appreciated, the concepts have now permeated much of our thinking about vertebrate as well as invertebrate social evolution. Studies of mating system evolution in insects and vertebrates have similarly undergone long periods of independent development. Much of the theory developed for vertebrates has only recently been applied to insects, and many of the important insect examples are still unfamiliar to vertebrate workers. It is also true that several major contributors to mating system theory (e. g. Jeff Parker) have begun with insect perspectives and too rarely have their ideas been applied to vertebrate systems. Clearly, we are now in a period when researchers on mating system evolution in the two groups of taxa are comparing notes and the field is undergoing rapid change as a result (e. g. Baker, 1983; Thornhill and Alcock, 1983; and others cited in later sections below). Since this Symposium is very much in accord with such a period of contrast and exchange, I would like to devote my contribution to considering similarities and differences in the evolution of one type of system present in both groups: Lek Mating. Leks have been the subject of several recent papers in the entomological literature (see below), and I have spent the last 10 years working on several avian and mammalian lek species. Recent insect work has certainly changed my views on the general rules which must govern lek evolution, and new developments in vertebrate examples should perhaps receive a bit more attention among entomologists.

2 Definitions of lek mating

In every field, one is inclined to categorize phenomena as a precondition to making contrasts. Despite the best intentions, we are all limited in identifying criteria by the examples at hand. A case in point is the definition of a «lek mating sytem». Most workers agree that an absence of male parental care is a necessary component in the definition of lek mating (Cf. Lack, 1968; Snow, 1963). Beyond this, there is considerable disagreement. Alexander (1975) was the first to distinguish between «resource based leks» and «non-resource based leks». In both cases, reproductively active males are clustered, but in the former case females visit the site at least in part to gain access to the resource, whereas only males (and matings) are available in the latter case. The distinction seems important because the cues used in mate choice are bound to differ in the two contexts and this in turn will influence the direction and intensity of sexual selection that results.

Based on work with birds and mammals, I suggested (Bradbury, 1977) that «lek» be restricted to cases where: a) there is no male parental care, b) males occur on territories which are spatially clustered within a population's accessible range, c) the territories of males contain no resources which influence female visits or mate choice, and d) once a female visits a male assembly, she has the freedom to select a mate. The first criterion was aimed at excluding storks or gulls where the other 3 conditions may be met, but mating is clearly monogamous. The second was intended to focus attention on those systems where the interactions of clustered males facilitated and affected mate choice by females. The third excluded Alexander's first category for the reason given above: one expected the form and dynamics of sexual selection to vary depending on whether resource quality or male phenotype was more important in female choice. Finally, the last category was intended to distinguish between typical vertebrate leks, in most of which females can make free choices, and many insect swarms where they apparently cannot (Downes, 1969; Eickwort and Ginsberg, 1980; Catts, 1982). Again, one would expect sexual selection to follow different trajectories in the two systems.

This restrictive definition, and even that of Alexander, have occasioned no little controvery. The recent field studies of «promiscuous» vertebrate species which fit criteria a, c, and d, but which show male clustering only when suitable habitat is patchy (e. g. ruffed grouse, some bower birds), suggest that the second criterion may unnecessarily deflect our attention away from forms which are otherwise under selective pressures similar to those of classical lek species (Gullion, 1976; Boag, 1965; Pruett-Jones and Pruett-Jones, 1982). Loiselle and Barlow (1978), working on fish mating systems, and Leuthold (1966), working on ungulates, suggest that the role of resources in effecting differential male mating success varies continuously and it is unhelpful to impose an artificial boundary between the two extremes. Baker (1983) makes a similar point for insect systems. Recent work on the provisioning of females by males in Mecopterans (Thornhill, 1976, 1981), empid flies (Downes, 1970; Alcock, 1973), and Orthopterans (via spermatophores; Cf. Leopold, 1976), many of which exhibit male clustering and/or active display, further confounds the issue of resource quality vs. male phenotype in mate selection. On top of these considerations, it is also now clear that even where males do not defend resources, the spatial disposition of resources needed by females may still determine indirectly the patterns of male dispersion (Parker, 1978; Bradbury, 1981; Bradbury and Gibson, 1983). Finally, several authors have argued either that some female insects DO have a chance to choose when approaching swarms, or else that they do not need to choose since preselection by the males themselves has occurred prior to female visits (Cf. Alcock and Thornhill, 1983 for most recent review).

I am now persuaded that the latter three criteria used in my definition of lek represent particular values for several orthogonal variables which can in principle vary independently with ecological conditions and taxonomic predilection. The lesson to be learned from contrasts between insect and vertebrate systems is that we need a broader framework in which both the classical leks of avian and mammalian species and the immense swarms of mayflies are obtained as special cases. If in retrospect, we find that the variables in such a framework are often correlated, it is at that point that one is justified in distinguishing between common patterns and setting down more formal definitions.

The framework I shall recommend is based on the notion of Parker (1978) and Thornhill and Alcock (1983) that males will distribute themselves nonrandomly according to the spatial dispersion of at least some of the potential sites where females can be encountered. Some potential encounter locations are clearly dependent on resource distributions: these include oviposition sites, female emergence sites, foraging sites, and microclimatic refuges (Parker, 1978). Other sites, as discussed below, are not close to resources but may yet be determined by resource distributions: these include transit bottlenecks and so-called «arbitrary» sites. I thus presume that for every species, some type of encounter site can be distinguished as a major determinant of male dispersion. Given the determination of male dispersion by one of these sites, males may vary in their proximity to the site: males aggregating around resources may actually set up territories on the resource (as in many vertebrates), or they may sit on adjacent trees and cliff faces (as in some insects). In short, the first set of parameters one must specify includes the type of encounter site (or sites) favored by a particular species, and the degree of proximity to these sites that males adopt. If one wanted to preserve the distinction of Alexander, one might collapse these two variables into one: consistent proximity to a resource demonstrably required by females. Notice that this shifts attention away fom the role of resources in biasing mate choice to their role in determining male dispersion.

The second required parameter would describe the actual spatial pattern of males which results from the adopting of a given type of encounter site and a given degree of proximity to such sites. These dispersions would vary with the encounter site dispersion, the degree of proximity to such sites adopted, and the rules of male settlement. Clearly, males which settle on dung pats could be clustered, random, or uniformly distributed depending on a variety of factors. The third and final parameter I would invoke would summarize the degree to which the female instead of the male determines the identity of successful males.

Leks as originally defined by myself and others constitute one set of values for these three variables. Whatever one wants to call the other variants, it seems fair to expect the general rules which predict classical leks under one set of conditions to be able to account for these other systems under different conditions. In the remainder of this paper, I shall use contrasts between insects and vertebrates to bracket the factors which appear to constrain these three parameter values and which may lead to significant correlations between them. In the end, I shall reconsider what we might want to mean when we call one system a «lek» and another something else.

3 Types of encounter sites and male proximity to them

Several entomological workers have enumerated the types of sites at which insect males might concentrate to encounter females (ALEXANDER, 1975; PARKER, 1978; ALCOCK, et al. 1978; BAKER, 1983; THORNHILL and ALCOCK, 1983). These include: emergence sites of virgin females, oviposition sites, foraging sites, sites with optimal microclimates, transit bottlenecks for females, major orientation cues for females, and arbitrary sites (vis., none of the above). The emphasis in the insect literature has been that males will go wherever they get the most effective access to females. If there are several types of sites, or if one type is dominated by a few males, then one may find several male strategies within the same species (e. g. ALCOCK, JONES and BUCHMANN, 1977; ALCOCK, 1979; CADE, 1979; review in THORNHILL and ALCOCK, 1983).

Vertebrate researchers have generated equivalent lists (e. g. EMLEN and ORING, 1977; WITTENBERGER, 1981; KREBS and DAVIES, 1981) and although the results are similar, there are some disjoint components. Male birds and mammals do not defend emergence sites since most species exhibit a significant delay between birth or hatching and sexual maturity: one does not find male hornbills waiting outside a tree hole to pounce on emergent fledglings. Males do aggregate at and often defend oviposition sites in a variety of fish, anuran, and avian species, and defense of parturition sites occurs in certain mammals. Defense of foraging resources is common in vertebrates. Defense of refuges with favorable microclimates for females occurs in bats. The degree to which vertebrate species which do not fit one of the above categories can be ascribed to male aggregation around female transit bottlenecks, over orientation cues, or due to arbitrary selection of cues is a topic of considerable discussion and will be reviewed below. One pattern common in mammals and fish, but rare in insects is the continued defense of a mobile group of females. The rarity of this behavior is not due to an inability of males to defend a single female: in fact, this behavior is quite common in a variety of species (e. g. PARKER, 1970). On the other hand, male insects very rarely defend entire groups of females at once. This is presumably because many insect females mate once after emergence and either do not form stable enough associations or do not live long enough to warrant male defense (THORNHILL and ALCOCK, 1983). In short, vertebrates and insects show considerable overlap in the sites at which males may encounter females, and thus in the available male strategies, but the lists are not identical.

Note that which type of site is used by a particular species of male is usually dependent on the dispersion of all potential encounter sites. For example, close proximity to foraging sites may be favored over other alternatives because foraging sites are easier to find, attract more females per «patch», or last longer than either oviposition or emergence sites. More subtly, when all critical resources for females are sufficiently scattered, an intermediate «cross roads» or node of female movements may be the better site. Two conditions of the resource distribution are thus necessary before proximity to transit bottlenecks or transit nodes are favored: a) that no resource itself is sufficiently concentrated that male settlement near the resource is a better strategy than near a node; and b) that the resources which control female movements are appropriately distributed to generate spatial heterogeneities in female movements. Female movements may be further concentrated if females must pass over or through some topographic feature to navigate between separated resource patches (e. g. BAKER, 1972, 1983).

In this regard, we do not know whether the so-called «arbitrary» sites are truly arbitrary, or whether thay also are modulated by resource distributions and female transit routes. There is certainly no theoretical reason why such arbitrary sites might not evolve. PARKER (1978) has suggested one scenario in which initial occupation of sites by males occurs because of some non-arbitrary benefit such as access to a major transit node or settlement in a location where female movements are more easily detected (e. g. hilltops, tall trees, etc.). This male preference could then lead to development of a female search image for such sites. The initial advantage to a female mutant with such an image would be its quicker ability to locate the most males and thus to reduce the costs of mate search and/or comparison. This female preference would then select for refined settlement choices by males. The resulting coevolutionary feedback between male site preference and female search image could easily shift the type of site from one of obvious utility to one of arbitrary specificity. The process is identical to that of sexual selection for other male traits by female choice (LANDE, 1981; KIRKPATRICK, 1982).

The question remains whether such sites are arbitrarily specific, remain tied to some specific location where females are more easily encountered anyway, or arise from some combination of these two processes. In some cases, sites which initially appeared to be arbitrary have later turned out to contain significant resources required by females (e. g. ponds used by bullfrogs, EMLEN, 1976; HOWARD, 1978). Certainly, additional cases in both taxa are likely to be found. Similarly, some sites which themselves are not resource or transit nodes may permit better surveillance of such adjacent sites and thus be of functional benefit to males (Cf. THORNHILL and ALCOCK, 1983). Even where there seems a strong correlation between sexual aggregations and arbitrary cues, it is not clear that this is strong evidence for high specificity in the cues (DOWNES, 1969; HÖLLDOBLER, 1976; DAVIDSON, 1982; ALCOCK, 1981). For example, in the *Pogonomyrmex* species in which the same Palo Verde trees are used by successive generations of alates for mating (HÖLLDOBLER, 1976; DAVIDSON, 1982), it would be of interest to contrast the effects on male and female visitation of tree form modification vs. tree relocation relative to perennial ant colonies in the area. The fact that the same trees are used in successive years does not require a highly specific search image. In fact, initially random settlement by males at a number of similar sites would usually result in higher local densities at points intermediate between the supplying colonies. The subsequent release of male secretions which attracted nearby males from less settled points would then exaggerate the initial settlement patterns and produce the perennial usage seen. In other words, the repeated use of the same trees need not imply a highly sophisticated search image, but could instead result from a general favoring of all large trees coupled with aggregative male settlement. Similar problems exist in vertebrate lek species where there are detectable similarities found among utilized lek sites, but not all sites meeting such conditions are occupied in a given year (Cf. BRADBURY, 1981; BRADBURY and GIBSON, 1983). The longer lives of vertebrates further confound the interpretation of perennial use of lek sites because males in many species tend to return to the same locations in which they were successful in prior years (e. g. Kruijt, et al., 1972). In short, in vertebrates there is no clear-cut evidence for fixed and arbitrary search images as the primary determinants of lek location; for some insect studies, the evidence seems more compelling (e. g. Lederhouse, 1982), but in others the necessary controls are lacking.

However rigorously we define arbitrary and transit sites, it still appears that vertebrates are much less likely to occupy sites distant from resources or female groups than are insects. Among the insects, many *Ephemeroptera* (Brittain, 1982), *Diptera* (Downes, 1969), *Hymenoptera* (Alcock, et al., 1978; Kimsey, 1980; Eickwort and Ginsberg, 1980), *Lepidoptera* (Shields, 1967; Lederhouse, 1972), *Trichoptera* (Tozer, Resh, and Solem, 1981), and *Coleoptera* (Hansen, et al. 1971; Hagen, 1962) appear to use sites at some distance from any emergence, oviposition, or foraging resource to await mates. As we shall note later, some of these species form «leks» in the restricted vertebrate sense (above), and others form swarms of varying configurations. Among vertebrates, such behavior is more rare. Leks, even counting more dispersed-male species, are uncommon in all of the vertebrate groups (Cf. Bradbury, 1981; Oring, 1982; Loiselle and Barlow, 1978). Swarming, even at resources, occurs rarely in vertebrates, the only clear examples being temperate bats (review in Bradbury, 1970), and possibly some cetaceans (Roger Payne, personal communication). By far the majority of vertebrates establish residence in close proximity to resources needed by females or defend the females themselves for long periods.

The conditions which ought to favor intermediate or arbitrary siting of males relative to resources instead of over or immediately adjacently to resources have been spelled out by a number of authors on both taxa (Emlen and Oring, 1977; Bradbury and Vehrencamp, 1977; Bradbury, 1981; Oring, 1982; Thornhill and Alcock, 1983). The basic argument is that scattered resources which are hard to find, occur in patches of restricted richness, or last for short periods may be insufficient attraction to enough females to make resource proximity a better strategy than occupation of some other site. The fact of the matter is that no one to date has actually quantified the cost of the alternatives and made a rigorous test of this notion. Whereas it might be possible to perform such a test using vertebrate species which show intraspecific variation in adoption of resource vs. «arbitrary» sites (e. g. topi, Montfort-Braham, 1975), it would perhaps be no more difficult to use parallel variation in insects, such as carpenter bees (Marshall and Alcock, 1981). In fact, it might be easier with an insect system to vary artificially food, emergence sites, and other relevant resources than in the vertebrate systems.

To understand the different emphasis given arbitrary and transit encounter sites by insects when compared to vertebrates, one would ideally do the above tests on both taxa. Perhaps one would find that even scaling for allometric differences, the resources are more finely dispersed for insects, or alternatively, that the costs for insects of search and defense go up faster with changes in resource dispersion than they do for vertebrates. There are a variety of factors which could generate the observed taxonomic biases in ways compatible with existing theory. However, until such economic contrasts are available, we shall have neither adequate tests of such theory nor the kinds of within-model refinement which make such theories more robust.

4 Patterns of male dispersion

Regardless of the type of encounter site adopted by males, the full range of male dispersion patterns may be encountered. Clearly, if one male can control an entire site, and sites are rare relative to males, other males may do best to occupy locations near to but not on the preferred site. If no male can control the site, then one may find a dense aggregation of males on the location. A variety of intermediates are of course possible. As suggested earlier, the observed pattern of males will depend on the relative densities of males and sites, the dispersion of the sites, and, perhaps most important, the settlement rules by which males distribute themselves.

Settlement rules have been proposed for both insects (PARKER, 1978) and vertebrates (BRADBURY and GIBSON, 1983). Both approaches assume that males have some idea about how many females are likely to pass through a particular encounter site. In both, the simplest cases presume that males occupy sites such that expected fitnesses are equal. This is an ideal free distribution in the sense of FRETWELL and LUCAS (1970). The difference between the two models is that PARKER treats each female as a fixed resource: that is, there is a fixed number of females likely to visit any given site and the site at which a given female is likely to mate is known. This leads to the prediction that if 10 females mate at site A and 4 females mate at site B, males ought to settle in the two sites in the ratio of 5 : 2. In Parker's model, settlement by a male at one site has no bearing on the fitness of males at an adjacent but separate encounter site.

In our model (called the «hotspot» model in prior papers), a female is allowed to move between encounter sites. We allow her an equal probability of mating at any of the occupied sites within her normal home range. This means that settlement of a male at one site within the home range of a given female devalues her worth (in fitness units) to males at other sites within her range. It does not affect the fitness of males at sites more than one female home range away fom the settled site. Settlement thus tends to equalize the summed fractional paternities of all males. Whereas PARKER's males settle all available sites in proportion to the number of females expected there, our males tend to ignore potential encounter sites intermediate between the best encounter sites. This generates more clustering than occurs in PARKER's example, and it produces a positive correlation between the sizes of female home ranges and the degree of male aggregation.

The simplest case for both models assumes that each male at a site has an equal chance of mating with any female there. Both models then go on to consider the case of despotic settlement in which early settlers can expect a higher fraction of matings than later males. In PARKER's model, the typical example presumes that male territoriality results in a non-equitable partitioning of matings, whereas in ours, either territoriality or differential attraction of females can generate the despotic mating distribution. In both models, the general settlement results are similar in despotic and free settlement, but later settlers tend to adopt less desirable encounter sites sooner in the despotic case. In our system, the correlation between female home range size and male aggregation is preserved in the despotic case, but the overall levels of male aggregation are reduced.

In short, PARKER's model maps males directly onto the dispersion of all acceptable encounter sites; our model maps males onto the same dispersion, but it either settles fewer of the sites or it increases the variance in males per site over that in PARKER's model. Perhaps more important, the dispersion of males in PARKER's analysis is invariant with female home range (within the constraint that these ranges are small relative to encounter site dispersion); in our model, two populations with exactly the same encounter site dispersion might exhibit quite different degrees of male aggregation due to differences in female home range size. The factor which determines which model applies is the ratio of female home range size and the average distance between potential encounter sites. Where encounter sites are discrete patches (which is usually the case for resources such as food, oviposition sites, emergence sites, microclimatic refuges, and TRUE arbitrary sites), such a contrast ought to be fairly straight-forward. Presumably female ranges are set by the most scattered of these sites; if males select this same type of site for mating, they will probably settle by Parker's rules. If in contrast, males adopt an encounter site which is more closely spaced than that setting the female range size, then our model will apply.

The situation is more complex for transit nodes and continously distributed resources. Here there is not a finite set of potential settlement sites against which the final male distribution can be compared. In fact, there is a surface which describes the probability of female encounter, and it is obvious that settlement at any one peak will devalue the surrounding points within a distance of one female home range. In general, such a model always shows some local devaluation and should thus be sensitive to female range size.

In many insects, foraging and oviposition occur at the same site. If emergence sites are not relevant (as females must mature after emergence), and if the foraging and oviposition sites are easily found and occupied, female ranges should match the distribution of occupied encounter sites and Parker's models should apply. This is apparently the case for dungflies and several other Dipterans which Parker has examined: in most cases, his predictions match very well with observed dispersions and behaviors (Parker, 1970, 1972, 1974). Unfortunately, no one has yet examined dispersions of insect males with regard to availability of encounter sites in species where one might expect a large difference between female range size and the dispersion of the occupied sites. Part of the problem is that identification of encounter site suitability must be made independent of the observed male distribution: in cases, such as arbitrary cues, where the male distribution is used to help define what a suitable site is (e. g. Alcock, 1983b; Davidson, 1982), it is impossible to perform the requisite contrasts. One contrast which could be made more easily is to look for correlations (either intra- or interspecifically) between female range size and male dispersion. As noted below, this must be done in such a way as to evaluate the possibilities of other mechanisms which might also generate such a correlation. Where the alternatives are unlikely, and the correlation holds, the hotspot model may be indicated.

In vertebrates, one is faced with similar problems. I have not yet tried to compare the two models for resource-oriented encounter site settlement in vertebrates. Other workers more familiar than I with such species would be better able to make these contrasts. However, within vertebrate lek species, there are a number of groups in which a good correlation between female range and male dispersion is obtained, and where alternative reasons for such a correlation have been disproved or are thought unlikely (Bradbury, 1981; Bradbury and Gibson, 1983). The variation in male dispersion seen in groups such as grouse, in which uniformly distributed males occur in small home range forms, exploded leks occur in moderate home range forms, and classical leks occur in large home range species, thus seems quite compatible with our proposed male settlement rules. There are a number of dispersed species, such as bower birds and the birds-of-paradise, which are currently under study and for which not only the correlations between home range and male dispersion, but also the mapping of males on known «hotspots», ought to be available soon for detailed tests of the model.

In short, male dispersion varies widely among both insects and vertebrates. At the moment, dispersion of available encounter sites seems sufficient to explain the dispersions of males in many forms; for the balance, in both taxa, female home range may be the additional parameter which is required to explain the observed variations.

5 Degrees of female choice

The final parameter of interest is the degree to which females are able to choose a mate, or instead have a mating imposed on them by males. At the phenomenological level, it is this parameter which seems to show some of the most marked differences between insects and vertebrates. There are several grades possible (modified from Borgia, 1979):

1. Grade 1: a female is free to visit several males and select a mate from them using phenotypic cues of males and without harassment or sacrifice of access to critical resources.
2. Grade 2: a female is free to visit several males, but the latter restrict female access to a critical resource in exchange for mating favors.
3. Grade 3: a female is approached and chased or harassed by aggressive males, but she remains able, perhaps at no small cost, to refuse matings.
4. Grade 4: a female is immediately chased and mounted by one or more males and has no overt choice of a mate. Any non-randomness in male success is strictly due to intrasexual interactions between males.

One might expect *a priori* that the degree of female choice might not be independent of either the type of encounter site adopted by males or their disperion. Where males occupy resource sites, the costs to female fecundity or survival of avoiding a mating with a particular male may be quite high. This need not be true for arbitrary sites unless they are very widely dispersed. Similarly, where males are more widely dispersed on large territories, it should in general be easier for females to avoid particular males in favor of other ones (assuming female ranges are large enough to contain several males). Where males are densely clumped and territorial boundaries do not exist or are poorly respected, continuous interference by adjacent males could greatly limit female choice.

One of the points now clear to me from recent work is that these correlations, while reasonable, may not be binding. For example, male dungflies aggregated over dung pats, (a food and oviposition site), generate at best a Grade 3 and more usually a Grade 4 situation (Parker, 1970; Borgia, 1981, 1982). In contrast, many *Drosophila,* which also aggregate over combined food and oviposition sites, exhibit elaborate courtship by males to females, and although there may be some jostling between males, the situation seems closer to Grade 1 than to other grades. Insect species with dispersed or loosely aggregated male territories in arbitrary sites show a tendency to be Grade 3 or Grade 4: examples where females can visit several territories, but are rushed and jumped by males include tarantula hawks (Alcock, 1981), and the flies *Physiphora demandata* (Alcock and Pyle, 1979) and *Cuterebra spp.* (Catts, 1982). On the other hand, females of the black swallowtail butterfly are able to pass through the territories of several such males and thus to shop for a mate (Lederhouse, 1982). Euglossine bees also show loosely clustered male territories at which females appear to have reasonable freedom of choice (Kimsey, 1980). In contrast to vertebrates where Grade 1 is more likely for such male dispersions, swallowtails and euglossines appear to be uncommon exceptions to a general insect pattern of reduced female choice where males are loosely aggregated on arbitrary territories.

The same taxonomic bias seems to recur in the degree of female choice for highly clustered males over arbitrary or transit sites. Whereas most insect swarms of males show Grade 4 behaviors, most vertebrate leks allow the female relatively unconstrained choice of a mate (Downes, 1969; Alcock, et al., 1978; Eickwort and Ginsberg, 1980; Thornhill and Alcock, 1983; Bradbury and Gibson, 1983). There are some exceptions on both sides. Among the insects, the lekking *Drosophila* are very similar in their behavior to classical leks of vertebrates (Spieth, 1974; Parsons, 1977, 1978). The several forms of lekking fireflies appear to have retained a considerable amount of female leverage in refusing unwanted mates (Lloyd, 1979). Among the vertebrates, some birds-of-paradise (LeCroy, et al., 1980), small leks of grouse (Hamerstrom and Hamerstrom, 1960) and buff-breasted sandpipers (Myers, 1979) exhibit high levels of male interference or dominance on leks. Females seem to exercise no mate choice in some temperate bats where the nearest analog to the observed male dispersion is probably an insect swarm (Cf. review in Bradbury, 1970).

It has become fashionable recently to argue that many female insects which visit mating swarms have more choice of a mate than meets the eye. Davidson (1982) recently argued that females approaching *Pogonomyrmex* swarms were likely to be retaining some degree of choice because there was positive assortative mating according to size. In fact, the pattern of size association she reports is more likely explained by male preferences for larger females and subsequent competition between males for such females, than for joint choice by both sexes (Parker, 1983). Thornhill and Alcock (1983) suggest that female honeybees may be able to discourage or allow additional matings after a preceeding copulation by retaining or removing, respectively, the plug formed by the prior male's detached genitalia in their own genital orifice. A similar mechanism may apply in some midges (Downes, 1978). This method, if it occurs, does not of course give the female any control over her first mate. Even if one allows that there is SOME residual female choice in insect swarms, the degree of choice is clearly less than that in classical vertebrate lek systems.

In short, for each type of male dispersion and encounter site, there are obvious exceptions to the simple notions of facilitated avoidance suggested earlier. Overall, insects seem to exhibit reduced female choice even when males are territorial and only loosely clustered; vertebrates, particularly at transit or arbitrary encounter sites, tend to retain considerable female choice even when males are highly aggregated. Two classes of explanations for these patterns have been suggested. The first compares the benefits and costs for each sex of persisiting when a male and female disagree on whether a mating should occur. The most general case occurs in species where males are totally indiscriminate and it is the female which may want to avoid a mating (Cf. Thornhill and Alcock, 1983).

Parker (1979, 1983) has developed several models outlining options and thresholds for female abandonment of choice. I shall not repeat these here, but the sense of them can be seen in the following simple contrast. Consider a female which could either expend some cost, C, to locate a male with some phenotypic cue value B, or she could select males at random at no additional costs and obtain a male of mean cue value M. Let p be the probability that with investment C, she actually locates or is able to distinguish a male of trait B from the average male. Let W(B) be the number of grand-offspring this female achieves, and W(M) be the grand-offspring of a non-discriminating female. For choice to be favored, it must be the case that

$$p[W(B)-C] + (1-p)[W(M)-C] > W(M)$$

or rewriting, that

$$[W(B)-W(M)]/C > 1/p$$

This basically says that the gain in fitness through choice, relative to the costs of that choice, should be larger than the reciprocal of the discrimination probability for that cost. For fixed B, M, C, and function W(B), a female with poor discrimination abilities for a given cost C requires a higher relative benefit of choice to justify the behavior than one with good discrimination abilities at the same cost. The cost to find a male with cue B when the mean is M will of course depend on the variance in male cues in the population, how far the female must travel between males, and how such travel, exposure, and time are scaled into losses of fecundity and survival. Densely aggregated males might lead to lower search costs, but they may concomitantly increase the costs of resisting unwanted mates. Such tradeoffs may be complex. The numerator of the left side of the equation is more complex yet. For cases where males provide some environmental benefit to females, such as access to a resource, the difference between W(B) and W(M) might be quite large. Where the only benefits are genetic, such as the likelihood that a more actively displaying male will produce sons who are themselves more active displayers, the achieved difference between the mated father and the mean will be devalued in the female's sons by a) the degree to which display is heritable, and b) the degree to which sons with a higher cue value are exposed to greater survival risks prior to mating. To justify choice for a given C and p, a low heritability and strong normalizing selection on males will have to be compensated by a very steep W function. In contrast, a shallow W function will require mild selection on males and a higher heritability. As several workers have pointed out, this may lead to a paradox since the execution of strong choice will reduce the amount of heritable variation in male cues until the benefit of choice no longer meets the inequality (BORGIA, 1979; TAYLOR and WILLIAMS, 1982; PARKER, 1983). However, LANDE (1977) has argued that if the trait is polygenic, there will be sufficient back-mutation and recombination to maintain a steady level of heritable male variation; continued female choice would then be favored.

If this model is appropriate, it suggests that the left hand side of the inequality above is more likely to be smaller than the right in insects than it is in vertebrates. This could be true if a) insects were less able to discriminate between males for a given cost than vertebrates; b) the cost of discrimination increased more rapidly for a given (B-M) than in vertebrates; c) insects had lower levels of heritable variation in cues than vertebrates; or d) the dependence of fitness on male cues is steeper in vertebrates. Quite frankly, we have no numbers as yet to test any of these possibilities. One would think that once arriving at a dense aggregation of males insects, the cost to a female of finding sufficent options would be small and thus choice favored. The difficulty may be that whereas a vertebrate female can leave a lek and either return later or go to another lek if harassed, the average female insect does not have sufficient time or energy to exercise this option. The alternative option of remaining at the site and fending off unwanted males can also become expensive. Still, why should *Drosophila* favor female choice both over foraging and oviposition sites and at arbitrary lek sites?

One factor may be that males, as well as females, are somewhat discriminating in some species. PARKER (1978) considers such cases in his general treatment. The gist of the argument, as with females, is that persistence pays for a male only when the chances of finally mating with the female exceed those of searching for and mating with an alternative female. As pointed out by THORNHILL and ALCOCK (1983), the latter probability depends greatly on the operational sex ratio. They suggest that males in many swarms over arbitrary sites are visited by females only at slow rates and most of these females mate only once. Since male longevity in such a swarm is limited, males which aggressively mount any visiting female are favored. The drosophilids and a few other taxa of insects are unusual in that females are cyclically receptive and each female thus makes several visits to mating sites. This effectively increases the ratio of encountered females per male and *a priori* would increase the benefits of males abandoning any one unreceptive female for a more receptive one. As THORNHILL and ALCOCK (1983) suggest, some male discrimination between receptive and non-receptive females would then be favored. Were females to outstrip males in the arms race between male discrimination and the ability of receptive but unwilling females to feign non-receptivity, the emphasis on courtship and a more Grade 1 mating system would result (Cf. PARKER, 1983). Perhaps this is what has occurred in drosophilids and the few other insects with Grade 1 mate choice. It certainly could apply to many of the vertebrate leks where Grade 1 choice appears to be the rule. In summary, if male and female generally disagree over mating preferences, the most likely explanations for the taxon-specific (and exceptional) distributions of different grades of mate choice will be obtained by careful economic contrasts between the sexes in the benefits and costs of persistence. These in turn will be affected by the dispersions, life histories, heritabilities, and other extrinsic factors which are common to each taxon and differ among related species.

The second type of explanation for the absence of female choice in either taxon rests on the notion that where females do not appear to contest male advances, they must be in favor of them. In other words, the behaviors of males among themselves and the choices of females, if they were exercised, would be exactly congruent. In such a case, females do not need to exercise any choice: the males do it for them. The notion that females benefit by mating with the most aggressive or dominant males pervades much of the recent insect literature (e. g. Borgia, 1979; Davidson, 1982; Parker, 1983; Thornhill and Alcock, 1983), and until recently, seemed a reasonable interpretation of the vertebrate lek literature (Williams, 1966; Trivers, 1972; Zahavi, 1975; Wiley, 1974; Borgia 1979). The link between aggressive males in insect swarms and successful males in vertebrate leks was generated by a presumed preference of females in the latter case for central territories on leks and a consequent competition by males for such sites. In the insect swarms, males fought for the females directly; in the latter, they fought for sites which females preferred. Even in some insects, males fight for central territories and females appear to select males on the basis of the resulting patterns of ownership (Lederhouse, 1982). Thus both systems really could be using the same mechanism, and the apparent differences in female choice are illusory.

It now appears that this link was fragile at best. The evidence for center-preferences in vertebrate leks was inferred from a) the correlation between male reproductive success and relative location, and b) competition by males to obtain territories closest to the lek centers (Wiley, 1974; Ballard and Robel, 1974; Buechner and Roth, 1974). The problem is that both supportive observations could also be generated by centripetal competition among less successful males to set up territories near males which were successful for reasons other than location. One assumption of the first interpretation is that there is some arbitrary convention which defines the site preferred by both sexes. If so, one expects the location of successful males to be as stable as the location of the arbitrary cues, and males which replace central males should ascend to high mating status. It now appears that available data for the more carefully-studied vertebrate leks either goes against these predictions, or is equivocal (Cf. review by Bradbury and Gibson, 1983). In fact, there is increasing support for female discrimination among males on the basis of male display performance and subsequent clustering of unsuccessful males around successful ones (Hartzler, 1972; de Vos, 1983; Kruijt, de Vos, Bossema, and Bruinsma, unpublished; Bradbury and Gibson, 1983 and unpublished observations).

Why then might female insects accept the outcomes of aggressive male encounters in most species, but most vertebrate females and certain exceptional insects utilize other phenotypic cues of males to exercise choice? One interpretation by some workers is that the cues used by females in either case are those most likely to confer general adaptive benefits on both sons and daughters of discriminating females (Borgia, 1979; Thornhill, 1980; Thornhill and Alcock, 1983; Parker, 1983). As an example, suppose that foraging ability were a heritable trait that was expressed in daughters as higher fecundity or survival and in males as higher competitive abilities in access to mates. The cues used by females to select males might thus be body size or dominance status, where this was the best indicator of male foraging success, or male performance display, where this was a better indicator of foraging ability. The switch between accepting male aggression and using some other cue would thus depend on which type of male cue was best correlated with the output of both sons and daughters.

The problem with this notion is that the intervening variable of foraging ability adds an additional devaluation term to the numerator (above) in the female choice inequality. Using the same example, the critical heritability is that for foraging ability, and the devaluations are the correlation coefficients between male foraging ability and competitive ability, and female fecundity or survival and foraging ability. It is well known that the correlations between status and some trait such as body size, stamina, and the like need not be very high even when dominance status is very clear-cut and stable (LANDAU, 1951). Thus even if female choice of males was easier (since males did the work of choosing), and even if daughters gained somewhat by that choice, the net gain in fitness would be devalued by the imperfect correlations between traits as well as the levels of heritability. If devalued enough, the gain would not warrant adopting this form of choice. Perhaps more pertinent, such choice might not be resistant to invasion by an alternative form of female choice which did not operate under the constraint of optimizing son and daughter fitness, but instead focused on traits in males which were more closely correlated with current male competitive success and thus benefitted sons only (Cf. KIRKPATRICK, 1982; BRADBURY and GIBSON, 1983).

If one insists that female insects are accepting male decisions because that is the best genetic choice open to them, and if the type of adaptive cue scenario in the preceeding paragraph is deemed not operating, is there any reasonable way to explain the differences between the taxa in the observed grades of female choice? If there is, it is not yet clear to me. I could imagine a sensory argument which assumed that insects were less able than vertebrates to recognize bizarre mutant male displays which had no adaptive value to daughters but which tried to capture the attention of receptive females. It is possible that once such a cue was established by chance or sensory predisposition, it could «runaway» and govern all subsequent mate choice (LANDE, 1981; KIRKPATRICK, 1982). The difficulty with such an argument is the number of exceptional insects which do show elaborate courtship at arbitrary locations with Grade 1 female choice. In lieu of such an argument, my own feeling is that the economic contrasts appropriate to contests between male and female persistence (noted earlier) will provide the most reasonable explanations of the patterns we have been discussing.

6 Conclusions

Do insects have leks? How should we define leks? I am persuaded by my review of both literatures that there may turn out to be specific combinations of type of encounter site, male dispersion, and degree of female choice which recur repeatedly in both vertebrates and insects. There are certainly insects with aggregated males, no nearby resources, and Grade 1 female choice which could be called «leks» by the classical vertebrate criteria. Despite this, I think we either need to formulate a whole host of additional definitions (e. g. no-choice swarms, no-choice dispersed territories, etc.), or else remain with the notion that these three variables can evolve sufficiently independently that any finite set of fixed definitions will always leave some species outside our dictionary. Both approaches are valuable. The first is so because it stresses the clear correlations which must arise in some variable combinations: males cannot as easily adopt a Grade 2 approach when aggregating over arbitrary sites as they can over a resource. Once correlations appear, one must examine their generality across taxa. This forces refinement and/or rejection of the basic categories. With regard to my old definition, it is now obvious that it is not the absence of resources within territories which should distinguish lek from non-lek species, but instead the degree to which males regulate access to those resources to obtain matings.

Bell birds and Epauletted Bats may both court females from within territories which contain resources needed by females; they do not however use controlled access to these resources as leverage in obtaining females (B. SNOW, 1973; BRADBURY, 1981).

The second approach often has as much value in explaining what we see as in what we don't: no one is likely to define a mating system category which doesn't appear in their favored organisms. Using a multivariable format, we can see what parts of the parameter space are never occupied and ask why not. This then allows one to add new taxa as they are described: perhaps they will fill the empty parameter space or perhaps they will only reinforce any existing clusters of points. In this chapter, I have tried to accomodate insects and vertebrates into a more common framework. The result has been both a reinforcement of some existing patterns common to both groups, and the occupation of parameter spaces not shared by the two larger taxa. Next year, maybe someone else can tackle snails and ctenophores.

References

Alcock, J. (1973): The mating behavior of *Empis barbatoides* Melander and *Empis poplitea* Loew (Diptera: Empidae). J. Nat. His. *7,* 411–420.

Alcock, J. (1979): The behavioral consequences of size variation among males of the territorial wasp *Hemipepsis ustulata* (Hymenoptera: Pompilidae). Behaviour *71,* 322–335.

Alcock, J. (1981): Lek territoriality in the tarantula hawk wasp *Hemipepsis ustulata* (Hymenoptera: Pompilidae). Behav. Ecol. Sociobiol. *8,* 309–317.

Alcock, J. (1983a): Territoriality by hilltopping males of the great purple hairstreak, *Atlides halesus* (Lepidoptera, Lycaenidae): convergent evolution with a pompilid wasp. Behav. Ecol. Sociobiol *13,* 57–62.

Alcock, J. (1983b): Constancy in the relative attractiveness of a set of landmark territorial sites to two generations of male tarantula hawk wasps (Hymenoptera: Pompilidae). Anim. Behav. *31,* 74–80.

Alcock, J., E. M. Barrows, G. Gordh, L. J. Hubbard, L. Kirkendall, D. Pyle, T. L. Ponder, and F. G. Zalom (1978): The ecology and evolution of male reproductive behavior in the bees and wasps. J. Linn. Soc. London, Zool. *64,* 293–326.

Alcock, J., C. E. Jones, and S. L. Buchmann (1977): Male mating strategies in the bee *Centris pallida* Fox (Hymenoptera: Anthophoridae). Amer. Nat. *111,* 145–155.

Alcock, J. and D. W. Pyle (1979): The complex courtship behavior of *Physiphora demandata* (F.) (Diptera: Otitidae). Zeit. für Tierpsych. *49,* 325–335.

Alexander, R. D. (1975): Natural selection and specialized chorusing behavior in acoustical insects. pp. 35–77, in Insects, Science and Society. D. Pimentel, ed. New York, Academic Press.

Baker, R. H. (1972): Territorial behaviour of the nymphalid butterflies, *Aglais urticae* and *Inachis io.* J. Anim. Ecol. *41,* 453–469.

Baker, R. H. (1983): Insect territoriality. Ann. Rev. Entomol. *28,* 65–89.

Ballard, W. B. and R. J. Robel (1974): Reproductive importance of dominant male greater prairie chickens. Auk *91,* 75–86.

Barrows, E. M. (1976): Mating behavior in halictine bees. II. Microterritorial and patrolling behavior in males of *Lasioglossum rohweri.* Zeit. für Tierpsych. *40,* 377–389.

Boag, D. A. (1965): Population attributes of blue grouse in southwestern Alberta. Can. J. Zool. *44,* 799–814.

Borgia, G. (1979): Sexual selection and the evolution of mating systems. pp. 19–80, in Sexual Selection and Reproductive Competition in Insects. M. S. Blum and N. A. Blum, eds. New York, Academic Press.

Borgia, G. (1981): Mate selection in the fly *Scatophaga stercoraria:* female choice in a male-controlled system. Animal Behaviour *29,* 71–80.

Borgia, G. (1982): Experimental changes in resource structure and male density: size-related differences in mating success among male *Scatophaga stercoraria.* Evol. *36,* 307–315.

Bradbury, J. (1970): Social organization and communication. pp. 1–72, in Biology of Bats, Vol. III. W. Wimsatt, ed. New York, Academic Press.

Bradbury, J. (1977): Lek mating behavior in the hammerheaded bat. Zeit. für Tierpsych. *45,* 225–255.

Bradbury, J. (1981): The evolution of leks pp. 138–169, in Natural Selection and Social Behavior: Recent Research and Theory. R. D. Alexander and D. W. Tinkle, eds. New York, Chiron Press.

Bradbury, J. and R. Gibson (1983): Leks and mate choice. pp. 109–138, in Mate Choice. P. P. G. Bateson, ed. Cambridge, Cambridge University Press.

Bradbury, J. and S. L. Vehrencamp (1977): Social organization and foraging in emballonurid bats. III. Mating systems. Behav. Ecol. Sociobiol. *2,* 1–17.

Brittain, J. E. (1982): Biology of mayflies. Ann. Rev. Entomol. *27,* 119–147.

Buechner, H. K. and R. Schloeth (1965): Ceremonial mating behavior in Uganda kob *(Adenota kob thomasii).* Zeit. für Tierpsych. *22,* 209–225.

Cade, W. (1979): The evolution of alternative male strategies in field crickets. pp. 343–379, in Sexual Selection and Reproductive Competition in Insects. M. S. Blum and N. A. Blum, eds. New York, Academic Press.

Catts, E. P. (1979): Hilltop aggregation and mating behavior in *Gasterophilus* (Diptera: Gasterophilidae). J. Med. Entomol. *16,* 461–464.

Catts, E. P. (1982): Biology of New World botflies: Cuterebridae. Ann. Rev. Entomol. *27,* 313–338.

Davidson, D. W. (1982): Sexual selection in harvester ants (Hymenoptera: Formicidae: Pogonomyrmex). Behav. Ecol. Sociobiol. *10,* 245–250.

Doolan, J. M. (1981): Male spacing and the influence of female courtship behaviour in the bladder cicada, *Cystosoma sundersii* Westwood. Behav. Ecol. Sociobiol. *9,* 269–276.

Downes, J. A. (1969): The swarming and mating flight of Diptera. Ann. Rev. Entomol. *14,* 271–298.

Downes, J. A. (1970): The feeding and mating behaviour of the specialized Empidinae (Diptera): observations on four species of *Rhamphomyia* in the high arctic and a general discussion. Canad. Entomol. *102,* 769–791.

Downes, J. A. (1978): Feeding and mating in the insectivorous Ceratopogoninae (Diptera). Memoirs Entomol. Soc. Can. *104.*

Eickwort, G. C. and H. S. Ginsberg (1980): Foraging and mating behavior in Apoidea. Ann. Rev. Entomol. *25,* 421–446.

Emlen, S. T. (1976): Lek organization and mating strategies in the bullfrog. Behav. Ecol. Sociobiol. *1,* 283–313.

Emlen, S. T. and L. W. Oring (1977): Ecology, sexual selection, and the evolution of mating systems. Science *197,* 215–223.

Frankie, G. W., S. B. Vinson, and R. E. Colville (1980): Territorial behavior of *Centris adani* and its reproductive function in the Costa Rican dry forest (Hymenoptera: Anthophoridae). J. Kansas Entomol. Soc. *53,* 837–857.

Fretwell, S. D. and H. L. Lucas (1969): On territorial behavior and other factors influencing habitat distribution in birds. I. Theoretical development. Acta Biotheoret. *19,* 16–36.

Gullion, G. W. (1976): Re-evaluation of «activity clustering» by male grouse. Auk *93,* 192–193.

Hagen, K. S. (1962): Biology and ecology of predacious Coccinellidae. Ann. Rev. Entomol. *7,* 289–326.

Hamerstrom, F. N. and F. Hamerstrom (1951): Mobility of the sharp-tailed grouse in relation to its ecology and distribution. Amer. Midl. Nat. *46,* 174–226.

Hamilton, W. D. (1964): The genetical evolution of social behavior. I, II. J. Theor. Biol. *7,* 1–52.

Hanson, F. E., J. F. Case, E. Buck, and J. B. Buck (1971): Synchrony and flash entrainment in a New Guinea firefly. Science *174,* 161–164.

Hartzler, J. E. (1972): An Analysis of Sage Grouse Behavior. Ph. D. Dissertation Thesis. Missoula, Montana: University of Montana, 234 pp.

Holldobler, B. (1976): The behavioral ecology of mating in harvester ants (Hymenoptera: Formicidae: *Pogonomyrmex*). Behav. Ecol. Sociobiol. *1,* 405–423.

Howard, R. D. (1978): The evolution of mating strategies in bullfrogs, *Rana catesbeiana.* Evol. *32,* 850–871.

Janetos, A. C. (1980): Strategies of female mate choice: a theoretical analysis. Behav. Ecol. Sociobiol. *7,* 107–112.

Kimsey, L. S. (1980): The behavior of male orchid bees (Apidae, Hymenoptera, Insecta) and the question of leks. Anim. Behav. *28,* 996–1004.

Kirkpatrick, M. (1982): Sexual selection and the evolution of female choice. Evol. *36,* 1–12.

Krebs, J. R. and N. B. Davies (1981): An Introduction to Behavioural Ecology. Sunderland, Massachusetts, Sinauer Associates.

Kruijt, J. P., G. J. de Vos, and I. Bossema (1972): The arena system of black grouse. pp. 399–423 in, Proc. XVth Intl. Ornithol. Congr. Leiden, E. J. Brill.

Kruijt, J. P., G. J. de Vos, I. Bossema, and O. Bruinsma (unpubl.): Sexual selection in black grouse. Unpublished manuscript.

Lack, D. (1968): Ecological Adaptations for Breeding in Birds. London, Chapman and Hall.

Lande, R. (1977): The influence of mating system on the maintenance of genetic variability in polygenic characters. Genetics *86,* 485–498.

Lande, R. (1981): Models of speciation by sexual selection on polygenic traits. Proc. Nat. Acad. Sci., USA *78,* 3721–3725.

Landau, H. G. (1951): On dominance relations and the structure of animal societies, I. Effect of inherent characteristics. Bull. Math. Biophysics *13,* 1–19.

LeCroy, M., A. Kulupi, and W. S. Peckover (1980): Goldie's bird-of-paradise: display, natural history, and traditional relationships of people to the bird. Wilson Bull. *92,* 289–301.

Lederhouse, R. C. (1982): Territorial defense and lek behavior of the black swallowtail butterfly, *Papilio polyxenes.* Behav. Ecol. Sociobiol. *10,* 109–118.

Leopold, R. A. (1976): The role of male accessory glands in insect reproduction. Ann. Rev. Entomol. *21,* 199–221.

Leuthold, W. (1966): Territorial behavior of Uganda kob. Behaviour *27,* 255–256.

Lloyd, J. E. (1971): Bioluminescent communication in insects. Ann. Rev. Entomol. *16,* 97–122.

Lloyd, J. E. (1979): Sexual selection in luminescent beetles. pp. 293–342, in Sexual Selection and Reproductive Competition in Insects. M. S. Blum and N. A. Blum, eds. New York, Academic Press.

Loiselle, P. V. and G. W. Barlow (1978): Do fishes lek like birds? pp. 31–76, in Contrasts in Behavior, E. S. Reese and F. J. Lighter, eds. New York, Wiley and Sons.

Marshall, L. and J. Alcock (1981): The evolution of the mating system of *Xylocopa varipuncta* (Hymenoptera: Anthophoridae). J. Zool. *193,* 315–324.

McAlpine, D. K. (1979): Agonistic behavior in *Achias australis* (Diptera, Platystomatidae) and the significance of eyestalks. pp. 221–230, in Sexual Selection and Reproductive Competition in Insects. M. S. Blum and N. A. Blum, eds. New York, Academic Press.

Montfort-Braham, N. (1975): Variations dans la structure sociale du topi *Damaliscus korrigum* Ogilby, au Parc National de l'Akagera, Rwanda. Zeit. für Tierpsych. *39,* 332–364.

Myers, J. P. (1979): Leks, sex, and buff-breasted sandpipers. Amer. Birds *33,* 823–825.

Oring, L. (1982): Avian mating systems. pp. 1–92, in Avian Biology, Vol. VI. D. Farner, J. King, and K. Parkes, eds. New York, Academic Press.

Parker, G. A. (1970): The reproductive behaviour and nature of sexual selection in *Scatophaga stercoraria* (L.) (Diptera: Scatophagidae). II. The fertilization rate and the spatial and temporal relationships of each sex around the site of mating and oviposition. J. Anim. Ecol. *39,* 205–228.

Parker, G. A. (1972): Reproductive behaviour of *Sepsis cynipsea* (L.) (Diptera: Sepsidae). I. A preliminary analysis of reproductive strategy and its associated behaviour patterns. Behaviour *41,* 172–206.

Parker, G. A. (1974): The reproductive behaviour and the nature of sexual selection in *Scatophaga stercoraria* (L.) (Diptera: Scatophagidae). VIII. The behaviour of searching males. J. Entomol. (A) *48,* 199–211.

Parker, G. A. (1978): Evolution of competitive mate searching. Ann. Rev. Entomol. *23,* 173–196.

Parker, G. A. (1979): Sexual selection and sexual conflict. pp. 123–166, in Sexual Selection and Reproductive Competition in Insects. M. S. Blum and N. A. Blum, eds. New York, Academic Press.

Parker, G. A. (1982): Phenotype limited evolutionarily stable strategies. pp. 173–201, in Current Problems in Sociobiology. B. R. Bertram, T. H. Clutton-Brock, R. I. M. Dunbar, D. I. Rubenstein, and R. Wrangham, eds. Cambridge, Cambridge University Press.

Parker, G. A. (1983): Mate quality and mating decisions. pp. 141–166, in Mate Choice. P. P. G. Bateson, ed. Cambridge, Cambridge University Press.
Parsons, P. A. (1977): Lek behavior in *Drosophila* (Hirtodrosophila) polybori Malloch-an Australian rainforest species. Evol. *31,* 223–225.
Parsons, P. A. (1978): Habitat selection and evolutionary strategies in *Drosophila:* an invited address. Behav. Genet. *8,* 511–526.
Pruett-Jones, M. A. and S. G. Pruett-Jones (1982): Spacing and distribution of bowers in MacGregor's bowerbird *(Amblyornis macgregoriae).* Behav. Ecol. Sociobiol. *11,* 25–32.
Shields, O. (1967): Hilltopping. J. Res. Lepid. *6,* 69–178.
Snow, B. (1973): Notes on the behavior of the white bellbird. Auk *90,* 743–751.
Snow, D. (1963): The evolution of manakin displays. pp. 553–561, in Proc. 13th Intl. Ornith. Cong., Vol. I.
Spieth, H. T. (1974): Courtship behavior in *Drosophila.* Ann. Rev. Entomol. *19,* 385–405.
Taylor, P. D. and G. C. Williams (1982): The lek paradox is not resolved. Theoret. Pop. Biol. *22,* 392–409.
Thornhill, R. (1976): Sexual selection and nuptial feeding behavior in *Bittacus apicalis* (Insecta: Mecoptera). Amer. Nat. *110,* 529–548.
Thornhill, R. (1980): Competitive charming males and choosy females: was Darwin correct? Florida Entomol. *63,* 5–30.
Thornhill, R. (1981): *Panorpa* (Mecoptera: Panorpidae) scorpionflies: systems for understanding resource-defense polygny and alternative male reproductive efforts. Ann. Rev. Ecol. System. *12,* 355–386.
Thornhill, R. and J. Alcock (1983): The Evolution of Insect Mating Systems. Cambridge, Mass.: Harvard University Press. 547 pp.
Trivers, R. L. (1972): Parental investment and sexual selection. pp. 136–179, in Sexual Selection and the Descent of Man, 1871–1971. B. G. Campbell, ed. Chicago, Aldine Press.
Tozer, W. E., U. H. Resh and J. O. Solem (1981): Bionomics and adult behavior of a lentic caddisfly, *Nectopsyche albida* (Walker). Amer. Midl. Nat. *106,* 133–144.
Velthuis, H. H. W. and J. M. F. de Camargo (1975): Observation on male territories in a carpenter bee, *Xylocopa (Neoxylocopa) hirsuitissima* Maidl (Hymenoptera, Anthophoridae). Zeit. für Tierpsych. *38,* 409–418.
de Vos, G. J. (1983): Social behavior of black grouse: an observational and experimental field study. Ardea *71,* 1–103.
de Vries, P. J. (1980): Observations on the apparent lek behavior in Costa Rican rain forest *Perrkhybis pyrrha* (Pieridae). J. Res. Lepid. *17,* 142–144.
Wiley, R. H. (1974): Evolution of social organization and life history patterns among grouse (Aves: Tetraonidae). Quart. Rev. Biol. *49,* 209–227.
Williams, G. C. (1966): Adaptation and Natural Selection. Princeton, N. J.: Princeton University Press. 207 pp.
Wittenberger, J. F. (1981): Animal Social Behavior. Boston, Duxbury Press.
Zahavi, A. (1975): Mate selection – a selection for a handicap. J. Theor. Biol. *53,* 205–214.

IV. Social Organization

Fortschritte der Zoologie, Bd. 31 · Hölldobler/Lindauer (Hrsg.): Experimental Behavioral Ecology
G. Fischer Verlag · Stuttgart · New York · 1985

From solitary to eusocial: need there be a series of intervening species?

CHARLES D. MICHENER

Departments of Entomology and of Systematics and Ecology, and Snow Entomological Museum, University of Kansas, Lawrence. Kansas 66045, USA

Abstract

Bees of the genus *Ceratina* are ordinarily solitary but certain species have social tendencies. In some the mother regularly breaks down partitions between cells, inspects and cleans them, and then reconstructs the partitions. Such behavior is probably not related to social interactions among adults. Perhaps all species form prereproductive assemblages of the young adults reared in a nest, often with their mother since females of some species may live for three years. Interactions in these assemblages include direct transfer of nectar. Foraging individuals may also deposit food on the nest walls from which it is eaten by others. Probably in such assemblages mutual tolerance and other behaviors evolved that permitted formation of colonies of adults.

Most such colonies contain two bees. Within a population colonies are less common than nests occupied by single females. A colony probably usually consists of sisters or of a mother and daughter. Sometimes the members are mated females eith enlarged ovaries (probably quasiosocial) but sometimes only one bee is mated and has enlarged ovaries (queen) while the other has slender ovaries and an empty spermatheca (worker). Queens are ordinarily larger than workers and do not forage while workers are smaller and do the foraging. If the colony consists of sisters it is semisocial; if it consists of a mother and daughter it is eusocial. Eusociality could arise within a population of basically solitary bees by fixation of eusocial behavior already sometimes expressed; there need not be a series of intervening species.

In halictine bees most species are either solitary or eusocial; once eusocial tendencies arise they probably quickly become fixed. This may be because halictids nest in the ground where nests can be found in a two-dimensional search by predators, while *Ceratina* and allodapines nest in stems and twigs where search must be three-dimensional. The advantage of a guard at the nest entrance, impossible for solitary forms but common for colonies of two or more, would be greater for halictids.

1 Introduction

A traditional method of thinking about the evolution of an attribute, particularly if an ancestral and a derived state are easily recognizable, is to look for organisms intermediate for the attribute, i. e., links among existing forms. Such intermediates are then placed with the extremes in a logical sequence from ancestral to derived, and one postulates that the evolution of the attribute followed that sequence. This comparative method of reconstructing evolution, i. e., looking for species whose attributes constitute links in a chain, has been in use since the time of DARWIN and has led to numerous evolutionary scenarios, including those that most of us use, to one degree or another, in explaining the evolution of eusocial insects from their solitary ancestors. WHEELER (1923) postulated seven steps from solitary insects that scatter their eggs to eusocial insects. Evans (1958) envisioned an ethocline of 13 steps from parasitic pompilid wasps to the most highly eusocial vespids. Similar sequences have been postulated for bees. Thus MICHENER (1958) discussed the evolutionary routes followed by bees from solitary to eusocial, and WILSON (1971, p. 99) explained these routes in some detail, reporting on the «evolutionary steps to eusociality in the bees» as envisioned by MICHENER (1969).

Caution in acceptance of the evolutionary interpretations suggested by such sequences has been voiced. Thus MICHENER (1974a, p. 47) noted that while it is tempting to look at the ethoclines as evolutionary sequences, the primitively eusocial mode of life has probably arisen not only from subsocial and from semisocial ancestral species, as suggested by earlier studies, but may also have arisen in some cases directly from solitary antecedents within a species. [Explanations of some terms, adjusted for use with *Hymenoptera,* are as follows: A *solitary* female mass provisions cells in her nest, closes them, and has no contact with her growing offspring. A *subsocial* group consists of immature offspring (eggs, larvae, etc.) and their mother who cares for them progressively. A *quasisocial* colony consists of adult females working on the same cells without caste differentiation or division of labor; all lay eggs. *Semisocial* and *eusocial* colonies have at least behaviorally and reproductively differentiated castes, workers and queens; those of semisocial colonies are of the same generation, those of eusocial colonies are not (e. g., mother and daughters). For further explanation see MICHENER (1974). The term eosocial was introduced by SAKAGAMI and MAETA (1977) for quasisocial colonies involving two generations. A *colony,* for this paper, is a group of nest-making adult females; subsocial groups and prereproductive assemblages (see below) are not considered colonies.]

The present paper is intended to emphasize the viewpoint that while the scenarios usually cited, involving a series of species intermediate between the solitary and the eusocial conditions, may in some cases be correct, they are sometimes wrong. There is increasing evidence that the primitively eusocial way of life can arise, among bee taxa having appropriate potentials, without intervening stages in the sense of *species* that are semisocial, subsocial, etc. The potential exists for direct development of eusociality involving individuals within a hitherto solitary bee population.

2 *Ceratina*

2.1 Life history; solitary behavior

The genus *Ceratina* is ordinarily considered to consist of solitary bees. Studies of European and North American species (MALYSHEV, 1913; KISLOW, 1976) have shown that each nest is made by a single female. The adults are long lived, and overwintering groups of adults of both sexes are common in the old nest burrows. The bees disperse and mate in the spring. (One of the Japanese species, *C. flavipes,* disperses in the autumn but mating is in the spring.) In spring each female excavates her own new nest (in the two species studied in detail by KISLOW), or in some species such as *C. japonica* old nest burrows are often reused (SAKAGAMI and MAETA, 1977 and in press). Even in such cases, although one female may remain in the old nest, and although she might start provisioning a cell before all of her associates (sisters or daughters) have left, only one working female is usually present in each nest that is being provisioned.

Each nest is a burrow in the pith of a dead stem, subdivided by partitions of pith fragments to form a series of cells, each of which is mass provisioned by the mother and provided with a single egg. When the young bees emerge as adults, they crowd their way toward the nest entrance, past younger siblings, destroying the partitions and often (in some species nearly always) rebuilding them to reform the cells. Ultimately, when all the young have matured, the partitions are cleared away and a group of adults, siblings and sometimes their mother, is found together in the burrow. They remain there through the autumn and winter (or disperse in the case of *C. flavipes*), thus completing the cycle. Such groups in temperate species have been called prehibernation assemblages but a more appropriate term is prereproductive assemblages, considering tropical as well as temperate species. Most species that have been studied in detail have only one generation per year but *C. iwatai* in Japan is bivoltine and some species in the tropics are believed to have several generations annually, as does *C. okinawana* in southern Japan and *C. dallatorreana* in California. With modifications required by the voltanism and climate, the life cycle as outlined above is characteristic of the many species of *Ceratina*, a genus of numerous subgenera and considerable morphological diversity found in all continents.

KISLOW (1976) and SAKAGAMI and LAROCA (1971) both give good reviews of behavioral and life history observations of *Ceratina* species in all parts of the world. While the major studies have been made in temperate areas, *Ceratina* nests in the tropics also are ordinarily provisioned by individual bees. Likewise the related genera *Pithitis* and *Manuelia* are, so far as known, solitary at the time of nest construction and provisioning. In short, there are no species of *Ceratina* or of closely related genera that consistently work as colonies of adults.

Solitary nest construction and provisioning (and egg laying) as outlined above have been verified by my examination of many nests of *Ceratina (Zadontomerus) calcarata* and *C. (Z.) strenua* in the vicinity of Lawrence, Kansas. In no case did there seem to be two or more bees working together. More impressive are the hundreds of nests of the same species examined weekly and frequently X-rayed to determine their internal conditions by KISLOW. She reported no instance of bees working together in making or provisioning cells. The presumably similarly solitary *C. callosa* (probably really *C. chalybea;* DALY, 1983) studied by MALYSHEV is in the subgenus *Euceratina*. Excellent studies of several Japanese species of the subgenera *Ceratina* and *Ceratinidia* by SAKAGAMI and MAETA (1977, 1982, and in press) also show that the majority of nests were built, provisioned, and the eggs laid by lone females.

Ceratina species show a number of features unusual for solitary bees. Two of these deserve emphasis here. First, long life of the adults has already been mentioned. KISLOW as well as SAKAGAMI and MAETA (1977) demonstrate that the mothers may protect immature progeny from natural enemies. Moreover, mothers are often associated for a time with their adult progeny. In the species studied by KISLOW and MALYSHEV such mothers die in the autumn without further reproductive activity at an age of about one and a quarter years. SAKAGAMI and MAETA (1977) found that some females of *C. (Ceratinidia) japonica* and *C. (C.) flavipes* survive and reproduce in a second summer, thus living for more than two years, and a few hibernate the following winter; one in captivity laid a single egg in her third summer when she would have been nearly three years old. KATAYAMA and MAETA (1979) reported that some females of *C. (Ceratina) megastigmata* survive for three years, i. e., function in three brood-rearing seasons; two females reared broods of offspring in the third year (in captivity).

Second, the prereproductive assemblages of young adults of both sexes, sometimes with their mother, are well known. The bees are more or less active, and often interact with one another for several months of late summer and autumn in temperate species. They tolerate their siblings although there is a certain amount of butting and pushing according to KISLOW. There is also food transfer among them. KISLOW reports seeing a female bring a little pollen into the burrow on her scopae and place it on the wall, from which it was eaten by a male bee. SAKAGAMI and MAETA (1977) found such behavior to be commonplace; in late summer and autumn both mothers and older juvenile females foraged and returned with pollen loads. Such pollen was consumed by the nestmates. Small amounts of nectar have also been seen on burrow walls (DALY, 1966). KISLOW reports one incident in which a bee appeared to take food from the mouthparts of another individual. Such behavior (nectar transfer) is of regular occurence in prereproductive assemblages of *C. japonica.*

2.2 Colonies of mature adults

There has been scattered evidence of small colonies of adults in some nests of various unrelated species of *Ceratina*. Under seasonally anomalous laboratory conditions CHANDLER (1975) reported eusocial colonies of *Ceratina calcarata*. I examined many nests of *Ceratina (Calloceratina) laeta* near Kourou, French Guiana, at different times of year. Of these 25 contained an egg or young larva or a partially provisioned cell. Most such nests contained only one adult bee; a few contained one or more additional adult females that were unworn juveniles presumably not yet dispersed from the earlier use of the stem by a prereproductive assemblage. One nest, however, contained a cell being provisioned, a large adult female with sperm cells in the spermatheca and enlarged ovaries, and a smaller female with somewhat worn wings, showing that she was not a juvenile adult as yet not dispersed. She had slender ovaries and no sperm cells in the spermatheca. There are no behavioral data but it was tempting to think of the first as queenlike and the second as workerlike.

I made similar casual observations of occasional nests containing a workerlike individual, or even two such individuals, in *Ceratina (Ceratina) braunsiana* in South Africa and *Ceratina* of the subgenus *Ceratinidia* in Java. Dissections, in the few cases where they were made, showed the smaller bee to have slender ovaries and often to be unfertilized. A single nest of *C. australiensis,* however, was found to contain a cell just provisioned and two adult females, both with enlarged ovaries (MICHENER, 1962). Again there were no behavioral data.

Recent studies of Japanese species (SAKAGAMI and MAETA, 1977, 1982, and in press) have added greatly to the available information on colonies in *Ceratina*. Among over 3000 field nests of *C. (Ceratinidia) flavipes,* nearly all contained solitary mothers. Note that this is the species which disperses in autumn, so that in spring when nests are being established, the females are mostly isolated from one another. In the closely related *C. japonica,* however, the prereproductive assemblages persist through the winter and dispersal sometimes seems to be incomplete in the spring; at least this may well be the explanation of the frequency of colonies in this species. Of 230 new nests 1.3 percent contained more than one female, but among 203 reused nests (built the preceeding year and presumably used for prereproductive assemblages) the equivalent figure is 31.0 percent. For all nests, an estimated 19.2 percent were inhabited by two or more mature females. Of these, 82.6 percent contained two females, the rest, three except for one nest with four. Of 103 reused nests of *C. (Ceratina) okinawana,* 12.6 percent contained two females. A few colonies have also been recorded for *Ceratina iwatai* and *C. megastigmata,* both in the subgenus *Ceratina.*

The same authors were able to produce artificially a few colonies of *C. japonica* by crowding – liberating numerous females in a cage with few potential nests. Colonies were formed more readily with females of different sizes than with females of about the same size. Ordinarily the occupant of a nest defends it against would be joiners. This helps to explain the rarity of colonies among new nests in the field.

The females within colonies of *C. japonica* are diverse in their relationships. For simplicity of expression the following comments are in terms of two-bee colonies. In over 81 percent of the field colonies, both bees matured the preceding summer. The rest contained one such bee along with an old bee, in its second reproductive summer. In about half the colonies, the two bees had similar ovarian development, commonly both enlarged and capable of producing eggs. Such a colony is quasisocial and may consist of bees belonging to the same generation (commonly sisters) or to different generations (?mother and daughter). In the remaining colonies, one of the bees had enlarged ovaries while the other had markedly less enlarged or slender ovaries. This suggests division of labor (see below); if the bees are of the same generation (?sisters) the colony is semisocial, if they are of different generations (?mother and daughter) the colony is eusocial.

Twenty-one field collected, semisocial or eusocial nests each contained at least one inseminated bee with enlarged ovaries and one uninseminated individual with slender ovaries; only two contained an inseminated individual with slender ovaries or an uninseminated one with enlarged ovaries. Thus as in most social insects, egg-layers (queens) are usually mated and those that do not lay or that lay less (workers) are usually not mated.

The correlates of caste are similar to those in common social Hymenoptera. Among females from colonies, there is a positive relationship between body size, ovarian enlargement, and insemination. In most semisocial or eusocial colonies, the queen is larger than the worker. When the bees are reproductively more or less equivalent, as in quasisocial associations, size differences are usually less.

Reproductivity, measured by the number of completed cells, is greater for colonies than for lone bees. Reproductivity per female, however, remains constant. Thus two-bee nests are usually deeper than solitary nests and contain about twice as many cells when finished. This is so whether there is reproductive nonequivalence and apparent division of labor as in eusocial and semisocial associations or reproductive equivalence as in quasisocial colonies.

Sakagami and Maeta were able to make detailed behavioral observations of lone as well as colonial *C. japonica* in observation nests. A lone female on sunny days commonly forages in the mornings and early afternoons, then guards (= rests with posterior end toward nest entrance) and manipulates the pollen in the cell being provisioned until late afternoon or early evening when she lays an egg. In two-female nests with division of labor, the smaller female often behaves much like a lone individual. However, when the smaller bee returns, the larger one often faces outward and buccal transfer of nectar occurs before the forager goes to deeper parts of the nest. During most of the day the larger female spends much time guarding but in late afternoon she frequently eats her nestmate's egg on the food mass, remanipulates it, and lays her own egg. In various nests such successive ovipositions, always with oophagy, were observed seven times. Except in one nest containing two bees of about the same size, prolonged daytime guarding in twelve colonies was always by the large individual. In the same colonies no foraging by large individuals was observed, again with the exception of the colony with little size difference. Eggs were laid either by the large individual only or by both, but more often by the large one. (There are few data on behavior in colonies of similar-sized bees which might have been quasisocial).

SAKAGAMI and MAETA (1982) also produced artificial observation colonies of two other species of *Ceratina* by the same methods used with *C. japonica*. For *C. okinawana*, which occasionally forms colonies in the field, the results are not very different from those obtained with *C. japonica*. *C. flavipes*, however, almost never forms colonies in the field. Of 13 multifemale nests produced in a cage by crowding, 10 were unstable and produced no brood; food transfer was not observed and the bees soon dispersed or died. Interestingly, guarding occured in all cases and was mainly by the larger bee of each pair, as in *C. japonica*. In three nests two bees did remain together for 71 to 88 days; colonies were formed, in two of which cells were constructed and provisioned. In the nest that produced no cells the large bee foraged and also transferred food to the small one. In the other two nests, one of which was semisocial, the other eusocial, passing of individuals in the burrow was less easy than in the case of *C. japonica*, food transfer in one nest was seen from large to small but in the other it was the reverse, foraging was by both bees or only by the large one, guarding was by both or only the small one. In one nest foraging by the small individual and guarding by the large one were established, as in *C. japonica*. Thus interactions were considerably more mixed, and not so suggestive of ordinary eusocial Hymenoptera, as in *C. japonica*. The only exception to this statement concerns egg laying, which in both nests was primarily by the larger individual.

2.3 Subsocial *Ceratina*

Although all *Ceratina* females mass provision their cells, some species are subsocial in the sense of contacting and actively caring for their young. This is in contrast to solitary bees which mass provision, oviposit in, and then abandon their cells or at least have no contact with their developing progeny. Some *Ceratina* species, likewise, are fully solitary and do not care for their larvae except to remain in and presumably guard the nest entrance. For example, SAKAGAMI and MAETA (1977) found that in *C. iwatai* cell partitions, larval feces, etc., are not removed until the beginning of the prereproductive phase. In *C. megastigmata* removal of partitions and feces occurs soon after the youngest larva has finished feeding. In both species a minority of the nests contains colonies of adults. From this I judge that colony formation is not causally related to the continuing care of larvae. Such care, now fully established for some *Ceratina* species, is most unusual among bees; it reinforces our understanding of the tendency of these bees to interact with conspecifics.

Females of various species, during and after nest provisioning, regularly open the cells containing their progeny. KISLOW, with the help of X-ray photographs, found mothers of two *Zadontomerus* species deep in their nests and associated with every stage from egg to pupa. SAKAGAMI and MAETA (1977), using observation nests, were able to observe directly mothers of two species of the subgenus *Ceratinidia* re-entering their cells and going deep in the nests, and report more such activity associated with older immatures than with eggs and young larvae. Obviously a female can work her way down the nest burrow without damaging delicate stages such as eggs and young larvae. Of course such activity involves destruction of partitions between cells. Although partitions are occasionally omitted, they are normally rebuilt as the mother moves up and down the nest. In both *Zadontomerus* and *Caratinidia* she regularly cleans cells by incorporating the fecal pellets of larvae and other debris into the bottom of the burrow or into the partitions between cells. SAKAGAMI and MAETA found that a mother may return repeatedly to a given cell, removing recently voided feces with each visit. KISLOW found that she responds to a dead egg by shortening the cell, filling in around the pollen mass with particles of pith; by completely destroying the affected cell, breaking up the pollen mass and incorporating it into cell partitions to produce one long plug (i. e., thick partition) between cells; or rarely by merely removing a partition to produce a long cell containing an unused food mass and an immature bee.

Inspection, cell cleaning, cell modification, destruction and reconstruction of partitions between cells, and movement of immature stages for short distances within the nest are characteristic of various unrelated species. The subgenera *Zadontomerus* and *Ceratinidia,* in which such behavior is fully documented, are not closely related; fragmentary data suggest similar behavior in species of other subgenera. Thus lack of partitions between older immature stages and empty cells containing larval feces before emergence of young adults have been reported for various species (Sakagami and Laroca, 1971) and presumably are results of the mother's attention to immature stages.

3 Comparative comments

3.1 Xylocopinae

This subfamily is divisible into two tribes, the Xylocopini (large carpenter bees) and the Ceratinini (small carpenter bees). The latter includes the genera already mentioned, *Ceratina, Pithitis,* and *Manuelia.* In addition it includes the allodapine bees, or alternatively these are put in a separate tribe, Allodapini. Allodapine bees are morphologically quite different from other Ceratinini and have many derived characters; they might have arisen from the *Ceratina* group but could not be ancestral to it. Thus there are strong reasons to believe that the solitary or subsocial mode of life of *Ceratina* is ancestral for the genus, derived from the solitary mode of life of most other bees, and not from that of any ancestor that lived in colonies of adults.

Allodapines nest in pithy stems as does *Ceratina* but the burrow is not divided into cells by partitions; in nearly all genera the immature stages live in frequent contact with one another, and are moved about by the adults and fed progressively, not by mass provisioning. Thus a nest containing a mother and her young progeny is obviously subsocial, unlike subsocial *Ceratina* whose partitions and mass provisioning obscure the subsociality. (For reviews see Michener, 1971, 1974a, b). It is now apparent that there is more similarity in behavior between allodapines and *Ceratina* such as *C. japonica* than had been realized. The following is a list of some major features of *Ceratina* and allodapine biology.

(1.) Nests of all *Ceratina* and allodapine species are either always or usually made by lone females.

(2.) In some *Ceratina* species and probably all allodapine species a minority of the nests contains colonies of two or more females. In the allodapine *Allodapula dichroa* only 3 percent of the nests contained colonies of mature adults; at the other extreme, in *Braunsapis facialis* and *B. leptozonia* 34 and 38 percent of the nests contained such colonies (Michener, 1971).

(3.) In *Ceratina* such colonies are usually in old nests and therefore probably consist of bees that were in the same prereproductive assemblage – probably usually sisters or a mother and her daughter(s). The same is probably true of allodapines but prereproductive assemblages are less obvious in the tropics; also colonies probably arise when young females remain in their mothers' nests as workers, without an obvious prereproductive phase.

(4.) Much less commonly colonies are found in new nests. In these cases the bees may have the same average relationship to one another as the bees in the local population. However, they may well be able to recognize former nestmates or their relatives; at least some halictid bees have such abilities (Greenberg, 1979; Smith, 1983). If this is true for Ceratinini, even colonies in new nests may commonly consist of close relatives.

(5.) Adult females in colonies may be similar in reproductive ability. In this case they are probably similar behaviorally and in size, although data supporting this statement are sparse. Such quasisocial colonies are uncommon or temporary in allodapines.

(6.) In other colonies adult females may be quite different in reproductive ability. One may have enlarged ovaries with an oocyte ready to become an egg while the other has slender ovaries without or with few enlarged oocytes. Such bees differ behaviorly and in size. The bee with larger ovaries, the queen, at least in *Ceratina japonica,* is the principal egg-layer, the principal guard, the minimal forager, and is usually inseminated. The bee with smaller ovaries, a worker, lays fewer eggs or none, rarely guards, is the principal forager, and is often uninseminated. The queen is larger than the worker. Bees in such a colony may be of the same generation (semisocial) or of different generations (eusocial). The same statements probably apply to allodapines, whose nests are initially occupied by a mother and her young (subsocial); when young females mature one or a few may remain and become workers while the mother becomes a queen (eusocial); if she dies one of her daughters presumably becomes the new queen with other daughters being workers (semisocial); as workers die off they may be replaced by porgeny of the new queen (again eusocial); etc.

(7.) In ovarian development, behavior and size, there are all intermediates between queens and workers.

(8.) Workers may be reproductively disadvantaged and hence prone to join other bees, but at least some of them in *Ceratina japonica* and probably in allodapines are capable of provisioning and laying by themselves if orphaned. Items 9 to 11 likewise are based on *C. japonica* but probably apply also to allodapines.

(9.) Loss of a worker leads to foraging by the former queen, now alone. She behaves like any other lone individual.

(10.) There is little obvious agonism between colony members. However, queens sometimes eat the eggs laid by workers. The reverse has not been seen.

(11.) If the adults in a colony with division of labor differ noticably in age, the older one is ordinarily the queen.

(12.) Food transfer by regurgitation between colony members occurs frequently.

(13.) In *Ceratina,* although the cells are mass provisioned, adult females of some species open cells containing immature stages, remove feces, and reconstruct partitions between cells. Thus a solitary nest contains a group (the mother and her immature offspring) that can be regarded as subsocial because of the direct and continuing care of the young by their mother. In allodapines there are no cells and the adults feed the larvae progressively, remove feces, etc.

(14.) Such care of growing young is not universal in *Ceratina.* For example in *C. iwatai* the mother does not open, inspect, or clean the cells during the growth of her progeny. Nests of this species inhabited by single females would be solitary, not subsocial, by all standards. Probably no allodapine is solitary by all standards; a lone female caring for her young is subsocial.

(15.) In both *Ceratina* and allodapines in late summer and autumn adult males and females remain together in their nest burrows. In the tropics similar associations are presumably briefer and less obvious. Such prereproductive assemblages probably consist largely of siblings, often with the mother as well. Contact among the individuals and limited food transfer occur.

(16.) Most and perhaps all types of interactions seen among adults in colonies of *Ceratina* can also be seen in prereproductive assemblages. The same is likely to be true for allodapines.

(17.) No storage of significant amounts of food for adult consumption occurs in the nests or in prereproductive assemblages except that in the genus *Allodapula* significant amounts of pollen are sometimes stored.

This brief review shows the similarity in nesting and social behavior between allodapines and *Ceratina* such as *C. japonica.*

The Xylocopini or large carpenter bees share with the *Ceratina* group of Ceratinini many biological features. There are similar prereproductive assemblages. Except for these, some species (e. g., *Xylocopa pubescens)* are solitary. So far as known no species opens cells and inspects or cleans the growing young; thus none is subsocial. But some species have some nests that contain two to several working adult females instead of only one. Food transfer occurs much as in *Ceratina.* Often the relationships are communal, with different females using different branches of the nest, something that is impossible for *Ceratina* because narrow stems do not permit branching nests. In other species, e. g. *X. sulcatipes,* quasisocial and semisocial relationships have been reported and since mothers often survive and are associated with their adult progeny, eusocial associations are possible. Data are sparse and difficult to obtain because the nests are in wood; the best information comes from X-ray examination of nests in boards (Gerling, Hurd and Hefetz, 1983). Few ovarian and spermathecal examinations have been made.

3.2 Apidae

This family, which may have had a common origin with the Xylocopinae (possibly they are sister groups), needs no detailed attention here because early stages in its social evolution are extinct or inadequately studied. The subfamily Euglossinae contains both solitary and slightly social species, but details are largely lacking. The subfamily Bombinae has initially subsocial associations which become primitively eusocial colonies with maturation of the first workers, and the Meliponinae and Apinae consist of the only highly eusocial bees.

3.3 Halictini

The only remaining group of bees in which eusocial behavior has evolved is the Halictidae. This family in not at all related to the xylocopine-apid complex, except that both are bees (superfamily Apoidea). Nearly all of the eusocial halictids are in the tribe Halictini and the following remarks concern that group only. In the Halictini, as in the Xylocopinae, the question arises as to whether eusocial behavior evolved via a series of species exhibiting progressively higher social levels, or alternatively, directly from a solitary or polymorphic species. [There are few subsocial species, for brood cells, once closed, are not opened for addition of food or for cleaning except in a few of the most fully eusocial species of *Lasioglossum* (subgenus *Evylaeus*). Their eusociality is certainly a derived condition in a limited group of *Evylaeus* and has nothing to do with early stages of evolution of halictine eusocial life.]

In the Halictini there are many solitary species as well as many eusocial species. Nests of the latter commonly start with a single female (later to become the queen of a colony) rearing a female-biased or all female brood, most or all of which become workers. There are no regularly quasisocial species and only one that is reportedly semisocial. There may be groups of nest foundresses that are quasisocial or semisocial, as in *Polistes,* but these conditions are temporary and for most species unusual; single foundress colonies are more common. There are also other temporary situations in which the colonies are not eusocial. For example, if the queen dies and is replaced by one of her workers, as happens regularly in *L. (Dialictus) zephyrum,* there is necessarily a stage when the colony is semisocial before the old workers die and are replaced by the porgeny of the new queen, returning the colony to the eusocial condition.

Considering more specifically the American species of the subgenus *Dialictus* of *Lasioglossum,* a large group of morphologically monotonous, closely related species, one finds solitary as well as eusocial species. The solitary ones are similar in mode of life to other solitary burrowing bees. The better known eusocial ones such as *L. (D.) zephyrum* and *imitatum* commonly live in colonies of several females (for a review see Michener, 1974). Such forms shed little light on the actual origin of eusociality unless one considers the ontogeny of colonies relevant. However, species of *Dialictus* exist that regularly live in minimal eusocial groups. The best known examples are *L. (D.) breedi* and *seabrai* (Michener, Breed, and Bell, 1979). As in most species, nests are started by lone females, but so far as known all that survive obligately develop colonies. The average number of adult females per nest is about 1.5; in nests containing colonies of two or more bees, the average number is 3.0 and 2.3 respectively, for the two species. Even in species with such small mature colony sizes (the maximum number of females found in a nest was 4), the organization is usually eusocial; the queens have developed ovaries and are inseminated, the workers have more slender ovaries and usually are uninseminated. No significant size difference between the castes was found. In a few nests there were two bees with more or less equally developed ovaries – perhaps quasisocial colonies that developed after loss of a queen, allowing ovarial enlargement of two workers. Foraging in *L. breedi* is largely by uninseminated individuals, regardless of ovarian size, suggesting that queens tend to remain in their nests as in other eusocial colonies.

4 Discussion

Important contributors to the potential for colony formation, at least in Xylocopinae, must have been the prevalence of prereproductive assemblages and of long lived adults. Prereproductive assemblages occur even in the fully solitary species of *Ceratina* and *Xylocopa,* suggesting that mutual tolerance and defense, food exchange, foraging for nestmates, etc., evolved in connection with such assemblages, before and probably prerequisite to evolution of nesting colonies of adults. Once evolved, the behavior associated with assemblages could also function in reproductive colonies of adults, where longevity also would be important in permitting co-occurence of adults of different ages and generations. Beyond the stage in which mature females of various ages could live together, there seems to have been no necessary sequence of stages leading to eusociality.

The importance of the *Ceratina* data for an understanding of the origin of eusocial behavior is that when they live in colonies, these bees lack uniformity in social relationships. It is as though whatever two females remain together or join one another, whether they are sisters, mother and daughter, or perhaps unrelated bees, interactions result. Within the same species there exists a whole gamut of social categories. The colonies can be quasisocial, semisocial, or eusocial, all at low frequency in the same species of bee. For the first, in which both bees (in a 2-bee colony) are fully reproductive, one only needs to imagine the evolution of the ability of the bees to tolerate one another during the nest making phase, as well as during the prereproductive phase. For semisocial and eusocial arrangements, however, one female is the principal reproductive and the other is less reproductive or even nonreproductive. Within a predominantly solitary or subsocial species one must imagine the evolution of the potential for the reproductively less competent individual to function as a worker and the reproductively more competent individual, as a queen.

The differences between the worker and queen morphs in *Ceratina japonica* and probably in other species of diverse subgenera, and certainly in various allodapines, are remarkably similar to those found in familiar eusocial insects. Large body size, large ovaries, insemination, reduced foraging, receipt of food in the nest, and probably a tendency to destroy eggs laid by a worker, all go with the *Ceratina* queens; small size, small ovaries (hence reduced egg laying activities), frequently lack of insemination, active foraging, and food transfer to the queen in the nest are all attributes of workers. Most of the same features differentiate the castes in colonies of allodapine bees. Among allodapines caste differences seem least in those species that only rarely produce colonies of mature adults. That all these caste differences go hand in hand suggests that they must have evolved through selection to form the *associations* of characteristics. They go hand in hand not only in these bees but in eusocial insects generally: hence the association of features of the castes hase occurred repeatedly, must therefore be important, and begins at least in bees at an extraordinarily early stage in the evolution of sociality.

It is not difficult to envision evolution of such castes as a result of increased inclusive fitness of the worker, control by the queen, mutual advantage, or a combination of these. The high frequency with which uninseminated females of *Ceratina japonica* become workers suggests that reproductively disadvantaged bees can increase their (inclusive) fitness by joining relatives. However, in each of the three nests in which the order of joining of bees to form a colony is known, a large bee (ultimately a queen) joined a small bee (ultimately a worker) that had already started the nest. It may or may not be legitimate to interpret this as a large bee forcing itself like a parasite on a small one and thereafter controlling it.

In spite of the censuses indicating that reproductivity per *Ceratina* is the same whether alone or in colonies, the result might be quite different if all nests started could be followed throughout a nesting season. Many lone females may be destroyed by predators, so that they do not appear in censuses, i. e., guards in colonies may reduce the number of nests destroyed. If this is true, both bees in a colony must have higher fitnesses than they would if they were alone in spite of the fact that the number of cells per bee in completed nests is the same.

Evolution of castes in Xylocopinae has occurred in situations in which cooperative nesting must not be a major advantage; if it were, there would be Xylocopinae with colony behavior fixed. Instead, eusocial life is a minority condition in all Xylocopinae in which it is known. Selection has somehow been effective enough to assemble the potentials for the attributes of each caste without leading to high frequency or fixation of colonial life. Presumably the association of caste attributes, such that a queen, for example, always has a given list of characteristics, is genetically specified. But the particular caste into which a given individual will develop is under environmental control. This is true of most eusocial insects, and evidently of *Ceratina japonica* also, since the functions (hence caste) of an individual can vary according to its associates or lack of them.

The existence of a minority of nests containing colonies, some of them with castes (i. e., semisocial and eusocial), in various species of *Ceratina,* in most allodapines, and perhaps also in *Xylocopa,* suggests that this polymorphism or at least a potential for it arose in a remote common ancestor of the modern species and has persisted, without ever proceeding to fixation. Presumably from such polymorphic ancestral types, forms arose in the Apidae that consistently showed similar social behavior, so that species rather than individual colonies could be regarded, for example, as eusocial.

For the Halictini the virtual absence of subsocial, communal, quasisocial, semisocial, etc., species, especially in American *Dialictus,* supports the view that, as in the *Ceratina* group, there is no need to postulate the existence of a series of species occupying intermediate social levels from solitary to eusocial. Indeed there is evidence that such series do not exist. The species exhibiting, in maximal colony development, the intermediate social levels such as semisociality, mostly belong to different phyletic lines (Eickwort and Sakagami, 1979).

It is more difficult to envision the evolution from solitary to eusocial in Halictini than in the xylocopine-apid complex because halictine nests are usually inhabited either by solitary or by eusocial bees. Selection for eusocial attributes in Halictini must be sufficiently strong to associate the various features of each morph and also to carry the process to fixation so that populations and species quickly become characterized by the existance of the morphs, i. e., by eusociality. Again, of course the morphs are not genetically specified; in the absence of a queen a worker can become a queen and intermediates between the casts are common. But the potential to develop into one morph or the other must have evolved and after the young or juvenile adult period, most individuals can be identified by behavior and dissections as members of one caste or the other. (For details on one species see Brothers and Michener, 1974).

Perhaps nest sites influence the evolution of sociality in *Ceratina* and in Halictini. *Ceratina* nests are normally dispersed because the dead broken stems which they occupy are dispersed. The nest entrances are widely scattered in three-dimensional space and wingless parasites and predators must search in a maze of stems and twigs most of which do not lead to nests. It follows that parasite and predator pressure on *Ceratina* is probably only moderate (although the advantages of guards have been demonstrated) and that the advantages of colonial life may not be enough to lead to fixation of polymorphism and sociality. Presumably such advantages are countered by disadvantages, at least for a worker capable of independent life; perhaps they are sufficient to prevent fixation of the polymorphism, and in some forms like *Ceratina flavipes* may prevent colony formation entirely. Halictini, on the other hand, usually nest in the ground. On the two-dimensional surface, especially if suitable sites are few or if for other reasons the nests are aggregated, parasite and predator pressure is often high. For this reason guards may be more important than in the Ceratinini and probably the tendency is to go all the way toward colony formation as a fixed way of life once the process starts.

Acknowledgments

I am indebted to S. F. Sakagami and Y. Maeta for the use of their paper in press on *Ceratina japonica* and for reading and commenting upon a manuscript version of my section on *Ceratina*. C. J. Kislow provided a copy of her dissertation with excellent photographic prints of figures. For useful comments on the manuscript I am indebted to William T. Wcislo and David W. Roubik. This paper is a product of NSF grant BNS82-00651.

Contribution number 1883 from the Department of Entomology, University of Kansas.

References

Brothers, D. J. and C. D. Michener (1974): Interactions in colonies of primitively social bees III. Ethometry of division of labor in *Lasioglossum zephyrum*. Jour. Comp. Physiol., *70,* 129–168.

Chandler, L. (1975): Eusociality in *Ceratina calcarata* Robt. Proc. Indiana Acad. Sci., *84,* 283–284.

Daly, H. V. (1966): Biological studies of *Ceratina dallatorreana,* an alien bee in California. Ann. Entom. Soc. Amer., *59,* 1138–1154.

– (1983): Taxonomy and ecology of Ceratinini of North Africa and the Iberian Peninsula. Syst. Entom., *8,* 29–62.

Eickwort, G. C. and S. F. Sakagami (1979): A classification of nest architecture of bees in the tribe Augochlorini, with description of a Brazilian nest of *Rhinocorynura inflaticeps*. Biotropica, *11*, 28–37.

Evans, H. E. (1958): The evolution of social life in wasps. Proc. 10th Internat. Congr. Entom. (Montreal), *2*, 449–457.

Gerling, H. E., P. D. Hurd, Jr., and A. Hefetz (1983): Comparative behavioral biology of two middle east species of carpenter bees (*Xylocopa* Latreille). Smithsonian Contr. Zool., no. *369*, i-iii + 1–33 pp.

Greenberg, L. (1979): Genetic component of bee odor in kin recognition. Science, *206*, 1095–1097.

Katayama, E. and Y. Maeta (1979): Brood development and adult activities of a small carpenter bee, *Ceratina megastigmata*. Kontyu, *47*, 139–157.

Kislow, C. J. (1976): The comparative biology of two species of small carpenter bees, *Ceratina strenua* F. Smith and *C. calcarata* Robertson. Ph. D. dissertation, Univ. of Georgia, Athens, Georga, USA, iii + 1–221 pp.

Malyshev, S. (1913): Life and instincts of some *Ceratina*-bees. Horae Soc. Entom. Rossicae, *40(8)*, 1–58, pl. 3 (in Russian, English summary).

Michener, C. D. (1958): The evolution of social behavior in bees. Proc. 10th Internat. Congr. Entom. (Montreal), *2*, 441–447.

– (1962): The genus *Ceratina* in Australia, with notes on its nests. Jour. Kansas Entom. Soc., *35*, 414–421.

– (1969): Comparative social behavior of bees. Ann. Rev. Entom., *14*, 299–342.

– (1971): Biologies of African allodapine bees. Bull. Amer. Mus. Nat. Hist., *145*, 219–302.

– (1974a): The social behavior of the bees. Cambridge, Mass.: Harvard Univ. Press.

– (1974b): Polymorphismus bei allodapinen Bienen, pp. 246–256, *in* G. H. Schmidt, Sozialpolymorphismus bei Insekten, Stuttgart: Wissenschaftliche Verlagsgesellschaft MBH.

Michener, C. D., M. D. Breed, and W. J. Bell (1979): Seasonal cycles, nests, and social behavior of some Colombian halictine bees. Revista Biol. Tropical, *27*, 13–34.

Sakagami, S. F. and S. Laroca (1971): Observations on the bionomics of some neotropical xylocopine bees, with comparative and biofaunistic notes. Jour. Fac. Sci., Hokkaido Univ. (6, Zool.) *18*, 57–127.

Sakagami, S. F. and Y. Maeta (1977): Some presumably presocial habits of Japanese *Ceratina* bees, with notes on various social types in Hymenoptera. Insectes Sociaux, *24*, 319–343.

Sakagami, S. F. and Y. Maeta (1982): Further experiments on the artificial induction of multifemale associations in the principally solitary bee genus *Ceratina*, pp. 171–174, *in* M. D. Breed, C. D. Michener, and H. E. Evans (eds.), The Biology of Social Insects, Proc. 9th Congr., Internat. Union Study Soc. Insects., Boulder, Colo., Westview Press.

Sakagami, S. F. and Y. Maeta (in press): Multifemale nests and rudimentary castes in the basically solitary bee *Ceratina japonica*. Jour. Kansas Entom. Soc., *57*.

Fortschritte der Zoologie, Bd. 31 · Hölldobler/Lindauer (Hrsg.): Experimental Behavioral Ecology
G. Fischer Verlag · Stuttgart · New York · 1985

The principles of caste evolution

Edward O. Wilson

Museum of Comparative Zoology, Harvard University, Cambridge, Massachusetts 02138, USA

Abstract

Caste and division of labor are at the core of colony organization in social insects, because they determine the efficiency of energy utilization and resource allocation. In this review I present some of the generalizations that have emerged during the study of these phenomena, especially those encountered in ants during the past ten years.

Castes in ants and other advanced social insects are either physical, marked by anatomical differences, or temporal, based on programmed changes in behavior as a function of aging. The percentages of these castes existing in each colony are largely the outcome of adaptive demography; that is, the schedules of natality and survivorship appear to have been shaped by natural selection at the colony level. Thus the first adult worker brood of the leafcutter ant *Atta cephalotes* consists of workers that just span the minimum range required to collect fresh vegetation and raise fungi on it. This demographic property maximizes the number of individuals that the founding queen produces and presumably reduces to a minimum the chance that any size-group will go extinct before the appearance of the second brood.

Another basic principle of caste evolution in ants is that all physical subcastes (majors, minors, and intermediates) have been generated during evolution by relatively simple modifications in allometric growth. The 44 genera with polymorphic species (the remaining 219 are entirely monomorphic) can be placed within a straightforward array of evolutionary grades ranging from mild allometry and unimodal size-frequency distributions to complete dimorphism. In all species so far studied, allometry is accompanied by alloethism, the size-dependent change in response thresholds and frequency of performance of certain social behaviors.

The division of labor among these various physical and temporal castes is often tightly regulated by feedback loops, the nature of which has only recently begun to receive systematic attention in experimental studies. In *Pheidole dentata*, for example, the major workers are activated for colony defense by the minor workers when the latter report the presence of fire ants or thief ants (genus *Solenopsis*) by means of odor trails combined with other chemical cues. The magnitude of the majors' reaction, and with it the overall pattern of colony response, changes with the intensity of signals received by both minors and majors during the invasion. On the other hand, other colony-level responses are now known in which feedback and regulation are lacking. Thus the percentage of major workers produced by *P. dentata* colonies is independent of the frequency and intensity of attacks on them by *Solenopsis;* there appears to be no way to adjust defense expenditures to the degree of stress.

Optimization studies, employed to identify the selective agents and to evaluate their effectiveness, are in the earliest stages of development. An example is presented of an experimental study on *Atta,* where the foraging (and leaf-cutting) size group was found to be the one that provides the maximum net energetic yield, as opposed to the maximization of other conceivable performances – or no maximization at all.

As such analyses increase in number, and can be synthesized into a more nearly complete picture of colony functioning, it will become useful again to treat the colony as a superorganism and to build exact models based on the interplay of individual- and colony-level selection.

1 Introduction

It is possible to distinguish two very general problems in the study of social insects, to which all other topics and theoretical reflection play a tributary role. The first is the origin of social behavior itself. Entomologists are especially interested in the advanced state called eusociality, which characterizes all of the ants and termites and a small percentage of aculeate wasps and bees. This evolutionary grade is defined as the combination of three features, namely care of the young by adults, overlap in generations, and a division of labor into reproductive and nonreproductive castes.

The last of the traits, the existence of a subordinate or even completely sterile worker caste, is the rarest of the three. It is also by far the most significant with reference to further evolutionary potential, for when individuals can be turned into specialized working machines, an intricate division of labor can be achieved and a relatively complicated social organization becomes attainable even with a relatively simple repertory of individual behavior. It does not matter, to use a rough metaphor, if the separate pieces of a clock are relatively simple in construction, so long as they can be shaped to serve particular roles and then fitted together into a working whole. You cannot construct a proper clock if every little piece is also required to keep some sort of time on its own. By the same token natural selection cannot readily form a complex colony if every member must complete a full life cycle on its own. But reproductive sterility and the origin of a worker caste imply substantial altruism of the kind most difficult to achieve in evolution, since it is directed not to offspring but to nephews, nieces, brothers, and sisters. There is still a great deal of uncertainty concerning the conditions under which the twelve or so phyletic lines of insects (eleven in the aculeate Hymenoptera, one in the Isoptera) have worked against the main evolutionary currents to attain this kind of altruism. What are the relative contributions of kin selection and parental manipulation? What ecological constraints and nest site preferences are most likely to add to the inclusive fitness of the altruistic workers? These questions are the focus of a great deal of continuing theoretical and experimental work, which has reached a quite sophisticated level during the past decade.

The second major problem in the study of social insects, with which I will be concerned in this essay, can be called *colony design,* or, more fully, the strategic design of colonies. Once a sterile worker caste is in place, that is, once the equivalent of the less-than-independent levers and wheels of the clock can be manufactured, the important consideration becomes the best arrangement of castes and division of labor for the functioning and reproduction of the colony as a whole. In an earlier study (WILSON, 1968) I argued that the colony can be most effectively analyzed if treated as a factory within a fortress. Natural selection operates so as to favor colonies that contribute the largest number of mature colonies in the next generation, so that the number of workers per colony is only of incidental importance. In other words, the key measure is not how big, or strong, or aggressive a colony of a particular genotype can become but how many successful new colonies it generates. Hence the functioning of the workers in gathering energy and converting it into virgin queens and males is vital. This part of colony functioning constitutes the factory. But the colony is simultaneously a tempting target for predators. The brood and food stores are veritable treasure houses of protein, fat, and carbohydrates. As a consequence colonies must have an adequate defense system, which often takes the form of stings and poisonous secretions and even specialized soldier castes. This set of adaptations constitutes the fortress.

The ultimate currency in the colony fitness equations is energy. The workers appear programmed to sweep the nest environs in such a way as to gain the maximum net energetic yield. Their size, diel rhythms, foraging geometry, recruitment techniques, and methods of food retrieval are the qualities most likely to be shaped by natural selection. But even as energy is being collected and distributed to the queen and brood, the colony is subject to predation. A certain number of individuals must be sacrificed in periodic defenses of the colony and during the riskier steps of the daily foraging expeditions. The loss of energy required to replace them is entered on the debit side of the fitness ledger.

Colony design then becomes essentially a problem in economics. I have suggested use of the similar expression «ergonomics» to acknowledge that work and energy are the sole elements of calculation, and nothing like human-like transactions with credits and money is involved (WILSON, 1968; OSTER and WILSON, 1978).

The core of ergonomic organization is division of labor, based upon a system of roles and castes. This is the topic on which I have concentrated my own research during the past ten years and will partly summarize here. I propose to cover the main principles of ergonomics and caste evolution, but because the literature documenting them is now so large the examples presented will be taken principally from my own research. This has the additional advantage of bringing with it a greater degree of familiarity, but in making the restriction it is not my intention to imply that the excellent research being conducted by others is any less important; see for example the reviews by SCHMIDT et al. (1974), OSTER and WILSON (1978).

2 Caste, task, role

The aim of any ergonomic (as well as any human-economic) analysis is the prescription of a complete balance sheet, and to this end it is necessary to define as precisely as possible the agents entailed. A *caste* is a set of colony members, smaller than the total colony population, that specialize on particular tasks for prolonged periods of time. Ordinarily, and perhaps in the social insects invariably, the set is distinguished by some kind of additional marker – a larger size, some other kind of anatomical feature, a different age, or even some less apparent physiological trait. To make the classification somewhat more formal, a physical caste is one distinguished not only by its form of labor but also by one or more distinctive anatomical traits, while a temporal caste is set apart by a combination of particular behavior patterns and age. So far as known, all physical castes within a given colony of ants have the same genotype. The differences among them are based on variation in size or some other incidental quality determined during individual development, and not upon differences in genes. A *task* is a set of behaviors that must be performed to achieve some purpose of the colony, such as repelling an invader or feeding a larva. A set of closely linked tasks can be defined as a *role,* even if the acts are otherwise quite different. It is generally true, for example, that in ants the act (task) of grooming the queen is closely linked with the very different acts of regurgitating to the queen and removing freshly deposited eggs. All of these responses can be considered part of the single role of queen care, and it is conceivable that the role might be filled by one, two, or more castes, depending on the species.

The linkage between castes and roles is close, and in the case of the more complex forms of morphological and behavioral variation, it is sometimes necessary to employ roles to distinguish castes. In the leaf-cutter ant *Atta sexdens,* for example, size variation is continuous, and all of the size groups together perform a total of 22 tasks. The physical castes cannot be separated as discontinuous groups from one another on the basis of any known set of discrete anatomical traits. Instead, four castes have been defined on the basis of «role clusters», which are more or less segregated as distinctively shaped sets of activity curves. They are, respectively: gardener-nurses, within-nest generalists, forager-excavators, and defenders (see Figure 1). This discretization is only partial, and finer, arbitrary subdivisions at the lower end of the size scale are possible.

3 Adaptive demography

A key principle of caste evolution is the adaptive nature of colony demography (Wilson, 1968). Ordinary demography, of the kind found in nonsocial organisms, in social vertebrates, and in more primitively social insects such as subsocial wasps and bees, is a function of the individual parameters of growth, reproduction, and death. A large amount of documentation from free-living and laboratory populations supports the general belief that growth and natality schedules are direct adaptations shaped by natural selection at the level of the individual (May et al., 1981; Krebs and Davies, 1981). On the other hand, the size and age structure of the population as a whole are epiphenomena, in the sense that they reflect the individual-level adaptations but do not constitute adaptations in their own right. Thus a sharply tapered age distribution in many species of fishes and birds results

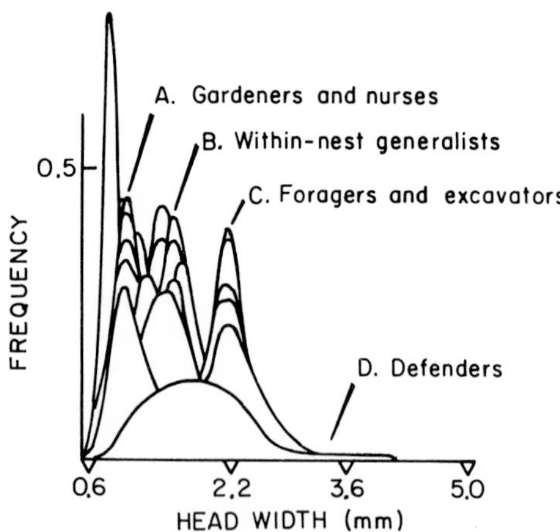

Fig. 1: The four physical castes of the leaf-cutter ant *Atta sexdens,* defined on the basis of size and role clusters. The frequency curves, which have been smoothed by eye, are of workers of different size that perform each of 22 tasks; the frequency curves fall into four groups called role clusters. (From Wilson, 1980a).

from a high birth rate and a high mortality schedule throughout the life span, but in itself does not contribute to the survival of the population or the individual members. The exact reverse is the case of the eusocial insect colony defined as a population. The demography, not its causal parameters, is directly adaptive. The workers are for the most part sterile; their birth and death schedules have meaning only with reference to the survival and reproduction of the queen. Hence the unit of selection is the colony as a whole. The traits of the colony are the larger features of demography, the age and size frequency distributions. What matters are such higher-level traits as the number of very large adults available to serve as soldiers, the number of small young adults functioning as nurses, and so on through the entire caste roster and behavioral repertory. Each species has a characteristic age-size frequency distribution, and the evidence is strong that this colony-wide trait is not an epiphenomenon, but has been shaped by natural selection, constituting a direct adaptation.

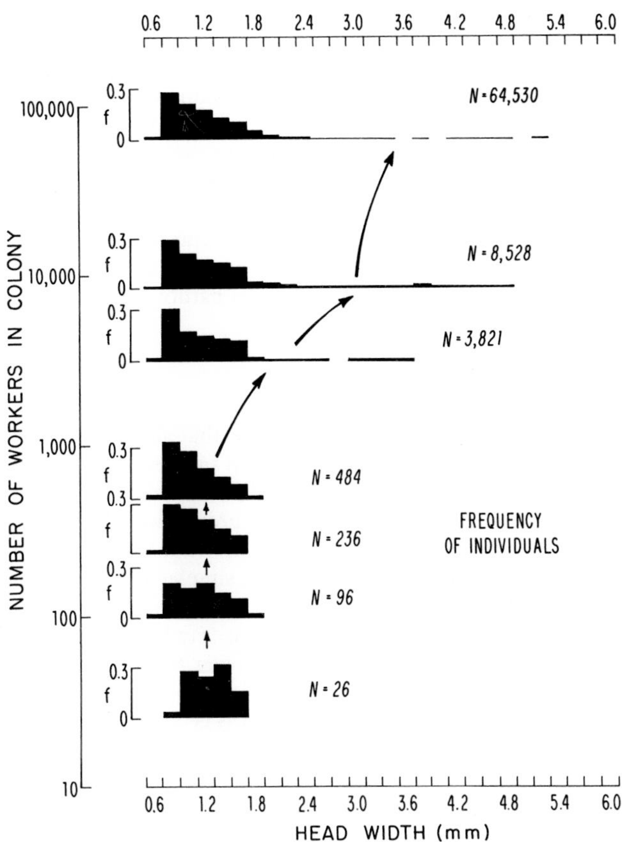

Fig. 2: The ontogeny of the caste system in the leaf-cutter ant *Atta cephalotes* is reflected strikingly in the changes in the size-frequency distribution that occur with colony growth. Shown here are seven representative colonies from the field in Costa Rica and author's laboratory. *N*, the number of workers in each colony, based on complete censuses; *f*, the frequency of individuals according to size class. (From Wilson, 1983b).

A striking example of adaptive demography is provided by developmental changes in the caste system of the leaf-cutter ant *Atta cephalotes* (Wilson, 1983b). Beginning colonies, started by a single queen from her own body reserves, have a nearly uniform size-frequency distribution across a relatively narrow head-width range of 0.8–1.6 mm. The key to the arrangement is that workers in the span 0.8–1.0 mm are required as gardeners of the symbiotic fungus on which the colony depends, while workers with head widths of 1.6 mm are the smallest that can cut vegetation of average toughness. This range also embraces the worker size groups most involved in brood care. Thus, remarkably, the queen produces about the maximum number of individuals which together can perform all of the essential colony tasks. As the colony continues growing, the worker size variation broadens in both directions, to head width 0.7 mm or slightly less at the lower end to over 5.0 mm at the upper end, and the frequency distribution becomes more sharply peaked and strongly skewed to the larger size classes (see Figure 2).

A more physiological question immediately arises from the observed colony ontogeny of *Atta:* which is more important in determining the size-frequency distribution, the size of the colony or its age? In order to learn the answer, I selected four colonies 3–4 years old and with about 10,000 workers and reduced the population of each to 236 workers, giving them a size-frequency distribution characteristic of natural young colonies of the same size collected in Costa Rica. The worker pupae produced at the end of the first brood cycle possessed a size-frequency distribution like that of small, young colonies rather than larger, older ones. Thus colony size is more important than age.

The *Atta* example is just an extreme case of what appears at least on the surface to be programmed colony demography among ant species. However, the physiological control mechanisms remain almost entirely unexplored. The «rejuvenation» effect in *Atta cephalotes* indicates that a feedback loop of some kind is involved, as opposed to an irreversible maturation of the size-frequency distribution; but its nature has not been investigated.

4 Allometry and alloethism

Another basic principle of caste evolution in ants is that all worker subcastes appear to have originated ultimately from allometric growth. In simplest terms, the different forms, designated usually as minors, medias, and majors, differ from one another by size, and the size variation is accompanied by differences in body proportions. Thus majors, the largest workers, typically possess heads that are larger relative to the rest of the body than is the case in the minors and medias. In ant species with simpler forms of differentiation the relation can be approximated by the power function $y = bx^a$, where x is one dimension (say width of the pronotum), y is another dimension (say head width), and a and b are fitted constants characteristic of the species. The power function is in fact the defining property of allometry in the original, strict sense. It is of considerable importance that the pattern of allometry that characterizes a given colony is genetically prescribed, but the size attained by individual workers is not.

Of the 263 living genera of ants known worldwide, only 44 possess polymorphic species, in which the worker caste is differentiated into easily recognized subcastes. By standard techniques used in phylogenetic systematics, it has been possible to infer with some confidence which are primitive caste systems and which are derived. The following sequence has occurred repeatedly and independently in many separate evolving lines of ants, as deduced from the comparison of related species showing different stages of advancement (Wilson, 1953). First, the size variation of the workers within colonies broadens and is accompanied by simple allometry, usually including a disproportionate widening of the head, a decrease in the total behavioral repertory, and some degree of specialization – in most cases for defense. Next, the size-frequency distribution is skewed to the larger size classes; then it becomes bimodal, with peaks corresponding to what are now easily recognizable minor and major castes. In some but not all of the species occupying this evolutionary grade, the allometry has become diphasic or triphasic; that is, the allometric curve breaks at one or two points, so that the slope (a in the equation $y = bx^a$) changes value once or twice. This advance either speeds up or slows down the shifts in proportion, and it complicates the overall pattern of subcaste differentiation. Finally, in some of the most advanced species, in genera such as *Acanthomyrmex*, *Oligomyrmex*, *Pheidole*, and *Zatapinoma*, the medias drop out, the size-frequency curve is divided into two discrete curves, and the two allometric curves are separated and come to possess different slopes. Paradoxically, this last step simplifies the caste system somewhat by changing its base from continuous size variation and allometry over a broad segment of the size range to a straightforward two-state polymorphism.

The paramount quality of allometric evolution in ants is that all evolving lines (so far as can be reasonably inferred) started with a single allometric curve and stayed with it until the final stage of complete dimorphism, when at last the curve was broken and the two segments evolved in an outwardly independent manner. No phylogenetic group in ants has

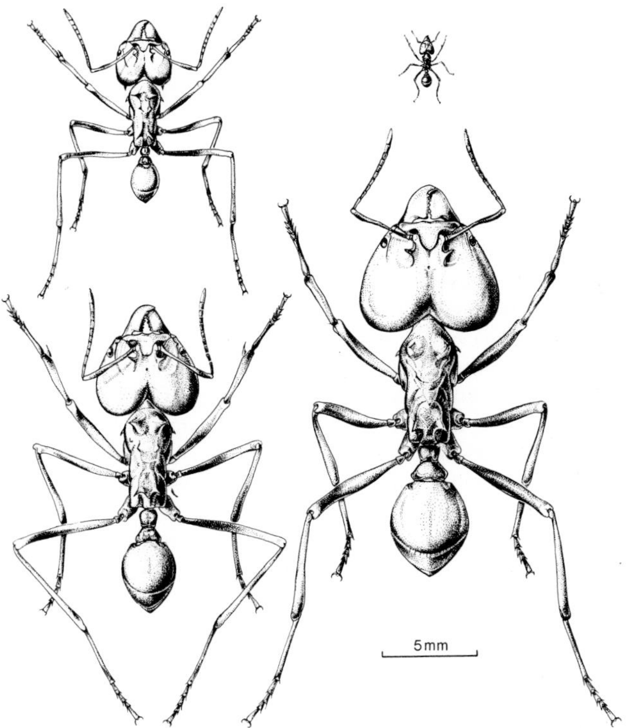

Fig. 3: The caste system of the genus *Atta*, here exemplified by *A. laevigata*, is among the most extreme found in ants, yet it is still based upon single allometry curves combined with size variation. (From Oster and Wilson, 1978).

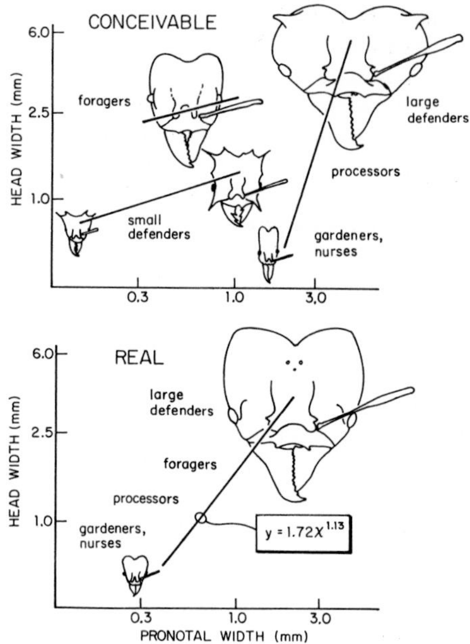

Fig. 4: A conceivable array of *Atta* castes is compared with the actual array in *A. sexdens,* in order to illustrate the principle that the evolution of ant castes is based on a single allometric curve and hence constrained in a fundamental manner. (After Wilson, 1980a).

relied on multiple allometric lines. If it had, this property alone would have permitted the rapid evolution and fine-scale adjustment of a much larger number of specialized worker castes from the outset. In Figures 3 and 4 I have illustrated this principle with a depiction of what actually exists in the highly polymorphic *Atta sexdens,* in contrast to an imaginary example of what *might* have evolved with the addition of only one more allometric curve.

Closely coordinated with the allometry of the physical caste systems is *alloethism,* the regular change in behavior patterns as a function of size. In both species on which measurements have been taken, the fire ant *Solenopsis geminata* and leaf-cutter *Atta sexdens,* the curves were found to possess steeper slopes than the corresponding allometric curves of the anatomical structures employed in the behavior (Wilson, 1978, 1980a). Part of the pattern in the fire ant species is illustrated in Figure 5.

An important feature of both allometry and alloethism in ants is the clearly adaptive function in all cases thus far carefully studied. The extreme, sometimes bizarre anatomical productions are not incidental to some general process of differential growth, nor are the remarkable shifts in behavior with increases in body size neutral by-products of larger or smaller brains. On the contrary, each species has a distinct morphological and behavioral program in its caste differentiation which, in all cases so far analyzed, has proved closely related to its principal life habits.

The correlation between the caste system and colony natural history is strikingly illustrated in the fire ants. The more primitive pattern of *Solenopsis invicta* is typical of almost all of the fire ant species, which comprise the subgenus *Solenopsis* of *Solenopsis* (as opposed to the thief ants, which comprise the subgenus *Diplorhoptrum*). The allometry of

S. invicta is very slight: the majors have heads that are only somewhat shorter, broader, and more quadrate with reference to the remainder of the body (see Figure 6). The behavioral differences are also modest, since the majors perform the same tasks as the minors with the exception of caring for the eggs and young larvae. Workers in these larger size classes simply handle pieces of brood, prey, and soil particles that are correspondingly larger in size. Hence transport and nest excavation appear to be enhanced by the polymorphism, although that subjective impression has not yet been put to a quantitative test. The more important point is that *Solenopsis geminata* has added something entirely new. Allometry is steep, so that the larger medias and majors are grotesque creatures possessing massive heads and blunt, toothless mandibles. These two features together contribute to a substantial increase in the crushing power of the ants. I was able to show that the largest workers are in fact specialized for milling seeds, the majors so much so that they have become restricted to the smallest behavioral repertory thus far recorded in the social insects – milling and self-grooming. *Solenopsis geminata* has undergone this striking revision in its caste system and division of labor as part of its trend toward granivory, which is the most extensive known among the fire ant species.

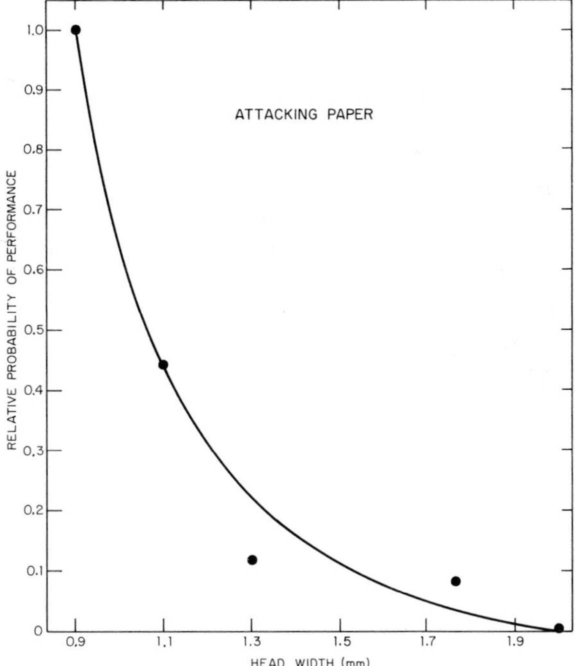

Fig. 5: Part of the alloethic curve for aggressive response by workers of the fire ant *Solenopsis geminata* to a disturbance of the nest. «Relative probability» is defined as the fraction of workers responding in a size group relative to the highest fraction responding in any size group (in this case the group head width 0.9 mm). (From Wilson, 1978).

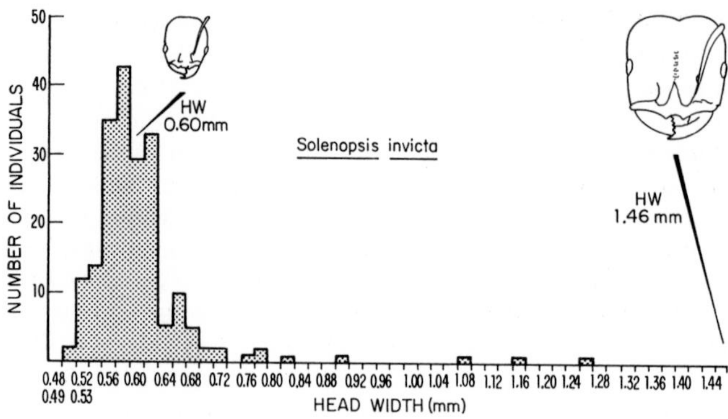

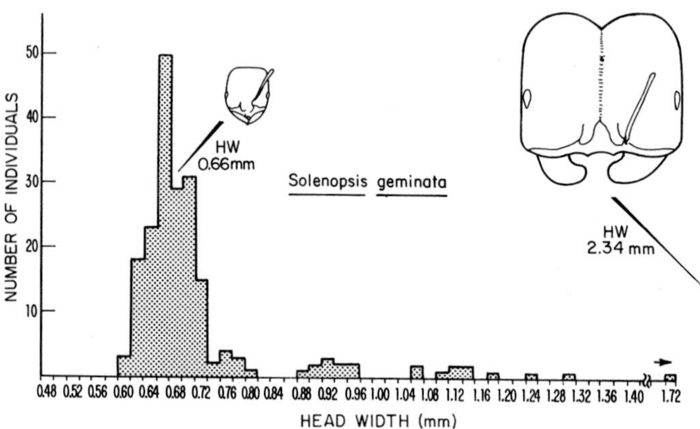

Fig. 6: The worker caste systems of two fire ant species, the red imported fire ant *Solenopsis invicta* and the native granivorous form *S. geminata*. The outlines shown are the heads of workers at the two extremes of size variation; that is, a small minor worker contrasted with a large major worker. Also presented are the size-frequency distributions of random samples of workers from large laboratory colonies of the two species. (From Wilson, 1978).

5 The superorganism

From about 1900 to 1950, many biologists and philosophers developed a keen interest in holism and emergent evolution. They often turned to colonies of social insects as prime examples of superorganisms, by which is meant aggregates of individuals so well organized that they take on the attributes of true organisms. These early expositions were almost entirely limited to recognizing various sorts of analogies. For example, the queen and spermatheca were compared with the reproductive organs and the workers with the variously differentiated soma, trophallaxis was cited as the equivalent of circulation, soldiers were thought the parallel of the immune system, and so on. While initially stimulating, the entire elaborate exercise eventually exhausted its possibilities. The limitations of a primarily analogical approach became increasingly apparent when investigators turned more to the fine details of communication and caste, and the expression «superorganism» was all but dropped from the technical literature by the 1960s.

The time may be at hand for a revival of the superorganism concept. There are two reasons, both stemming from the increased information and technical competence acquired by entomologists during the analytic phase of the 1960s and 1970s. The first is the beginning of a sound developmental biology of the insect colony, and the second is the rapid improvement of optimization analysis in behavioral ecology and sociobiology.

The new developmental biology entails an understanding of the way in which castes are determined, their ratios regulated, and their actions coordinated through communication. There is a prospect now of drawing analogies at a deeper level, consisting in the precise comparison of these organizational processes with their equivalents in the growth and differentiation of tissues in organisms. I do not believe it too much to suggest that insect sociobiology will in fact contribute to a future general theory of biological organization based upon quantitatively defined principles of feedback, quality control, amplification and cascading effects, and optional spatial arrangements. As WILLIAM M. WHEELER pointed out in 1927, the ant colony offers a notable advantage in studies of a physiological nature over conventional organisms. Its parts – the workers and other colony members – can be temporarily separated from one another and reassembled without severe damage to the remainder of the colony. It is as though a conventional physiologist could remove a hand, dissect its parts, record its physiology and behavior while separated from the arm, and then reconstruct and reattach it with only a modest amount of disturbance. Furthermore, as I will report shortly, this procedure offers the great additional advantage of being repeatable with the very same colony, so that other potentially complicating factors, such as genetic structure, age, and colony size, are kept tightly controlled.

6 Feedback and spatial arrangement

As an example of a complex and precise form of feedback, consider the phenomenon of enemy specification in *Pheidole dentata*. This colony-level defensive reaction has certain behavioral parallels with the immune response of the vertebrate body that make it all the more interesting. *P. dentata,* like most other members of its genus, is completely dimorphic. The minor workers possess body proportions that are ordinary for myrmicine ants, and they participate in all of the 22 tasks recorded for entire *P. dentata* colonies, including defense (WILSON, 1976a, b). The majors have exceptionally large heads, filled mostly with massive adductor muscles that pull the sharp mandibles together like miniature wire clippers. Their overall repertory is much more limited than that of the minors, consisting in the performance of only five tasks.

When scouts of most ant species approach close to a *P. dentata* nest, they are typically met and dispatched by the minor workers. But if the territorial invaders are fire ants or thief ants, in other words, members of the genus *Solenopsis,* they evoke an entirely different response. Minor workers first touch the intruders, often grappling with them briefly, and then run back and forth between the site of contact and the nest, laying an odor trail from the poison gland of the sting. The combination of the trail pheromone and *Solenopsis* odor adhering to the bodies of the minor workers galvanizes the majors into action. They follow the trail out to the intruder and meet them in combat, sometimes succumbing to the bites and the potent *Solenopsis* venom but just as often eliminating their opponents by cutting them to pieces. Often only a single *Solenopsis* worker is sufficient to draw out ten or more majors, which then continue to patrol the area for an hour or longer. This diligence helps to «blind» nearby *Solenopsis* colonies to the presence of the *Pheidole* colonies. The maneuver is a distinct advantage to a *Pheidole* colony, since the workers of a *Solenopsis* colony outnumber its own workers by as much as a hundred to one and move in quickly to destroy and consume the *Pheidole* colony when alerted to its presence by successful scouts.

The entire alarm-defense response is based upon feedback loops and differential response to varying levels of severity in the challenge from the *Solenopsis*. Laboratory studies with fire ants have shown that in the earliest stages the number of majors recruited increases with the number of intruders encountered, and hence depends on the amount of trail pheromone detected and the number of minor workers reporting back that are laden with the distinctive *Solenopsis* scent. When the fire ants invade in larger numbers the strategy changes. Now fewer trails are laid, and the *Pheidole* majors fight closer to the nest along a shorter perimeter. When the invasion becomes still more intense, a truly radical shift in tactics occurs: the minors abscond with their brood and scatter out of the nest in all directions, leaving the fighting majors to their fate. Later, if the fire ants vacate the scene, the survivors may return to the site, rear a new crop of majors, and carry on as before.

Similar feedback occurs in labor allocation in ant colonies, and the nature of the loop can sometimes be detected by experimentally removing castes that are specialized for a particular task. In *Atta* colonies most leaf cutting is performed by media workers with head widths ranging from 1.8 to 2.4 mm (with a mode within 2.0–2.2mm), although workers as small as 1.6 mm or as large as 3.6 mm also participate. In one experiment on four colonies of *A. cephalotes,* I removed more than 90 percent of the workers in the size group from 1.8 through 2.2 mm. Contrary to expectation, the colonies did not respond by adding workers from the adjacent size classes to the leaf-cutter force; in other words, there was no increase in workers smaller than 1.8 mm or larger than 2.2 mm. Yet the rate of leaf harvesting remained unaffected, due to the fact that excess workers in the adjacent size classes were already present on a stand-by basis in the foraging area. Moreover, the few survivors in the prime group increased their individual activity by approximately five times. In unmodified colonies, the prime foragers tend to displace others from the edges of the leaves, where most cutting takes place. When these individuals are removed, the auxiliaries participate more freely (WILSON, 1983a).

I have found it useful to make a distinction between two kinds of flexibility on the part of social insects, and according to level of organization. Both responses contribute to social homeostasis. A colony as a whole shows a certain degree of *resiliency* in each task. Thus, even though most of the prime foragers of *Atta cephalotes* were removed in the experiment just cited, the colony continued to perform as well as ever – although at approximately one-third less energetic efficiency. The colony was therefore almost totally resilient in mechanical performance, although with a substantial loss in net energetic yield.

Colony-level resiliency depends on the degree of individual behavioral *elasticity*, which is defined as the amount by which the repertory of individual workers is modified when the colony is placed under stress. In the case just cited, the surviving *Atta* workers proved elastic enough in behavior, increasing their level of activity by as much as five times, to confer nearly complete resiliency to the depleted foraging force.

A far more striking example of elasticity is found in the Neotropical ant *Pheidole guilelmimuelleri*. The major workers of this species are proportionately very large compared to the minor workers, even by the standards of the highly dimorphic *Pheidole*. Their heads are massive, thickwalled, and extremely modified for cutting and crushing. In an undisturbed colony the majors display very low levels of activity and a restricted repertory, confined mostly to self-grooming, occasional bouts of seed milling, and participation in colony defense. However, when all of the minor workers were experimentally removed, the majors showed a quite surprising degree of behavioral expansion. They increased their repertory by 4.5 times, from 4 to 18 kinds of acts, now performing most of the tasks limited previously to the minor workers, while their overall rate of activity, in tasks per individual per unit time, grew almost 20-fold. Another Neotropical species, *P. pubiventris*, displayed a similar but slightly less marked shift under the same experimental conditions (WILSON, 1984).

Students of social insects still know very little about such feedback mechanisms and the magnitude of homeostasis in most categories of behavior and adaptive demography. They have also just begun to examine the geometry of labor allocation within nests and its relation to caste differentiation. A central aspect of this category of organization is the pattern of change in the behavior of individual workers – in other words the differentiation of temporal castes – and its relation to the location of the workers in and around the nests. In theory, two extreme possibilities in the evolution of temporal castes exist. First, workers can undergo changes in responsiveness to various kinds of stimuli in a strongly discordant manner as they grow older, so that each task is addressed by a distinctly

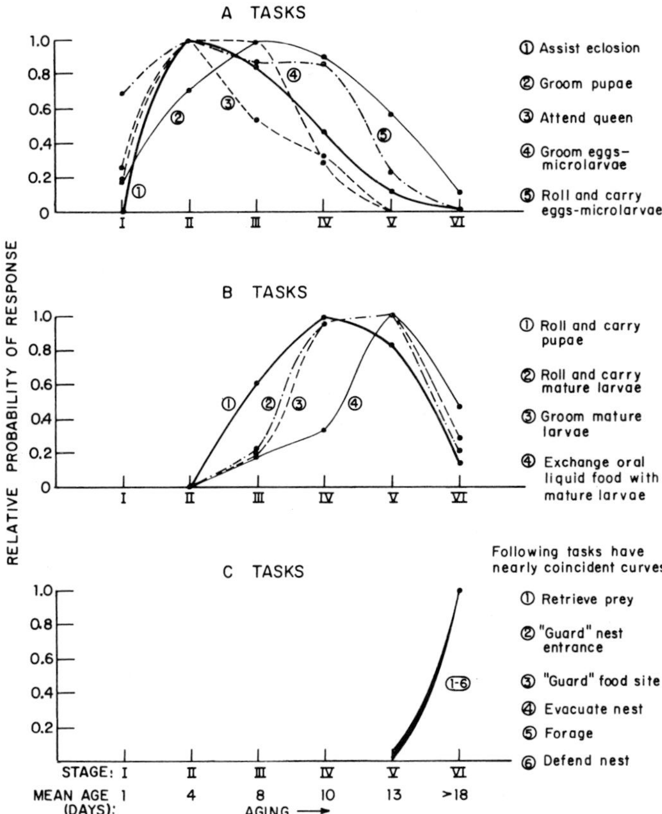

Fig. 7: The discrete partitioning of clusters of tasks by workers of *Pheidole dentata* according to age. The relative probabilities of response (definition: see caption of Figure 5) are given for 15 tasks; stages I–VI are different ages estimated from the degree of exoskeleton darkening and calibrated with individuals marked at the time of eclosion. (From WILSON, 1976b).

different frequency distribution of workers belonging to different age groups. Because these age-frequency distributions change almost gradually from one task to another in covering many such tasks, the resulting temporal caste system is referred to as «continuous.» At the opposite extreme, the aging worker can undergo changes in responsiveness to different stimuli in a highly concordant manner, so that all of the tasks are attended by one or relatively few frequency distributions of workers belonging to different age groups. The resulting temporal caste system has been referred to as «discrete», and the evolutionary process leading to it has been called behavioral discretization (WILSON, 1976b).

In one of the first species studied with this distinction in mind, *Pheidole dentata,* the caste system proved to be closer to the discrete pattern, although not extreme in form (see Figure 7). A close examination of the data displayed in the figure shows that the particular manner in which *P. dentata* colonies are discretized results in an increase in their spatial efficiency. For example, it is energetically more efficient for a particular ant grooming a larva to regurgitate to it as well, or for the worker standing «guard» at the nest entrance to join in excavation when the entrance is buried by a cave-in. The other juxtapositions documented in Figure 6 make equal sense when the spatial arrangement of the colony as a whole is considered. The queen, eggs, first instar larvae (microlarvae), and pupae are typically clustered together and apart from the older larvae, although the positions are constantly being shifted and pupae in particular are often segregated for varying periods of time well away from other immature stages. Thus the *A*-ensemble of workers can efficiently care for all of these groups, moving from egg to pupa to queen with a minimum of travel. The mean free path of a patrolling worker, to put the matter another way, is minimized by such versatility; it utilizes the least amount of energy in travelling from one contingency to another. It makes equal sense for *A* workers to assist the eclosion of adults from the pupae, since the latter developmental stage is already under their care.

7 Caste optimization

The simple fact of the occurrence of evolution is easy to document at the level of both single genes and complex traits controlled by genes at multiple loci. It is also relatively easy to demonstrate the adaptive nature of the evolving traits in most cases, by careful analyses of the function of the phenotypic traits and their effect on survival and reproduction. But it is an entirely different matter to judge such cases of evolution with reference to optimization. In practical terms, we are required to ask how well a species has performed in the course of its evolution, and whether there exists an attainable adaptive peak toward which the species is still moving. Such questions, despite their abstract flavor, are not merely philosophical exercises but truly scientific in nature, that is, soluble by means that can be objectively repeated and verified. The answer to them, if well formulated, can help to predict such important phenomena as the rate and direction of evolution, the impact of a newly imported species on its host environment, and the outcome of competitive interactions between pairs of species.

Under most circumstances, and in the biology of most species, the measure of optimization is obviously going to be technically forbidding. In biology as in engineering, optimization implies an optimum, a goal with reference to which the system may or may not have been ideally designed. Only when the nature of the goal is precisely identified may we presume to use such familiar terms as «suboptimum», «transitional,» and «maladaptive» with any degree of confidence. The complete analysis of optimization, or strategic analysis as it might equally well be called, requires the following four components (OSTER and WILSON, 1978): (1) a state space, designating the conceivable parameters such as size and behavioral response and their relevance to the goal; (2) a set of conceivable strategies, such as foraging procedures, caste structure, division of labor, and pattern of colony development; (3) the goal itself, consisting in one or more optimization criteria, or fitness functions, including net energetic yield during foraging; (4) a set of constraints, defined by those states and strategies beyond the reach of the species and hence outside the scope of analysis (we know, for example, not to expect ants that weigh ten kilograms or have metal jaws).

In most studies of behavioral ecology and sociobiology, optimization models are very difficult to design in a way that satisfies the formal and rigorous demands of the four criteria. But this is not an unusual circumstance for biological disciplines in general. It is equally impracticable to test Mendelian genetics with redwoods or to advance the cellular biology of learning with butterflies. The study of animal behavior seeks paradigmatic species that are easily managed in the field and laboratory, as well as phenomena that are relatively unambiguous with reference to the question asked. The biologist should always keep in mind August Krogh's rule of biological research: for every problem there exists an organism ideally suited to its solution. Moreover, an often unappreciated bonus awaits the successful completion of an optimization study. By appropriate experimental design it is possible to learn which fitness criteria actually hold in nature, in other words, which selection pressures have been most active in shaping the trait under consideration. Indeed, the data may indicate that *no* feasible criterion is approached closely, so that the trait can be identified with some confidence as maladaptive or at least substantially suboptimal. So much for the baseless charge sometimes heard that natural selection theory is «panglossian,» circular in logic, and not susceptible to rigorous testing.

Ant castes are exceptionally well suited for optimization studies in sociobiology. The reason is that individual worker ants are full organisms with ordinary, whole patterns of social behavior, yet they are also clearly specialized for particular well-defined tasks. This restriction is extreme in certain highly specialized castes and in particular the major workers of the dimorphic species, which may perform only one or several tasks during their entire lifetime and which possess bodies and behavioral repertories clearly modified to that end. Thus well-defined anatomical structures and behavioral acts can be more readily assayed with reference to the four elements of optimization models; we really are able to specify a restricted and relatively easily defined state space, a set of strategies, testable fitness functions, and measurable constraints.

Consider the case of the prime foraging caste of the leaf-cutter ants of the genus *Atta*, which I recently studied with reference to ergonomics (Wilson, 1980b). As noted, these insects, which possess head widths from 1.8 to 2.4 mm (and a mode of 2.0–2.2 mm), comprise less than ten percent of the work force and do little else except forage and cut vegetation. Here, then, was an opportunity to learn whether the *Atta* colony as a whole, and the leaf-cutting size class in particular, is optimized with reference to foraging. To do so, I began by conceiving three alternative a priori criteria of evolutionary optimization that are consistent with my own understanding of *Atta* biology. These are the reduction of predation by means of evasion and defense during foraging, the minimization of foraging time through skill and running velocity during foraging, and energetic efficiency, which must be evaluated on the basis of both the energetic construction costs of new workers and the energetic cost of maintenance of the existing worker force.

In order to measure the performance of various size groups within the *A. sexdens* worker caste in isolation, I devised a «pseudomutant» technique. At the start of each experiment, groups of foraging workers were thinned out until only individuals of one size class were left outside the nest. Measurements were then made of the rate of attraction, initiative in cutting, and performance of each size group at head-width intervals of 0.4

mm. Other needed measurements were made in body weight, oxygen consumption, and running velocity for each of the size classes.

The pseudomutant technique has the great advantage of permitting the same colony to be used over and over again, with the experimenter modifying it in a precise manner as though it were a mutant in a certain trait, but with little or no alteration of the remainder of its traits. In other words, there is no genetic noise or pleiotropism to confuse the analysis. The procedure used in the *Atta* experiments is roughly comparable to an imaginary study of the efficiency of various conceivable forms of the human hand in the employment of a tool. In the morning we painlessly pull off a finger and measure performance of the four-fingered hand, then restore the missing finger at the end of the day; the next morning we cut off the terminal digits (again painlessly) and measure the performance of a stubby-fingered hand, restoring the entire fingers after the experiment; and so on through a wide range of variations and combinations of hand form. Finally, we are able to decide whether the natural hand form is near the optimum.

The data from the pseudomutant scanning revealed that the size-frequency distribution of the leaf-cutter caste in *Atta sexdens* conforms closely to the optimum predicted by the energetic efficiency criterion for harder varieties of vegetation, such as thick, coriaceous leaves. The distribution is optimum with reference to both construction and maintenance costs (see Figure 8). It does not conform to other criteria conceived prior to the start of the experiments.

I next constructed a model in which the attraction of the ants to vegetation and their initiative in cutting were allowed to «evolve» genetically to uniform maximum levels. The theoretical maximum efficiency levels obtained by this means were found to reside in the head-width 2.6–2.8 mm size class, or 8 percent from the actual maximally efficient class. In the activity of leaf cutting, *A. sexdens* can therefore be said not only to be at an adaptive optimum but also, within at most a relatively narrow margin of error, to have been optimized in the course of evolution. In other words, there appears to be no nearby adaptive peak that is both higher and attainable by gradual macroevolution.

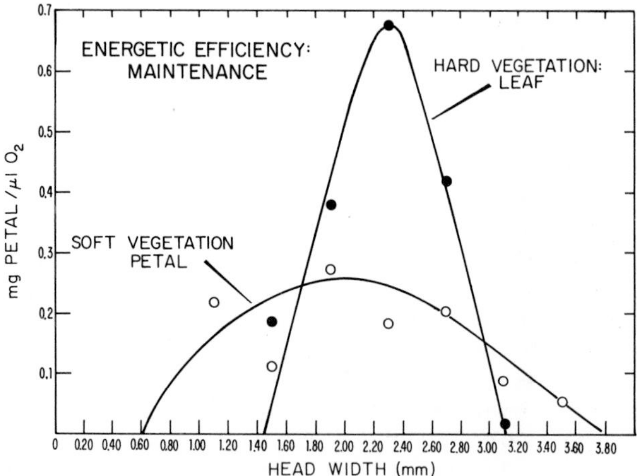

Fig. 8: The principal foraging caste of leaf-cutter ants *(Atta sexdens)*, comprising workers in the size range 1.8–2.4 mm, is also the most efficient in net energetic yield during foraging. The cost criterion used in the measurements depicted here is oxygen consumption, reflecting the energetic cost to maintain workers of various sizes within the colony. (From Wilson, 1980b).

Not all such studies reveal optional responses or even adaptive mechanisms of the sort intuitively expected. For example, what would be the physiological response of an *Atta* colony, as opposed to the purely behavioral reaction described earlier, if a large part of its foraging specialists were suddenly removed by some catastrophe outside the nest? One prudent response on the part of a colony would be to manufacture a higher proportion of workers in the 1.8–2.2 mm size class during the next brood cycle to make up the deficit. However, when I removed over 90 percent of this group in experimental colonies, no differential increase in the production of 1.8–2.2 mm workers could be detected in comparison with sham-treated control colonies. As a consequence, this group remained underrepresented in the foraging arenas by about 50 percent at the end of the first brood cycle, a period of eight weeks.

Another striking example of a feedback loop that did not evolve is stress-induced change in the production of major workers in the ant *Pheidole dentata*. What happens when a colony is attacked repeatedly over a long period by fire ants, the enemy to which its alarm-defense system is specially tuned? An obvious adaptation would be to raise the production of majors (increase defense expenditures, to put it in more familiar human terms) until the stress is relieved. But this response could not be induced by Ardis Johnston and myself in laboratory colonies. We stressed *P. dentata* colonies heavily through four brood cycles by regularly forcing them to fight and destroy *Solenopsis invicta* workers. Contrary to our own expectation, the proportion of majors did not change significantly from colonies stressed with another species of ant *(Tetramorium caespitum)* or from the proportion that prevailed prior to the experiments, when the colonies were free of any pressures through many brood cycles (Johnston and Wilson, 1984).

8 Future directions in caste research

The examples I have cited from my own research, which are paralleled in some instances by the work of other investigators, suggest the richness of caste and division of labor among the 10,000 or more living species of ants. But they also reveal how little we have learned of the underlying control mechanisms. It seems probable that as research proceeds, feedback loops of various kinds, including inhibitory pheromones, facilitation, competitive sequestering of nutrients, competitive displacement of one caste by another at particular tasks, and yet others, will sometimes be discovered in circumstances where they are least expected, and just as often found lacking where they were intuitively most expected. The pattern that appears might first resemble an irrational patchwork of mechanisms, as indeed it seems thus far in *Atta*. Yet there is no reason to suppose that meaningful trends will fail to emerge as studies become more detailed.

Lumsden (1982) has recently reexamined the superorganism concept in a more formal and systematic manner that can serve as a useful framework for additional empirical research. He points out that three key features in insect societies need to be linked in models in order to make quantitative predictions about the optimization of caste systems and division of labor. They are the form of the caste system, including the often large differences in energetic costs of different size groups; the quality of interaction among members of the various size groups; and the remarkably short-term, stochastic nature of the colony's memory in terms of past nutritional states, pheromone levels, and the like. Of special immediate interest is the second feature, the modes of interaction. Each species must possess an ergonomic interaction matrix, such that the mutual effects of individuals belonging to different size groups can be cooperative, interfering, or both. The matrix will have measurable effects on the net energetic yield due to foraging and processing of food, as well as defense.

In LUMSDEN's conception, natural selection will tend to shape the ergonomic interaction matrix in a way that maximizes the net energetic yield of the colony as a whole. The ultimate superorganismic traits are the matrix and the yield. The proximate mechanisms are the various modes of communication and feedback which, as I have stressed in this summarization of principles, are just beginning to come under close scrutiny.

It is encouraging to look to the more distant future, when empirical studies might be joined closely to theoretical modeling of the kind advanced by OSTER and WILSON (1978) and LUMSDEN (1982). We might then find it possible to marry the emerging generalizations of caste evolution with those of developmental biology, creating a deeper and more heuristic analogy between organism and superorganism and bringing the study of social insects into still closer alignment with the remainder of biology.

Acknowledgment

The author's personal research reported in this review has been supported by a series of grants from the National Science Foundation, the latest of which is NSF BSR-8119350.

References

Johnston, A., and E. O. Wilson (1984): Correlates of variation in the major/minor ratio of the ant *Pheidole dentata* (Hymenoptera: Formicidae). Ann. Ent. Soc. Amer. in press.

Krebs, J. R., and N. B. Davies (1981): *An introduction to behavioural ecology.* Sinauer Associates, Sunderland, Mass. x + 292 pp.

Lumsden, C. J. (1982): The social regulation of physical caste: The superorganism revived. J. Theoret. Biol. *95,* 749–781.

May, R. M., ed. (1981): *Theoretical ecology: Principles and applications,* second edition. Sinauer Associates, Sunderland, Mass. ix + 489 pp.

Oster, G. F., and E. O. Wilson (1978): *Caste and ecology in the social insects* (Monographs in Population Biology No. 12). Princeton University Press, Princeton, N. J. xvi + 352 pp.

Schmidt, G. H., ed. (1974): Sozialpolymorphismus bei Insekten: Probleme der Kastenbildung im Tierreich. Wissenschaftliche Verlagsgesellschaft MBH, Stuttgart. xxiv + 974 pp.

Wheeler, W. M. (1927): *Emergent evolution and the social.* Keegan Paul, Trench, Trubner, London. 57 pp.

Wilson, E. O.(1953): The origin and evolution of polymorphism in ants. Quart. Rev. Biol. *28,* 136–156.

– (1968): The ergonomics of caste in the social insects. Amer. Nat. *102,* 41–66.

– (1976a): The organization of colony defense in the ant *Pheidole dentata* Mayr (Hymenoptera: Formicidae). Behav. Ecol. Sociobiol. *1,* 63–81.

– (1976b): Behavioral discretization and the number of castes in an ant species. Behav. Ecol. Sociobiol. *1,* 141–154.

– (1978): Division of labor in fire ants based on physical castes (Hymenoptera: Formicidae: *Solenopsis*). J. Kans. Entomol. Soc. *51,* 615–636.

– (1980a): Caste and division of labor in leaf-cutter ants (Hymenoptera: Formicidae: *Atta*). I. The overall pattern in *A. sexdens.* Behav. Ecol. Sociobiol. *7,* 143–156.

– (1980b): Caste and division of labor in leaf-cutter ants (Hymenoptera: Formicidae: *Atta*). II. The ergonomic optimization of leaf cutting. Behav. Ecol. Sociobiol. *7,* 157–165.

– (1983a): Caste and division of labor in leaf-cutter ants (Hymenoptera: Formicidae: *Atta*). III. Ergonomic resiliency in foraging by *A. cephalotes.* Behav. Ecol. Sociobiol. *14,* 47–54.

– (1983b): Caste and division of labor in leaf-cutter ants (Hymenoptera: Formicidae: *Atta*). IV. Colony ontogeny of *A. cephalotes.* Behav. Ecol. Sociobiol. *14,* 55–60.

– (1984): The relation between caste ratios and division of labor in the ant genus *Pheidole* (Hymenptera: Formicidae). Behav. Ecol. Sociobiol. in press.

Fortschritte der Zoologie, Bd. 31 · Hölldobler/Lindauer (Hrsg.): Experimental Behavioral Ecology
G. Fischer Verlag · Stuttgart · New York · 1985

Termite polygyny: the ecological dynamics of queen mutualism

Barbara L. Thorne

Museum of Comparative Zoology, Harvard University, Cambridge, Mass. 02138 USA

Abstract

Polygyny, the existence of more than one fertilized and egg-producing queen within a social insect colony, is a mechanism by which an insect society, and its individual queens, can side-step ecological and evolutionary constraints encountered by closely related, obligately monogynous species. Colony growth rate, probability of survival, age of first reproduction, resistence to predation, and ability to exploit resources may all be enhanced by the coexistence of multiple queens. Although polygynous females must share reproductive output and may not survive as the reigning queen as the colony ages, there are significant mutualistic advantages to joining multiple queen associations. This paper surveys the phyletic distribution of polygyny in termites, and focuses on a comparison of the ecological dynamics of polygyny in the New World species *Nasutitermes corniger* (Motschulsky) and the African termite *Macrotermes michaelseni* (Sjöstedt). Finally, ecological and evolutionary convergences between multiple queen associations in the phylogenetically independent orders Isoptera and Hymenoptera are examined.

1 Introduction

Social behavior enables insect colonies to escape some of the morphological, physiological, and neurological limits faced by closely related solitary insects. The increased biological flexibility of an insect society also results in expanded ecological adaptability. Castes are one example of behavioral dynamics mediating such ecological resilience. Division of labor within a colony enables specialization, and also permits concurrent rather than sequential accomplishment of tasks (Oster and Wilson 1978). Queens, specializing in egg production, are tended and fed by workers, and can therefore allocate their entire time and augmented energy budgets to reproduction. This reproductive specialization, coupled with the ability of many social insect colonies to replace dead or senescent reproductives, enables insect societies (and in many cases individual queens) to function as perennials while most of their solitary relatives are confined to an annual life cycle. As perennials queens achieve a much higher fecundity than as annuals, and their fertile progeny are not all dispersed into the same «basket» of time and space. Labor castes and group living also enable many social insect colonies to build elaborate nests which render their living environment relatively homeostatic, allow food storage or cultivation, and/or give protection from predators. Solitary insects are rarely able to invest in intricate nest construction, and they use simple or existing cavities or shelters for refuge, oviposition, and brood care.

Polygyny is a condition in which multiple fecund queens coexist and are cared for by nonreproductive castes within a single colony. This is another behavioral method by which an insect society, and its individual queens, can evade ecological and evolutionary constraints faced by closely related, obligately monogynous species. The cumulative productivity of multiple queens confers a mutualistic advantage to each contributor (Thorne 1984). In both Hymenoptera and Isoptera, growth rates of polygynous colonies are higher than those possible for monogynous colonies of the same age (e.g. Metcalf and Whitt 1977; Gibo 1978; West-Eberhard 1978a; Thorne 1982a, 1984; Darlington 1984). Increased colony population size can result in increased fecundity and probability of survival for both the colony and individual queens (see Gibo 1974, 1978; Gamboa 1978; Thorne 1984). Larger societies may be able to exploit available resources more effectively. For a colony, multiple queens insure that the death of one fertile female will not kill the colony; in this sense, polygyny is a mode of risk minimization (e.g. West-Eberhard 1978b; Bartz and Hölldobler 1982; Thorne 1982a; Darlington 1984). Thus social behaviors and organization, including polygyny, increase the ecological flexibility and resilience of an insect colony as compared with a solitary relative.

This paper focuses on polygyny in termites, first surveying its phyletic distribution and then comparing dynamics in the only two species in which multiple primary queens have been studied, *Nasutitermes corniger* (Motschulsky) in Central America (Thorne 1982a, b, c; 1984) and *Macrotermes michaelseni* (Sjöstedt) in Kenya (Darlington 1984). Analysis shows that even though the two species appear different, both show (1) a mutualistic advantage to multiple queen association, and (2) the advantages of polygyny in allowing escape from ecological constraints. Finally, I examine ecological and evolutionary convergences between polygynous associations in the independent orders Isoptera and Hymenoptera.

2 Phyletic distribution of multiple primary queens in termites

Termite reproductives are of three morphological forms depending upon developmental origin. «Primary» or «first-form» queens are derived from alates, and have conspicuous plates on the dorsal thorax marking abcission points of the nuptial wings. After the mating flight, alates pair, search for a nest site, mate, and initiate new colonies. In Higher termites (family Termitidae) the female ovarioles enlarge and the abdomen becomes physogastric. These primary reproductives have the highest potential fecundity of the three reproductive forms.

«Nymphoid» and «ergatoid» reproductives, derived from nymphs and workers respectively, are termed ‹supplementary› or ‹neotenic› reproductives. They normally differentiate only upon death or senescence of the original primary queen, and they obligatorily remain in the natal nest. Multiple supplementary reproductives are relatively common in colonies of some termite species, particularly in the Lower termite families. However, because female and male primary forms are the only potentially independent, dispersive reproductives, they are regarded as the true queens and kings, and will be the focus of this report.

Although generally considered rare in termites, literature accounts report multiple alate-derived queens in colonies of 38 species (15 genera) of Isoptera, representing all 4 subfamilies of the phylogenetically advanced family Termitidae (Table I). This list is undoubtedly conservative. Extensive field surveys of termite colony reproductive status have been few because royal compartments in many species are not readily accessible (e.g. mound, log, or subterranean nests). Since polygyny appears facultative, numerous colonies must be sampled to estimate its frequency in any particular species and habitat.

Most of the entries in Table I are accounts of a single colony housing multiple primary queens. It is therefore premature to compare polygynous species, but some patterns are suggested. Although the number of associated queens is variable among and within taxa, the Nasutitermitinae have the largest number of polygynous species and the highest queen counts. Multiple primary kings are found commonly in the royal chambers of polygynous colonies in all subfamilies. Queens within an association are generally of approximately the same size and pigmentation. Exceptions to this pattern appear to be cases of mother-daughter affiliations (reported in several species of *Macrotermes* and *Odontotermes*). Multiple reproductives are almost always found in the same queen cell, often sharing single chambers within the cell.

«Adultoid» reproductives are primary replacement reproductives: in some species, mature alates may remain within the natal nest and assume the reproductive position(s) from dead or senescent parents. Experimental removal of established queens with subsequent observation of one or more young primaries documents this replacement route in *Macrotermes natalensis, Odontotermes badius, O. latericius* (Coaton 1949), *Astalotermes quietus* (Noirot 1956), and *Macrotermes michaelseni* (Darlington 1978, 1984; Sieber and Darlington 1982). The mode of formation of polygynous associations in most species is unknown. Primary foundresses and adultoid replacement reproductives are morphologically indistinguishable. In a number of cases (particularly within the Apicotermitinae, Termitinae, and Macrotermitinae), however, colony age or direct observations suggest multiple adultoid replacements. Such replacements are siblings (or possibly half-siblings if the colony was originally polygynous).

Comparison of polygyny in *Nasutitermes corniger* and *Macrotermes michaelseni*

Polygyny has been investigated in only two termite species. Fortunately, these two focal species are phylogenetically and geographically disparate: *Nasutitermes corniger* (Nasutitermitinae) is Neotropical while *Macrotermes michaelseni* (Macrotermitinae) is an Old World termite. Juxtaposing a comparable phenomenon in such dissimilar species is likely to be more striking than comparisons between closely related species. Many gaps exist in our understanding of the ontogeny of polygynous associations in each species, but available data suggest that some of the patterns and processes of multiple queen colonies are quite different in *Nasutitermes corniger* and *Macrotermes michaelseni.*

Nasutitermes corniger is an ecologically and economically dominant termite throughout rainforest, gallery forest, shrublands and disturbed areas of Central America. It is reported to range south into Columbia, Ecuador and Venezuela (Thorne 1980). Colonies of up to one million individuals are composed of two worker castes, nasute soldiers, and a seasonal reproductive brood. The arboreal carton nests have a network of intercalated galleries surrounding the dense, fist-sized royal cell. Data reported here were collected by the author in areas of young second growth near the Panama Canal, Republic of Panama.

Macrotermes michaelseni is a common termite in semi-arid grasslands in Kenya. Major and minor mandibulate soldiers and major and minor workers staff colonies of up to 5.25 million termites (Darlington 1982). The termites forage on grass litter, and culture fungus within the mound. *M. michaelseni* builds earthen mounds, with a closed ventilation system. Fungus combs surround nursery chambers, which embed a capsule-like royal cell positioned centrally in the hive (Darlington 1977). Observations and data summarized

Table I

	Site	# Primary Queens	# Primary Kings	Queens Approx. Same Size?	Same Queen Cell?	Adultoids?	Reference
Apicotermitinae							
Apicotermes desneuxi Emerson	Africa	2	?	?	?	?	Bouillon 1964 p. 183
Apicotermes kisantuensis Sjöstedt	Africa	2	2	?	?	?	Bouillon 1964 p. 183
Astalotermes quietus (Silvestri)	Africa	3–9	?	yes	?	yes	Noirot 1956 p. 146, and pers. comm.
Ateuchotermes tranquillus (Silvestri)	Africa	11[a]	2	?	?	yes	Silvestri 1914–15 p. 54
Termitinae							
Cubitermes aemulus Silvestri	French Guinea	4	2	?	?	yes	Silvestri 1914–15 p. 100
Cubitermes curtatus Silvestri	French Guinea	7	?	?	?	yes	Silvestri 1914–15 p. 103
Cubitermes sankurensis Wasmann	Africa	2	2–4	?	?	?	Bouillon and Lekie 1964 p. 209
Microcerotermes amboinensis Kemner	Indonesia	1	?	?	?	?	Weyer 1930 p. 378
Microcerotermes parvus (Haviland)	South Africa	2	2	?	yes	?	Warren 1909 p. 125
	Ivory Coast	6+[b]	2[b]	yes	no[d]	yes	Noirot 1956 p. 150, and pers. comm.
Microcerotermes sikorae (Wasmann)	Madagascar	2	?	?	yes	?	Sjöstedt 1914 p. 9
Paracapritermes primus Hill	Australia	5[c]	1	?	?	yes	Hill 1942 p. 418
Pericapritermes heteronotus Silvestri	Guinea	3	1	?	yes	?	Silvestri 1914–15 p. 139
Protocapritermes krisiformis (Froggatt)	Australia	12	?	?	?	yes	Hill 1942 p. 416
Termes hospes (Sjöstedt)	Ivory Coast	10+[a,e]	> 1	yes	?	yes	Noirot 1956 p. 151, and pers. comm.
Termes kraepelinii (Silvestri)	Australia	several	?	?	?	yes	Hill 1942 p. 393
Termes panamaensis (Snyder)	Panama	9	1	yes	?	?	NMNH collection
Nasutitermitinae							
Araujotermes parvellus (Emerson)	Brazil	24	2	yes	no[f]	?	A. E. Mill, pers. comm.
Nasutitermes amboinesis (Kemner)	Indonesia	16	?	yes, no[g]	no?[h]	?	Weyer 1930 p. 376, 378

	Site	# Primary Queens	# Primary Kings	Queens Approx. Same Size?	Same Queen Cell?	Adultoids?	Reference
Nasutitermes corniger (Motschulsky)	Panama	2–33–55?[i]	2–17	yes	yes	sometimes (w/budding)	Dudley & Beaumont 1889a p. 61; 1890 p.158, 161, 167–8. 1889b p. 88, 106–7; Dietz & Snyder 1923 p. 295; Molino & Zetek (NMNH collection); Thorne 1982a, b;
	Costa Rica						Prestwich, Rettenmeyer (pers. comm.)
Nasutitermes costalis (Holmgren)	West Indies	7–8	4–7	?	?	?	NMNH collection
N. morio Banks[j]	Cuba	9	1	yes	?	?	Barbour & Brooke, MCZ collection
N. sanchezi Banks[j]	Jamaica	2–5	2–6	yes	?	?	Hubbard, NMNH collection
Nasutitermes ephratae (Holmgren)	Panama	5	?	yes	?	?	Zetek & Molino, NMNH
	Mexico	6	> 1	24–32 mm	yes	?	Becker 1961 p. 81
	Brazil	2	2	?	yes	?	Mathews 1977 p. 166
	Costa Rica	12	?	?	yes	?	H. Jacobson, pers. comm.
Nasutitermes ripperti (Rambur)	Jamaica	4	2	yes	implied	?	Andrews 1911 p. 202–4
Nasutitermes sp. (near *latus*)	Philippines	9	5	?	yes	?	C. K. Starr, pers. comm.
Nasutitermes sp.	Uruguay	6, 7	2	yes	?	?	Parker, NMNH collection
Macrotermitinae							
Hypotermes obscuriceps (Wasmann)[k]	Ceylan	2	2	yes	yes	?	Escherich in Hegh 1922 p. 174
Macrotermes bellicosus (Smeathman)	Erythrée	2	?	no	yes	?	Silvestri in Hegh 1922 p. 173
Macrotermes gilvus subsp. *malayanus* (Haviland)	Malaya	2–8	6	?	yes	?	Ridley 1910 p. 157

	Site	# Primary Queens	# Primary Kings	Queens Approx. Same Size?	Same Queen Cell?	Adultoids?	Reference
Macrotermes michaelseni (Sjöstedt)	Kenya	2–7	1–3	usually	yes[l]	often	Darlington 1978 p. 22–53; 1984
Macrotermes natalensis (Haviland)	Zaire[m]	3	?	yes	yes	?	Bequaert 1921 p. 194–6
		4[n]	2	?	?	?	Hegh 1922 p. 173
	South Africa	2	1	no	yes, no	yes	Coaton 1949 p. 149
Microtermes havilandi Holmgren	South Africa	3–4	?	?	?	?	Warren 1909 p. 125
Odontotermes badius Holmgren	South Africa	2–4	?	no	yes	yes	Coaton 1949 p. 338
Odontotermes bangalorensis Holmgren	India	6	1	yes	yes	?	Holmgren 1913 p. 107, 109
		1	2				
Odontotermes brunneus Hagen	India	3	?	?	no	probably	Holmgren 1912 p. 785
Odontotermes latericius Holmgren	South Africa	4–6	2	no	no	yes	Coaton 1949 p. 376–8
		2–3	?	?	yes	?	Warren 1909 p. 121
Odontotermes obesus (Rambur)	India	2	2	no	yes	implied	Roonwal & Gupta 1952 p. 293–4
		4	2	yes	yes	?	Roonwal & Chhotani 1962 p. 975–6
Odontotermes redemanni Wasmann	India	2	2	yes	yes	?	Escherich in Hegh 1922 p. 174
		2+	?	yes	yes	?	Mukerji & Raychauduri 1942 p. 175
Odontotermes vulgaris (Haviland)	South Africa	2	2	?	yes	?	Warren 1909 p. 123
Odontotermes wallonensis (Wasmann)	India	2	2	?	yes	?	Holmgren 1913 p. 110
		3	2	yes	yes	?	Mathur & Chotani 1960 p. 623

a Incomplete coloration.
b Additional primary reproductives present, not counted or collected.
c Implication that females are very young; not necessarily egg-laying.
d No defined queen cell(s); queens found relatively close together.
e Colony only partially sampled; may have been additional reproductives.
f Queens found in proximate cells.
g Different colonies.
h Queens found wandering, possibly from disturbance during nest dissection, therefore status of queen cell occupation is questionable.
i Dudley & Beaumont do not mention whether the "55 queens" are physogastric or dealate imagoes. They are not in the MCZ collection with Beaumont's other specimens.
j Synonymous with *N. costalis* according to T. E. Snyder (1949).
k Formerly *Odontotermes obscuriceps*; nomenclature revised according to Ruelle 1970.
l One exception (2 queen cells) of 350 nest observations (J.P.E.C. Darlington, pers. comm.).
m Because of locality, species may be M. [illegible]

here are based on extensive sampling by J. P. E. C. DARLINGTON around Kajiado, in the Rift Valley Province of Kenya (DARLINGTON 1984 and pers. comm.)

Research methods for the two studies were similar. Colony population size and caste ratio data were derived from complete colony dissections and volumetric subsampling of removed termites. *Nasutitermes* nests were inactivated by refrigeration; *Macrotermes* colonies were fumigated with methyl bromide. All sampled *Nasutitermes* colonies were completely dissected and sampled for population size. Nest population was assessed in only a small proportion of *Macrotermes* colonies. Most nests were opened live for other purposes, and the royal cell was excavated to determine queen status.

At least in the sampled habitats, multiple primary queens are relatively common in both termite species. Of 361 mature queenright *Macrotermes michaelseni* mounds, 84 (23 %) contained 2 or more physogastric queens. In *Nasutitermes corniger,* 25 of 76 (33 %) queenright colonies housed more than one queen.

3.1 Similarities

There are a number of similarities between the two species. First, with the exception of a single colony in each species, all queens are found within a single royal cell in the termitaria. In *Macrotermes,* all females share the same chamber (one exception). *Nasutitermes* queens commonly occupy a single chamber, although in some nests chambers containing queen groups may be stacked like apartment tiers within the royal cell. Second, associated queens appear to be of the same age, as judged by pigmentation, degree of physogastry, and fresh weight. In *Macrotermes michaelseni* there are a few exceptions to this pattern (N=3 of 90 observations, DARLINGTON 1984), but these rare cases appear to be transient eclipses of queen replacement (a senescent mother being succeeded by one or more young daughters). Third, polygyny is not restricted to immature nests. Multiple queens are frequently present in alate-producing colonies (see THORNE 1983). Fourth, boosted by the cumulative egg production of multiple queens, polygynous colonies grow larger than a conspecific monogynous colony headed by a queen of the same age. At least in *N. corniger,* such polygynous colonies also grow faster (THORNE 1982, 1984). Fifth, multiple primary kings are occasionally found in colonies of both *Macrotermes* and *Nasutitermes*, but in each case the usual situation is one male even if there are multiple females. Multiple kings are sometimes found in immature *N. corniger* colonies, or as an apparently transient phase of king replacement in mature colonies of both species (THORNE 1982a, b, 1984; DARLINGTON 1984). Finally, no aggression has been observed among cohabiting queens in either species. However, neither of these studies focused specifically on queen-queen behaviors.

3.2 Differences

Differences between polygyny in *Macrotermes* and *Nasutitermes* are in some cases quite dramatic. First, size, pigmentation, and morphological data (attrition of antennal segments) indicate that polygynous *Nasutitermes* queens are relatively young individuals (THORNE 1984). The weight distribution of individual *Nasutitermes* queens in large groups is markedly skewed toward the smaller size classes (Fig. 1). Monogynous queens (stippled bars) are more evenly distributed through all weight classes, and many reach sizes never attained by multiple queens. Polygynous queens are always a light cream color, while monogynous females frequently have darker pigmentation characteristic of older individuals. In contrast, *Macrotermes* queen associations may have young or old queens (although females *within* any one group are apparently of the same age). Polygynous

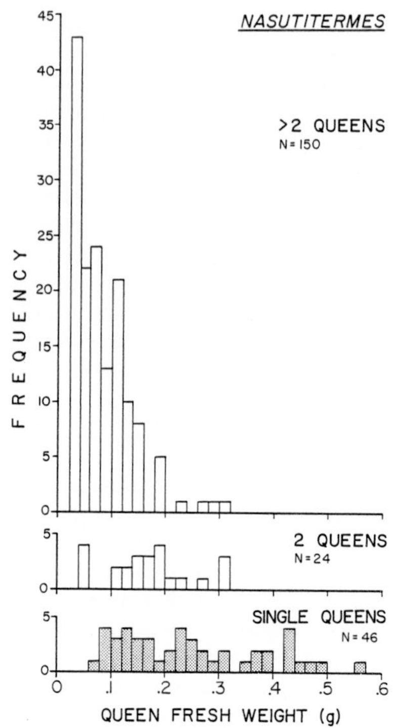

Fig. 1: Histogram of fresh weight distribution of monogynous *N. corniger* queens (stippled bars), queens in two queen groups, and queens in larger polygynous associations.

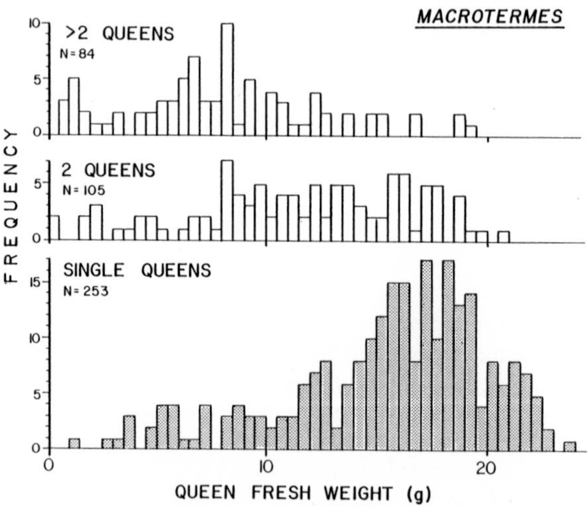

Fig. 2: Histogram of fresh weight distribution of monogynous *M. michaelseni* queens (stippled bars), queens in two queen groups, and queens in larger polygynous associations. Redrawn from Darlington 1984.

queen weights are distributed over most of the range covered by single queens, with no modes to suggest that only large, small, or intermediate weight queens are found in groups (Fig. 2). Further, some polygynous *Macrotermes* queens have darkly pigmented abdomens indicating that they are elderly (Darlington 1984).

Relative queen weight is a second major difference between the species. The largest *Nasutitermes corniger* queen found was just over 1/2 gram. *Macrotermes* queens range up to 24 grams (abcissas, Figs. 1 and 2). An average *Macrotermes* queen is nearly 100 times heavier than an average *Nasutitermes* queen.

A third difference between the two species is that the largest number of *Macrotermes* queens found together was 7 (N=10 of 84 (12 %) of the polygynous groups had more than 3 queens), while up to 33 cohabiting females were found in *Nasutitermes corniger* (N=9 of 25 (36 %) of the groups housed more than 3 queens). *Nasutitermes* colonies therefore have much larger numbers of queens, in a considerably higher frequency, than in *Macrotermes*.

A fourth difference between *Macrotermes* and *Nasutitermes* is in the mode of formation of polygynous colonies. Figure 3 is a diagramatic representation of possible derivations of polygyny in termites, and transition routes between monogyny and polygyny (a comparative scheme for ants is presented in Hölldobler and Wilson 1977). In both *N. corniger* and *M. michaelseni*, observations and experiments suggest that multiple queen colonies can be formed by pleometrosis after an alate flight. In *Nasutitermes corniger*, mature colonies can actively bud and seed the satellite daughter nest with alates from the parental colony (Thorne 1982a, b). Budding has not been observed in *Macrotermes michaelseni*. If alate brood are present, a *Macrotermes* colony whose queen dies can replace her (and the king if necessary) with female and male alate progeny (Sieber and Darlington 1982). Such queen replacement may be observed by experimentally removing a queen and resampling the colony a year later, or recent replacement may be inferred when a large termitaria is excavated and found with only small or lightly pigmented young reproduc-

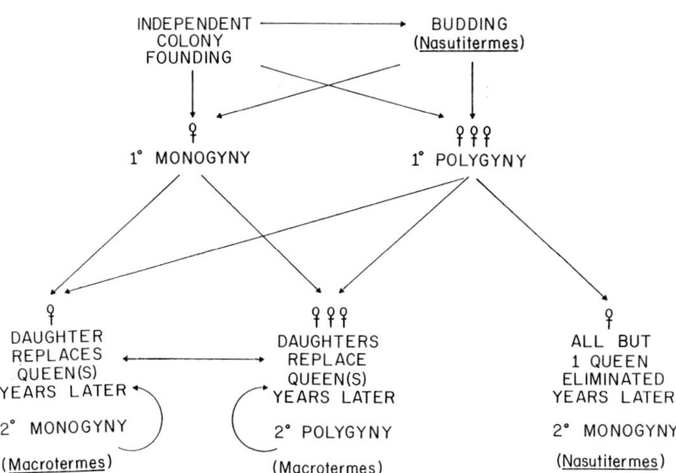

Fig. 3: Possible derivations of polygyny in termites, and transition paths between monogyny and polygyny in *Nasutitermes corniger* and *Macrotermes michaelseni*. When a transition route is thought to be used primarily by one species and not the other, the generic name of the species favoring the option is given in parentheses.

tives in a colony of mature soldiers and workers. Queen replacement appears to be common in *Macrotermes michaelseni.* Such requeening may occur in *Nasutitermes corniger,* but it has not been observed. The present study focused on young nests in recently disturbed habitats. Queen replacement might be more common in older *N. corniger* colonies.

Multiple queens could arise if one or more daughters join their mother and become egg producers. In such a case, colony dissection would reveal a large, darkly pigmented queen accompanied by one or more lightly pigmented, probably smaller, females. This has never been observed in *Nasutitermes corniger,* and has been found only rarely (N = 3 cases) in *Macrotermes michaelseni.* In all cases the older queen appeared senescent and in poor health, so it is likely that overlap of reproductive generations is temporary and that such maturation of daughters is in anticipation of replacing their mother. Note that all methods of polygynous colony formation excepting pleometrosis after an alate flight definitely result in associations of sister or half-sister queens. Cofounders grouping after nuptial flights may or may not be sibs.

Observations suggest that nearly all cases of *Nasutitermes corniger* polygyny are the result of pleometrosis following the alate flight, or as a result of parental colony budding. In contrast, the majority of *Macrotermes michaelseni* multiple queens are apparently replacement reproductives occupying large, established mounds (DARLINGTON 1984 and pers. comm.). If this general pattern is confirmed, it infers that accelerated colony growth rates are advantageous to young *N. corniger* colonies while the cumulative egg production boost of multiple queens is a benefit to large, mature *M. michaelseni* societies.

A fifth interspecific difference is that in *Macrotermes* colonies there is a very high correlation between the number of queens in a and their cumulative weight (Fig. 4A). No such relationship holds among *Nasutitermes corniger* queens (Fig. 4B). Thus egg production of *Macrotermes* colonies appears to jump significantly with queen number, while productivity of *Nasutitermes* depends on both queen number and size.

Furthermore, DARLINGTON (1984) reports that productivity of two small *Macrotermes* queens is functionally equivalent to that of a single large queen. The population size of a

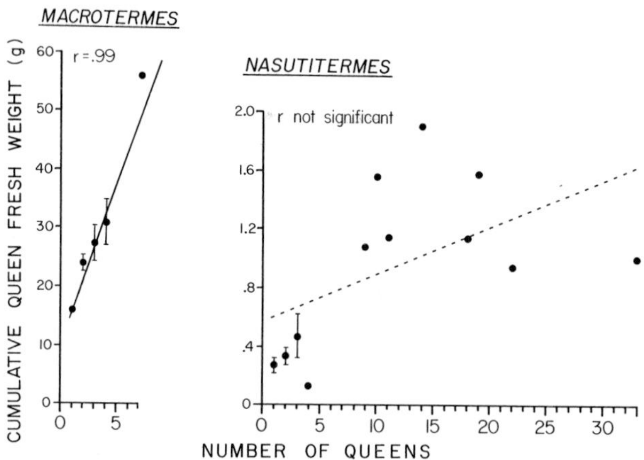

Fig. 4: Correlations between number of queens in a polygynous association and their combined fresh weight in *A. Macrotermes michaelseni* and *B. Nasutitermes corniger.* Standard errors are shown in error bars. (Graph *A* redrawn from DARLINGTON 1984).

colony headed by two 10 gram queens is comparable to that of a colony housing a single 20 gram queen. No such pattern holds for *Nasutitermes corniger*. Data for egg production appear in THORNE (1982a, 1984); data for neuter population size are presented in Figure 5. 95 % confidence limits of the monogynous and polygynous principal axis correlation lines do not overlap, indicating that, for a given total weight, one large queen is more productive than two smaller ones. (It should be noted, however, that through partnership in a polygynous association, a small queen can be part of a large, rapidly growing colony – an impossibility for a small solitary female).

Cumulative productivity differences between *Macrotermes* and *Nasutitermes* may be due to the striking differences in queen weight. Because all *Nasutitermes* queens are tiny compared with any *Macrotermes* queen, a larger percentage of their body weight is head and thorax. *Macrotermes* queen weight is essentially all abdomen. Therefore, adding *Macrotermes* queen weights may give a more representative index of potential egg productivity than adding *Nasutitermes* queen weights.

Finally, a sixth difference between the two species is in the stability of polygyny. In *Nasutitermes corniger,* evidence suggests that queen number is reduced over time, eventually resulting in monogyny (THORNE 1982a, b, 1984). In *Macrotermes michaelseni,* polygynous associations may be long-lived and stable (DARLINGTON 1984).

3.3 The ecology and evolution of polygyny in *N. corniger* and *M. michaelseni*

Underlying these similarities and differences in details of multiple queen dynamics, however, the ecological constraints changed by polygyny are analogous in *Nasutitermes corniger* and *Macrotermes michaelseni.* Multiple queens enable these termite societies to evade risks and restrictions posed by physiological limits on growth rate and body size of individual queens (including limits to queen growth rate due to a low energy surplus returned by a small foraging force), and by any one queen's eventual mortality. In *N. corniger,* polygyny appears most common early in the colony life cycle when an amplified growth rate rapidly boosts a young society out of the vulnerable «incipient colony» size class. In *M. michaelseni,* polygyny is found among all age classes, with a significant proportion of cases appearing to be replacement reproductives in established mounds.

Using LESLIE matrix projections and the example of *N. corniger,* THORNE (1984) showed that with either a slight increase in the probability of survival during early age classes or with a decrease in the age of first reproduction, females in polygynous groups have an equal or higher expected fitness than queens attempting to initiate colonies, and boost them to the reproductive threshold, alone. Such projections account for the probability of elimination from the association as time progresses. The chance of being the one surviving queen is «worth» the risk of elimination. Colonies with single foundresses grow more slowly and therefore may have a lower probability of survival in the precarious incipient colony size class. A bigger termite colony with an increased buffer provided by larger numbers of soldiers and workers can withstand biotic and abiotic disturbances (an ant raid, an ant-eater attack, drought, famine) with more resilience than a smaller counterpart. Furthermore, single foundress colonies take longer for their neuter population to reach a size at which the colony can successfully rear alate brood. Thus each queen in a polygynous group benefits from the mutualism via a higher probability of survival and/or an increased probability of reproduction in early age classes. An adaptive value of polygyny is not dependent on kin selection; it can be modeled effectively based solely on mutualism.

The same arguments may also apply to pleometrotic *M. michaelseni* colonies, but replacement reproductives in mature colonies require a different explanation. DARLINGton (1984) argues that multiple *M. michaelseni* queens enable an absolute increase in a mature colony's population size. For example, the largest colony excavated in her study (5.25 million termites) was headed by 7 fully mature (middle aged) queens, the largest association found in *M. michaelseni* (Darlington 1978, pers. comm.). The combined fresh weight of these queens was 55.9 g, far above the possible weight (and thus egg production capacity) of any single queen (Darlington 1984).

A complete explanation for the evolution of polygyny, however, must address the question of why potentially independent young queens would forego autonomous colony foundation and instead join a multiple association. In a polygynous colony, energy devoted to rearing the fertile brood must be divided, equally or unequally, among progeny of all queens. A monogynous queen does not have to share reproductive output.

Why might a *M. michaelseni* female alate become a replacement queen within her natal nest rather than attempt to found her own colony following a dispersive nuptial flight, and how might an increased mature population size in a *M. michaelseni* mound favor the queens within? Replacement reproductives acquire several advantages over foundresses. By assuming the throne of an established colony, they receive a 'dowry' of a soldier and worker support staff, a pre-constructed termitaria, and deed to a proven microhabitat (the mound is known to be in an adequate resource environment because at least one alate brood was reared successfully). This inheritance translates into a substantial savings in time and energy. Resources which an incipient colony (and its queen) must allocate towards these basics can, in an established colony, be channeled directly into reproduction. Thus replacement queens probably receive the same major advantages as pleometrotic queens, albeit with even greater assurance. They are likely to have a higher probability of survival (once they obtain a reproductive position), and they can expect an earlier age of first reproduction. Furthermore, growth rates of individual replacement queens are ex-

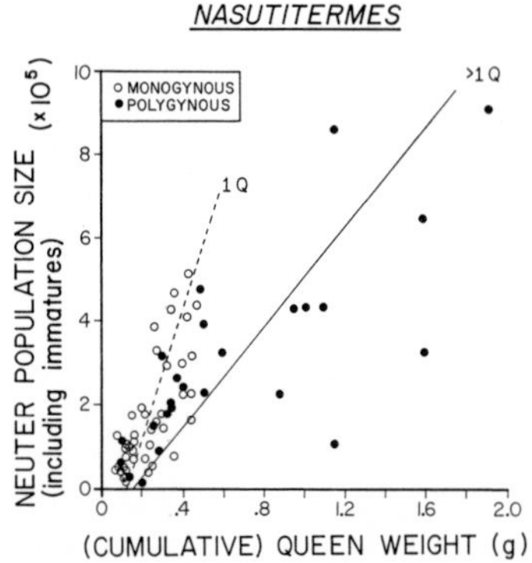

Fig. 5: Principal axis lines of queen fresh weight vs colony neuter population size correlations for monogynous *N. corniger* colonies (open circles; dashed line), and cumulative queen fresh weight vs neuter population size for polygynous *N. corniger* colonies (solid circles; solid line). For single queen colonies, y = 1,575,532 × – 202,807; r = 0.77. 95 % confidence limits on the slope of the principal axis are 1,278,339 and 2,079,079. The polygynous correlation line is y = 602,870 × – 87,376; r = 0.74. 95% confidence limits on the slope are 446,235 and 928,844. Non-overlap of the 95 % confidence limits indicates that the lines are distinct, and that monogynous queens are more productive than an equivalent biomass of smaller queens.

pected to be higher than for either haplo- or pleometrotic foundresses because the existing foraging force can immediately tend and feed the new reproductives (SIEBER and DARLINGTON 1982).

All of the above arguments hold for single as well as for multiple replacement reproductives. Some *M. michaelseni* colonies house only one replacement; others requeen with 2 or 3 females (DARLINGTON 1978, 1984). Presently, we lack data on the process of queen replacement. How and when are replacements «chosen» amidst the throngs of alate princesses, or do candidates compete for queen status? Nonreproductive colony members would favor multiple replacements if they boost colony reproductive output. Although a single female might «prefer» to reign as the sole replacement, the cost of defending that position may be high enough that she tolerates one or two partners. Multiple replacement reproductives are close kin (sisters if the natal nest was monogynous). Costs of shared reproductive output are thus minimized by the inclusive fitness benefits of producing nieces and nephews.

Cumulative egg production from multiple replacement queens increases the population size of their colony. If the proximate resource environment is sufficiently rich, more foragers can more effectively explore and exploit the habitat. This increased energy input might result in a larger fertile brood (alate) output. Such escalated productivity coupled with inclusive fitness payoffs could, on a per queen basis, compensate for the shared origin of the alate brood.

If polygyny confers such advantages to a colony and to individual queens within the association, why are the minority of dissected *N. corniger* or *M. michaelseni* colonies polygynous? Variability in the number of replacement *M. michaelseni* queens is discussed above. Current evidence from *N. corniger* suggests that polygyny is restricted to relatively young colonies. Sampling from a broader age class distribution, one expects that many societies will have already made the transition to monogyny. Furthermore, polygyny in incipient colonies appears to be facultative. A queen which founds singly must build her colony's population alone, but if successful, she also assumes the uncontested throne of a mature, alate-producing nest. Polygyny is expected to be most common in young *N. corniger* colonies located in habitats with resources rich enough to support rapid growth, and with predator and/or competitor pressure strong enough to give large colonies a survival and reproductive advantage. Even in such habitats, however, not all colonies will be polygynous because – in the hurried quest for nest sites – some tandem alate pairs may not have time to search for additional cofounders, especially if the density of alates landing in the areas is low. In incipient *Macrotermes* colonies, these same influences may explain facultative polygyny.

Thus while the ontogeny of polygynous *N. corniger* and *M. michaelseni* colonies differ, some of the ecological and evolutionary advantages of multiple queens appear similar. At the colony level, polygyny insures against incidental death of any one female. To the participating queens, polygyny confers a mutualistic advantage in that they are part of an ensemble which, cumulatively, can be more productive on a per queen basis.

4 Polygyny in the Isoptera and Hymenoptera: A case of convergent evolution?

Polygyny occurs in many groups of eusocial Hymenoptera, including polybiine wasps (WEST-EBERHARD 1969, 1973, 1978a) and a number of ant genera (reviewed in BUSCHINGER 1974, HÖLLDOBLER and WILSON 1977). Multiple egg-layers (usually foundresses) are prevalent in many species of primitively social bees and wasps (e. g. MICHENER 1974, WEST-EBERHARD 1978a). Polygyny in species of the phylogenetically independent orders Isoptera and Hymenoptera may have envolved in response to similar ecological constraints.

Polygyny in eusocial Hymenoptera originates 1) through pleometrosis; 2) via budding or swarming from a mature colony; or 3) by secondary polygyny (addition of or replacement by daughters) (reviewed in Hölldobler and Wilson 1977, West-Eberhard 1978a, b, 1982). These are the same pathways known to generate associations of multiple primary queens in termites.

Wasp cofoundresses are frequently close kin (e. g. West 1967, Gibo 1974, 1978, Gamboa 1978, West-Eberhard 1978a, 1982). Associated termite queens are definitely related if they colonize a bud, or if they replace a parent in the natal nest. Queens may or may not be close relatives in termite colonies founded by pleometrosis following alate flights.

Cofounding Hymenoptera queens often develop intra-group dominance and gonad size hierarchies. Behavioral rank normally parallels gonad size hierarchies (e. g. Waloff 1957, West 1967, Jeanne 1972, Hölldobler and Wilson 1977, West-Eberhard 1978b, Bartz and Hölldobler 1982). Such hierarchies result in differential egg production among females, and many hymenoptera colonies become literally or functionally monogynous before producing the first alate brood (e. g. West 1967, Gamboa 1978, Gibo 1978, West-Eberhard 1978b, Bartz and Hölldobler 1982). Termite queens within a polygynous *Nasutitermes corniger* or *Macrotermes michaelseni* colony are normally of the same size, and no behavioral dominance hierarchy has been observed. Data suggest, however, that polygynous *N. corniger* colonies do make a transition to monogyny, although not always before production of a fertile brood. Such long term polygyny originating from pleometrosis is unknown in wasps and has not been clearly demonstrated in ants. Polygynous associations in *Macrotermes michaelseni* (in many cases replacements) appear stable over time.

In cases where it has been evaluated, the probability of survival of a polygynous Hymenoptera colony (and therefore of successful queens) is higher than for sympatric monogynous conspecifics (e. g. Waloff 1957, Gibo 1974, 1978, Gamboa 1978, Bartz and Hölldobler 1982, Tshinkel and Howard 1983). Enhanced survivorship can be due to a decreased predator or parasite pressure because more females are present to guard the incipient colony, a decreased probability of a nest being usurped by a foreign queen, an accelerated growth rate enabling escape from the vulnerability of a small colony, or to a combination of these factors.

Production of non-reproductive progeny is important for colony survival, and also as support staff for rearing alate brood. West-Eberhard (1978b) described «temporary queens» in *Metapolybia aztecoides:* multiple females who lay worker eggs, thus boosting the number of nurses and foragers. The number of queens is then reduced to one prior to laying eggs destined to become fertile progeny. This single queen, supported by the enlarged worker force contributed by her polygynous predecessors, can produce many more reproductive progeny than would have been possible had the nest been continuously monogynous. *Metapolybia aztecoides* colonies cycle between polygyny and monogyny (single queens are replaced by groups of inseminated offspring). Many eusocial Hymenoptera colonies, and the termite *Nasutitermes corniger,* maintain a single queen after the transition to monogyny. Like *Metapolybia aztecoides,* colonies profit from the polygynous pulse, but only in the early age (size) classes. In *N. corniger,* multiple queens may produce alates before the transition to monogyny. Cofounding queens assume the risk of not reigning as the sole survivor, but they may survive long enough to produce some fertile progeny. The payoff is substantial for the single successful queen. Membership in a stable *Macrotermes michaelseni* queen association permits each queen to be part of large, productive colony. The mean weight (and thus egg laying capacity) of polygynous queens is less than that of single queens (Fig. 2), but replacement reproductives likely grow more quickly than foundresses, and are thereby compensated by an earlier age of reproduction.

Thus the ecological factors of early age class probability of survival and expected per queen reproductive output appear to have influenced the evolution of polygyny in the phylogenetically independent Isoptera and Hymenoptera. Parallels in social behavior, colony organization, caste polymorphism, life histories, and foraging behavior between the two orders have been of marked interest: polygyny is a further example of convergent patterns.

Acknowledgements

Special thanks to Dr. J. P. E. C. DARLINGTON of the National Museums of Kenya for allowing me to cite her extensive data on polygyny in *Macrotermes michaelseni*, and for helpful discussion of some of the points presented in this paper.

H. JACOBSON, A. E. MILL, C. NOIROT, G. D. PRESTWICH, C. RETTENMEYER, and C. K. STARR generously provided personal observations of multiple primary queen associations, enabling me to expand the species and geographic lists in Table I.

I appreciate useful comments by J. P. E. C. DARLINGTON, B. HÖLLDOBLER, S. C. LEVINGS, and K. P. SEBENS on earlier versions of this manuscript.

Portions of this research were supported by Harvard University, the Smithsonian Tropical Research Institute, NSF Dissertation Improvement Grant DEB-80-16415, and the American Association of University Women.

References

Andrews, E. A. (1911): Observations on termites in Jamaica. Journal of Animal Behaviour *1*, 193–228.

Bartz, S. H. and B. Hölldobler (1982): Colony founding in *Myrmecocystus mimicus* Wheeler (Hymenoptera: Formicidae) and the evolution of foundress associations. Behavioral Ecology and Sociobiology *10*, 137–147.

Becker, G. (1961): Beobachtungen und Versuche über den Beginn der Kolonie-Entwicklung von *Nasutitermes ephratae* Holmgren (Isoptera). Zeitsch. Angew. Ent. *49*, 78–94.

Bequaert, J. (1921): Insects as food. How they have augmented the food supply of mankind in early and recent times. Natural History *21*, 191–200.

Bouillon, A. (1964): Etude de la composition des sociétés dans trois espècies d'*Apicotermes* Holmgren. Pages 181–196 *in* A. Bouillon, editor. Etudes sur les termites africanis. Masson, Paris.

Bouillon, A. E. & R. Lekie (1964): Populations, rythme d'activite diurne et cycle de croissance du nit de *Cubitermes sankurensis* Wasmann (Isoptera, Termitinae). Pages 197–213 *in* A. Bouillon (ed.) Etudes sur les termites africanis. Masson, Paris.

Buschinger, A. (1974): Monogynie und Polygynie in Insektensozietäten. In: G. H. Schmidt (ed.) Sozialpolymorphismus bei Insekten. Wissenschaftliche Verlagsgesellschaft, Stuttgart. pp. 862–896.

Coaton, W. G. H. (1949): Queen removal in termite control. Farming in South Africa *24*, 335–338.

Darlington, J. P. E. C. (1977): Nest structure and distribution of the population within the nest of the termite *Macrotermes subhyalinus*. Proc. 8th Int. Congr. IUSSI, Wageningen.

Darlington, J. P. E. C. (1978): Populations of nests of *Macrotermes* species in Kajiado and Bissell. Annual Report, International Centre of Insect Physiology and Ecology 6, 22–23.

Darlington, J. P. E. C. (1982): Population dynamics in an African fungus-growing termite. In: Breed, M. V., C. D. Michener & H. E. Evans (eds). The Biology of Social Insects. Westview Press, Boulder, Colorado. pp. 54–58.

Darlington, J. P. E. C. (1984): Multiple primary reproductives in the termite *Macrotermes michaelseni* (Sjöstedt). Insect Science and its Application, in press.

Dietz, H. F. and T. E. Snyder (1923): Biological notes on the termites on the Canal Zone and adjoining parts of the Republic of Panama. United States Department of Agriculture Journal of Agricultural Research *26*, 279–302.

Dudley, P. H. and J. Beaumont (1889a): The termites or so-called «white ants» of the Isthmus of Panama. Journal of the New York Microscopical Society *5*, 56–70, 111–112.

Dudley, P. H. and J. Beaumont (1889b): Observations on the termites, or white-ants of the Isthmus of Panama. Transactions of the New York Academy of Science. *1*, 85–114.

Dudley, P. H. and J. Beaumont (1890): Termites of the Isthmus of Panama. Part II. Transactions of the New York Academy of Science. *9*, 157–180.

Gamboa, G. J. (1978): Intraspecific defense: Advantage of social cooperation among paper wasp foundresses. Science *199*, 1463–1465.

Gibo, D. L. (1974): A laboratory study on the selective advantage of foundress associations in *Polistes fuscatus*. Canadian Entomologist *106*, 101–106.

Gibo, D. L. (1978): The selective advantage of foundress associations in *Polistes fuscatus* (Hymenoptera: Vespidae): A field study of the effects of predation on productivity. Canadian Entomologist *110*, 519–540.

Hegh, E. (1922): Les Termites. Brussels.

Hill, G. F. (1942): Termites (Isoptera) from the Australian region. Council for Scientific and Industrial Research, Commonwealth of Australia. 479pp.

Hölldobler, B. and E. O. Wilson (1977): The number of queens: An important trait in ant evolution. Naturwissenschaften *64*, 8–15.

Holmgren, N. (1912): Termites from British India (Bombay) collected by Dr. J. Assmuth, S. J. Journal of the Bombay Natural History Society *21*, 774–793.

Holmgren, N. (1913): Termites from British India (near Bombay, in Gujerat and Bangalore), collected by Dr. J. Assmuth, S. J. Part II. Journal of the Bombay Natural History Society *22*, 101–117.

Jeanne, R. L. (1972): Social biology of the neotropical wasp *Mischocyttarus drewseni*. Bulletin of the Museum of Comparative Zoology *144*, 63–140.

Mathews, A. G. A. (1977): Studies on termites from the Mato Grosso State, Brazil. Academia Brazileira de Ciencias, Rio de Janeiro, Brazil.

Mathur, R. N. and O. B. Chhotani. 1960. Three queens in mounds of *Odontotermes wallonensis* (Wasmann) (Termitidae: Isoptera) Indian Forester *86*, 623–624.

Metcalf, R. A. and G. S. Whitt (1977): Relative inclusive fitness in the social wasp *Polistes metricus*. Behavioral Ecology and Sociobiology *2*, 353–360.

Michener, C. D. (1974): The Social Behavior of the Bees. Belknap Press of Harvard University Press, Cambridge, Mass.

Mukerji, D. and S. Raychaudhuri (1942): Structure, function, and origin of the exudate organs in the abdomen of the physogastric queen of the termite *Termes redemanni* Wasmann. Indian Journal of Entomology, New Dehli *4*, 173–199.

Noirot, C. (1956): Les sexués de remplacement chez les termites supérieurs (Termitidae). Insectes Sociaux *3*, 145–158.

Oster, G. F. and E. O. Wilson (1978): Caste and ecology in the social insects. Princeton University Press, Princeton, New Jersey.

Ridley, H. N. (1910): A termite's nest with 8 queens. Journal of the Straits Branch of the Royal Asiatic Society, Singapore. *54*, 157.

Roonwal, M. L. and O. B. Chhotani (1962): Termite «*Odontotermes obesus* (Rambur)»: royal chamber with four queens and two kings. Journal of the Bombay Natural History Society *59*, 975–976.

Roonwal, M. L. and S. D. Gupta (1952): An unusual royal chamber with two kings and two queens in the Indian mound-building termite *Odontotermes obesus* (Rambur) (Isoptera, Termitidae). Journal of the Bombay Natural History Society *51*, 293–294.

Ruelle, J. E. (1970): A revision of the termites of the genus *Nasutitermes* from the Ethiopian region (Isoptera: Termitidae). Bulletin of the British Museum (Natural History) Entomology *24*, 363–444.

Sieber, R. and J. P. E. C. Darlington (1982): Replacement of the royal pair in *Macrotermes michaelseni.* Insect Science and its Application *3,* 39–42.
Silvestri, F. (1914–1915): Contribuzione alla conoscenza dei Termitidi e Termitofili dell' Africa occidentale. I. Termitidi. Bollettino del Laboratorio di Zoologia Generale e Agraria della R. Scuola Superiore dAgricoltura in Portici *9,* 3–145.
Sjöstedt, Y. (1914): Termiten aus Madagaskar eingesammelt von Hern. Dr. W. Kaudern 1911–1912. Arkiv for Zoologi *8,* 1–19.
Snyder, T. E. (1949): Catalog of the termites (Isoptera) of the world. Smithsonian Institution Publication 3953, Washington, D. C.
Thorne, B. L. (1980): Differences in nest architecture between the Neotropical arboreal termites *Nasutitermes corniger* and *Nasutitermes ephratae.* Psyche *87,* 235–243.
Thorne, B. L. (1982a): Polygyny in termites: multiple primary queens in colonies of *Nasutitermes corniger* (Motschulsky) (Isoptera: Termitidae). Insectes Sociaux *29,* 102–117.
Thorne, B. L. (1982b). Reproductive plasticity in the Neotropical termite *Nasutitermes corniger. in* P. Jaisson, editor. Social Insects in the Tropics. Vol I. Université Paris-Nord, Paris. pp. 22–29.
Thorne, B. L. (1982c): Multiple primary queens in termites: phyletic distribution, ecological context and a comparison to polygyny in Hymenoptera. *in* M. D. Breed, C. D. Michener & H. E. Evans (eds) The Biology of Social Insects. Westview Press, Boulder, Colorado. pp. 206–211.
Thorne, B. L. (1983): Alate production and sex ratio in colonies of the Neotropical termite *Nasutitermes corniger* (Isoptera; Termitidae). Oecologia *58,* 103–109.
Thorne, B. L. (1984): Polygyny in the Neotropical termite *Nasutitermes corniger:* Life history consequences of queen mutualism. Behavioral Ecology & Sociobiology *14,* 117–136.
Thorne, B. L. and C. Noirot (1982): Ergatoid reproductives in *Nasutitermes corniger* (Motschulsky): Isoptera, Termitidae. International Journal of Insect Morphology and Embryology *11,* 213–226.
Tshinkel, W. R. and D. F. Howard (1983): Colony founding by pleometrosis in the fire ant, *Solenopsis invicta.* Behavioral Ecology and Sociobiology *12,* 103–113.
Waloff, N. (1957): The effect of the number of queens of the ant *Lasius flavus* (Fab.) (Hym. Formicidae) on their survival and on the rate of development of the first brood. Insectes Sociaux *4,* 391–408.
Warren, E. (1909): Notes on the life histories of Natal termites, based on the observations of the late G. D. Haviland. Annals of the Natal Museum. Vo. II, Part I. pp. 113–128.
West, M. J. (1967): Foundress associations in polistine wasps: dominance hierarchies and the evolution of social behavior. Science *157,* 1584–1585.
West-Eberhard, M. J. (1969): The Social Biology of Polistine Wasps. Miscellaneous publications, Museum of Zoology, University of Michigan *140,* 1–101.
West-Eberhard, M. J. (1973): Monogyny in «polygynous» social wasps. Proceedings VII International Congress IUSSI, London, 396–403.
West-Eberhard, M. J. (1978a): Temporary queens in *Metapolybia* waps: Nonreproductive helpers without altruism? Science *200,* 441–443.
West-Eberhard, M. J. (1978b): Polygyny and the evolution of social behavior in wasps. Journal of the Kansas Entomological Society *51,* 832–856.
West-Eberhard, M. J. (1982): The nature and evolution of swarming in tropical social wasps (Vespidae, Polistinae, Polybiini). In: P. Jaisson (ed), Social Insects in the Tropics, Vol. I. Proc. IUSSI symposium, Mexico 1980. pp. 97–128.
Weyer, F. (1930): Beobachtungen über die Entstehung neuer kolonien bei tropischen Termiten. Zool. Jahrb (Syst) *60,* 327–380.

Fortschritte der Zoologie, Bd. 31 · Hölldobler/Lindauer (Hrsg.): Experimental Behavioral Ecology
G. Fischer Verlag · Stuttgart · New York · 1985

The honeybee queen and the social organization of her colony

H. H. W. Velthuis

Laboratory of Comparative Physiology, University of Utrecht, The Netherlands

Abstract

This paper deals with the mechanisms by which the honeybee queen dominates her workers and how laying worker bees compare to the queen. Queen pheromones are produced by the mandibular glands, the tergal glands and the Arnhard glands. The chemicals in the mandibular gland secretion mainly belong to the fatty acids in the C_6–C_{10} range. The various effects of the queen on the workers constitute one syndrome of reactions; there are no separate stimuli causing specific effects. This complex stimulation of the workers is still unsufficiently known. Learning plays a major role.

The various grades of laying workers can be considered as a series of developmental stages in adult life from workers towards queenlike individuals. Abiotic factors, colony conditions and individual ability determine success or failure in this process towards reproduction. Its competitive nature is reflected in the trophallactic interactions, leading to food exchange. Laying workers are offered food more frequently due to their being distinct. The chemical composition of their mandibular gland secretion has been studied. Some changes in the blend of chemicals occur with the emergence of functional groups of bees, but in general these changes do not explain the differentiation of laying workers. Apart from the mandibular glands other discriminators are present. Among these are probably the tergal glands.

1 Introduction

Thirty years ago the important discovery was made that queen honeybees exercise control over their workers by means of chemical emanations. Simultaneously, Butler (1954), Pain (1954) and de Groot and Voogd (1954) published papers in which they maintained that extracts from queens could affect bees in the same manner as did a living queen. The interpretations of their results differed from the very beginning, and several of the questions that arose at that time are still relevant and in part unsolved. Butler used the term ‹queen substance› to denote a single component and some authors still use it as the popular name for 9-oxo-trans-2-decenoic acid (9-ODA), while others include in the term all the chemicals that are involved in the regulation of the worker by the queen.

The discovery that several sources produced different chemical components of this queen substance stimulated the discussion on their mode of action. Initially, the idea that found most support was that each of these substances had its own particular way of entering into the workers' system and subsequently had its own, specific effect. At present it seems generally accepted that these substances together constitute a complex stimulus, leading to a syndrome of reactions. The complexity of the relation between queen and worker concerns the processing of the information the worker receives from the queen and its subsequent translation into responses, rather than the multifunctional properties of the substances involved. These problems focus on the stimulus-response relationship between queen and individual worker. But both queen and worker are members of the colony which, as a unit, responds to environmental factors. In this ecological setting the queen is no longer the exclusive sovereign but is strongly influenced by the actions of the workers. Very little is known about this adaptation.

The organization of the colony should be understood in the sense that its function is to produce sexuals. The intriguing interweaving of cooperation and competition, so apparent in many features of social insects, becomes increasingly important. On the one hand the honeybee colony functions as a unit in which the queen produces the eggs and the workers tend the brood and care for the nest, and on the other hand each worker possesses a pair of ovaries and would be able to produce eggs, were she not prevented from doing so by the queen. The analysis of this situation has been a matter of interest to physiologists and ethologists, and its ultimate properties have also become an area for theoretical investigations. Therefore, I shall deal at length with laying workers, which are of special interest, since in several social bees reproduction by workers is a normal event, whereas in others, like the honeybee, it is only an aberrant development without any actual consequences for the gene pool.

2 The stimulus complex of the queen

2.1 Chemical identity

Three sources of queen pheromones are known: the mandibular glands (Butler and Simpson, 1958), the tergal glands (Velthuis, 1970b; Vierling and Renner, 1977) and the Arnhard glands on the tarsi (Lensky and Slabezki, 1981). The morphology of these glands has been studied by Altenkirch (1962), Lukoschus (1956, 1962), Nedel (1960), Renner and Baumann (1964), Cruz-Landim (1967), Hemstedt (1969) and da Costa (1982). Of these sources, only the mandibular gland pheromones have been studied chemically. The secretion is characterized by the presence of several classes of substances, the most prominent of which are the fatty acids ranging from 6 to 10 carbon atoms. The known substances are listed in Table 1. An interesting aspect is the great variability of the secretion both in its composition and in the relative amounts of the substances reported. I consider this variation as essential for the understanding of social relationships.

The two most important substances are the 9-ODA, predominant in the secretion of mature queens, and 10-hydroxy-decenoic acid (10-HDA), which is the only or the major component found in worker mandibular glands, especially in nurse bees. Crewe (1982) was able to demonstrate race-specific variation in the ratio 9-ODA/9-HDA (9-hydroxy-decenoic acid) in queens, and a considerable variation in the ratio 10-HDA/10-HDDA (10-hydroxy-decanoic acid) in worker mandibular glands, and royal jelly has been reviewed by Boch and Shearer (1982).

Of the many substances listed in Table 1a pheromonal function could be attributed only to the 9-ODA. The latter is found in queens of all four species of *Apis* (Shearer et al., 1970) and functions as a sex attractant. It is from this function that the social one, the relation between queen and workers, was probably derived (Velthuis, 1976, 1977;

Table 1: The mandibular gland secretion of queens and worker bees. (sources are 1: Barbier et al., 1960; 2: Pain et al., 1960; 3: Callow et al., 1964; 4: Ruttner et al., 1976; 5: Boch et al., 1979; 6: Crewe and Velthuis, 1980; 7: Saiovici, 1983).

substance	reported to be present in		
	queen by	queenright worker by	laying worker and false queen by
major components			
9-oxo-trans-2-decenoic acid	1		4
9-hydroxy-trans-2-decenoic acid	3	6	6
10-hydroxy-2-decenoic acid	3	2	6
10-hydroxydecanoic acid	6	5	6
8-hydroxyoctanoic acid	6	6	6
methyl-p-hydroxybenzoate	3		
minor components			
hexanoic acid	5	5	
octanoic acid	5	5	
2-octanoic acid	5	5	
nonanoic acid	3		
decanoic acid	3		
9-oxo-decanoic acid	3		
9-hydroxy-decanoic acid	3		
sebacic acid		5	
p-hydroxybenzoic acid	3		
4-hydroxy-benzyl alcohol			7
4-hydroxy-benzylaldehyde			7

Michener, 1977) when 9-ODA became involved in the several responses of the workers to the queen (court formation, swarm attraction, inhibition of queen rearing, reduction of ovary development and foraging activity; see Verheijen-Voogd, 1959; Pain, 1961; Butler and Fairey, 1964; Velthuis and van Es, 1964; Boch et al., 1975 and Simpson, 1979). None of the other substances from the mandibular gland, when tested separately, showed any influence on any of the worker responses tested. Some of the substances may perhaps be precursors to real pheromones and others may play a role as keeper substance (see Boch et al., 1975). On the other hand it should be remembered that the real stimulus probably consists of a blend of components rather than of several substances acting on their own. Research has been concentrated on synergists of 9-ODA, but none of them except the 9-HDA has such a function. Butler and Simpson (1967) claimed that 9-HDA was involved in the attraction of a swarm by the queen, but this effect has been denied (Boch et al., 1975). Ferguson and Free (1981) report a stimulating effect of 9-HDA on the Nasonof pheromone release by workers at the hive entrance where 9-ODA is not effective. Indeed, 9-HDA is more volatile and could be more useful than 9-ODA for distant communication.

While 9-ODA mimics satisfactorily the attractiveness of a queen for drones, its activity in relation to worker responses is a poor imitation of the mixture secreted by the queen. Indeed, Pain (1961) claimed already that unstable volatile compounds are involved, which she called »Pheromone II«. Much later, others came to the same conclusion (Boch et al., 1975; Simpson, 1979; Ferguson et al., 1979; Ferguson and Free, 1981; Velthuis,

1982). In contrast to what is generally believed I maintain that, in spite of the attention that has been paid to queen substance, we do not know its chemical nature in the broad sense.

2.2 Biophysical properties

The probability that the chemical nature of the stimulus is a rather complex one gives rise to questions concerning the physical aspects of the message and how it reaches the worker bee: some substances may be hardly volatile and only by touching the queen will the worker be able to smell her; other substances are more volatile, permitting detection from a distance. The importance of bodily contact was stressed long ago (VERHEIJEN-VOOGD, 1959; VELTHUIS, 1970 b, 1972), so was the importance of odour perception at short range (VAN ERP, 1960; BUTLER and FAIREY, 1963). By electrophysiological methods EAG's and single-cell responses were obtained by puffing air, carrying 9-ODA, over worker and drone antennae (KAISSLING and RENNER, 1968; BEETSMA and SCHOONHOVEN, 1966; ADLER et al., 1973). Within the hive, however, the active space of a volatile substance may become rather limited if the body surface of queen and worker absorb it; then again, it is only by contacting such a surface that the worker can obtain its information. Indeed, BUTLER et al. (1973) were unable to detect movements of prospective court bees towards the queen when they were on the combs. One and the same substance, therefore, may have merely an arrestant function inside the hive but will have an attractant function outside.

In fact several biophysical mechanisms are involved. We know this from the observations of JŬSKA (1978), who found an exponential decline of the arrestant effect of traces left behind by a queen in a cage. A similar exponential decline characterizes the attention paid to a court bee by other workers after her stay with the queen (JŬSKA et al., 1981). The time constants of the exponential functions are of the order of magnitude of 10–15 min. This quick disappearance contrasts with the long-lasting effectivity of pure 9-ODA and extracts from queens and of queens from museum collections, in relation to court behaviour and ovarial inhibition. There are apparently at least two different dispersal mechanisms.

2.3 Production and reception

Little is known about the important quantitative aspect of production and release. Measurements have been restricted to the determination of amounts of 9-ODA present within the mandibular gland; about 100–300 μg is generally present in a queen and quantities of up to 1500 μg have been found occasionally (PAIN et al., 1967, 1974; PAIN and ROGER, 1976; SHEARER et al., 1970; BOCH et al., 1975). How much of this material is used on a daily basis or during certain behaviours? Probably not much, for after swarming queens still contain much material; furthermore, BUTLER et al. (1973) and BOCH et al. (1975) report a positive rather than a negative correlation between attraction (of workers and drones, resp.) and the amounts 9-ODA present later on. Therefore, the signal emitted will not change considerably in time.

With regard to the nature of the stimulus it is of importance that bees distinguish between virgin and mated queens (BUTLER and SIMPSON, 1956) although the amount of 9-ODA in queen's glands is not necessarily different. Furthermore, bees are perfectly able to recognize their own queen in choice tests (VELTHUIS and VAN ES, 1964; BOCH and MORSE, 1974; AMBROSE et al., 1979), indicating that both innate and learned components play a role. Individual odour blends emanating from the queen and the general colony odour may be involved. Even the learned odour of a paint mark may elicit behaviour originally directed towards the queen (VELTHUIS, 1972; BOCH and MORSE, 1979). It is this versatile character of the stimulus complex that makes it difficult to give a satisfactory and operational definition of ‹queen substance›.

3 Some responses

3.1 Retinue behaviour

The queen patrols in that area of the colony where brood is reared and where young bees tend the brood. It seems obvious that reactions to the presence or absence of the queen will be found primarily in the young bees. Workers respond either by a quick retreat or by turning their head towards her. Bees give way as soon as she moves and if she passes by, they are unable to follow. The release of retinue is apparently bound up with the body surface of the queen and her immediate surroundings.

The queen is hardly touched by the outstretched worker antennae, which barely move. The mandibles of the workers are generally open, pointing downwards. Often a droplet is held between the mandibles, supported by the basal part of the otherwise withdrawn tongue, irrespective of the position relative to the queen. ALLEN (1955) states that only 1–5 feedings occur per egg produced. The mean duration of a feeding is about 45 sec. A court bee hardly ever refuses to feed the queen.

The extension of the tongue by workers is another aspect of court behaviour. This precludes, of course, food offering. The frequency is rather variable. BUTLER (1954) regarded tongue extrusion as an essential element of retinue behaviour whereby bees would obtain queen substance by licking the queen, but later (1973, 1974) he retracted that argument.

A behaviour that resembles court behaviour is the »balling« of a queen. Balling occurs mainly as a reaction to human interference. It implies an increase in the number of bees surrounding the queen, sometimes to over 60. These bees are therefore very close to each other, forming a tight ball of several layers of bees, preventing the queen from moving. Those that are closest to the queen touch her with the full length of their antennae and with their tongue and mandibles. Some bees may seize her wings, legs, antennae or tongue if food begging has occurred. Such aggressive bees may even climb on to the queen and may attempt to sting, and sometimes actually do so. During most of this balling the queen remains motionless. If undisturbed, balling usually changes into retinue behaviour. Because of this gradual transition and because of the relatedness of several behavioural elements I am inclined to interpret balling and retinue as two intensities of one and the same system.

WALTON and SMITH (1969, 1970) and YADAVA and SMITH (1967a, 1971b) made a thorough study of balling, especially in relation to virgin queens introduced into cages with workers from a colony. They demonstrate the involvement of the queens' mandibular glands. But queens whose antennae are removed are not balled; they apparently react to a stimulus provided by the workers examining the queen.

3.2 Queen rearing

Preparatory to swarming, new queens are reared in swarm cells. First of all queen cups are built, in which the queen deposits an egg. Such eggs often disappear, as do the larvae developing from them (ALLEN, 1956, 1965). Swarm cells are thought to be constructed as a response to insufficient production or distribution of queen pheromones.

Emergency queen cells are constructed only as a response to the sudden disappearance of the queen. Some worker cells are transformed into queen cells. First of all, the outer rim of the horizontal, hexagonal cell is changed into a round rim and from that a vertical, hanging queen cell is constructed. The worker larva in the cell is fed like a queen larva as soon as the rim of its cell is transformed; this occurs about 7–10 hrs after the removal of the queen. Nurse bees probably react to the different cell types in order to deliver the appropriate food.

The causal organization of queen rearing from swarm cells and from emergency cells may be identical; its occurrence is more subject to a gradual change in the colony. The process, therefore, is less predictable. An important factor is whether the queen is able to reach the cell and to kill the larva. If this is prevented, as in some commercial breeding methods, many queens can be reared in a queenright colony. The queen apparently controls only the first differentiation of the larva.

In the colony only a limited number of queen cells are constructed. This may be related to the production of pheromones from the queen pupae (BOCH, 1979; KOENIGER and VEITH, 1983).

Once the cells are sealed, the colony has to decide whether to use the cells for supersedure or swarming, or to tear them down again, which happens frequently. Here an interplay of external factors and colony conditions occur. In the latter the queen and her pheromones are again involved.

3.3 The workers' ovaries

Generally, there is only a limited number of bees whose ovaries are activated and then only the initial stages of oogenesis occur. However, once swarming preparations are being made, workers often have well developed oocytes and may occasionally lay an egg. Highly developed ovaries may also occur in wintering bees (MAURIZIO, 1954).

Removal of the queen enhances this process but the queen is not the only factor that governs the state of the ovary of the individual worker. Race differences and seasonal influences occur related to nutrition in the larval stage; in the adult worker the balance between the availability of protein and the colony's demands determines how much protein is left for oogenesis. Furthermore, workers do affect each other and as a result of this, group size is a variable.

What determines which bee out of a group is to activate its ovaries? Age is the most important factor; most probably it is the younger workers whose ovaries are to be activated. Their physiology is adapted for the maintenance of a high protein level in the haemolymph (ENGELS, 1974). Although there is a correlation between degree of exposure to queen pheromones and ovary development at the population level (VELTHUIS, 1970 a, 1976), it is of interest that this is not true at the individual level. This points to the preponderant effect of exchange – probably of a nutritive and of a communicational nature – on the process of oogenesis. There is also social control of the actual egg laying. PEREPELOVA (1929) already distinguished between ‹anatomical› and ‹functional› laying workers. It is still not known, however, what determines this difference.

3.4 The integration at colony level

The foregoing responses cannot be explained solely by regular contact or absence thereof between queen and each individual worker. Apparently workers that encounter their queen and stay in her court for a sufficient period of time subsequently transmit a message to other colony members (BUTLER, 1954; VERHEIJEN-VOOGD, 1959). The message is derived from the mandibular glands of the queen (VELTHUIS, 1972). Worker bees, while contacting the queen, obtain some of her pheromones, either via their antennae or their tongue. In subsequent contacts with nestmates their message is passed on as long as the substance remains on their body. Such workers, called messenger bees by SEELEY (1979), are most often engaged in antennal contacts with nestmates, walk more, thereby dispersing the message, and perform fewer labour acts than their nestmates. The amount of substance on their bodies must be extremely small if the message is 9-ODA; Seeley estimated this to be less than 0.1 ng! FERGUSON and FREE (1980) claim that the first workers that a messenger bee meets also obtain sufficient queen substance to be perceived by nestmates. Like the messenger bee these secondary messengers have an increased number of contacts. In contrast to SEELEY, FERGUSON and FREE found that both messenger bees and secondary messengers became engaged in an increased number of food transfers. These effects of participation in the court ceased, however, within 5 min, rather a short time in comparison to the times reported by VELTHUIS (1972) and SEELEY (1979). This is probably connected with the fact that their colony is much smaller; their bees are therefore much more in contact with the queen.

Although this system of distribution is of importance, the queen herself distributes the information by moving within the brood area. We can deduce this from the fact that a queen without mandibular glands, and thus unable to provide the messenger substance, can head a colony in an apparently perfect way and a considerable period of time (NEDEL, 1960; VELTHUIS, 1970b). Such queens have a court because of their tergal glands (VELTHUIS, 1970 b; VIERLING and RENNER, 1977). LENSKY and SLABEZKY (1981) discovered that the Arnhard glands on the tarsi provide another pheromone. These foot-prints are involved in the inhibition of queen cup construction. The model devised by LENSKY and SLABEZKY for the regulation of this behaviour is, however, primitive in that it does not include the other means by which this and other behaviours are regulated.

4 The laying worker

4.1 Its occurrence in the colony

If a colony is dequeened and rearing a new queen is prevented, the colony will produce laying workers. Not only the queen, but also the brood (JAY, 1970, 1972) and the reciprocal effect between workers (VELTHUIS et al., 1965) interfere with oogenesis. The moment the queen is removed the once stable colony sets forth on a developmental pathway along which brood production ceases and ageing of its members takes place. Against this social background individual worker bees compete for a chance to reproduce. Each individual has a restricted period of life during which it is optimally equipped to become a laying worker. Therefore, certain age classes produce many more laying workers than others, since only a few per cent of the individuals will actually lay eggs.

The development of laying workers is accompanied by the occurrence of aggressiveness. Although aggressive bees are characterized by having, to some extent, activated ovaries, there is no direct correlation between laying eggs and being aggressive (VELTHUIS, 1976). We can distinguish 3 groups of workers: those that lay eggs or behave as if they do so, those that are aggressive towards each other and towards the laying workers, and the remaining majority of the colony, which we may call the undifferentiated ones. Laying worker bees were once aggressive bees; they need some days for the transition, during which period they are rather inactive. When they become egg layers, especially at the beginning, they often go through the behavioral sequence of egg laying without producing the egg. The duration of their laying worker phase is very much influenced by the conditions inside and outside the colony. Sometimes they are very severely attacked and may be killed. Then laying workers disappear within a few days of becoming laying workers. In other observation hives, for a number of bees this phase may last much longer; we observed laying workers being reproductive for periods of 25 and 30 days. They always disappear suddenly; only once did we find a former laying worker that became a forager for 2 days. Therefore, the developmental process of a laying worker and of a forager can be considered as alternative pathways, which may diverge already early in adult life, for we found bees becoming aggressive at an age of 5 days and egg laying may start at an age of 7–10 days.

4.2 False queens

This special type of laying worker (Sakagami, 1968) deserves our attention because it attains the social position of the queen: the false queens have a court, overt aggressiveness in the colony ceases and the false queens strongly inhibit other workers from becoming egg layers. The number of false queens is always rather restricted. Their behaviour is distinct from that of laying workers. They walk over the comb like queens and lay their eggs singly in the centre of the cell (laying workers place eggs irregularly and often put many together in the same cell). False queens have a prolonged life. They occur only rarely in nature, but we are able to produce them experimentally by mixing different races of bees. Especially the *A. m. capensis* worker may become a false queen if placed amidst young workers of other races. The most noteworthy peculiarity of *A. m. capensis* workers is the fact that they produce diploid eggs by automictic parthenogenesis (Onions, 1912; Jack, 1916; Ruttner, 1976), their daughters being genetically identical (Verma and Ruttner, 1983).

We followed the development of a freshly emerged *capensis* worker into a false queen by video-recording her contacts with the *mellifera* nestmates. In her first days she was not very distinct but gradually her behaviour became different. She moved more rapidly over the comb, bumping against other workers, walking over and under them. *Mellifera* workers on the other hand tend to stop and wait until their nestmates move aside or make a detour. There was also a change in the nature of the head-to-head contacts. Generally, bees offer and ask for food which, in a certain percentage of contacts, leads to food transmission. *Capensis* workers offer food only when very young; when they are older, the *mellifera* workers chiefly offer food to them. At an age of 5–7 days the *capensis* may have a court and may start ovipositing. Ruttner, Koeniger and Veith (1976) have shown that she produces 9-ODA by that time.

These observations led us to ask two other questions. First, does trophallaxis play a role in the ontogeny of a common laying worker and secondly, what is the nature of the communication between worker bees when they enter upon trophallaxis?

4.3 Trophallaxis

One difficulty in the analysis of trophallaxis in a honeybee colony is that the partners of individuals that are studied generally remain anonymous as far as their own history is concerned. At best, age classes can be distinguished. Such studies, therefore, tend to accentuate the randomness of food exchange and neglect individual differentiation.

In order to avoid these problems we (Korst and Velthuis, 1982) reduced the colony to 2–5 bees of the same age. With 2 animals together a difference often develops between them; one specializes in asking for food, the other in offering. The frequency with which these behaviours lead to food transfer is rather low. Perhaps the performance has a function on its own. But even with a low rate of response one individual gains more from its partner than it gives. In groups of 5 bees essentially the same process has been found, but bees have here the possibility to select or avoid certain partners. It was also possible to demonstrate that bees, when taking the initiative to ask or to offer, take into account the identity of a partner (Fig. 1).

In some, but not in all groups distinct hierarchies were found. It is interesting to note that the animal that receives the most food has a hindgut that contains hardly any pollen grains and its ovaries are best developed. It probably got its protein from the other individuals. At the same time, the animal with the lowest behavioural rank has a hindgut brimfull of pollen and has resting ovaries. It could well be that this animal performed the digestion for the others.

Let us now return to the queenless observation hive and see how trophallactic interactions relate to aggressive bees and laying worker bees. We placed individually marked bees, aged 1–2 days, in a queenless observation hive. As soon as they differentiated into aggressive workers their trophallactic contacts were sampled on a daily basis. This sampling continued for those that became laying workers later on. As has been explained already, this transition takes a few days, and during this period the bees are rather inactive. This made it possible for us to recognize future laying workers and to concentrate on them during this period.

Interesting results were obtained, especially after the laying workers had been separated into two groups: the more active laying workers (the prominent ones, generally characterized by a long period of egg laying) and those that laid for a brief period only (the minor laying workers). As Table 2 shows, the prominent laying workers have high relative values for asking compared to offering, for being offered compared to being asked, and they experience trophic advantage. The minor laying workers have high negative values for

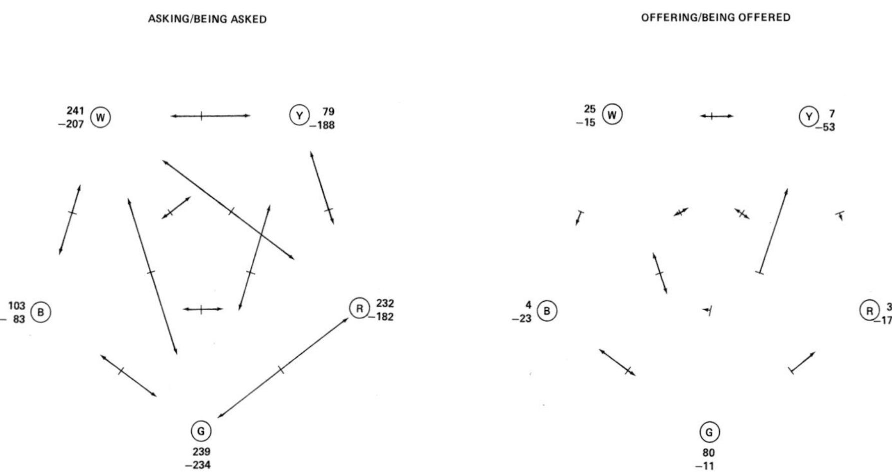

Fig. 1: The frequency and the distribution pattern of the initiatives *asking* and *offering* within a group of five *A. m. capensis* worker bees. The length of each arrow between two bees indicates the number of initiatives taken by and directed at each of them. The total score for each type of initiative taken by and directed at each individual bee is given by positive and negative numbers resp. The figure illustrates the non-randomness of these interactions.

Table 2: Trophallactic interactions of aggressive and laying workers. These interactions are given in the ratio $\frac{\text{asking} - \text{offering}}{\text{asking} + \text{offering}}$ for their initiatives and in the ratio $\frac{\text{being offered} - \text{being asked}}{\text{being offered} + \text{being asked}}$ for the initiatives taken by their partners.

		major LWB	minor LWB	AWB
$\frac{A-O}{A+O}$	before egg laying period	+0.55	−0.27	−0.05
	during egg laying period	+0.58	−0.33	−0.07
$\frac{B_o-B_a}{B_o+b_a}$	before egg laying period	+0.22	−0.30	−0.11
	during egg laying period	+0.19	−0.45	+0.02

these ratios, whereas the aggressive bees, as a group, are intermediate. The disadvantages experienced by the minor laying workers probably contributed to their early disappearance.

Not in every case does such a differentiation in the laying workers occur. As already mentioned, in some years only minor laying workers developed. In their trophallactic contacts they gained from the undifferentiated workers, but lost even more to the aggressive ones. We consider environmental factors such as weather and an abundance of flowers to be the most important determinants for the duration of egg-laying.

4.4 Mandibular gland content

MONTAGNER and PAIN (1971) suppose that pheromones are involved in the regulation of trophallactic behaviour and KORST and VELTHUIS (1982) suggest that pheromonal differences could be involved in the labour distribution within the colony. When one observes the decisiveness of an aggressive bee approaching a laying worker and the suddenness of its attacks even if the laying worker seems to be at rest, it is relatively easy to assume that there is pheromonal recognition. As a first possible source the mandibular gland of the worker needs to be investigated.

In young bees, functioning as nurse bees, the gland produces 10-HDA and some other substances, which are also found in worker and royal jelly. Therefore the gland is thought to be involved in larval nutrition and eventually in caste determination. Essential components for queen development do not stem from this gland, however, since PENG and JAY (1977, 1979) were able to rear queens using workers deprived of their mandibular glands. In older bees, functioning as guards or foragers, the gland produces 2-heptanone, an alarm pheromone. Laying workers and false queens are able to produce 9-ODA (RUTTNER et al., 1976; HEMMLING et al., 1979; CREWE and VELTHUIS, 1980) but the laying condition is not always associated with the production of this substance. The presence of 9-ODA does not imply being treated as a queen by nestmates, and in small groups of queenless *capensis* workers they may all produce this substance, but only a few will become laying workers. In fact the mandibular gland secretion of workers ranges from a most workerlike composition (principally 10-HDA) to a most queenlike one (containing 9-ODA as the major component). Perhaps the extent to which the secretion is queenlike is of greater importance than the presence or absence of certain substances.

In an effort to elucidate this matter, we raised bees of different races under different circumstances and sampled their mandibular glands as the groups became older. The composition of the secretion was studied by means of GC and MS. A first group consisted of *capensis* workers that were to become false queens. A second group consisted of the descendants of a *capensis* queen mated to *carnica* drones. Some of them were left with their mother queen in an observation hive, others were placed among young *mellifera* bees so that they could become laying workers. The third group of bees were sister *mellifera* workers; some were placed in the queenright colony, others in a queenless portion of it where differentiation into aggressive and laying workers occurred.

The composition of the mandibular glands of these bees is summarized in Table 3. *Capensis* false queens produce large amounts of 9-ODA as well as several compounds that have not yet been identified. The relative occurrence of known substances does not change considerably with age. In the F_1, *capensis* × *carnica*, queenright bees demonstrate very little effect of age on the mixture of substances but if they are orphaned a change occurs. This is rather a sudden occurrence involving an increase in 9-HDA and 8-HDA and the appearence of 9-ODA at an age of 8–12 days. Before and after this change the secretion is of rather constant composition. Although the acquisition of a reproductive function is accompanied by a change in the mandibular gland secretion, this change probably does not precede oogenesis, and the causal relation seems weak.

Table 3: The composition of mandibular gland secretion in workers of various descend and physiological condition (average percentages).

		1OHDA	1OHDDA	9HDA	9ODA	8HOA	other known	unknown
capensis								
false queens,	8 days	3.6		7.0	69.0	3.5	5.0	12.0
	24 days	2.6		5.0	45.0	4.2	2.2	42.0
hybrid **capensis × carnica**								
laying workers,	8 days	41.0	4.5	12.0		6.5		35.0
	25–30 days	25.5	2.5	27.0	19.5	10.5		15.0
in colony with queen,	8 days	65.5	9.5	7.0		4.0		14.0
	25 days	64.5	15.5	4.5		4.5		11.0
mellifera								
in colony with queen,								
	7–17 days	73.0	17.0	2.0		3.5		
	20–36 days	76.0	15.0	1.5		2.0		
in queenless hive,								
undiffer.	7–17 days	77.0	17.0	2.0	[1]	2.0		
	20–36 days	73.0	18.5	1.5		3.0		
aggress.	7–17 days	75.5	15.5	1.0		4.0		
	20–36 days	70.0	20.0	1.0		5.0		
egg layer	7–17 days	77.0	17.0	1.5		2.5		
	20–36 days	75.0	17.0	2.0		4.0		

[1] in several bees, irrespective of the category, under queenless conditions traces of 9ODA have been detected.

In the *mellifera* there is hardly any qualitative difference, either in relation to age, or in relation to presence or absence of their queen. Especially for this race an age- or function-dependent type of secretion, serving communication, seems of little importance, at least in the present set of data. Furthermore, as in the queen, the eventual involvement of the mandibular gland in communication between workers is sufficiently supported by other, and probably more important, mechanisms. This can be concluded from a study by SAIOVICI (1983), who removed mandibular glands from young *capensis* workers. This did not interfere with the bees becoming egg layers in groups of *mellifera*, nor did it reduce their inhibitory effect on the *mellifera*'s ovaries. Preliminary observations by BILLEN (unpubl.) show that laying *capensis* workers possess active tergal glands, which so far have been considered as exclusive to queens.

5 Concluding remarks

The foregoing observations on the origin and the continued existence of laying workers have been made with the expectation that mechanisms will become known which may possibly also govern the division of labour under queenright conditions. This goal has not yet been achieved. Instead, we are aware of the many gaps in our understanding of the communication mechanisms and of the consequences of such communication for behaviour – from food exchange to egg laying. Indeed remarkably little is yet known about the more classic problems related to source, chemistry and biological functions of the queen pheromones and to the labour distribution between the workers as a result of these pheromones; in addition, the way in which the colony responds to, the environmental fluctuations is completely unknown. Only by knowing about these mechanisms will it be possible for us to distinguish whether and when a worker bee has the choice between being selfish or altruistic, if such a distinction has any significance at all at a causal level.

It seems to me as if we really have only a rather superficial knowledge of the honeybee, in the sense that only the very general correlations are known but not the causal relations leading to them. It is, however, through these relationships that we may gain further insight into the evolution of the social structure of the honeybee.

Acknowledgements

It is with pleasure that I mention the kind cooperation of Miss F. VAN VLIET, Mrs A. VAN DER KERK, Mrs S. M. MC NAB, M. A., L. THEUNISSEN and M. VAN GEFFEN in the various stages of this study.

References

Adler, V. E., Doolittle, R. E., Shimanuki, H. & Jacobson M. (1973): Electrophysiological screening of queen substance and analogues for attraction to drone, queen, and worker honey bees. J. econ. Entomol. *66*, 33–36.

Allen, M. D. (1955): Observations on honeybees attending their queen. Brit. J. anim. Behav. *3*, 66–69.

– (1956): The behaviour of honeybees preparing to swarm. Brit. J. anim. Behav. *4*, 14–22.

– (1965): The production of queen cups and queen cells in relation to the general development of honeybee colonies, and its connection with swarming and supersedure. J. apicult. Res. *4*, 121–141.

Altenkirch, G. (1962): Untersuchungen über die Morphologie der abdominalen Hautdrüsen einheimischer Apiden (Insecta, Hymenoptera). Zool. Beiträge (n. F.) *7,* 161–238.

Ambrose, J. T., Morse, R. A. and Boch, R. (1979): Queen discrimination by honey bee swarms. Ann. Entomol. Soc. Amer. *72,* 673–675.

Barbier, M. and Lederer, E. (1960): Structure chimique de la ‹substance royale› de la reine d'abeille *(Apis mellifera).* C. R. Acad. Sci. Paris *250,* 4467–4469.

Beetsma, J. and Schoonhoven, L. M. (1966): Some chemosensory aspects of the social relations between the queen and the worker in the honeybee (*Apis mellifera* L.). Proc. K. Ned. Akad. Wet. C *69,* 645–647.

Boch, R. (1979): Queen substance pheromone produced by immature queen honeybees. J. apicult. Res. *18,* 12–15.

– and Morse, R. A. (1974): Discrimination of familiar and foreign queens by honey bee swarms. Ann. Entomol. Soc. Amer. *67,* 709.

– and – (1979). Individual recognition of queens by honey bee swarms. Ann. Entomol. Soc. Amer. *72,* 51–53.

– and Shearer, D. A. (1982): Fatty acids in the heads of worker honeybees of various ages. J. apicult. Res. *21,* 122–125.

–, – and Young, J. C. (1975): Honey bee pheromones: field tests of natural and artificial queen substance. J. Chem. Ecol. *1,* 133–148.

–, – and Shuel, R. W. (1979): Octanoic and other volatile acids in the mandibular glands of the honeybee and in royal jelly. J. apicult. Res. *18,* 250–253.

Butler, C. G. (1954): The method and importance of the recognition by a colony of honeybees *(Apis mellifera)* of the presence of its queen. Trans. Roy. Entomol. Soc. London *105,* 11–19.

– (1973): The queen and the ‹spirit of the hive›. Proc. Roy. Entomol. Soc. London C *37,* 59–75.

– (1974): The world of the honeybee. Collins, London, 2nd ed.

–, Callow, R. K., Koster, C. G. and Simpson, J. (1973): Perception of the queen by workers in the honeybee colony. J. apicult. Res. *12,* 159–166.

– and Fairey, E. M. (1963): The role of the queen in preventing oogenesis in worker honeybees. J. apicult. Res. *2:* 14–18.

– and Simpson, J. (1956): The introduction of virgin and mated queens directly and in a simple cage. Bee World *37* 105–114, 124.

– – (1958): The source of the queen substance of the honey-bee (*Apis mellifera* L.). Proc. Roy. Ent. Soc. London A *22,* 120–122.

– (1967): Pheromones of the queen honeybee (*Apis mellifera* L.) wich enable her workers to follow her when swarming. Proc. Roy. Ent. Soc. London A *42,* 149–154.

Callow, R. K., Chapman, J. R. and Paton, P. N. (1964): Pheromones of the honeybee: chemical studies of the mandibular gland secretion of the queen. J. apicult. Res. *3,* 77–89.

Costa Leonardo, A. M. (1982): Ciclo de desenvolvimento das glândulas mandibulares de *Apis mellifera* L. (Hymenoptera: Apidae) e a regulação social na colônia. Diss. Inst. Biociências, São Paulo.

Crewe, R. M. (1982): Compositional variability: the key to the social signals produced by honeybee mandibular glands. In (M. D. Breed, C. D. Michener and H. E. Evans, Eds.) The biology of social insects, 318–322. Westview, Boulder.

– and Velthuis, H. H. W. (1980): False queens: a consequence of mandibular gland signals in worker honeybees. Naturwissenschaften *67,* 467–469.

Cruz-Landim, C. da (1967): Estudo comparativo de algumas glândulas das abelhas (Hymenoptera, Apoidea) e respectivas implicações evolutivas. Arq. Zool. (São Paulo) *15,* 177–290.

Engels, W. (1974): Occurrence and significance of vitellogenins in female castes of social Hymenoptera. Amer. Zool. *14,* 1229-1237.

Erp, A. van (1960): Mode of action of the inhibitory substance of the honeybee queen. Insectes Sociaux *7,* 207-211.

Ferguson, A. W. and Free, J. B. (1980). Queen pheromone transfer within honeybee colonies. Physiol. Entomol. *5,* 539-366.

– (1981): Factors determining the release of Nasonov pheromone by honeybees at the hive entrance. Physiol. Entomol. *6,11* 15–19.

– –, Pickett, J. A. and Winder, M. (1979): Techniques for studying honeybee pheromones involved in clustering, and experiments on the effect of Nasonov and queen pheromones. Physiol. Entomol. *4,* 339–344.

Groot, A. P. de and Voogd, S. (1954): On the ovary development in queenless worker bees (*Apis mellifera* L.) Experientia *10,* 384.

Hemmling, C., Koeniger, N. and Ruttner, F. (1979): Quantitative Bestimmung der 9-oxodecensäure im Lebenszyklus der Kapbiene (*Apis mellifera capensis* Escholtz). Apidologie *10,* 227–240.

Hemstedt, H. (1969): Zum Feinbau der Koschewnikowschen Drüse bei der Honigbiene *Apis mellifica* (Insecta, Hymenoptera). Z. Morph. Tiere *66,* 51–72.

Jack, R. W. (1916): Parthenogenesis amongst the workers of the Cape honey-bee: Mr. G. W. Onions' experiments. Trans. Ent. Soc. Lond. *64,* 396–404.

Jay, S. C. (1970): The effect of various combinations of immature queen and worker bees on the ovary development of worker honeybees in colonies with and without queens. Can. J. Zool. *48,* 169–173.

– (1972): Ovary development of worker honeybees when separated from worker brood by various methods. Can. J. Zool. *50,* 661–664.

Jûska, A. (1978): Temporal decline in attractiveness of honeybee queen tracks. Nature, London *276,* 261.

–, Seeley, T. D. and Velthuis, H. H. W. (1981): How honeybee queen attendants become ordinary workers. J. Insect Physiol. *27,* 515–519.

Kaissling, K. E. and Renner, M. (1968): Antennale Rezeptoren für Queen Substance und Sterzelduft bei der Honigbiene, Z. vergl. Physiol. *59,* 357–361.

Koeniger, N. and Veith, J. J. (1983): Identification of a glyceride (glyceryl-1, 2-dioleata-3-palmitate) as a brood pheromone of the honey bee (*Apis mellifera* L.) Apidologie *14,* 59.

Korst, P. J. A. M. and Velthuis, H. H. W. (1982): The nature of trophallaxis in honeybees. Insectes Sociaux *29,* 209–221.

Lensky, Y. and Slabezki, Y. (1981): The inhibitory effect of the queen bee (*Apis mellifera* L.) footprint pheromone on the construction of swarming queen cups. J. Insect Physiol. *27,* 313–323.

Lukoschus, F. (1956): Untersuchungen zur Entwicklung der Kasten-Merkmale bei der Honigbiene (*Apis mellifera* L.). Z. Morph. Ökol. Tiere *45,* 157–197.

– (1962): Zur Entwicklung von Leydigschen Hautdrüsen bei der Honigbiene. Z. Morph. Ökol. Tiere *51,* 261–270.

Maurizio, A. (1954): Pollenernährung und Lebensvorgänge bei der Honigbiene. Landwirtsch. Jb. Schweiz *69,* 115–182.

Michener, C. D. (1977): Aspects of the evolution of castes in primitively social insects. Proc. 8th Int. Congress IUSSI, Wageningen, 2–6.

Montagner, H. and Pain, J. (1971): Etude préliminaire des communications entre ouvrières d'abeilles au cours de la trophallaxie. Insectes Sociaux *18,* 177–192.

Nedel, J. O. (1060): Morphologie und Physiologie der Mandibeldrüse einiger Bienen-Arten. Z. Morph. Ökol. Tiere *49,* 139–183.

Onions, G. W. (1912): South African ‹fertile› worker bees. Agric. J. Un. S. Afr. *3,* 720–728.

Pain, J. (1954): Sur l'ectohormone des reines d'abeilles. C. R. Acad. Sci. Paris *239,* 1869–1870.

– (1961): Sur la phéromone des reines d'abeilles et ses effets physiologiques. Ann. Abeille *4,* 73–153.

–, Barbier, M. and Roger, B. (1967): Dosages individuels des acides céto-9-décène-2 oique et hydroxy-10 décène-2 oique dans les têtes des reines et des ouvrières d'abeilles. Ann. Abeille *10,* 45–52.

– and Roger, B. (1976): Variation de la teneur en acide céto-9 décène-2 oique en function de l'âge chez les reines vierges d'abeille (*Apis mellifica ligustica* S.) C. R. Acad. Sci. Paris *283,* 797–799.

–, – and Theurkauff, J. (1974): Mise en évidence d'un cycle saisonnier de la teneur en acides céto-9 et hydroxy-9 décène-2 oique des têtes de reines vierges d'abeille. Apidologie *5,* 319–355.

Peng, Y.-S. and Jay, S. C. (1977): Larval rearing by worker honey bees lacking their mandibular glands. I. rearing by small numbers of worker bees. Can. Entomol. *109,* 1175–1180.

– (1979): Larval rearing by worker honey bees lacking their mandibular glands II. rearing by larger number of bees. Can. Entomol. *111,* 101–104.

Perepelova, L. I. (1929): Laying workers, the egg-laying activity of the queen, and swarming. Bee World *10,* 69–71.

Renner, M. and Baumann, M. (1964): Über Komplexe von subepidermalen Drüsenzellen (Duftdrüsen?) der Bienenkönigin. Naturwissenschaften *51*, 68–69.
– and Vierling, G. (1977): The secretion of the tergite glands and the attractiveness of the queen honeybee to drones in the mating flight. Behav. Ecol. Sociobiol. *2*, 329–338.
Ruttner, F. (1976): Die Bienenrassen Afrikas. Verh. 25. Int. Bienenz. Kongr. Apimondia, 334–364.
– , Koeniger, N. and Veith, H. J. (1976): Queen substance bei eierlegenden Arbeiterinnen der Honigbiene (*Apis mellifera* L.). Naturwissenschaften *63*, 434.
Saiovici, M. (1983): 9-Oxodecenoic acid and dominance in honeybees. J. apicult. Res. *22*, 27–32.
Sakagami, S. F. 1968): The false-queen: fourth adjustive response in dequeened honeybee colonies. Behaviour *13*, 280–296.
Seeley, T. D. (1979): Queen substance dispersal by messenger workers in honeybee colonies. Behav. Ecol. Sociobiol. *5*, 391–416.
Shearer, D. A., Boch, R., Morse, R. A. and Laigo, F. M. (1970): Occurrence of 9-oxodec-*trans*-2-enoic acid in queens of *Apis dorsata, Apis cerana,* and *Apis mellifera.* J. Insect Physiol. *16*, 1437–1441.
Simpson, J. (1979): The existence and physiological properties of pheromones by which worker honeybees recognize queens. J. apicult. Res. *17*, 233–249.
Velthuis, H. H. W. (1970a): Ovarian development in *Apis mellifera* worker bees. Ent. exp. et appl. *13*, 377–394.
– (1970b): Queen substances from the abdomen of the honeybee queen. Z. vergl. Physiol. *70*, 210–222.
– (1972): Observations on the transmission of queen substance in the honey bee colony by the attendants of the queen. Behaviour *41*, 103–129.
– (1976): Egg laying, aggression and dominance in bees. Proc. XV Int. Congr. Entomol., Washington 436–449.
– (1977): The evolution of honeybee queen pheromones. Proc. 8th Int. Congr. IUSSI, Wageningen, 220–222.
– (1982): Communication and the swarming process in honeybees. In: (M. D. Breed, C. D. Michener and H. D. Evans, Eds.) The biology of social insects, 323–328. Westview, Boulder.
– and Es, J. van (1964): Some functional aspects of the mandibular glands of the queen honeybee. J. apicult. Res. *3*, 11–16.
–, Verheijen, F. J. and Gottenbos, A. J. (1965): Laying worker honeybee: similarities to the queen. Nature *207*, 1314.
Verheijen-Voogd, C. (1959): How worker bees perceive the presence of their queen. Z. vergl. Physiol. *41*, 527–582.
Verma, S. and Ruttner, F. (1983): Cytological analysis of the thelytokous parthenogenesis in the Cape honeybee (*Apis mellifera capensis* Escholtz). Apidologie *14*, 41–48.
Vierling, G. and Renner, M. (1977): Die Bedeutung der Tergittaschendrüsen für die Attraktivität der Bienenkönigin gegenüber jungen Arbeiterinnen. Behav. Ecol. Sociobiol. *2*, 185–200.
Walton, G. M. and Smith, M. V. (1969): Balling behavior of worker honey bees. Am. Bee J. *109*, 300–301, 305.
– (1970): Effect of mandibular gland extirpation on acceptance of the queen honey bee. J. econ. Entomol. *63*, 714–715.
Yadava, R. P. S. and Smith, M. V. (1971a): Aggressive behaviour of *Apis mellifera* L. workers towards introduced queens. I Behavioural mechanisms involved in the release of worker aggression. Behaviour *29*, 212–236.
– (1971b): Aggressive behavior of *Apis mellifera* L. workers towards introduced queens. II Role of mandibular gland contents of the queen in releasing aggressive behavior. Can J. Zool. *49*, 1179–1183.

Fortschritte der Zoologie, Bd. 31 · Hölldobler/Lindauer (Hrsg.): Experimental Behavioral Ecology
G. Fischer Verlag · Stuttgart · New York · 1985

Cooperative breeding strategies among birds*

Stephen T. Emlen and Sandra L. Vehrencamp

Animal Behavior in the Division of Biological Sciences, Cornell University, Ithaca, NY 14853, USA, and Dept. of Biology at the University of California, San Diego, La Jolla, CA 92039, USA

Abstract

In this paper we address several fundamental questions pertaining to the evolution of altruism by examining the phenomenon of cooperative rearing of young among brids.

We argue that an individual bird should cooperate in the rearing of offspring other than its own *only* when its opportunities for reproducing successfully on its own are severely curtailed. Ecological constraints arising from prolonged habitat saturation or from large, unpredictable fluctuations in resources critical for reproduction can restrict breeding opportunities and lead to the evolution of cooperative breeding societies. Under such circumstances, the lifetime inclusive fitness of non-breeding individuals which aid close relatives rear offspring may exceed that of individuals which abandon nesting activities altogether.

Even in cooperative societies, competition between helpers (hopeful reproductives) and breeders (actual reproductives) can be intense. We model the conflicting interests as well as the available options of breeder and helper alike and conclude that the *form* of the cooperative breeding system – whether highly asymmetrical (as in helper-at-the-nest species) or largely equitable (as in some communal nesting species) – is dependent upon the difference in dominance levels of the participants. In the past, cooperatively breeding birds have been portrayed as living in totally harmonious, aid-giving societies. Contrary to this portrayal, we suggest that they represent societies faced with severe ecological hardships which restrict the opportunities for younger, subordinate individuals to breed on their own. They also represent societies in which competitive conflicts of interest between different group members reach extreme limits. Viewing the various forms of cooperative breeding in terms of the balance of interests between conflicting group members, with dominants having greater behavioral leverage than subordinates, provides a useful framework for understanding each system and the evolutionary forces that shaped it.

---

* This article is modified from a chapter by the same title which appeared in Perspectives in Ornithology, pages 93–120, A. H. Brush and G. A. Clark, eds. Cambridge University Press 1983.

1 Introduction

The cooperative rearing of young is a topic of considerable interest to both biological and social scientists. Such behavior reaches its extreme development in many eusocial insect societies, where vast numbers of individuals live their entire lives as sterile workers, rearing young but never themselves becoming reproductives (Wilson, 1971). Such sterile castes have not yet been reported among vertebrates. But there are numerous instances in which individuals of vertebrate species (most of them avian) forego breeding for a significant portion of their adult lives, and spend such time helping to rear young that are not their own offspring. These extra individuals care for their foster offspring by bringing food and by helping to alert and protect the young against predators. Avian helpers also may play roles in nest construction and incubation. Recent reviews (Brown, 1978; Emlen, 1978) have compiled evidence showing that pairs breeding with the assistance of such «helpers» generally achieve a significantly higher reproductive success than pairs breeding alone.

There are three fundamental questions surrounding the topic of cooperative breeding in animals. First, what role have ecological factors played in promoting the development of such aid-giving societies? Second, how can such seemingly altruistic behavior be explained in terms of natural selection theory? And third, what behavioral tactics will members of such societies adopt to maximize their own fitness when interacting with others?

In the following article, we attempt to address each of these three topics. Before doing so, however, it is necessary to define our terms. We define an *auxiliary* as any mature, non-breeding member of a reproducing group. It may or may not provide aid in the rearing of young. *Cooperative breeding* refers to any case where individuals in addition to the two genetic parents provide care in the rearing of young. These cases can be subdivided into two types. In *Helper-at-the-nest* systems, auxiliaries contribute physically, but not genetically, to the young being reared. Auxiliaries serve as helpers, but they do not engage in sexual activity with the breeding pair. *Communal breeding* systems are those in which parentage of the offspring is shared. In the case of shared paternity, more than one male has a significant probability of fathering some of the offspring. With shared maternity, multiple females have a significant probability of contributing eggs to a communal clutch. Although this subdivision is by no means absolute, we feel that it is useful in modelling questions about ecological determinants, as well as behavioral tactics, that are important for the evolution and maintenance of cooperative breeding systems.

2 The ecological determinants of helping-at-the-nest

Approximately 5 percent (300) of the living avian species are known to exhibit aid-giving behavior during breeding, and the vast majority of these are of the helper-at-the-nest type. The list includes species belonging to a wide variety of taxonomic groups, inhabiting a broad range of geographic locations and habitats, and filling the spectrum of ecological niches. This diversity has stifled the search for common ecological denominators, and has caused some workers to conclude that no common thread underlies the parallel evolution of cooperative breeding in birds. While agreeing that aid-giving behaviors have undoubtedly evolved independently many times in the class Aves, and that each species can be understood fully only when studied in conjunction with its own, unique, ecology and history, we nevertheless propose that certain broad ecological generalizations can be made about the necessary and sufficient conditions that have led to the evolution of cooperative breeding systems.

In most helper-at-the-nest species, the auxiliaries are younger than the breeders; indeed they are usually grown offspring of the breeders (Brown, 1978; Emlen, 1978). Let us consider any post-breeding group of birds comprising parents and maturing offspring. There are two behavioral options available to the maturing individuals: they can disperse and attempt to become established as independent breeders in the upcoming nesting season, or they can remain with their parents in their natal groups through the next season. In the latter case, they face a second subset of options: to help or not to help the group's breeders in rearing the next generation of nestlings. Our first question is: «Under what conditions will grown offspring postpone dispersal and remain in their natal groups»?

There is a large body of literature discussing the gains that accrue to individuals by virtue of living in groups. The two most cited benefits include a) increased alertness and protection against predators, and b) enhanced efficiency in localization and exploitation of patchy, ephemeral food resources (Alexander, 1974; Bertram, 1978).

Surprisingly, neither of these factors usually is implicated in studies of cooperative breeders. Instead, field workers consistently stress the difficulties that younger birds have in finding vacant territories, obtaining mates, or breeding successfully alone. We suggest that the selective pressures favoring group living in most helper-at-the-nest species are fundamentally different from those favoring grouping in other types of avian aggregations such as nesting colonies, foraging flocks, or gregarious roosts.

2.1 Constraints on the option of independent breeding

If foraging and/or anti-predation benefits from the primary reason for gregariousness in cooperative breeders, then the average fitness of individual group members should increase as some function of increasing group size up to an optimum size, and decrease thereafter. Koenig (1981 a) applied this logic to an analysis of the published data available for cooperatively breeding birds. He found that annual, *per-capita* reproductive success showed no consistent pattern with group size. Of 16 cases analyzed, pairs were the most productive units in 7, pairs were roughly equivalent to larger groups in 5, and groups were most successful (on a per-capita basis) in only 4. Koenig further predicted that if cooperative groups formed because of inherent benefits of grouping, then the most frequently observed group sizes should coincide with the most productive group size as determined by per-capita success. Such was not the case: although in 12 cases pairs were equally or more productive than larger groups, in all cases the observed mean group size exceeded two. From these analyses, Koenig (1981 a) concluded that most cooperatively breeding birds live in groups because they are «forced» to do so by severe ecological constraints that limit the option of younger birds to become established as independent breeders.

The idea is not new; it was first formulated by Selander (1964) during his study of helpers in tropical wrens. It was generalized by Brown (1974) and recently it has been expanded and developed into more specific models by Woolfenden and Fitzpatrick (1978), Gaston (1978 a), Koenig and Pitelka (1981), and Emlen (1981, 1982 a). What is new is the realization that the ecological constraints concept can be generalized to encompass all categories of cooperative breeders. Furthermore, data are becoming available which, for the first time, allow a test of the major predictions of the model.

In its most concise form, the model states that when ecological constraints exist which severely limit the possibility of personal, independent reproduction, selection will favor delayed dispersal and continued retention of grown offspring within their natal units. It is the restriction upon independent breeding more than any inherent gain realized by grouping per se that leads to the formation of cooperative breeding units. Differing proximate factors can be responsible for limiting the option of personal reproduction in different species. To date, three categories of constraining factors have been suggested:

2.1.1 *Shortage of territory openings*

Many cooperative breeders are permanently territorial species that inhabit stable or regularly predictable environments. Further, many have specific ecological requirements such that suitable habitat is restricted. All available high-quality habitat becomes filled or «saturated». Unoccupied territories are rare, and territory turnovers are few. As the intensity of competition for space increases, fewer and fewer individuals are able to establish themselves on quality territories. Assuming that occupancy of a suitable territory is a prerequisite for reproduction, the option of breeding independently becomes increasingly limited. A new individual can become established as a breeder only (a) when it challenges and defeats a current breeder on an occupied territory, (b) when it competes to fill a vacancy that results from the death of a nearby breeder, or (c) when it buds-off or inherits a portion of the parental territory itself. The non-breeder must wait until it attains sufficient age, experience and status to enable it to obtain and defend an independent territory.

The notion of habitat saturation does not explain why auxiliaries do not disperse into more marginal habitats where competition is less. Koenig and Pitelka (1981) propose that a second factor is necessary for the evolution of cooperative breeding in permanently territorial species. Not only must optimal habitat be saturated, but marginal habitat must be rare. When this is the case, a maturing individual is severely constrained either from establishing itself as an independent breeder in the optimal habitat or successfully surviving and breeding in an outlying area. By this model, cooperative breeding is predicted to occur in those species whose ecological requirements are sufficently specialized that marginal habitat is rare, or which occupy habitats that are physically restricted or relict in distribution.

2.1.2 *Shortage of sexual partners*

A parallel argument can be made emphasizing the shortage of sexual partners rather than spatial territories. Many species of cooperatively breeding birds have a skewed sex ratio, with an excess of males (Rowley, 1965; Ridpath, 1972; Dow, 1977; Reyer, 1980).The reason for such skewing is poorly understood, but its effect is to increase competition for mates, leading to a demographic constraint on the option of becoming established as an independent breeder.

2.1.3 *Prohibitive costs of reproduction*

A major stumbling block to the acceptance of the generality of the breeding constraints model lies in the realization that many species of cooperatively breeding birds reside in areas where the concepts of ecological saturation or shortage of marginal habitat simply do not apply. This is especially true of arid and semiarid environments in Africa and Australia. Not only are cooperative breeders common in such environments, but many are either nomadic or inhabit areas subject to large scale, unpredictable, fluctuations in environmental quality (Rowley, 1965, 1976; Harrison, 1969; Grimes, 1976). The carrying capacity in such environments changes markedly and erratically from year to year. Avian populations, with their relatively low intrinsic rates of increase, cannot keep pace with these changes. Consequently, the degree of habitat saturation (if any) changes dramatically across seasons. We cannot speak of any consistent shortage of territory openings as the driving force in the evolution of cooperative breeding. But we can still speak of constraints on the option of independent breeding.

In variable and unpredictable environments, erratic changes in the carrying capacity create the functional equivalents of breeding openings and closures (EMLEN, 1981; 1982 a). As environmental conditions change from year to year so, too, does the degree of difficulty associated with successful breeding. In benign seasons, abundant food and cover decrease the costs to younger, less experienced individuals of dispersing from their natal groups and breeding independently. In harsher seasons, the costs associated with such reproductive ventures increase, eventually reaching prohibitive levels. As conditions deteriorate, breeding options become more constrained, and the constraints hit first at the younger, more subordinate individuals. The predicted outcome is the continued retention of such individuals in the breeding groups of older (usually parental) individuals.

2.2 Tests of the constraints model

The ecological constraints model predicts that the frequency of occurrence of nonbreeding auxiliaries will vary directly with (a) the degree of difficulty in obtaining a territory (for permanently territorial species), (b) the degree of skew in the population sex ratio (for species facing a shortage of mating partners), and (c) the level of environmental harshness (for species in erratic, unpredictable habitats). Relevant data for testing these predictions are available for three species of cooperative breeders. We have analyzed and plotted them in Figures 1 a–c.

The first graph shows the effect of territory constraints on the retention of offspring in the Acorn Woodpecker, *Melanerpes formicivorous*. This species lives in permanently territorial groups of from 2–15 individuals. During breeding, only a single nest is tended at any one time, and most or all group members help to incubate the eggs and feed and defend the young. Yearling individuals either emigrate and attempt to become independent breeders, or remain with their natal groups. Those that remain postpone breeding themselves and play full roles in their group's cooperative breeding attempts. Figure 1a plots the annual incidence of yearling «helpers» as a function of the proportion of all territories that became vacant during the preceeding year (the annual turnover rate). The figure incorporates data reported by MAC ROBERTS and MAC ROBERTS (1976), STACEY and BOCK (1978) and STACEY (1979), as well as unpublished results kindly provided by P. STACEY and W. KOENIG, R. MUMME and F. PITELKA. Acorn Woodpecker populations from Arizona, New Mexico, and coastal California, are pooled in the diagram.

The second example comes from studies of IAN ROWLEY on cooperatively breeding *Malurus* wrens in Australia (1965, 1981). These are classic helper-at-the-nest species, and

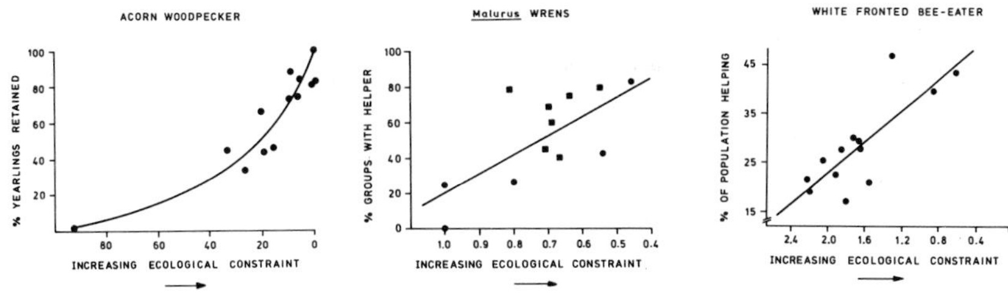

Fig. 1: The occurrence of helpers plotted as a function of the severity of ecological constraints for three species of cooperative breeders. A. Acorn Woodpeckers *(Melanerpes formicivorous)*. B. Malurus wrens (● = *M. cyaneus* and ■ = *M. splendens*). C. White fronted Bee-eaters *(Merops bullockoides)*. See text for details.

breeding units consist either of simple pairs or trios (the helper virtually always being a male). Figure 1b graphs the percentage of breeding groups with a (male) helper as a function of the shortage of sexual partners (using the ratio of females to males in the population as the index of demographic constraint).

The third example is taken from Emlen's (1981, 1982 a) work on the White-fronted Bee-eater, *Merops bullockoides*. This is an example of a colonial, cooperative breeder that inhabits the erratic, unpredictable environment of the Rift Valley of Kenya. The survival of nestlings is strongly influenced by the food supply, and reproductive success varies greatly across different seasons[1]. The food supply, in turn, is highly dependent upon the pattern of local rainfall. Consequently, Figure 1c plots the percentage of the population serving as helpers in each of 13 colonies, using a measure of rainfall as the index of ecological constraint (the log of total rainfall occurring in the month preceeding breeding).

Although the proximate factors responsible for the constraint upon independent breeding differ among these three species, in each case the intensity (magnitude) of the constraint is a good predictor of the incidence of helpers.

The retention of non-breeding auxiliary individuals in their natal groups is the usual first step in the evolution of cooperative breeding. By itself, however, it is insufficient to explain the development of actual helping behavior. We must also consider the question: «When should a retained auxiliary help in the rearing of the next generation of nestlings?»

3 The adaptive advantage of helping

Helping behavior is an intriguing phenomenon from the standpoint of evolutionary theory because of its seemingly altruistic nature. The auxiliary that feeds and defends nestlings other than its own not only incurs the costs and risks of alloparental care, but it also improves the reproductive success of other individuals in the population relative to its own. Evolutionary theory predicts that helping should only evolve when it actually benefits the individual helper. The focus of much current research on helper-at-the-nest species is on identifying the ways in which helpers might benefit from helping.

In formulating a testable hypothesis for investigating the selective advantage of helping, it is necessary to establish an alternative behavioral strategy as a reference point for comparison. The obvious alternative to helping is not helping. However, there are few cases of retained auxiliaries that do not also aid the breeding pair in some way. Rather, non-helpers typically disperse and attempt to breed independently. Therefore, the usual alternative strategy is dispersal from the natal territory at the age of sexual maturity and independent breeding thereafter. The evolutionary hypothesis specifically predicts that the fitness of birds that help during their lives will be equal to, or greater than, the fitness of birds that attempt early dispersal and breeding. Such a comparison incorporates the fitness effects of both delayed reproduciton and helping. Thus the question: «Why do helpers provide aid at the nest of others?» is closely tied to the question of why the auxiliaries remain on their natal territory.

How might helping behavior be adaptive to the helper? We discuss five general types of advantages that may accrue to helpers. The first deals with the potential benefits of delayed reproduction, while the second through fifth deal specifically with the question: «Why help?».

---

[1] Unpublished data of S. T. Emlen indicate nestling starvation rates range from 25–45% of hatchlings.

3.1 Survivorship advantage from delayed breeding

As the prospects for breeding independently become increasingly poor, it will become advantageous for certain individuals to forgo reproduction until conditions are better. Younger, subordinate birds will be more influenced by constraining factors of the types discussed in the previous sections than older, experienced individuals. By remaining on the parental territory and delaying reproduction, a young bird can increase its chances of surviving through the difficult waiting period and thereby improve its chances of becoming a breeder in the future when conditions are more favorable.

If the short-term cost of not reproducing is more than compensated for by improved longevity, then the survivorship advantage alone is sufficient to explain the evolution of delayed breeding. However, improved survivorship may be only one component of the total benefits that result in a higher lifetime reproductive success for delayed reproducers compared to non-delayers.

3.2 Breeding experience

Several studies of helpers-at-the-nest species have indicated that experienced breeders are more successful in rearing offspring than inexperienced breeders, and that older helpers provide more effective aid than younger ones (Woolfenden, 1975; Lawton and Guindon, 1981). These findings suggest that one advantage of helping during the first year or two of life may be improved breeding expertise. In order for breeding experience to be the sole explanation for the helping phenomenon, however, the amount of improvement in a helper's subsequent reproductive success as a breeder would have to outweigh the cost of its not reproducing during the helping years. This will only be true if helping is a more effective learning process than actual breeding. While helping may be a less risky method of learning, it seems unlikely that it is a more effective one.

Consequently, the attainment of breeding experience is unlikely to be the sole advantage of helping. However, it may well be an important contributory factor. The most likely benefit of breeding experience occurs in conjunction with the survival advantage of remaining on the natal territory when ecological conditions for independent breeding are constraining. The breeding experience of helping provides a bonus to auxiliaries who are otherwise forced to remain with their parents. In this case the improvement in a helper's subsequent reproductive success as a breeder need only outweigh the cost of the aid for the helping to be advantageous.

3.3 Parentally manipulated helping

Helping may function to assure the young bird of continued membership in the group. Youngsters that remain on their natal territory may impose a cost on their parents. The young may not only reduce the level of food on the territory and compete with the new generation for food, but they also may disrupt the breeding pair's reproductive efficiency by interfering with nest construction, interrupting copulation and incubation, and attracting predators to the nest (Zahavi, 1974). Helping to rear nestlings may be the «price» the youngsters must pay to be allowed to remain on the parental territory (Gaston, 1978 a). Parents should be selected to expel young that do not help. Parents are in a position to demand such aid, because it is the youngster that stands to benefit most from retention (Emlen, 1982 b). If the amount of aid an auxiliary provides is determined by parental pressure, then it should be the case that the greater the benefit of retention to the helper, the greater the helping effort the parents can demand (see next section).

In most cooperative species, helpers do improve the reproductive success of the breeders (Brown, 1978; Emlen, 1978). However, helpers usually do not work as hard as the parents themselves, and this is reflected by the fact that per capita reproductive success usually declines with increasing group size (Koenig, 1981 a). In most species, there is variation in the amount of help given as a function of age, sex, and group size. Proof of the parental manipulation hypothesis for the «advantage» of helping will rest upon its ability to predict these variations in helper effort.

3.4 Laisons for the future

Given that auxiliaries are temporarily constrained from breeding and remain on the natal territory as «hopeful reproductives», helping behavior may function in part to speed up the auxiliary's ascendency to breeding status by cementing bonds and creating laisons for future cooperative ventures. There are several mechanisms by which this might work.

(1.) Helping may heighten the dominance position of the helper relative to other group members and improve its position in the queue for breeding slots on the parental territory or on neighboring territories.

(2.) Helping may produce potential laisons that later help the auxiliary to attain breeding status. In some helper-at-the-nest species, coalitions of siblings or half-siblings disperse together. Such groups of dispersers, it has been claimed, have a greater chance of winning territorial disputes than do solitary individuals (Ridpath, 1972; Ligon and Ligon, 1978; Gaston, 1978 b; Koenig, 1981 b). To the degree that the social bonds formed during helping itself are important for the development of these coalitions, we may conclude that helping in one season may increase a helper's prospects for breeding in future years (Ligon and Ligon, 1978).

(3.) By helping cooperate in territorial defense, an auxiliary may bring about an expansion of the natal territory. Cases of such expansion being followed by the auxiliary budding off a portion of the natal area as its own have been reported in jays and babblers (Woolfenden and Fitzpatrick, 1978; Gaston, 1978 a). If an auxiliary remains on its natal territory until the death of one or both of its parents, it may take over the entirety of the territory. This is a widespread practice among both cooperatively breeding birds and mammals. By such an «inheritance» process, ownership of a high quality territory can be passed down through generations, remaining in the same family lineage well beyond the lifespans of individual members.

3.5 Increased inclusive fitness via collateral kin

All of the above mentioned benefits of helping increase the fitness of a helper via the direct or personal component of inclusive fitness, i. e., lifetime production of offspring (West-Eberhard, 1971; Brown, 1980). A second type of genetic gain inherent in helping behavior involves that component of inclusive fitness attributable to gene copies through collateral kin[2] (termed the indirect (Brown, ibid.) or kin (West-Eberhard, ibid.) component). The debate over the importance of kin selection has led to an unfortunate and, in our opinion, unproductive controversy.

It is true that for the vast majority of helper species for which data are available, the helpers are predominantly grown offspring that remain with their parental groups and help to rear full or half sibs. Thus the indirect component of inclusive fitness is large and could prove to be a major factor in the evolution of helping behavior. Some workers,

[2] Collateral kin are genetic relatives other than dircect descendants.

however, have mistakenly taken this correlation of group relatedness with helping as the sole evidence to build a case for the essentiality of kin selection in the evolution of seemingly altruistic behaviors. Other workers have erred in the opposite direction. By finding ways in which individual helpers gain personally through helping (i. e., points A, B and D above), they have concluded that kin selection is either unimportant to understanding helping behavior, or that kin selection does not even exist.

Both approaches are too narrow in outlook and miss the point. Evolution is the process of changing gene frequencies in a population throught time. Natural selection does not distinguish between a gene copy produced by a direct descendent and one fostered by a collateral kin. All gene copies are tallied, irrespective of who or what was responsible for their survival in the population. The interesting question, then, is not whether kin selection exists or not, but rather whether collateral kin interactions (as opposed to personal offspring production) have been an important or essential component in the evolution of helping behavior.

To take this debate out of the realm of semantic argument and into the realm of quantitative biology, VEHRENCAMP (1979) has devised a simple index for determining the relative importances of direct and indirect selection to the current evolutionary maintenance of a behavioral trait. I_k, the kin index, calculates the proportion of the total gain in inclusive fitness that is due to the indirect component of selection for any type of cooperation among kin, compared to the non-cooperative situation:

$$I_k = \frac{W_{RA} - W_R)\, r_{ARy}}{(W_{AR} - W_A)\, r_{Ay} + (W_{RA} - W_R)\, r_{ARy}}$$

where $(W_{RA} - W_R)$ is the change in lifetime reproductive success of the recipient, R, when it is aided by the donor, A,

$(W_{AR} - W_A)$ is the change in lifetime reproductive success of the donor, A, when it provides aid to R,

r_{ARy} is the relatedness of A to R's young, and

r_{Ay} is the relatedness of A to its own young.

The index is applicable only to situations in which the net change in inclusive fitness of A (i. e., the denominator) is positive. When I_k is greater than 1, there is a net cost to A's direct fitness and pure kin selection is acting. When I_k is less than zero, then A is manipulating R at a net cost to R's direct fitness. When I_k is between 0 and 1, both the direct and indirect components are increased by the cooperation, and the value of the index gives the proportion that is due to the indirect component.

Rowley has calculated I_k for the Splendid Wren, *(Malurus splendens)*, an Australian helper-at-the-nest species (1981). He found that the value of I_k ranged from 0.35 to 0.51 for a typical male helper. This suggests that both the direct and indirect component are increased via the helping behavior, and that the direct component is slightly more important than the indirect component. Preliminary calculations on other cooperative breeders also yield values of I_k between 0.4 and 0.5 (VEHRENCAMP, 1979).

If one considers these five points together, helping behavior ceases to present the evolutionary paradox that it first appeared to do. Severe ecological constraints often eliminate or greatly restrict the option of independent breeding. When viewed against the backdrop of this constraint, a number of possible adaptive functions can be proposed for helping behavior. These range from direct improvements in a helper's lifetime reproductive success to improvements in the indirect component of fitness by aiding close collateral kin. The differing importance of such factors for each type of cooperative system remains one of the principal challenges of future work on cooperative birds.

4 The communal breeding perspective: balancing the within-group conflict

Despite the fact that most cooperatively breeding avian species are characterized by one-sided helping behavior, one cannot dismiss the communal breeding systems. In these species, egg-laying and/or fertilization are shared, so that the fitness benefits as well as the «parental» care costs are more equitably distributed among group members. Many workers have conceived of the communal breeders as belonging to a totally separate and distinct category from the helper species. However, communal and helper systems differ in only one key parameter: the probability that a helper simultaneously will be a reproductive. We refer to this as the degree of bias in current personal reproductive benefits. The degree of bias is a continuously varying characteristic, ranging from no bias (i. e. equal division of reproductive benefits) to a very high bias (i. e. zero personal reproductive benefits to some members). Intermediate levels are found in many of the so-called equitable communal breeders.

This perspective on cooperative breeding now forces the investigator of helper-at-the-nest species to address a third question: «Why don't helpers reproduce communally along with their parents?». Once the communal breeding strategy is established as an alternative hypothesis to helping, the focus of research is shifted away from just the helper. The breeder/helper *relationship* is now viewed as a general dominant/subordinate interaction in which the interests and strategies of the two parties diverge to different degrees. This view forces us to recognize that the cooperatively breeding group is composed of competing individuals with conflicting strategies. Different individuals hold different options and different leverages for imposing their «will» on the group. The social result of these conflicts, in terms of the degree of bias and type of cooperative society, must represent the balance among the various competing interests.

4.1 Types of conflicts

Emlen has employed this approach in evaluating the nature of behavioral interactions in helper species (Emlen, 1982 b). In modelling the conflict between helpers and breeders, he used the inclusive fitness equations of Trivers (1974), West-Eberhard (1975) and Vehrencamp (1979) to establish the zones of conflict. The direction of the conflict, i. e. whether the breeder or the helper had the greatest leverage, changed depending on two critical variables: (a) the severity of the ecological constraints on younger, subordinate individuals, and (b) the magnitude of the advantage (or disadvantage) of group living.

The results of his analysis are summarized graphically in Figure 2 a and b. Each figure is a two-dimensional schematic representation of the fitness consequences of continued retention of an auxiliary. The fitness of a breeder increases from left to right while that of an auxiliary increases from bottom to top. The vertical and horizontal lines passing through the center of each figure represent positions at which retention of the auxiliary results in a zero change in *direct fitness* of breeder and helper respectively. Thus the direct fitness of a breeder is increased in all areas to the right of the vertical line while its *inclusive fitness* is increased in all areas to the right of the line marked B-B. Similarly, the *direct fitness* of an auxiliary is increased by the association with a breeder in all areas above the horizontal line, while the auxiliary's *inclusive fitness* is increased in all areas to the right of the line marked H-H.

When retention and helping benefit both helper and breeder, there is clearly no conflict and helping will evolve (speckled zone). Similarly, when retention and helping are disadvantageous to both parties, helping will not evolve (white zone). But, because of certain asymmetries in the inclusive fitness equations, there are two zones of breeder/helper con-

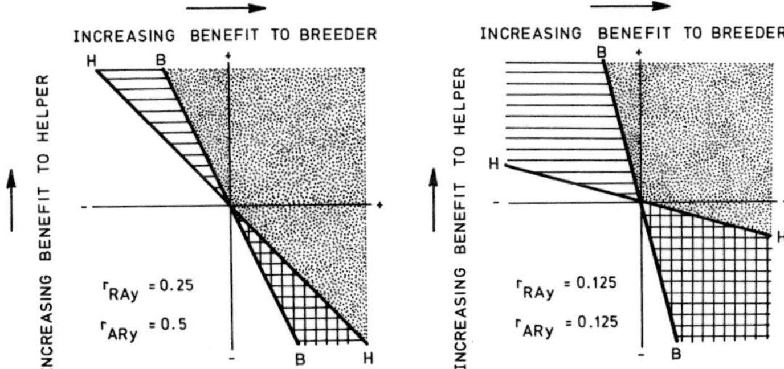

Fig. 2: Fitness space diagrams for breeders and helpers of two different degrees of relatedness. r_{AR_y} = coefficient of relatedness between an auxiliary and the young of the breeder (whom it usually helps to rear). r_{AR_y} = coefficient of relatedness between a breeder and the young of the auxiliary. The inclusive fitness of a breeder is increased through the retention of a helper in all areas to the right of line B-B; the inclusive fitness of an auxiliary is increased by such an association in all areas to the right of line H-H. Speckled areas indicate zones where continued association is mutually beneficial. Striped and cross-hatched areas represent zones of conflict, in which one party gains, but he other loses, by the association. Two distinct types of breeder-helper conflict, and the behaviors expected with each, are described in the text. Note that as the degree of relatedness between breeder and helper decreases, the zones of conflict increase (modified from EMLEN, 1982 b).

flict over the issue of helping (EMLEN, 1982 b). The size of these conflict zones increases as the degree of relatedness between breeder and helper decreases (compare Figs. 2 a and b).

Type I Conflict (striped zone), typifies helper-at-the-nest societies. Severe ecological constraints effectively prohibit dispersal and independent breeding by yearlings, yet retention of such auxiliaries results in a loss to the personal fitness of the breeder via competition, interference, greater conspicuousness to predators, etc. This means that the auxiliary gains, but the breeder loses, from retention. Assuming that the breeder is dominant over the auxiliary, the breeder will either expel the subordinate from the group or demand that the subordinate aid in the care of its young so as to increase the fitness of the breeder above the breeder retention tolerance line (labelled B-B). Notice that retention of related auxiliaries can evolve even when there is a cost in direct fitness to the breeder, since the improved survival chances of a retained auxiliary can increase the indirect component of the breeders' inclusive fitness.

In Type II Conflict (cross-hatched zone), the breeder gains from the presence of the auxiliary but the auxiliary loses from retention. This will occur when ecological constraints against novice breeders are slight but there is a direct benefit of group living *to breeders*. Even though the breeder would still benefit from the presence of the helper, the breeder has little leverage to prevent helper dispersal. The only option the breeder has is to entice the auxiliary to stay by «forfeiting»[3] some of its fitness benefits via shared fertilization, shared egg-laying, or reciprocation of helping (EMLEN, 1982 b). The dominant breeders must allow the helper a large enough genetic contribution to the clutch to increase the helper's fitness above *its* retention tolerance line (labelled H-H). *This leads to a communal breeding system with a lower degree of bias.*

[3] The term «fitness forfeiting» is used in the sense of ALEXANDER (1974).

This model illustrates the nature of conflicts within groups, and shows how the options and leverages of the different group members change as a function of changing environmental conditions. Furthermore, it specifically shows that the strongly biased helper-at-the-nest system can evolve only when the helper stands to gain more from retention in the long run than the breeder does, whereas lower-biased communal systems evolve when the breeder gains more than the helper.

4.2 How much bias?

Following up this conflict approach, we can now pursue the question of how much fitness the dominant must forfeit to keep the helper in the group. A short-term model of inclusive fitness costs and benefits is required and the boundary conditions for such a model are that both the helper and breeder must break even, i. e., all solutions must lie within the speckled area of the fitness space of Figure 2. Furthermore, this model assumes that, within the limits set by the subordinate's options, the dominant individual has the greatest leverage to push the fitness advantage in its favor. Given these assumptions, it is a simple matter to solve for the optimal degree of bias (Vehrencamp, 1983).

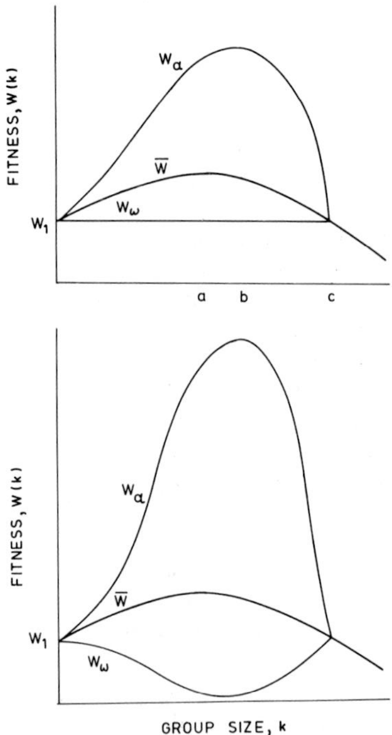

Fig. 3: The maximum degree of bias in fitness that can be generated between dominants and subordinates in groups containing (top) unrelated individuals, and (bottom) individuals related by a coefficient of relatedness r = .25. W is the average or per capita fitness of group members, W_α is the fitness of the dominant, and W_ω is the fitness of a subordinate. The distance between the W_α and W_ω curves represents the amount of bias possible at that group size. See text for further explanation.

Take the simplest case of a group of unrelated individuals where grouping benefits all group members on average. The function relating the per capita reproductive success to group size, k, is given by $\overline{W}$ (k) (Figure 3 top). For each unit of fitness that the dominant can usurp from a subordinate, one unit of fitness accrues to the dominant. For simplicity we also assume that there is only one dominant in the group and that all subordinates are affected by the same amount. The dominant cannot lower the fitness of any one subordinate below the fitness of a solitary breeder, $\overline{W}$ (l), otherwise subordinates will be selected to leave and breed solitarly. The lowest possible fitness of a subordinate in a group, $W_{\omega}(k)$, is therefore depicted as a horizontal line intersecting $\overline{W}$ (l). The fitness increment of the dominant, $W_{\alpha}(k)$, is the difference between $\overline{W}$ (k) and $\overline{W}$ (l) times the number of subordinates. The bias in fitness is therefore reflected in the degree to which the fitness of dominant and subordinate diverge. Clearly the greater the average benefit to grouping compared to solitary living, the greater the bias can be.

For groups of related individuals, a similar graph can be derived, but in this case the subordinates leave the group when their *inclusive fitness* would be greater if they bred solitarily (Figure 3 bottom). Here, the direct fitness of the subordinate can go well below the direct fitness of a solitary breeder, and the higher the degree of relatedness, the greater the bias can be. As in the unrelated case, grouping and biasing can only occur when there is a net benefit to grouping, when $\overline{W}$ (k) > $\overline{W}$ (l).

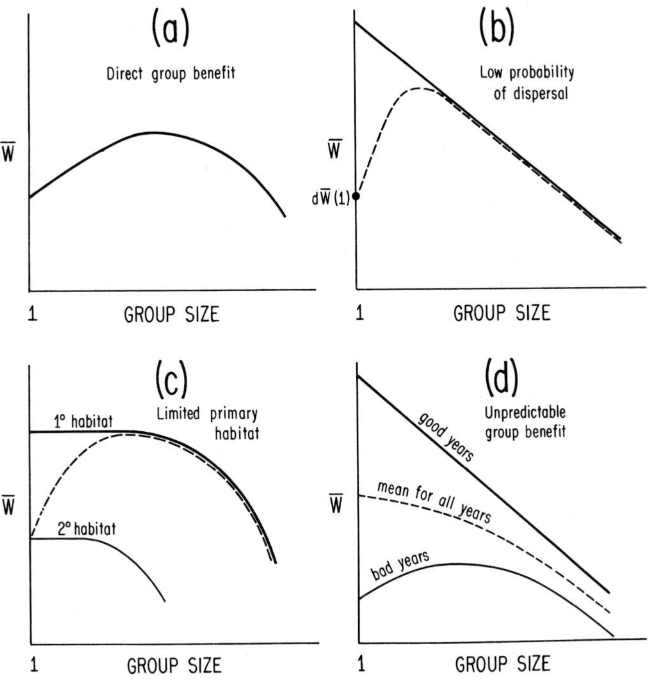

Fig. 4: Four ways to generate a benefit to grouping: (A) direct group benefit, (B) low probablity of dispersing to the optimal solitary situation, (C) limited primary habitat with no disadvantage due to grouping, and (D) direct benefit to grouping occurring unpredictably in poor years only. See text for further explanation.

Given that parents are dominant over their offspring and can control dispersal, group composition, and the degree of bias, certain sets of ecological conditions can be shown to lead to specific types of cooperatively breeding societies, e. g. egalitarian (communal) breeding among relatives, strongly biased breeding with relatives, or moderately biased breeding among non-relatives (Vehrencamp, 1983).

As we saw earlier, there are several situations in which grouping can be advantageous. These conditions are depicted graphically in Figure 4. When group breeding leads to a per capita increase in fitness compared to solitary breeding, there is a direct benefit to group living (Fig 4 a). Communal cooperative breeding will result if there are no strong ecological constraints. The White-winged Chough, *Corcorax melanorhanphus,* provides an example of such a society (Rowley, 1978).

In another case, grouping could lead to net decrease in per capita fitness compared to solitary breeding (Fig 4 b, solid line). However, if the constraints against becoming an independent breeder are high, the intrinsic benefit of independent breeding must be devalued accordingly. This leads to an effective «hump» in the $\overline{W}$ (k) curve and grown offspring may be «forced» to remain on their natal territories (Fig 4 b, dashed line). As discussed previously, this appears to have been the most common cause of grouping among helper-at-the-nest species. The probability of dispersing successfully, here denoted as d, can be calculated for permanently territorial species as the number of breeding slots available relative to the number of auxiliaries competing for them and is an inverse measure of the level of habitat saturation or ecological constraint. This condition can only lead to groups of close relatives since the groups form by retention of natal members. The florida Scrub Jay, *Aphelocoma coerulescens*, is a well-studied example (Woolfenden, 1975).

Another situation in which grouping might be favored is illustrated in Fig. 4 c. Here there is a primary, high quality, habitat which is limited, and a secondary habitat which can support only pairs but which is not limited. The two types of habitats must be close by or arranged in a mosaic so that dispersers have a choice between them. Large groups will occur on good territories, small groups or pairs on the poorer territories. This situation is graphically similar to the Verner-Orians polygamy model (Orians, 1969) and can lead to communal groups of non-relatives. Cooperative breeding in the Grove billed Ani, *Crotophaga sulcirostris,* has been explained in this manner (Vehrencamp, 1978).

Finally, the conditions for grouping may change unpredictably from year to year (Fig 4 d). If grouping is disadvantageous in good years and advantageous in bad years, then birds will hedge their bets by attempting to breed solitarily in most years, but will switch to groups if conditions deteriorate. If there is only a short time for breeding, some birds will initiate breeding but give up their nest to help others during bad times. The model predicts that only close relatives will be willing to accept this cost *and* be in a position to benefit from grouping via heightened inclusive fitness. This situation can lead to a facultative helper-at-the-nest system in which helping is most prevalent in poor years. The society of the White-fronted Bee-eater, *Merops bullockoides,* fits this pattern (Emlen, 1981, 1982 a).

In summary we have shown how a small set of ecological and social factors determine whether or not cooperative breeding is advantageous, and what form of cooperation will result. Typically, cooperatively breeding species have been described as living in harmonious, aid-giving groups. We believe this description to be exaggerated. Rather, such species live in societies faced with severe ecological hardships which restrict the opportunities for maturing young to become independent breeders. They also live in societies in which competitive conflicts of interest between different group members reach extreme limits. Viewing the various forms of cooperative breeding in terms of the balance of interest between conflicting group members, with dominants having greater behavioral leverage than subordinates, provides a more useful framework for understanding each system and the evolutionary forces that shaped it.

Acknowledgments

The research of the authors has been supported by the National Science Foundation through grants number BNS-76-81921 and BNS-79-24436 (STE) and DEB-78-25230 and DEB-80-022167 (SLV). Some of the ideas in this paper were developed while STE was a Visiting Fellow at the Center for Advanced Study in the Behavioral Sciences at Stanford.

References

Alexander, R. D. (1974): The evolution of social behavior. Annu. Rev. Ecol. Syst. *5*, 325–383.

Bertram, C. C. R. (1978): Living in groups: predators and prey. In J. R. Krebs and N. B. Davies, (eds.). Behavioral ecology: an evolutionary approach. pp. 64–96. Oxford, Blackwell Scientific Publications.

Brown, J. L. (1974): Alternative routes to sociality in jays – with a theory for the evolution of altruism and communal breeding. Am. Zool. *14*, 63–80.

Brown, J. L. (1978): Avian communal breeding systems. Annu. Rev. Ecol. Syst. *9*, 123–156.

Brown, J. L. (1980): Fitness in complex avian systems. In H. Markl. (ed.). Evolution of social behavior: hypothesis and empirical tests. Pp. 115–128, Verlag Chemie, Weinheim.

Dow, D. D. (1977): Reproductive behavior of the Noisy Miner, a communally breeding honeyeater. Living Bird *16*, 163–185.

Emlen, S. T. (1978): The evolution of cooperative breeding in birds. In J. R. Krebs and N. B. Davies (eds.). Behavioral ecology: an evolutionary approach, pp. 245–281. Oxford, Blackwell Scientific.

Emlen, S. T. (1981): Altruism, kinship and reciprocity in the White-fronted Bee-eater. In R. D. Alexander and D. Tinkle (eds.). Natural selection and social behavior: recent research and new theory. pp. 217–230. New York, Chiron Press.

Emlen, S. T. (1982 a): The evolution of helping. I. An ecological constraints model. Am. Nat. *119*, 29–39.

Emlen, S. T. (1982 b): The evolutuon of helping. II. The role of behavioral conflict. Am. Nat. *119*, 40–53.

Gaston, A. J. (1978 a): The evolution of group territorial behavior and cooperative breeding. Am. Nat. *112*, 1091–1100.

Gaston, A. J. (1978 b): Demography of the Jungle Babbler, *Turdoides striatus*. J. Anim. Ecol. *47*, 845–879.

Grimes, L. G. (1976): The occurence of cooperative breeding behavior in African birds. Ostrich *47*, 1–15.

Harrision, C. J. (1969): Helpers at the nest in Australian passerine birds. Emu *69*, 30–40.

Koenig, W. D. (1981 a): Reproductive success, group size, and the evolutuon of cooperative breeding in the Acorn Woodpecker. Am. Nat. *117*, 421–443.

Koenig, W. D. (1981 b): Space competition in the Acorn Woodpecker: Power struggles in a cooperative breeder. Anim. Behav. *29*, 396–409

Koenig, W. D. and F. A. Pitelka (1981): Ecological factors and kin selection in the evolution of cooperative breeding in birds. In R. D. Alexander and D. Tinkle (eds.). Natural selection and social behavior: research and new theory, pp. 261–180. New York, Chiron Press.

Lawton, M. F. and C. F. Guindon (1981): Flock composition, breeding success, and learing in the Brown Jay. *83*, 27–33.

Ligon, L. D. and S. H. Ligon (1978): The communal social system of the Green Woodhoopoe in Kenya. Living Bird *17*,159–198.

MacRoberts, M. H. and B. R. MacRoberts (1976): Social organization and behavior of the Acorn Woodpecker in central coastal California. Ornithol. Monogr. *21,* 1–115.

Orians, G. A. (1969): On the evolution of mating systems in birds and mammals. Am. Nat. *103,* 589–603.

Reyer, H.-U. (1980): Flexible helper structure as an ecological adaptation in the Pied Kingfisher *(Ceryle rudis).* Behav. Ecol. Sociobiol. *6,* 219–227.

Ridpath, M. G. (1972): The Tasmanian Native Hen, *Tribonyx mortierrii.* CSIRO Wildl. Res. *17,* 1–118.

Rowley, I. (1965): The life history of the Superb Blue Wren, *Malarus cyaneus.* Emu *64,* 251–297.

Rowley, I. (1976): Co-operative breeding in Australian birds. In H. Frith and J. H. Calaby (eds.). Proceedings of the XVI International Ornithological Congress (Canberra, Australia), p. 657-666.

Rowley, I. (1978): Communal activities among White-winged Choughs, *Corcorax melanorhamphus.* Ibis *120,* 178–197.

Rowley, I. (1981): The communal way of life in the Splendid Wren, *Malurus splendens.* Z. Tierpsychol. *55,* 228–267.

Selander, R. K. (1964): Speciation in wrens of the genus Campylorhynchus. Univ. Calif. Publ. Zool. *74,* 1–224.

Stacey, P. B. (1979): Habitat saturation and communal breeding in the Acorn Woodpecker. Anim. Behav. *27,* 1153–1166.

Stacey, P. B., and C. E. Bock, (1978): Social plasticity in the Acorn Woodpecker. Science *202,* 1297–1300.

Trivers, R. L. (1974): Parent-offspring conflict. Am. Zool. *14,* 249–264.

Vehrencamp, S. L. (1978): The adaptive significance of communal nesting in Groove-billed Anis *(Crotophaga sulcirostris).* Behav. Ecol. Sociobiol. *4,* 1–33.

Vehrencamp, S. L. (1979): The roles of individual, kin and group selection in the evolution of sociality. In P. Marler and J. G. Vandenbergh (eds.). Handbook of behavioral neurobiology, pp. 351–394. Vol. 3. New York, Plenum.

Vehrencamp, S. L. (1983): A model for the evolution of despotic versus egalitarian societies. Anim. Behav. *31:* 667–682.

West-Eberhard, M. J. (1975): The evolution of social behavior by kin-selection. Q. Rev. Biol. *50,* 1–33.

Wilson, E. O. (1971): The Insect Societies. Cambridge, Mass., Belknap Press.

Woolfenden, G. E. (1975): Florida Scrub Jay helpers at the nest. Auk *92,* 1–15.

Woolfenden, G. E. and J. W. Fitzpatrick (1978): The inheritance of territory in group-breeding birds. BioScience *28,* 104–108.

Zahavi, A. (1974): Communal nesting by the Arabian Babbler: a case of individual selection Ibis *116,* 84–87.

Fortschritte der Zoologie, Bd. 31 · Hölldobler/Lindauer (Hrsg.): Experimental Behavioral Ecology
G. Fischer Verlag · Stuttgart · New York · 1985

Helping, cooperation, and altruism in primate societies

Christian Vogel

Institut für Anthropologie, Georg-August-Universität, Göttingen, F.R. Germany

Abstract

Altruism and cooperation are among those theoretical concepts which have been fundamentally reevaluated by modern evolutionary biology in regard to their adaptive function and ultimate causation. This paper reviews various types of helping, mutual support, cooperation and altruistic acts within nonhuman simian primate societies following the perspective of their assumed evolutionary significance as well as their functional adaptation to the respective species-specific social settings. It is shown that helping, altruistic behavior, and cooperation are strongly biased toward close kin, particularly among females which in the majority of simian primates form the continuous and stable element of reproductive social units by remaining within their natal troop throughout their lifetime, whereas the males migrate. In those cases where males form the stable center of the social organization and females are transients, as for instance in chimpanzees, nepotistic behavior is accordingly more common among the males. Reciprocity, however, constitutes an additional motivation for altruistic acts among relatives as well as between unrelated individuals, especially among individuals occupying adjacent ranks within their troop's social hierarchy. As has been documented by several longterm longitudinal field-studies adequate rewarding of altruistic investments may be delayed for several months or even years under certain social circumstances. For reasons of controversial indications emerging from field observations and laboratory research the question of whether or not nonhuman primates possess true «empathetic» or «sympathetic» capabilities must at present remain unresolved.

1 Introduction

Thirty years ago, Hebb and Thompson (1954) noted: «There is a great deal of anecdotal material to suggest the existence of altruism and cooperation in primate societies –, but the evidence in general has been so poor that Zuckermann (1932) had to conclude that no altruism exists.»

However, the situation changed dramatically when evolutionary biologists and, in particular, sociobiologists began to analyze intensively the complementary concepts of competition and cooperation in order to uncover the modalities under which cooperation, helping, and altruistic behavior could have evolved. In part, by adapting these concepts to the respective requirements of argumentation, terms like altruism, cooperation, helping etc. have been redefined several times and, of course, sometimes in confusingly divergent ways.

Therefore, my first task will be to define these terms as I shall use them through this paper.

Helping, primarily, is an asymmetrical interaction within a dyadic context, one individual protecting another individual from being harmed, or one individual aiding, assisting, or supporting another individual to reach its goal or to complete its intended action. Helping may be reciprocated, it may be altruistic or selfish, it may or may not be consciously intended, and its motivation may be classified as nepotistic, opportunistic, hedonistic or something else.

Cooperation is a collaborative behavior of two or more individuals directed toward a common goal or to the production of some common behavioral effect. The partners of a cooperative act are operating simultaneously and are mutually coordinated, they may act in complementary roles («division of labor»), but not in «essentially different roles» (which would serve as a criterion of Kummer's, 1967, «tripartite relationships»). All partners gain benefit, whatever it may be, as a direct result of their joint action (compare Wrangham's, 1982, concept of «mutualistic interaction»). Note that we are neither referring to a particular type or quality of reward nor to the criterion of whether or not the partners actually gain more than they would if acting alone. Although the latter result might be intended, it should not by itself function as a component of the descriptive definition. It should further be noticed that our concept of cooperation is different from what has been called «cooperation» in the game-theoretical approaches like «prisoners dilemma», which includes, in particular, «reciprocal altruism» (e. g. Axelrod and Hamilton, 1981). By definition cooperation is here limited to non-altruistic interactions.

The most controversial concept of those dealt with here is undoubtedly *«altruism»*. In humanities and social sciences it has frequently been emphasized that «we have to know more about an incident than the behavioral outcome before we can decide whether or not it *is* altruistic. – The classical issues associated with altruism are *motivational*» (Krebs, 1978). We shall come back to this issue in the final part of this paper. However, here we adopt the restrictive definition coined by evolutionary biologists: *Altruism* is defined as behavior that benefits another individual at some cost to the altruist; costs and benefits being measured in terms of *individual fitness*. Note that by this definition a self-sacrificing parent or grandparent does not act altruistically, whereas a person who acts correspondingly in favor of a sibling or nephew etc. would be labeled as altruist on the basis of the above definition (see also Bertram, 1982).

In defining *«reciprocal altruism»* I follow Packer (1977): «Reciprocal altruism implies the exchange of altruistic acts between unrelated individuals as well as between relatives. If the benefits to the recipient of an altruistic act exceed the costs to the altruist, and if the recipient is likely to reciprocate at a later time, then the cumulative benefits for both individuals will have exceeded the cumulative costs of their altruism».

The counterpart of altruistic is *selfish:* a behavior is considered as being selfish if it serves the *personal fitness* of the actor, irrespective of whether or not it serves the fitness of others as well.

The last terminological differentiation I should make in advance is that between «alliance» and «coalition». For convenience I shall call an *alliance* any cooperative ad hoc formation of individuals acting together toward a common goal, whereas *coalitions* are long-term relationships between two or more particular individuals mutually reciprocating altruistic acts, irrespective of whether or not the benefits are delayed in time (see also Bertram, 1982).

The examples given in this paper are (a) confined to simian primates, (b) primarily focused on free-ranging conditions, and (c) limited to intra-group interactions. Further, I am not concerned here with the whole complex of parental cooperation in raising offspring (a cooperation which is particularly developed in several monogamous primate species). Finally, I do not refer to basic questions of general relationships between cooperation and competition with respect to the evolution of social life and social groupings in primates, units which are per se «cooperative» systems within a competitive setting, as for instance WRANGHAM (1980) argued on the basis of his «feeding competition model». Here I concentrate on actual, directly observable interactions between particular individuals or classes of individuals within social units.

In doing so I will not dwell on more or less singular, occasional, sometimes even anecdotal observations of one individual «helping» another, e. g. of a chimpanzee removing irritants from a conspecific's eye, or cleaning the recipient's teeth, but, I shall confine my examples to regularly observable interactions within the typical, sometimes even species-specific social settings.

2 Cooperation

Let us begin with two classical examples of cooperation. The first one deals with cooperative displays, which among primates are most perfectly developed in form of the famous mutually coordinated «duets» of some of the monogamous hylobatids (Genera: *Hylobates* and *Symphalangus*). These acustically as well as optically impressive pair-ceremonies, characterized by complementary role-taking, appear to serve a combination of distinct functions: (1) territorial advertisement and territorial spacing; (2) demonstration of pair solidarity toward neighbors as well as foreigners; and (3) if the «duet» is complete, to keep other potential sexual partners at a distance, and if the «duet» is incomplete, to attract potential complementary sexual partners (e. g. SERPELL, 1981). The common rewards are the maintenance or acquisition of nutritional as well as reproductive resources.

The second example concerns the obviously intentional cooperation during predatory episodes among chimpanzees *(Pan troglodytes)*. Mutually coordinated actions of adult males were observed with regard to stalking, cornering, and chasing the prey (TELEKI, 1973). Anticipatory capabilities of the actors have been inferred by the observers in all these contexts. The «hunting»-parties comprised up to 5 adult males, and the prey was immediately divided among the cooperating males, who then further distributed pieces of meat to other troop members (e. g. TELEKI, 1973; SUZUKI, 1975). It seems difficult to assess the substantial advantage (if any!) of «hunting» *cooperatively* compared to singly predating from the data published so far. In any case, this is true cooperation in the sense of our above definition.

These two examples are rather out of the ordinary illustrations of cooperation among nonhuman primates confined to highly specialized contexts. In general, cooperative and/or helping interactions among primates are directly or indirectly linked with reproductive strategies in specific ecological and/or social settings.

3 The «helper-at-the-nest» situation

By this metaphor a social and reproductive configuration has been characterized in which non-reproducing sons and/or daughters altruistically help their parents to raise their offspring. Among primates this classical pattern has been developed by the tiny marmosets (Family: Callitrichidae). The primarily insectivorous Callitrichids, owing to special ecological pressures, are adapted to an opportunistically «colonizing» reproductive style (see

EISENBERG, 1978; FORD, 1980). Marmosets produce twins or even triplets twice a year. The newborns weigh about one-fifth to one-quarter as much as their mother. Therefore, it appears plausible that this type of reproduction requires the active support of other family members (e. g. EPPLE, 1975, 1978; INGRAM, 1977). Usually this active support is provided by the breeding male of the strictly monogamous pair, i. e. the biological father. However, additional and demonstrably significant help (by carrying the babies, bringing them back to the mother for suckling, and even sharing food with them) ist frequently provided by elder siblings of the infants. From laboratory research we know that these other family members are inhibited in their own reproduction by the mere presence of the dominant breeding pair. Thus, even fully adult daughters do not ovulate, their sexual activities are completely suppressed as are those of adult sons. This is an almost classical example of the «helper-at-the-nest»-situation, the best we can find in nonhuman primates. In evolutionary terms the «helpers» may gain in *inclusive fitness* on the long run, but this has still to be proven by long-term field data.

Another frequently quoted, but much less unequivocal and less extreme type of helping others to rear their offspring has been described as «aunt behavior» (HINDE et al., 1964; HINDE and SPENCER-BOOTH, 1967; JAY, 1965) or «allomothering», «infant sharing» and «infant transfer», which in its most developed form seems to be a characteristic trait of colobine social behavior (e. g. HRDY, 1976, 1977; MCKENNA, 1979, 1981). During their first weeks of life, langur infants (*Presbytis entellus*) spend a considerable amount of day time with «allomothers» who carry them around, sometimes grooming and cradling them, but, sometimes undoubtedly exploiting and even severely mistreating the infants (see below). However, at least as far as elder siblings are involved, which have proven to be the most careful and assiduous allomothers (VOGEL, 1984), there is a slight approximation to the «helper-at-the-nest» situation, which may be explained in evolutionary terms of *kin selection* by increasing the inclusive fitness of the helpers.

4 Reciprocal relations

The most famous example of male-male coalitions reciprocating altruistic acts has been reported by PACKER (1977) among *Papio anubis*. By gestures one male solicits the support of another who may join him in threatening and attacking a third male who is consorting with an estrous female. «On six occasions the formation of a coalition directed against a consorting male resulted in the loss of the female by the opponent. In all six cases the female ended up with the enlisting male of the coalition; the solicited male generally continued to fight the opponent while the enlisting male took over the female. In each of those cases the solicited male risked injury from fighting the opponent while the enlisting male gained access to an estrous female» (PACKER, 1977). The female was taken over by the male who had solicited, and not necessarily by the dominant male of the pair (PACKER, 1979). More importantly, «thirteen different pairs of males reciprocated in joining coalitions at each other's request on separate occasions. Individual males which most frequently gave aid were those which most frequently received aid. These results suggest that preferences for particular partners may be partly based on reciprocation» (PACKER, 1977). It was at least unlikely that the coalition partners were close relatives, since anubis baboon males are typical transients. However, it seems likely that «the number of offspring that a male sired as a result of participating in reciprocating coalitions would be greater than if he did not» (PACKER, 1977). PACKER's observations fit well to those of DE VORE (1965) and HAUSFATER (1975) who had already reported that in *Papio anubis* as well as in *Papio cynocephalus* subordinate males sometimes gained access to estrous consort females of higher ranking males by cooperatively harassing the latter.

The counterpart, reciprocal female-female coalitions, has been described by DUNBAR (1979, 1980) in *Theropithecus gelada.* Among geladas dominance rank within the matrilines seems to be determined, at least in part, by the relative aggressiveness of individual females. However, since individual aggressiveness, and hence individual rank as well, have been shown to follow an inverted J-shaped function of age, and since the number of a female's offspring correlates positively with her dominance rank, females should try to reach their optimum position as early as possible, and then, should attempt to prevent rank decline through older ages. The author argued convincingly that these goals could best be reached by forming a stable age-graded female-female longterm coalition. This supposition could be verified by DUNBAR's observations. The age-graded relationship is reciprocal only in the long term, though in the short term it is asymmetrical. It has been shown «that females who form (such) coalitions gain a life-time reproductive advantage over those who do not because coalitions with younger females help to prevent the decline in rank that would otherwise occur in old age. It is argued that females prefer to form coalitions with daughters rather than with unrelated females because the mother-daughter relationship is the only bond of sufficient strength to provide the basis for an investment which is asymmetric in the short term and reciprocal only over the length of a life-time. Any benefits that accrue from kin selection are considered to be secondary» (DUNBAR, 1980). Of course, if these conclusions hold true, then we are dealing with one of the most impressive examples of «delayed benefit» (BERTRAM, 1982) which has been described so far in nonhuman primate societies.

Papio hamadryas provides another example of considerably delayed rewards for altruistic investments, in this case, of a male in favor of a female. An initial form of the hamadryas «one-male unit» consists of a young adult (or subadult) male having «adopted» a juvenile (or even infant) female, taking the role of the mother with every semblance of solicitous maternal care, supporting the juvenile female in any respect of daily life. The growing female herself «seems to transfer the mother's role into the male», and this situation persists into adulthood. «Throughout adult life, a female under extreme stress will cling to the male's back or be embraced by him in the way of the infant» (KUMMER, 1967). This peculiar relationship develops into a stable sexual and social consortship as the female matures. Thus, the male by adopting the mother's role invests altruistically into a longterm interindividual relationship which rewards him only months or even more than two years later (SIGG et al., 1982) in the currency of sexual favors and, ultimately, of reproductive success. This is another most intriguing example of «delayed benefit» (BERTRAM, 1982) in reciprocal relationships.

5 Help, support, and cooperative defense in reproductive contexts

In several species it has been observed that particularly estrous females do solicit and receive support from their respective male consort partners. For instance, SEYFARTH (1978 a) in a detailed study of a *Papio ursinus* troop found «that six of eight females were supported by adult males when they received aggression from other females proportionally more often during sexual consortship than at other times». These supports, however, never did induce any changes in female rank. On the other hand, SEYFARTH (1978 b) was also able to show that the majority of male aids occurred outside sexual consortship, and appeared to be unrelated to changes in female reproductive state. «Each male showed a clear preference in the distribution of these aids, giving the majority to the individual with whom he also shared most frequent proximity, grooming, etc.» Hence, immediate sexual favors are obviously not the main factor provoking male-female support.

Cooperative female support on behalf of preferred males has been observed in several primate species. Since, as Silk and Boyd (1983) noted, «the effectiveness of female mate choice may be influenced by male rank insofar as rank influences both male tenure and access to females, both males and females are expected to have an interest in the outcome of competitive interactions among adult males». Females, of course, act accordingly if preferred males are challenged. They are more likely to do so when a male is established in the troop and is already responsible for a number of their offspring. For instance, Hrdy (1977) as well as our team observed langur females *(Presbytis entellus)* jointly defending their established male leader against aggressive males from all-male bands, which tried to oust the leader. Hrdy argued that the main interest of these females supporting the father of their infants was to avoid infanticide by a new leader male (see below).

A particular type of (longterm) male-male support are the relationships between «leader» and «follower» within the reproductive units of *Papio hamadryas* first described by Kummer (1968). A young adult (or subadult) male enters the one-male unit by adopting submissive behavior that makes him acceptable to the harem-leader. The two males gradually develop something like a «working partnership» (Crook, 1971). The males adopt a sort of complementary role-taking, the leader keeping the part of a «decider» while the follower adopts the function of watching the surroundings and, sometimes, herding the females. As the older male ages, the younger male acquires sexual access to an increasing number of females of the harem; the leader, however, still retaining control of group movement and daily routine. A complex system of mutual «notification» is set up between them to avoid «misunderstandings». It seems difficult to explain why a healthy harem-leader should accept a competitor in the form of the follower. I think the only plausible reason could be that leader and follower are close relatives, as Abegglen (1984) suspected. An aging male could gradually diminish the competition on behalf of his own reproduction, if he is in an effective position to enhance substantially the reproductive chances of close relatives, and by this increase his own inclusive fitness.

Longterm coalitions between particular males conquering and defending bisexual troops, and so gaining or maintaining access to reproductive females, have also been described for *Alouatta seniculus* by Sekulic (1983).

There are also female-female «attack alliances» against particular males. For example, unrelated female rhesus monkeys *(Macaca mulatta)* band together in fights against adult male troopmates (Kaplan, 1977); alliances of female geladas *(Theropithecus gelada)* dominate their male harem-leader, who for his part dominates each single female of his unit (Bramblett, 1970); female alliances in vervet monkeys *(Cercopithecus aethiops)* are able to drive away otherwise dominant males (Seyfarth, 1980); and cooperative attacks of females against particular males preceded the latter's emigration in *Macaca mulatta* (e. g. Lindburg, 1971) and *Macaca fuscata* (e. g. Kurland, 1977). In *Presbytis entellus* the females of a troop have been observed jointly attacking intruding males in contexts of male «take-overs». They furiously defended mothers and infants against presumably or actually infanticidal males (Hrdy, 1974, 1977; Vogel and Loch, 1984; Sommer, 1984). If the participating females are nonrelatives this behavior might be explained in terms of potential reciprocal altruism, otherwise it could be considered as an altruistic act in defense of kin, and hence, of own inclusive fitness, whereas in the long run it is rather doubtful whether or not the females are really able finally to prevent infanticide (Hrdy; Sommer, 1984).

6 Cooperative hampering or even suppressing reproduction of other consexuals: intrasexual competition

This has been described for both sexes. For instance De Vore (1965) reported that even high ranking males of *Papio anubis* sometimes were prevented from completing copulation through persistent cooperative harassment and physical interruptions by lower ranking males of the troop. Killing the offspring of another male, of course, is a much more drastic and effective strategy to gain reproductive advantage at the expense of a competitor (e. g. Hrdy, 1977, 1979; Chapman and Hausfater, 1979; Hausfater and Vogel, 1982; Vogel and Loch, 1984; Sommer, 1984). This, however, is not a cooperative act.

Wasser (1981, 1983) has shown by his field data on *Papio cynocephalus* that females compete with each other on two levels: that of male choice and that of reproductive inhibition of competitors. The «ratio» behind the strategy: «a female is likely to increase her relative personal fitness each time she inhibits another female». Among Wasser's yellow baboons adult females «mediated reproductive inhibition in consexuals» by forming offensive alliances following the tactic: «attack those whose reproduction is easiest to inhibit». However, easiest to inhibit are the reproductive capabilities of females during the most stress-sensitive reproductive phases, i. e. during preovulatory estrus and during early pregnancy. Actually, premature termination of estruses as well as abortions induced by social harassment have now been reproted from several species of free-ranging primates (e.g. Wasser, 1981; Pereira, 1983; Sommer, 1984). The participation of particular females in these attack alliances should, therefore, depend on «common interests» among the allies, mediated by their reproductive state at the respective time of alliance formation, and on the dominance rank relations between allies and victims. Close relatives as victims should be excluded. These predictions were confirmed by Wasser's observations.

It is consequently not surprising that this type of female-female competition continues after parturition as well. One of the most effective ways of reducing the fitness of competitors would be to mistreat or harm infants of non-(or distantly)related females. This, though mostly overlooked, is an aspect of the so-called «aunt behavior» or «allomothering» (see Wasser and Barash, 1981). Hrdy (1976, 1977), Kurland (1977), Wasser (1981, 1983) and Vogel (1984) described incidents of apparently deliberately harming, endangering and even injuring infants of other females by allomothers among macaques, baboons, and langurs. Hence, Hrdy (1981) is correct in stating that allomothering, with all its complexities, provides a striking illustration of the fine line between cooperation and exploitation between a female helping a troop-mate and one helping herself at a troop-mate's expense. Real helping and «pseudohelping» are the two facets of allomothering (Vogel, 1984).

However, the complement to male infanticide is female infanticide, which has so far been described only in four cooperative incidents among the chimpanzees *(Pan troglodytes)* of the Gombe Reserve (Tanzania). Goodall (1977, 1979) reported that two adult females, mother and daughter, belonging to a high-ranking lineage cooperatively killed and cannibalized two infants of a particular low-ranking female and two infants of a particular middle-ranking mother respectively. It is not yet clear whether this behavior should be considered as abnormal and hence maladaptive, or whether it was prompted by mere need or appetite for meat and the two females just took the easiest prey (which seems unlikely, however), or whether mother and daughter pursued a particular reproductive strategy by eliminating the offspring of other non-related female troop members in order to reduce the number of competitors for their own offspring. In this connection it should be emphasized that in contrast to the majority of other nonhuman primate societies (in particular the «female-bonded societies», WRANGHAM, 1980) adult females in chimpanzee breeding units are most frequently not close genetic relatives, and so the victims were probably not related to the infanticidal females.

7 Helping and cooperation as tools to raise reproductive chances mediated through dominance rank and kinship

Particularly in female cercopithecines, there are strong correlations between kinship, dominance rank, and reproductive success. For instance, female macaques form stable linear rank orders according to three general rules: (1) females outrank all unrelated females outranked by their maternal relatives; (2) adult females rank immediately above their daughters (and their immature sons); (3) adult sisters rank in inverse order of their ages (e. g. SILK and BOYD, 1983).

KAPLAN (1977) observed in rhesus monkeys *(Macaca mulatta)* that «females aiding unrelated females against other unrelated females tended to help those animals ranked closer to them in the dominance hierarchy against animals who were more distantly ranked». He argued that supporting unrelated females of adjacent rank may stabilize the existing hierarchy among females, «the stable and central part of the rhesus monkey social group». However, since reciprocity generally seems to be infrequent among unrelated females, SILK (1982) admitted the question to be unresolved of why females form alliances to support unrelated females in the absence of reciprocity. DE WAAL (1978 a) claimed «redirected aggression» as an explanation of the prevailing «actor alliances» (in which the aggressive party of a dyadic encounter receives support) among unrelated female *Macaca fascicularis*, whereas SILK and BOYD (1983) argued that animals «form alliances in support of non-relatives in order to reduce the fitness of the aggressor».

Kinship, however, appears to be the primary matrix of helping and cooperation in primate societies. Since higher primates are characterized by a considerable longevity, social units can easily comprise three generations with all their possible ramifications in degrees of relatedness among troop members.

MASSEY (1977) found that within her study troop of *Macaca nemestrina* during agonistic «defender aids» (those aids provided on behalf of the victim of an ongoing aggressive interaction) the animals discriminated clearly between family members and non-family members. Moreover, they distributed their aids significantly according to the gradation of the coefficients of relatedness. «They chose to aid relatives of closer degrees of relatedness more often than relatives more distantly related». Reciprocity was common among family members (excluding mother-infant aids due to the highly unequal cost-benefit ratios involved). So true coalitions were observed only among members of the same genealogy. The author concluded «that kin selection and reciprocal altruistic selection may act conjunctively for family members which do reciprocate aids».

Within the «female-bonded» nepotistic social systems of ceropithecines acts of help and cooperation among females are in general strongly biased toward members of the own matriline. This has been documented e. g. by KAPLAN (1977, 1978) and BERMAN (1980) for *Macaca mulatta*; by KURLAND (1977) and WATANABE (1979) for *Macaca fuscata*; by MASSEY (1977) for *Macaca nemestrina*; by DE WAAL (1978 a) for *Macaca fascicularis*; by SILK (1982) for *Macaca radiata*; by PAUL (in prep.) for *Macaca sylvanus*; by CHENEY (1977) for *Papio ursinus;* and by FAIRBANKS (1980) for *Cercopithecus aethiops.*

In «female-bonded» societies males are transients, leaving their natal troop mostly (but not always) before, during, or shortly after reaching their full reproductive maturity. However, it has been observed that brothers or half-brothers later joined the same bisexual troop, and mutual support might be of essential importance in order to establish favorable positions within these reproductive settings.

For example, MEIKLE and VESSEY (1981) reported that among the free-ranging rhesus monkeys *(Macaca mulatta)* of La Cueva (Puerto Rico), young males frequently transferred into the same social group as their elder brothers. In these cases the authors observed that brothers spent more time close to each other than to other males, formed aggressive alliances with their brothers more frequently than with other males, and disrupted each other's interactions with estrous females less frequently than expected by chance. Males who had brothers within their non-natal troop remained within that troop significantly longer than males without brothers, and, of course, the length of time a male spent in a troop was positively associated with the male's dominance rank and, probably, also with his reproductive success. Therefore, the authors concluded that «HAMILTON's model of inclusive fitness remains a plausible explanation of nepotism among monkey brothers». SADE (1968) as well as BOELKINS and WILSON (1972) reported additional cases of fraternal associations outside the males' natal troops from rhesus monkeys of Cayo Santiago.

Similarly, in chimpanzees *(Pan troglodytes)* RISS and GOODALL (1977) described the successful challenge of the α-male by a team of two brothers, who were mutually supporting each other in raising their rank positions within the male dominance hierarchy. The authors suggested that the frequently observed longterm coalitions between pairs of chimpanzee males are regularly formed among brothers, which would be most effective since male chimpanzees generally remain in their natal troop. By «pushing» an elder brother into a dominant position the younger brother would not only benefit directly from reciprocation, but also through increasing his inclusive fitness, since there are indications that dominant males are able to inhibit successfully copulations of lower ranking males and thus, to enhance their own chances to inseminate estrous females (e. g. TUTIN, 1975; DE WAAL, 1978 b).

8 Helping and rank acquisition in «female-bonded» societies

There are several studies on helping and cooperation in relation to rank acquisition of adolescent females particularly in macaque and savannah baboon societies (e. g. BERMAN, 1980, for *Macaca mulatta*; CHENEY, 1977, for *Papio ursinus*; WALTERS, 1980, vor *Papio cynocephalus*).

For instance, CHENEY (1977) found in *Papio ursinus* that immatures received a considerable proportion of aids from those adult females who ranked adjacent to their mothers and with whom their mothers most frequently interacted. Offspring of high-ranking mothers were more frequently and more successfully aided, and consequently the existing rank hierarchy tended to be perpetuated across generations. CHENEY described a striking difference in alliance formation among immatures and coalitions between adult females. Immature animals formed aggressive alliances primarily with members of high-ranking matrilines, whereas adult females formed coalitions primarily with individuals of adjacent rank. It therefore seems necessary to assume «that over time the formation of alliances will be influenced by an additional factor: namely the animals will learn to cease to form alliances with those individuals who fail to reciprocate them», e. g. individuals of high-ranking matrilines on behalf of individuals of low-ranking matrilines. Hence during their ontogenetic development individuals will learn to ally primarily with animals who reciprocate, that is, with individuals of adjacent rank. «Reciprocal alliances between adjacently ranked animals might therefore be regarded at least in part as the result of compromises that animals adopt during development between the potential benefits to be derived from alliances with high-ranking individuals and the high cost entailed by lack of reciprocity». CHENEY's model, thus, predicts «that the pattern of alliances among adult females will be dependent upon the interaction between an individual's rank and the probability of reciprocity, and that over time reciprocal alliances will generally involve those of adjacent rank», which in savannah baboons, of course, are also most likely to be closely related.

Studying rank acquisition in *Papio cynocephalus*, WALTERS (1980) coined the concept of «targeted» vs. «non-targeted» females, in order to distinguish between females to which a certain adolescent female attempted to become dominant (targeted females), and those females to which the same adolescent female behaved submissively and always remained subordinate (non-targeted females). WALTERS' «examination of non-dyadic agonistic interactions revealed that adolescents frequently intervened against targeted females, but never against non-targeted females, but aided non-targeted females against adolescents. Interveners were unrelated adult females and other adolescents as well as kin. Targeting was determined by birth rank (the rank of the adolescent's mother at the time of the adolescent's birth) and did not depend upon interventions by the mother» (in some cases even the mother may become a targeted female of her own daughter, if the mother had declined in rank since the time of birth of her daughter!). By this mechanism each adolescent female finally reaches her «proper» position within the rank hierarchy, at least in part, by virtue of aid from other females, even if the mother had previously died. «Thus, who intervened for whom in interactions involving adolescent and adult females was governed more by rank relationships than by kin relationship or social bonds (e.g. grooming relationships). Animals were extremely reluctant to intervene against the existing or expected (based on birth rank) hierarchy», and by this they reinforced the transmitted dominance hierarchy among matrilines. «The great difference between the behavior of adolescents toward targeted and non-targeted females implies that young females recognize their position in their society, and the data indicate that this recognition is based on birth rank». Moreover, the other females too must have a perception of their own «proper» relationship with all adolescent female troop-mates. The social organization, in general, might be regarded as a conservative and rather stable compromise between the competetive strategies to maximize personal as well as inclusive fitness.

9 Helping, altrusim, and empathy

Let us return to the initially mentioned notion that the «classical issues associated with altruism are motivational» (KREBS, 1978). Everybody knows that for most «naive» people the concept of altruism is a part of the system of ethical values, and as such «truly altruistic behavior is evidence of Man as his best» (HATFIELD et al. 1978). So it ought to be absolutely unselfish. However, even among humanists it is controversial whether or not this moralistic measure might actually be applicable to man. Most psychologists, of course, would agree with COHEN's (1978) statement that «man operates, in a motivational sense, from point of view of self-interest» even if this self-interest might be as high-grade as to stand for «the desire of being morally superior» (WISPÉ, 1978). Following COHEN's definition: «altruism can be defined as an act of desire to give something gratuitously to another person or group because he, she, they, or it needs it or wants it», three significant elements arise: first, the desire to give something; second, the motivation to give it regardless of whether or not it will be rewarded; and third, the «empathetic» feeling for the needs of others. Thus, «empathy» is a central concept with respect to altruism. ARONFREED (1970) even defined «that any time a person's behavior is controlled by empathetic processes his behavior should be labeled ‹altruistic›». Hence, if we don't know anything about the «desire to give» and about «selfishness vs. unselfishness» of motivations in nonhuman primates, perhaps we might, at least, be able to decide whether or not they are capable of something like «empathy».

However, from field data published so far, there is not the slightest evidence that monkeys or even apes show signs of sympathetic empathy or «compassion» toward suffering, seriously wounded, or even dying conspecifics; not even reactions like warming, cleaning, protecting, or providing privileged access to food, similar to the so-called «invalid care» among dwarf mongoose *(Helogale undulata rufula)* as reported by Rasa (1976), have been found. The sad story of the male chimp's «Mc Gregor's» disease described by van Lawick-Goodall (1971) in her famous book «In the Shadow of Man» stands as a vivid illustration of the lack of any compassion among his familiar troopmates.

On the other hand, from laboratory experiments there are clear indications of empathetically mediated reactions and «altruistic» acts in apes and monkeys. For instance a chimpanzee (Hebb, 1949) and even rhesus monkeys (Stephenson, 1967) actively saved their companions from aversive stimuli, by brachially pulling them away from a place or object of risk. Moreover, experiments by Mirsky et al. (1958), Miller et al. (1963), Miller et al. (1967), and by Massermann et al. (1964) demonstrated that rhesus monkeys (if reared non-isolated!) react empathetically to communicative signals of distress (optical as well as acoustical signals) of their fellow monkeys and that their empathetic reactions mediated altruistic responses (Krebs, 1971).

In conclusion, the question of whether or not nonhuman primates possess «true» empathetic or sympathetic capabilities remains (for the time being) unresolved, at least in the light of definitions such as: «empathy is the self-conscious awareness of the consciousness of the other» (Wispé, 1968), or: sympathy designates the «tendency on the part of one social partner to identify himself with the other and so to make the other's goal to some extent his own» (Humphrey, 1976). Such conceptualisations presume other mental concepts, like «personal identity», «self-perception» or «self-consciousness» and «self-esteem» which are even more controversial with regard to their applicability to non-human creatures.

However, all the acts of helping and cooperation among non-human primates reported in this paper can be interpreted without any implications as to unselfish and empathetic motivations.

And finally, with regard to human morality I agree with Kummer (1978) that «obviously, deviation from selfish opportunism is neither a strictly necessary nor a sufficient criterion of morality».

Acknowledgements

I am grateful to Eckart Voland for valuable suggestions and to Paul Winkler for critical comments and technical help.

References

Abegglen, J.-J. (1984): On socialization in hamadryas baboons. Bucknell University Press, Lewisburg.

Aronfreed, J. (1970): The socialization of altruistic and sympathetic behavior: Some theoretical and experimental analyses. In: Altruism and helping behavior (ed. J. Macaulay & L. Berkowitz) Academic Press, New York, 103–126.

Axelrod, R., Hamilton, W. D. (1981): The evolution of cooperation. Science *211*, 1390–1396.
Berman, C. M. (1980): Early agonistic experience and rank acquisition among free-ranging infant rhesus monkeys. Internat. J. Primatol. *1*, 153–170.
Bertram, C. R. (1982): Problems with altruism. In: Current problems in sociobiology (ed. King's College Sociobiology Group) Cambridge University Press, Cambridge, 251–267.
Boelkins, R. C., Wilson, A. P. (1972): Intergroup social dynamics of the Cayo Santiago rhesus *(Macaca mulatta)* with special reference to changes in group membership by males. Primates *13*, 125–140.
Bramblett, C. A. (1970): Coalitions among gelada baboons. Primates *11*, 327–333.
Chapman, M., Hausfater, G. (1979): The reproductive consequences of infanticide in langurs: A mathematical model. Behav. Ecol. Sociobiol. *5*, 227–240.
Cheney, D. L. (1977): The acquisition of rank and the development of reciprocal alliances among free-ranging immature baboons. Behav. Ecol. Sociobiol. *2*, 303–318.
Cohen, R. (1978): Altruism: Human, cultural, or what? In: Altruism, sympathy, and helping (ed. L. Wispé) Academic Press, New York, 79–100.
Crook, J. H. (1971): Sources of cooperation in animals and man. In: Man and beast: Comparative social behavior (ed. J. F. Eisenberg & W. S. Dillon) Smithsonian Institution Press, Washington, 235–260.
De Vore, I. (1965): Male dominance and mating behavior in baboons. In: Sex and behavior (ed. F. A. Beach) J. Wiley & Sons, Chichester, 266–289.
Dunbar, R. I. M. (1979): Structure of gelada baboon reproductive units. I. Stability of social relationships. Behaviour *69*, 72–87.
Dunbar, R. I. M. (1980): Determinants and evolutionary consequences of dominance among female gelada baboons. Behav. Ecol. Sociobiol. *7*, 253–265.
Eisenberg, J. F. (1978): Comparative ecology and reproduction of New World monkeys. In: The biology and conservation of the Callitrichidae (ed. D. Kleiman) Smithsonian Institution Press, Washington, 13–22.
Epple, G. (1975): The behavior of marmoset monkeys (Callithricidae). In: Primate behavior, vol. 4 (ed. L. A. Rosenblum) Academic Press, New York, 195–239.
Epple, G. (1978): Reproductive and social behavior of marmosets with special reference to captive breeding. In: Primates in medicine, vol. 10 (ed. E. I. Goldsmith & J. Moor-Jankowski) S. Karger, Basel, 50–62.
Fairbanks, L. A. (1980): Relationships among adult females in captive vervet monkeys: Testing a model of rank-related attractiveness. Anim. Behav. *28*, 853–859.
Ford, S. M. (1980): Callitrichids as phyletic dwarfs, and the place of the Callitrichidae in Platyrrhini. Primates *21*, 31–43.
Goodall, J. (1977): Infant killing and cannibalism in free-living chimpanzees. Folia primatol. *28*, 259–282.
Goodall, J. (1979): Life and death at Gombe. National Geographic *155*, 592–621.
Hatfield, E., Walster, G. W., Piliavin, J. A. (1978): Equity theory and helping relationships. In: Altrusim, sympathy, and helping (ed. L. Wispé) Academic Press, New York, 115–140.
Hausfater, G. (1975): Dominance and reproduction in baboons *(Papio cynocephalus)*: A quantitative analysis. Contributions to primatology, vol. 7, S. Karger, Basel.
Hausfater, G., Vogel, C. (1982): Infanticide in langur monkeys (genus *Presbytis*): Recent research and a review of hypotheses. In: Advanced views in primate biology (ed. A. B. Chiarelli & R. S. Corruccini) Springer-Verlag, Berlin–New York, 160–176.
Hebb, D. O. (1949): Temperament in chimpanzees: I. Method of analysis. J. Comp. Physiol. Psychol. *42*, 192–206.
Hebb, D. O., Thompson, W. R. (1954): The social significance of animal studies. In: Handbook of social psychology, vol. 1 (ed. G. Lindzey) Addision-Wesley, Cambridge.
Hinde, R. A., Rowell, T. E., Spencer-Booth, Y. (1964): Behavior of socially living rhesus monkeys in their first 6 months. Proc. Zool. Soc. Lond. *143*, 609–649.
Hinde, R. A., Spencer-Booth, Y. (1967): The effect of social companions on mother-infant relations in rhesus monkeys. In: Primate ethology (ed. D. Morris) Weidenfield & Nicolson, London, 267–286.

Hrdy, S. B. (1974): Male-male competition and infanticide among the langurs *(Presbytis entellus)* of Abu, Rajasthan. Folia primatol. *22,* 19–58.

Hrdy, S. B. (1976): Care and exploitation of nonhuman primate infants by conspecifics other than the mother. In: Advances in the study of behavior, vol. 6 (ed. J. S. Rosenblatt, R. A. Hinde, E. Shaw & C. Beer) Academic Press, London–New York, 101–158.

Hrdy, S. B. (1977): The langurs of Abu. Harvard University Press, Cambridge.

Hrdy, S. B. (1979) Infanticide among animals: A review, classification, and examination of the implications for the reproductive strategies of females. Ethol. Sociobiol. *1,* 13–40.

Hrdy, S. B. (1981): The woman that never evolved. Harvard University Press. Cambridge.

Humphrey, N. K. (1976): The social function of intellect. In: Growing points in ethology (ed. P. P. G. Bateson & R. A. Hinde) University Press, Cambridge, 303–317.

Ingram, J. C. (1977): Interactions between parents and infants and the development of independence in the common marmoset *(Callithrix j. jacchus).* Anim. Behav. *25,* 811–827.

Jay, P. (1965): The common langur of North India. In: Primate behavior (ed. I. De Vore) Holt, Rinehart & Winston, New York, 197–249.

Kaplan, J. R. (1977): Patterns of fight interference in free-ranging rhesus monkeys. Am. J. Phys. Anthropol. *47,* 279–288.

Kaplan, J. R. (1978): Fight interference and altruism in rhesus monkeys. Am. J. Phys. Anthropol. *49,* 241–249.

Krebs, D. (1971): Infrahuman altruism. Psychol. Bull. *76,* 411–414.

Krebs, D. (1978): A cognitive-developmental approach to altruism. In: Altruism, sympathy, and helping (ed. L. Wispé) Academic Press, New York, 141–164.

Kummer, H. (1967): Tripartite relations in hamadryas baboons. In: Social communication among primates (ed. S. Altmann) Chicago University Press, Chicago, 63–71.

Kummer, H. (1968): Social organization of hamadryas baboons. Chicago University Press, Chicago.

Kummer, H. (1978): Analogs of morality among nonhuman primates. In: Morality as a biological phenomenon (ed. G. S. Stent) Abakon, Berlin, 35–52.

Kurland, J. A. (1977): Kin selection in the Japanese monkey. Contributions to primatology, vol. 12, S. Karger, Basel.

Lawick-Goodall, J. van (1971): In the shadow of man. W. Collins & Sons, London.

Lindburg, D. G. (1971): The rhesus monkey in North India: An ecological and behavioral study. In: Primate behavior, developments in field and laboratory research, vol. 2 (ed. L. A. Rosenblum) Academic Press, New York, 1–106 (1971).

Massermann, J. H., Wechkin, S., Terris, W. (1964): «Altruistic» behavior in rhesus monkeys. Am. J. Psychiatry *121,* 584–585.

Massey, A. (1977): Agonistic aids and kinship in a group of pigtail macaques. Behav. Ecol. Sociobiol. *2,* 31–40.

McKenna, J. J. (1979): The evolution of allomothering behavior among colobine monkeys: Function and opportunism in evolution. Am. Anthropol. *81,* 818–840.

McKenna, J. J. (1981): Primate infant caregiving behavior. In: Parental care in mammals (ed. D. J. Gubernick & P. H. Klopfer) Plenum Press, New York, 389–416.

Meikle, D. B., Vessey, S. H. (1981): Nepotism among rhesus monkey brothers. Nature (Lond.) *294,* 160–161.

Miller, R. E., Banks, J. H., Ogawa, N. (1963): Role of facial expression in «cooperative-avoidance conditioning» in monkeys. J. Abnormal Social Psychol. *67,* 24–30.

Miller, R. E., Caul, W. F., Mirsky, I. A. (1967): Communication of affects between feral and socially isolated monkeys. J. Personality Social Psychol. *7,* 231–239.

Mirsky, I. A., Miller, R. E., Murphy, J. V. (1958): The communication of affect in rhesus monkeys: I. An experimental method. J. Amer. Psychoanal. Ass. *6,* 433–441.

Packer, C. (1977): Reciprocal altruism in *Papio anubis.* Nature (Lond.) *265,* 441–443.

Packer, C. (1979): Male dominance and reproductive activity in *Papio anubis.* Anim. Behav. *27,* 37–45.

Pereira M. E. (1983): Abortion following the immigration of an adult male in yellow baboons. Am. J. Primatol. *4,* 93–98.

Rasa, O. A. E. (1976): Invalid care in the dwarf mongoose *(Helogale undulata rufula).* Z. Tierpsychol. *42,* 337–342.

Riss, D., Goodall, J. (1977): The recent rise to the alpha-rank in a population of free-living chimpanzees. Folia primatol. *27,* 134–151.

Sade, D. S. (1968): Inhibition of son-mother mating among free-ranging rhesus monkeys. Science & Psychoanalysis *12,* 18–38.

Sekulic, R. (1983): Male relationships and infant deaths in red howler monkeys *(Alouatta seniculus).* Z. Tierpsychol. *61,* 185–202.

Serpell, J. A. (1981): Duetting in birds and primates: A question of function. Anim. Behav. *29,* 963–965.

Seyfarth, R. M. (1978 a): Social relationships among adult male and female baboons. I. Behaviour during sexual consortship. Behaviour *64,* 204–226.

Seyfarth, R. M. (1978 b): Social relationships among adult male and female baboons. II. Behaviour throughout the female reproductive cycle. Behaviour *64,* 227–247.

Seyfarth, R. M. (1980): The distribution of grooming and related behaviours among adult female vervet monkeys. Anim. Behav. *28,* 798–813.

Sigg, H., Stolba, A., Abegglen, J.-J., Dasser, V. (1982): Life history of hamadryas baboons: Physical development, infant mortality, reproductive parameters and family relationships. Primates *23,* 473–487.

Silk, J. B. (1982): Altruism among female *Macaca radiata:* explanations and analysis of patterns of grooming and coalition formation. Behaviour *79,* 162–188.

Silk, J. B., Boyd, R. (1983): Cooperation, competition, and mate choice in matrilineal macaque groups. In: Social behavior of female vertebrates (ed. S. K. Wasser) Academic Press, New York, 315–347.

Sommer, V. (1984): Kindestötungen bei indischen Langurenaffen *(Presbytis entellus)* – eine männliche Reproduktionsstrategie? Anthropol. Anz. *42,* 177–183.

Stephenson, G. R. (1967): Cultural acquisition of a specific learned response among rhesus monkeys. In: Neue Ergebnisse der Primatologie/Progress in Primatology (ed. D. Starck, R. Schneider & H.-J. Kuhn) G. Fischer, Stuttgart, 279–288.

Suzuki, A. (1975): The origin of hominid hunting: A primatological perspective. In: Socioecology and psychology of primates (ed. R. H. Tuttle) Mouton, The Hague, 259–278.

Teleki, G. (1973): The predatory behavior of wild chimpanzees. Bucknell University Press, Lewisburg.

Tutin, C. E. G. (1975): Exceptions to promiscuity in a feral chimpanzee community. In: Contemporary primatology (ed. S. Kondo, M. Kawai & A. Ehara) S. Karger, Basel, 445–449.

Vogel, C. (1984): Infant transfer among common Indian langurs *(Presbytis entellus)* near Jodhpur: Testing hypotheses concerning the benefits and risks. In: Current primate researches (ed. M. L. Roonwal, S. M. Mohnot & N. S. Rathore) Scientific Publishers, Jodhpur.

Vogel, C., Loch, H. (1984): Reproductive parameters, adult male replacements and infanticide among free-ranging langurs *(Presbytis entellus)* at Jodhpur (Rajasthan), India. In: Infanticide: Comparative and evolutionary perspectives (ed. G. Hausfater & S. B. Hrdy) Aldine Publishing Co., New York, 237–255.

Waal, F. B. M. de (1978 a): Join-aggression and protective-aggression among captive *Macaca fascicularis.* In: Recent advances in primatology, vol. 1 (ed. D. Chivers & J. Herbert) Academic Press, London–New York, 577–579.

Waal, F. B. M. de (1978 b): Exploitative and familiarity-dependent support strategies in a colony of semi-free living chimpanzees. Behaviour *66,* 268–312.

Walters, J. (1980): Interventions and the development of dominance relationships in female baboons. Folia primatol. *34,* 61–89.

Wasser, S. K. (1981): Reproductive competition and cooperation: General theory and a field study of female yellow baboons *(Papio cynocephalus).* Ph. D.-Dissertation, University of Washington.

Wasser, S. K. (1983): Reproductive competition and cooperation among female yellow baboons. In: Social behavior of female vertebrates (ed. S. K. Wasser) Academic Press, New York, 349–390.

Wasser, S. K., Barash, P. P. (1981): The selfish «allomother»: A comment on Scollay and DeBold (1980). Ethol. Sociobiol. *2,* 91–93.

Watanabe, K. (1979): Alliance formation in a free-ranging troop of Japanese macaques. Primates *20,* 459–474.

Wispé, L. (1978): Introduction. In: Altruism, sympathy, and helping (ed. L. Wispé) Academic Press, New York, 1–9.
Wrangham, R. W. (1980): An ecological model of female-bonded primate groups. Behaviour *75*, 262–300.
Wrangham, R. W. (1982): Mutualism, kinship and social evolution. In: Current problems in sociobiology (ed. King's College Sociobiology Group) Cambridge University Press, Cambridge, 269–289.
Zuckerman, S. (1932): The social life of monkeys and apes. K. Paul, Trench, Trubner & Co., London.

V. Physiology and Societies

Fortschritte der Zoologie, Bd. 31 · Hölldobler/Lindauer (Hrsg.): Experimental Behavioral Ecology
G. Fischer Verlag · Stuttgart · New York · 1985

The social physiology of temperature regulation in honeybees

BERND HEINRICH

Zoology Department, Marsh Life Science Building, University of Vermont,
Burlington, VT 05405, USA

Abstract

It has been known for a long time that the internal temperature of honeybee colonies is regulated to within 0.5° of 35°C at both low and high ambient temperatures. The mechanisms of thermoregulation that have been identified include increased heat retention and heat production at low ambient temperatures, and fanning and evaporation of water at high temperatures. Similar thermoregulatory mechanisms are also used by most vertebrate homeotherms and in many insects as individuals. However, thermoregulating individuals have sensors that detect temperature changes, a centrally-located thermostat, and a neural communication system that acts to produce a coordinated response. Do the colonies of social insects have analogous mechanisms? Little work has been done to elucidate details affecting the apparently coordinated thermoregulatory responses of a bee society consisting of tens of thousands of individuals. I here review the data so far available and propose that responses observed in colonies as a whole could potentially be understood in terms of individuals acting behaviorally and physiologically to regulate their own body temperature, in the context of the proximal and evolutionary constraints and possibilities afforded by their social life. This model, although largely consistent with the data so far available, still needs considerable testing, and pertinent experiments are suggested.

1 Introduction

Sociality in bees is roughly correlated with ability to maintain a constant nest temperature. For example, as far as we know none of the solitary or semi-social bees regulate nest temperature, but primitively social bees, like bumblebees, regulate nest temperature within relatively wide limits, while the highly eusocial bees regulate nest temperature fairly precisely. The microclimate control may promote a greater sociality, although it is also possible that the greater sociality promotes better microclimate control. At the present time it may not be possible to separate cause and effect, but perhaps insights can be gained by examining the details of the thermoregulatory mechanisms.

Honeybee colony thermoregulation has been studied for over a century, but so far no one model of thermoregulation has proven adequate for winter and summer hives as well as swarm clusters that have left the hive. SIMPSON (1961) in his review concludes «In effect, the bees keep the center of their cluster at a constant temperature in summer and its periphery above the vital minimum in winter.» Recent work on thermoregulation in swarms (HEINRICH, 1981 a and b) is not fully compatible with previous models of colony thermoregulation. Although the regulation of both heat loss and heat production have repeatedly been identified as colony thermoregulatory mechanisms, very little is known about the role of the individual bees and how or whether they might be coordinated by communication to produce the appropriate colony response.

Honeybees are able to regulate their own body temperature by controlling both heat production by shivering (ESCH, 1960 CAHILL and LUSTICK, 1976) and heat loss by evaporating fluid from their mouthparts (HEINRICH, 1980), and the mechanisms of nest thermoregulation could conceivably be based on individual responses of bees which collectively affect nest temperature. On the other hand, it could involve a coordinated social response with information transfer and division of labor, or a combination of individual and colony responses. My intention in this paper is to explore the mechanisms of thermoregulation in honeybees in order to differentiate the possible alternatives, and to suggest critical experiments where data are so far lacking.

2 «Superorganisms,» and keeping warm in clusters

It is generally maintained that the honeybee colony acts like a «superorganism» in its thermoregulatory response (SOUTHWICK and MUGAAS, 1971, SOUTHWICK, 1982). The implication of such a view is that the thousands of individuals of the colony are not acting on their own, but rather that they subordinate their individual actions for a coordinated response, much like the cells and organs of an organism such as a mouse or a bird act under control of integration centers to produce a unified response.

There are numerous observations that seem to support the superorganism model of thermoregulation. For example, if one lowers the air temperature surrounding a swarm (HEINRICH, 1981 a) or a winter cluster of bees (SOUTHWICK and MUGAAS, 1971) core temperature tends to *increase* while the temperature of the cluster periphery is maintained above some minimum (Fig. 1). The physics of heat flow dictate that the higher the core temperature relative to the outside, the greater the rate of heat flow to the bees on the cluster periphery. The interpretation of these results and premises is therefore in apparent accord with the «superorganisms» model where the core bees are increasing their temperature in order to help defend the vital minimum temperature of their hivemates on the periphery. Conversely, the bees on the swarm periphery are preserving a minimum body temperature rather than the much higher body temperature they prefer when alone, and thus they help to economize the colony's often limited food supplies. A mouse or other homeotherm shows analogous responses, and these responses are also interpreted in terms of mechanisms and adaptive significance of the organism as a whole, rather than in terms of benefit to the various parts involved.

The mechanisms whereby the colony responses are achieved have been compared to those of vertebrate homeotherms. A mouse or other homeotherm, for example, releases or conserves heat from the body core by circulatory shunts and by varying its insulation. The swarm varies the circulation of air with heat through it to affect control of core temperature (HEINRICH, 1981 a). Also, as air temperatures decline bees in the swarm periphery crowd closely together and plug exit holes through which beat might escape from the core (Fig. 2). Like a mouse huddling into a ball in the cold, the swarm, too, contracts presenting a minimum surface area to the air for heat loss (Fig. 1).

Regulation of heat loss is only one potential mechanism of maintaining a stable temperature. Regulation of heat production to offset heat loss is another. Lowering of air temperature around a swarm causes an increase in its total rate of heat production (HEINRICH,

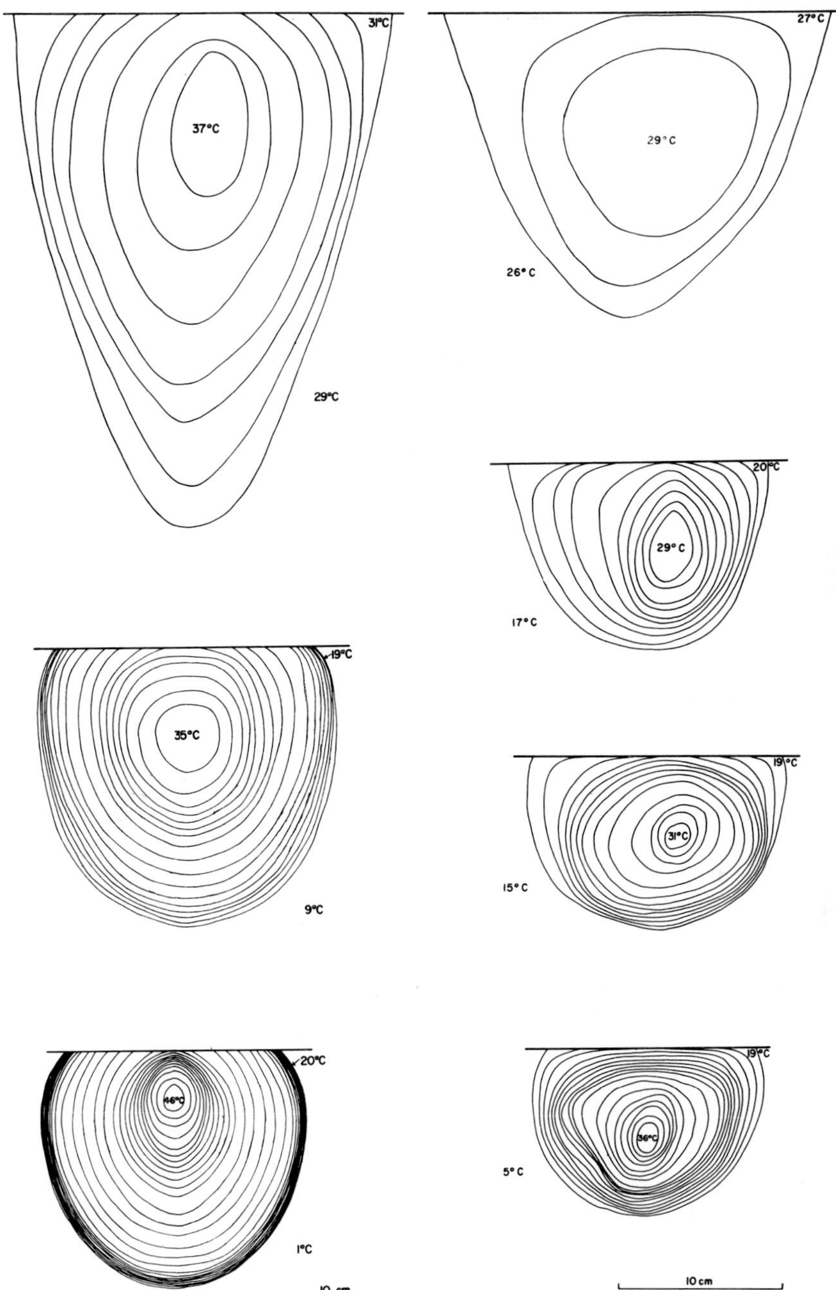

Fig. 1: Swarm shapes, and temperatures (in 1 °C isotherms) of one captive swarm of 16,600 bees (left) at 29 °C, 9 °C and 1 °C (HEINRICH, 1981 b) and of another captive swarm of 5,284 bees at 26 °C, 17 °C, 15 °C, and 5 °C (Heinrich, 1981 a).

1981 a). A vertebrate homeotherm behaves similarly. In a homeothermic vertebrate the thermoregulatory responses are controlled in the hypothalamic region of the brain. Sensory receptors on the organisms' periphery and in the hypothalamus itself monitor body temperature. When body temperatures decline below specific set-points located in the hypothalamus, neural commands are sent to the appropriate organs and these effect a coordinated response to the temperature challenge. Similarly, if a honeybee swarm or colony has bees on the periphery that act as receptors of temperature changes, which convey information about temperature to others that respond appropriately independently of their own needs, then the analogy of the «superorganism» might hold. If on the other hand bees receiving the temperature stimulus on the swarm periphery do not communicate the information to hivemates within the core so that they can respond appropriately to counteract the stimulus, then the «superorganism» model is inappropriate as a tool for promoting a greater understanding of the underlying mechanisms.

In examining the details of a swarm's responses, discrepancies between the vertebrate model and that of the swarm superorganism become apparent. First, a swarm can be divided, and both parts, with or without queen, continue to regulate their temperatures as before (HEINRICH, 1981 a) (It is so far not known if the response declines with time.). Secondly, swarm core temperature is often not regulated at one specific set-point. As other workers had observed with winter clusters (HIMMER, 1932; SOUTHWICK and MUGAAS, 1971; RITTER, 1978), I found that temperatures in the core of swarm clusters often tended to increase as air temperature decreased. The swarm periphery was maintained above 13°C, near the bees' chill coma temperature. Although these results do not comform to the model of thermoregulation by vertebrate homeotherms, they nevertheless do not contradict the superorganism model. Indeed, as other workers had interpreted for winter clusters (HIMMER, 1932; SIMPSON, 1961; SOUTHWICK and MUGAAS, 1971), I also initially attributed the increase in core temperature as the demonstration of a «superorganism» response, where the bees in the center of the swarm attempted to maintain the bees in the swarm periphery above their chill coma temperature by increasing the temperature gradient and therefore the heat flow to the periphery.

I designed experiments to test this model, but the idea that the core bees were responding to the temperatures felt by the mantle bees was not supported (HEINRICH, 1981 a). First, at low air temperatures individually marked bees on the mantle remained in place for hours, and days; they did not appear to exchange places with those in the center. Most of the exchange of bees between mantle and core took place at the higher air temperatures. When physical exchange of bees between core and mantle was prevented by a physical barrier (a screen sock), thermoregulation of the swarm remained unaltered. Clearly, there are no «messenger» bees shuttling about between the core and the periphery, telling the bees inside the swarm when to increase their heat production. Secondly, air (with presumptive pheromones) pumped from the center of one swarm located in the warm into one in the cold, and vice versa, failed to alter the thermoregulatory response, indicating that bees were not communicating possible thermal needs by pheromonal messages. Thirdly, perhaps the buzzing of chilled bees conveys information that is heeded by core bees. However, sound recordings from the periphery with a microphone about which the bees were clustered also had no effect on altering swarm core temperature. In summary, the core bees appear to play little active role in swarm thermoregulating and they do not respond to mantle bees.

Individual bees, if given a choice, seek out environmental temperatures near 35°C (HERAN, 1952), close to the thoracic temperature they regulate outside the hive while in flight (HEINRICH, 1979) while caged in small groups (CAHILL and LUSTICK, 1976), as well as while living in the summer hive (HIMMER, 1932; SIMPSON, 1961). However, in one large

Fig. 2: Photograph of bees on the mantle of a swarm at an air temperature of 25 °C (top) and at 3 °C (bottom). (From Heinrich, 1981 a).

swarm at an air temperature of 1°C I observed the bees at the core attain temperatures of 46°C (Fig. 1). Did the bees in the core of this swarm increase their metabolic rate to heat themselves up to near lethal temperatures in order to keep the bees on the swarm mantle from getting cold?

The metabolic rate of swarms increased at decreasing air temperatures (Fig. 3). However, there are at least two explanations for this observation: the bees may not only increase their metabolic rate to increase their body temperature, they could also be forced to have a higher metabolic rate (resting rate) if they are subjected to increasing body temperatures. It is difficult to differentiate these two alternatives. It is possible (and likely) that the mantle bees increase their active metabolism to try to keep warm, while the core bees are increasing their resting metabolism because they have no choice. It has so far not been possible to measure the metabolic rate of isolated bees *in situ* in the swarm. However, knowing the temperatures of the bees within the swarm one can calculate the total expected contribution by resting metabolism. Such calculations were made (Fig. 3) and the total metabolic rate of resting swarms was found too low to support the hypothesis that the bees were producing large amounts of heat by shivering to raise their body temperature. The total metablolic rate of swarms was no more than that expected from a composite of bees respiring at their body temparature-specific resting rates (HEINRICH, 1981 b). This observation suggested that although the bees in the core of the swarm may have had the highest respiratory rates, their increased metabolism was a passive function due to a higher body temperature that might have resulted from crowding, rather than from an active increase in respiration in order to keep the mantle bees warm.

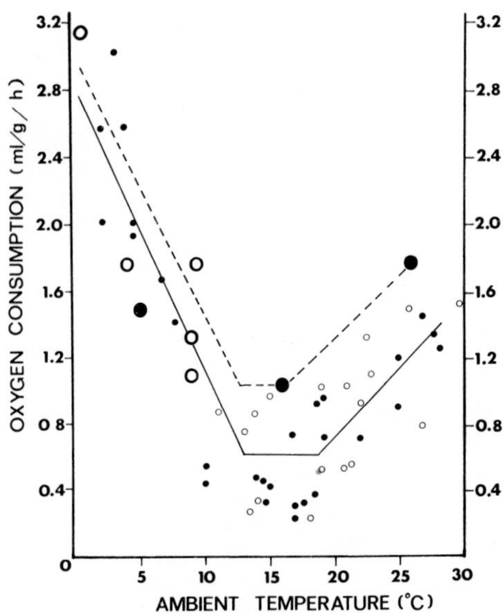

Fig. 3: Swarm metabolic rates as a function of ambient temperature. Open circles = oxygen consumption of swarms with 10,600–16,000 bees. Closed circles = oxygen consumption of swarms with 1,800–9,900 bees. Small circles represent measured rate of oxygen consumption (form HEINRICH, 1981 a). Large circles represent averages of calculated rates of resting bees produced by adding the expected resting rates over the observed isotherms of the two swarms (from HEINRICH, 1981 b).

In contrast to the «superorganism» model and in support of the above hypothesis, a series of experiments indicated that the bees on the swarm *periphery* affected core temperature. The mantle bees, by crowding inward when they experience low air temperature contract the swarm cluster. At low air temperatures, the bees were arranged shingle-like over the swarm surface (Fig. 2), as had been described for winter hive clusters (SIMPSON, 1961). The contracted swarms, unlike the expanded ones at air temperatures above 20°C, were apparently solid masses of bees without ventilatory channels (Fig. 4), as the bees from the exterior had crawled into the interior, plugging these channels as air temperature decreased. Heat produced by the core bees was then trapped within the swarm. When I transferred a swarm from 1°C to 16°C I observed an immediate *decline* in core temperature even as mantle temperature increased. It is likely that as the mantle bees became less crowded together as they warmed, air channels were created that caused heat loss from the core. At high air temperatures swarms routinely became elongated and less dense (Figs. 1 and 4). They contained channels for air flow, and air in the channels was cooler than the bees themselves.

If bees on the swarm mantle were responding to low air temperatures appropriately to conserve or release heat from the swarm core, the question remains: what is their stimulus for doing so? The data indicate that they respond, in large part, directly to the temperature they experience, and attempt to maintain their own body temperature at appropriate levels (HEINRICH, 1981 a).

At low air temperatures bees on the swarm mantle can warm themselves by crawling into the air passageways of the swarm, and by crowding their thoraxes close together and shivering when necessary. As the swarm structure becomes very tight, mantle bees are less able to force their way into the interior. When thus «trapped» on the periphery they may be subjected to very low temperatures where they become chilled and less able to shiver and to move, and presumably to replenish their food supplies from colony-mates. When taken to higher air temperatures where they can shiver vigorously (BASTIAN and ESCH,

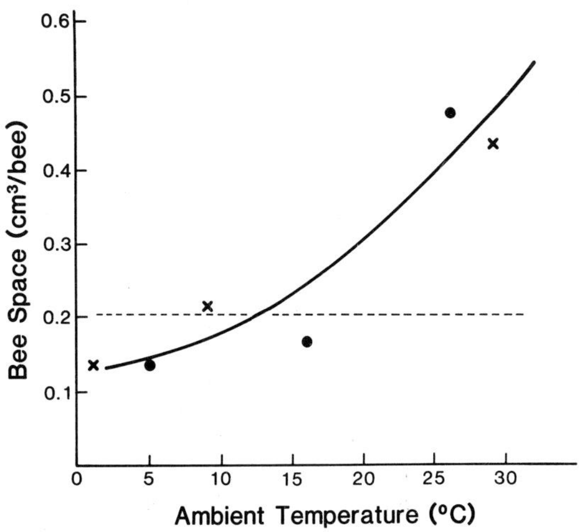

Fig. 4: Average crowding of bees within swarms calculated from swarm shape (Fig. 1) at different air temperatures. Filled circles = swarm of 5,284 bees. Crosses = swarm of 16,600 bees. Dotted line = a heap of 7,000 dead bees (HEINRICH, unpublished).

1970) they regulate a relatively high thoracic temperature (HEINRICH, 1981 a). Bees from the edge of winter clusters will immediately shiver and produce a large thoracic temperature increase if fed honey (ESCH, 1960). However, by being physically and energetically restricted from maintaining a continuous high body temperature at the edge of the cold bee cluster, the mantle bees are normally reduced to defending the minimum body temperature required to maintain sufficient motor control for crawling, and to remain attached to the swarm.

The core bees, however, may not be total victims of the actions of the mantle bees. If they become too heated by the tightly-packed mantle bees, they may counteract by crawling closer to the swarm periphery. By doing so they loosen the internal swarm cluster, which also cools the other bees that could potenially become overheated. I conclude that a major feature of this model of temperature control of swarms, where individual bees attempt to regulate their own temperature, does not fit the «superorganism» concept. However, presumably some of the bees' thermoregulatory responses may have evolved specifically in the social context, and there may be some features of the thermoregulatory behavior that are unique to the social context that do fit the superorganism concept. These presumed features have not yet been identified.

Thermoregulation of bees in their winter nests is poorly understood. However, like swarms, winter nests lack brood, and similarities in the mechanisms of thermoregulation can be expected.

As air temperature is lowered, the winter bee cluster contracts. As in swarms, during cluster contraction the outermost bees point their heads inward with their thoraxes close together, while their abdomens are exposed to the cold (SIMPSON, 1961). Conversely, as air temperature increases, the bee cluster expands. Obiously both the contraction of the cluster as well as increased heat production could increase the temperature of the core, but the two have never been adequately separated.

HIMMER (1933) observed that the edge of the cluster seldom declined below 10°C, the minimum temperature at which the bees could crawl and still hang onto the cluster. Since bees that cooled below this temperature would be lost from the hive, he speculated that mantle temperature (rather than that in the core of the bee cluster) is the most critical variable that the winter hive cluster tries to maintain.

As I later observed for swarms (HEINRICH, 1981 a), HIMMER (1933) and SOUTHWICK and MUGAAS (1971) also observed the curious behavior that as air temperature fell below 7–8 °C temperature on the cluster mantle remained independent of air temperature, while temperatures in the core often *increased.* (When air temperature increased above 10°C, however, the temperature of the mantle increased in parallel with air temperature). HIMMER and SOUTHWICK and MUGAAS all concluded that the central bees generated heat to keep the bees at the edge at the minimum (defined variously at 8 °C to 15 °C). HIMMER said «Es handelt sich also um eine typische Wärmeregulierung einer Tiergemeinschaft als biologische Einheit» (It is a matter of a typical thermoregulation of an animal society acting as a unit). In his review SIMPSON (1961) as well as SOUTHWICK and MUGAAS (1971) and SOUTHWICK (1982) in studies of the winter cluster also came to the same conclusion that the bee colony functions as a «superorganism» where «A honeybee cluster thermoregulates so as to maintain a constant surface temperature at low ambient temperature.»

Although the detailed studies performed on swarms have not yet been done on the winter cluster, the data at the present time suggest that the same basic mechanisms apply to both, provided certain important differences are taken into account.

Unlike in swarm clusters, which are temporary aggregations of a few hours or several days, bees in the winter may stay in a cluster for several weeks, moving periodically to feed at new combs with honey and then forming a new cluster. While moving, the swarm cluster is temporarily loosened or destroyed, potenially cooling all of the bees. Perhaps some of the moving may account for the large variation in temperature (14.5 °C–32 °C) that has been observed in winter hives (HIMMER, 1933. SOUTHWICK and MUGAAS, 1971). Circadian cycles of activity and heat production (KRONENBERG and HELLER, 1982; SOUTHWICK, 1982) also account for variation of hive temperature.

During the periods when bees move at low ambient temperature, there may be great increases in overall hive metabolism (SOUTHWICK, 1982). However, temporary increases of metabolism and food consumption are to be expected as many individuals, without the benefit of the cluster, are each attempting to regulate their own body temperature by increasing their rate of heat production (ALLEN, 1959; ESCH, 1960; CAHILL and LUSTICK, 1976). The metabolism of the tightly contracted winter cluster, in contrast, increases only slightly, even at sub-zero temperatures, and the metabolic rate of these winter clusters (SOUTHWICK, 1982) differs little from that obeserved in resting swarm clusters (HEINRICH, 1981 a). As reviewed by SIMPSON (1961), records of the amounts of food consumed by colonies in winter usually fail to show any definite correlation with outside temperature. As I have described for swarms, apparently much of the thermoregulatory response can therefore be explained by individuals crowding tightly together, trapping heat within the cluster.

One major difference between swarms and winter clusters, however, concerns the physiology of the individual bees. Following prolonged exposure to cold, bees acclimate (HERAN, 1952). Bees from the winter nest in temperature-preference tests in a temperature gradient, choose temperatures averaging 32.8 °C but as low as 28 °C, while summer bees choose temperatures near 35–38 °C (HERAN, 1952). Furthermore, winter bees have lower chill-coma temperatures than summer bees (FREE and SPENCER-BOOTH, 1960). Presumably these differences could alone account for both the lower core and mantle temperatures observed between winter hive clusters and summer swarms (Fig. 5).

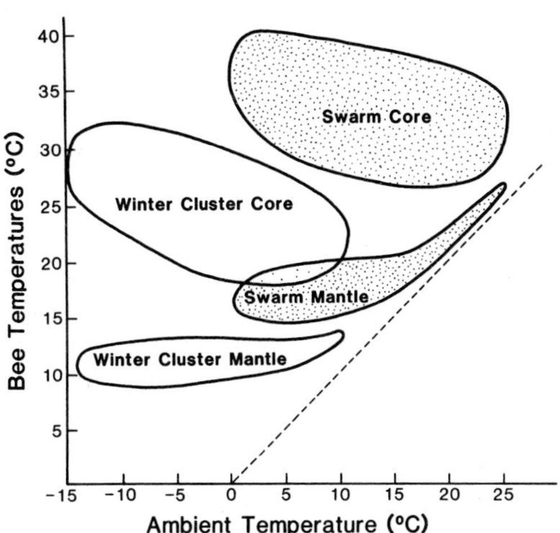

Fig. 5: Swarm and winter cluster temperatures as a function of air temperature. The cloud diagrams exclude the extremes but enclose at least 80 % of the observed temperatures. Data was derived from the following sources: Swarms (HEINRICH, 1981 a), winter cluster cores (SOUTHWICK and MUGAAS, 1971), winter cluster mantle (interpreted from review by SIMPSON, 1961).

3 «Superorganisms» and keeping cool

Honeybees not only elevate nest temperature at low air temperatures, they also prevent overheating of the nest at high ambient temperatures (Lindauer, 1954; Lensky, 1964). The classic experiments by Lindauer (1954) of honeybee colonies subjected to high temperatures suggest that nest cooling involves a highly coordinated social response. In part, the response is a modification of the bee's food storage behavior. Additionally, the thermoregulatory behavior could involve modifications of responses that aid the individuals to regulate their own body temperature, that secondarily affects hive temperature.

As a first response to high hive temperature, bees clustered upon the combs disperse. If they are not sufficiently cooled by dispersing they may leave the hive. Bees that leave the colony presumably lower their own body temperature, but in so doing they also provide space within, so that air circulation can occur and the hive can cool. Expansion of winter clusters or swarms at elevated temperatures has the same effect.

Lindauer (1954) observed further that bees began to fan and carry in water that evaporated and cooled the hive when it continued to be heat-stressed. As hive temperatures increased and the need for water for evaporative cooling increased, the foragers switched from collecting concentrated nectar to dilute nectar or to pure water. However, the switsch-over did not occur because the water carriers were aware of the need of water for temperature regulation within the hive. Rather, it occurred because other bees eagerly accepted the dilute nectar or water being brought even at the entrance of the colony, while those carrying concentrated nectar could not unload their cargo and consequently did not recruit. It cannot be presumed, however, that the bees were aware of the hive's needs and behaved accordingly. The bees from the nest interior that accepted the water regurgitated it and evaporated it from droplets on their extruded mouthparts (Lindauer, 1954), which we now know to be the primary mechanism whereby individual bees prevent overheating of their own body temperature (Heinrich, 1980). Evaporative cooling within the hive would act to reduce hive temperature, regardless of whether it occurs primarily off the comb, or from the individual bees. Thus, bees that accept water to cool themselves would also be acting to regulate hive temperature. So far it is not known if cool bees in a hot hive are willing to accept water for the hive, or if hot bees in a cool hive accept water.

In addition to individual bees holding and evaporating fluid, Lindauer (1954) observed the bees that received the water from water carriers to regurgitate onto the upper sides of the honeycombs, in the same way that dilute nectar is deposited so that it may be concentrated to become honey. Deposition of pure water cannot be accounted for on the basis of individual cooling, but the bees may not know what else to do with liquids. However, since the deposition of water in combs has taken on the function of cooling the hive, rather than the individual bee, it has assumed social significance. The behavior now aids the hive directly, rather than indirectly through what aids the individual hive members.

Temperature regulation of the hive also involves a third response – fanning. Bees fan after bringing dilute nectar or water into the hive possibly in response to high humidity. The bees also fan in response to other noxious stimuli, such as smoke, excessive CO_2, and invasions of small ants (personal observation).Whether or not they respond by fanning to high temperature (Neuhaus and Wohlgemuth, 1960), as such, is not clear. However, regardless of the various stimuli involved that induce fanning, the behavior affects hive temperature. Are there possibly individualistic behaviors that may have originally formed the basis for the hive's over-all response, and to what extent do these behaviors still form the basis of this overall response?

Bees aligning themselves in chains to create unidirectional air flow (Neuhaus and Wohlgemuth, 1960), appear to show an integrated colony response for optimal cooling of the hive. However, they might also be attempting to cool themselves, without reference to hive needs. Several kinds of small-bodied insects, such as butterflies (HEINRICH, 1972), syrphid flies (HEINRICH and PANTLE, 1975), and cicindelid beetles (KENNETH R. MORGAN, personal communication), fly when heat-stressed, thereby cooling themselves convectively. A bee fanning with its wings also necessarily drives air over itself for cooling. Like an individual insect seeking a drafty place, a second bee could enhance its own cooling by standing in the slip-stream of a second, etc. As a result, the bees would align themselves in chains and unidirectional flow would automatically be created, and cooling of the hive would be optimized. If nest fanning evolved from individuals trying to cool themselves one might expect that the hottest bees would fan first, and that they would initially align themselves outside the nest, pushing cool air over themselves and into it, rather than drawing warm air out of the hive and over themselves.

Many observations indicate that bees that fan do not respond to hive temperature, *per se*. NEUHAUS and WOHLGEMUTH (1960) observed that foragers, particularly those carrying pollen, often fanned before entering the hive. The temperature of the air in the vicinity of the antennae bearing the thermoreceptors (HERAN, 1952) is of no effect in stimulating fanning (NEUHAUS and WOHLGEMUTH, 1960). The body temperature of the bees, however, may be relevant. At least in the Arizona desert, most of the pollen carriers returning to the hive have empty honeycrops and, unlike nectar carriers, they almost never extruded liquid from their mouthparts (PAUL D. COOPER, personal communication). The pollen foragers had much higher thoracic temperatures (average = 43°C). Like NEUHAUS and WOHLGEMUTH (1960), Cooper reports that the pollen carriers «almost always stop and fan upon alighting before entering the hive,» drawing air from the outside of the hive to the inside. Since the fanners presumably had no knowledge of temperature within the hive, it is difficult to rationalize their behavior except in terms of response to their own body temperature. The returning pollen foragers only fanned for short durations. It would be of great interest to know what their body temperatures were when they stopped fanning. Whether or not the bee's fanning behavior is related to individual thermoregulation is so far unexamined. However, if the bees indeed fan to affect their own body temperature, it is significant that through their sociality, they have evolved to do it primarily in the vicinity of the hive, where it immediately benefits colony mates as well.

4 The effect of brood

Of major impact on colony thermoregulation is the presence of brood. The brood is incapable of regulating its own temperature (HIMMER, 1933) and it is sensitive to changes in temperature. In the summer nest the brood nest is maintained at 35°C ± 0.5°C (SIMPSON, 1961). Temperatures higher than 36°C caused developmental abnormalities, principally of the wings (HIMMER, 1933). How does the regulation of brood temperature relate to the model of bees acting primarily to affect their own body temperature? Have responses evolved to modify or channel the worker's «selfish» behavior so that the colony benefits?

Within a colony of tens of thousands of individuals, each is capable of shivering to produce heat (BASTIAN and ESCH, 1970). The heat produced as a by-product of the normal activity and resting metabolism of many individuals, and the colony-wide heating that results, frees the individuals for numerous tasks at the same time that the brood is deriving the benefit of the bees' metabolic heat. In order to heat the brood the workers need only remain in the vicinity of it. Thus, it is possible that the temperature control of brood could be explained in terms of individuals regulating their own body temperature, provided they did it precisely enough with the added behavioral response that the bees are *attracted* to the brood so that they stay in its vicinity sufficiently long to be cooled or heated by it, so that they can respond if temperature between the two differ.

Bees that are not in flight not only attempt to regulate their own thoracic temperature near 35°C by adjusting their rate of heat production (Cahill and Lustick, 1976) but they also choose temperatures near 35°C in a temperature gradient (Heran, 1952). Therefore, heat produced by the brood and the bees upon it should provide additional incentive for bees to crowd together upon the brood at low air temperatures, amplifying attraction based on pheremonal stimuli. Bees already attracted and producing heat would provide nuclei for the attraction of still more bees seeking warmth.

It has previously been thought that there were major differences between «winter» and «summer» thermoregulation (Simpson, 1961). It is likely, however, that the differences observed reflect primarily the presence or absence of brood and that without brood, clusters of summer bees behave similarly to those of winter (Ritter, 1978). Brood, indeed, has a strong attraction to the worker bees (Koeniger, 1978), and the shifting of the worker force to areas containing brood should predictably have a major effect on heat distribution within the hive. It would be of interest to know if bees are more attracted to warm brood where they could warm themselves, or to cold brood where they must shiver to keep themselves warm.

The inter-relationships between the bees' crowding upon the brood, regulating their own body temperature, and warming of the brood and stabilizing its temperature have so far not been elucidated. It remains unclear if the bees produce heat only to maintain their own body temperature elevated, or to elevate the temperature of the brood. Circumstantial evidence suggests that although the bees may act to regulate their own body temperature, they also respond specifically to brood temperature, *per se*. For example, the bees have numerous thermo-receptors on their antennae (Heran, 1952), which could serve to monitor brood temperature but which are unlikely to be involved in monitoring their own body temperature. In order to determine if bees respond to brood temperature, as such, Kronenberg and Heller (1982) allowed bees to build comb around a water-cooled device in order to control brood temperature, and found that lowering of brood temperature resulted in crowding of the bees upon the comb and in elevation of the colony's metabolic rate. This result suggests but does not prove that the bees regulate brood temperature as such, rather than merely attempting to regulate their own body temperature. As already mentioned, it is difficult to experimentally separate brood from body temperature, since experimental reductions of temperature of the brood upon which the bees are located would affect the bees upon the brood as well.

5 Conclusion

Thermorgulation of honeybee colonies involves many responses. Many if not most of these responses can be understood, in part, in terms of the individual bees regulating their own body temperatures. This perspective offers a single model to account for thermoregulation of not only winter and summer nests, which were previously thought to be qualitatively different, but also of swarms. Responses of individuals acting for their own survival were presumably selected because they also enhanced colony survival and reproduction. Modifications of these responses that aided the colony as a whole could theoretically be superimposed, so that the «superorganism» concept might apply in varying degrees to different functions.

The superorganism concept is obviously highly useful in understanding many aspects of honeybee colony organization. However, with respect to thermoregulation, the superorganism concept is a tool that applies only to a limited extent in framing questions on colony mechanisms. The concept that the individuals interact on the basis of their own thermoregulatory needs is a tool that has so far been applied only to a limited extent, and it promises to provide new insights. Finally, I speculate that regulation of brood temperature evolved as a by-product of individual temperature regulation. In a large social insect colony with thousands of metabolically active individuals, heat is produced that inevitably increases the temperature in the environs of the brood. As reasoned elsewhere for individual thermoregulation (HEINRICH, 1977) the brood must evolve to specialize its metabolic functions near the elevated temperature it normally experiences.

Acknowledgements

Writing of this paper was supported by NSF grant BSR-8116662. I thank the Akademie der Wissenschaften und Literatur of Mainz for sponsoring the symposium. I am also grateful to MARGARET HEINRICH and DR. JOAN HERBERS reading drafts of the manuscript and making many helpful comments and criticisms.

References

Allen, M. D. (1955): Respiration rates of worker honeybees of different ages at different temperatures. J. Exp. Biol. *36*, 92–101.

Bastian, J. and Esch, H. (1970): The nervous control of the indirect flight muscle of the honeybee. Z. vergl. Physiol. *67*, 307–342.

Büdel, A. (1958): Ein Beispiel der Temperaturverteilung in der Bienentraube. Z. Bienenforsch. *4*, 63–66.

Cahill, K. and Lustick, S. (1976): Oxygen consumption and thermoregulation in *Apis mellifera* workers and drones. Comp. Biochem. Physiol. *55A*, 355–357.

Esch, H. (1960): Über die Körpertemperaturen und den Wärmehaushalt von *Apis mellifera*. Z. vergl. Physiol. *43*, 305–335.

Free, J. B. and Spencer-Both. (1958): Observations on the temperature regulation and food consumption of honeybees *(Apis mellifera)*. J. Exp. Biol. *35*, 930–937.

Free, J. B. and Spencer-Booth, H. Y. (1960): Chill-coma and cold death temperatures of *Apis mellifera*. Ent. Exp. and Appl. *3*, 222–230.

Heinrich, B. (1972): Thoracic temperature of butterflies in the field near the equator. Comp. Biochem. Physiol. *43 A*, 459–467.

– (1977): Why have some animals evolved to regulate a high body temperature? Amer. Natur. *111*, 623–640.

– (1979):Thermoregulation of African and European honeybees during foraging, attack, and hive exits and returns. J. Exp. Biol. *80*, 217–229.

– (1980): Mechanisms of body temperature regulation in honeybees, *Apis mellifera*. I. Regulation of head temperature. J. Exp. Biol. *85*, 61–87.

– (1981 a): The mechanisms and energetics of honeybee swarm temperature regulation. J. Exp. Biol. *91*, 25–55.

– (1981 b): Energetics of honeybee swarm thermoregulation. Science *212*, 565–566.

– and Pantle, C. (1975): Thermoregulation in small flies (*Syrphus* sp.): Basking and shivering. J. Exp. Biol. *62*, 599–610.

Heran, H. (1952): Untersuchungen über den Temperatursinn der Honigbiene *(Apis mellifica)* unter besonderer Berücksichtigung der Wahrnehmung strahlender Wärme. Z. vergl. Physiol. *34*, 179–206.

Himmer, A. (1932): Die Temperaturverhältnisse bei den sozialen Hymenopteren. Biol. Rev. *7,* 224–253.

Koeniger, N. (1978): Das Wärmen der Brut bei der Honigbiene (*Apis mellifera* L.) Apidologie 9, 305–320.

Kronenberg, F. and Heller, H. C. (1982): Colonial thermoregulation in honeybees *(Apis mellifera).* J. Comp. Physiol. *148,* 65–76.

Lensky, Y. (1964): Comportement d'une colonie d'abeilles a des temperatures extremes. J. Ins. Physiol. *10,* 1–12.

Lindauer, M. (1954): Temperaturregulierung und Wasserhaushalt im Bienenstaat. Z. vergl. Physiol. *36,* 391–432.

Neuhaus, W. and Wohlgemuth, R. (1960): Über das Fächeln der Bienen und dessen Verhältnis zum Fliegen. Z. vergl. Physiol. *43,* 615–641.

Ritter, W. (1978): Der Einfluß der Brut auf die Änderung der Wärmebildung in Bienenvölkern *(Apis mellifera carnica).* Verh. Dtsch. Zool. Ges. 220, Gustav Fischer Verlag, Stuttgart, 1978.

Simpson, J. (1961): Nest climate regulation in honeybee colonies. Science *133,* 1327–1333.

Southwick, E. E. (1982): Metabolic energy of intact honeybee colonies. Comp. Biochem. Physiol. *71 A,* 277–281.

– and J. N. Mugaas. (1971): A hypothetical homeotherm: The honeybee hive. Comp. Biochem. Physiol. *40 A,* 935–944.

Fortschritte der Zoologie, Bd. 31 · Hölldobler/Lindauer (Hrsg.): Experimental Behavioral Ecology
G. Fischer Verlag · Stuttgart · New York · 1985

Feeding behaviour of nurse bees, larval food composition and caste differentiation in the honey bee *(Apis mellifera L.)*

J. BEETSMA[1]

Laboratory of Entomology, Agricultural University, Wageningen, Netherlands

JAN DE WILDE was scheduled to present a paper on «Problems and Issues in the Study of Caste Development in Social Insects». We were all shocked and deeply saddened when we learned of the sudden death of Professor DE WILDE, only two weeks before the commencement of this conference.

The editors are grateful to Dr. BEETSMA who, on very short notice, agreed to come to Mainz and present to us the latest findings on caste differentiation in honey bees.

1 Introduction

Caste development in social insects is induced by extrinsic factors. Climatic factors and consequently the availability of food outside the colony affect the conditions within the colony that ultimately regulate the development of castes.

In the social Hymenoptera both trophogenic factors and the activity of the corpora allata play an important role in caste development (DE WILDE and BEETSMA, 1982).

The most detailed understanding of caste differentiation concerns the honey bee.

According to DE WILDE (1975) the female castes of the honeybee can be considered as «ecomorphs». Genetically identical individuals are induced by environmental factors to develop into either queens or workers. «The intricate programmes of development lie in readiness from the very beginning of embryonic development, and need only to be «switched on» by one or a combination of extrinsic factors» (DE WILDE, 1976).

The role of juvenile hormone in the process of caste differentiation has been demonstrated by WIRTZ and BEETSMA (1972), WIRTZ (1973), GOEWIE and BEETSMA (1976), GOEWIE (1978) and EBERT (1978).

Juvenile hormone is considered to act as a «mediator» between the nutritional environment of the larva and the differentiation of caste characteristics.

Concerning the mechanism different hypotheses have been put forward (see review by BEETSMA, 1979).

The activity of the corpora allata (CA) in younger larvae seems not to be regulated by the medial neurosecretory cells (MNSC) in the brain. Both GOEWIE (1978) and REMBOLD and ULRICH (1982) found undifferentiated MNSC in these larvae and therefore they suggest that the activity of the CA is regulated by another system.

---

[1] Dedicated to the memory of my scientific-«father» JAN DE WILDE.

Different results were obtained concerning the moment of the initial activity of the MNSC in older larvae of both castes. REMBOLD and ULIRCH (1982) found that outgrowth of axons and incorporation of ^{3}H-uridine into RNA of the MNSC is delayed by one larval instar in worker larvae when compared to queen larvae. Therefore they suggest an earlier change from unknown system to the MNSC to regulate the activity of the CA in queen larvae.

Whereas REMBOLD and ULRICH (1982) found stainable neurosecretory granules in the MNSC only in the last instar queen larvae, GOEWIE (1978) demonstrated these granules already in 3-day-old larvae, irrespective of the transfer or non-transfer of larvae into queen cells.

In addition GOEWIE (1978) found considerable amounts of neurosecretory material in the CA-innervating axons from the MNSC only at the end of larval development. Therefore he suggested that the CA activity is inhibited by the MNSC.

2 Production of larval food

In the honeybee colony, queens and workers can be reared simultaneously both in the presence and in the absence of the queen. Queen rearing is released by the partial or complete absence of queen pheromones (BUTLER, 1954). Secondary releasers of this specific behaviour in nurse bees may be the shape, size and orientation of the queen cell and signals emanating from the larva. Until recently the discrimination between the nurse bees feeding queen and/or worker larvae was based on age of the bees and composition of larval food in relation to the content of the mandibular and hypopharyngeal glands of the bees.

Now a new approach of the problem has become possible by measuring the activity of the hypopharyngeal (HP) glands.

BROUWERS (1983) developed a method to measure the rate of incorporation of C^{14}-leucine into the proteins of extirpated HP glands *in vitro*. BROUWERS (1983) compared the size, protein content and activity of HP glands of bees of different ages with those of known nurse bees, collected from an observation hive.

It appeared that the activity of full-grown HP glands of summerbees showed conspicuous variation. In only one-third of these bees did the activity reach a level as found in the HP glands of bees collected when depositing food in a brood cell (> 800p.mol./h). Recent data indicate that the average activity of HP glands differs between nurse bees feeding queen and worker larvae.

3 Frequency and length of feedings

Both BROUWERS (Wageningen) and EBERT (Würzburg) have studied the feeding behaviour of nurse bees in worker cells. In addition BROUWERS (in preparation) recorded the offering of food in queen cells. Observations could be carried out by using video apparatus. The video camera was provided with a special recording tube that could be used at low intensities of red light. Using a time-lapse tape recorder the behaviour of the nurse bees could be recorded continuously over several days.

The frequency and length of feedings in queen and worker cells are completely different. Whereas in queen cells a bee offered food for a short time (< 50 seconds) about every five minutes throughout the whole period of larval development, in worker cells the frequency and duration of offering of food changed with the age of the larva.

In worker larvae after the offering of food at the moment of hatching it sometimes took 12 hours before the larva received food again. Older larvae were fed 1–2 times per hour. The length of feedings generally increased with the age of the larva (50 to 250 seconds), but after the age of about 70 hours short feedings (< 50 seconds) occurred in addition.

At this moment we can only guess about differences in the quality of food offered within a short or a long period. On the one hand it is obvious that secondary releasing factors of feeding behaviour present in the cell play a role. On the other hand, however, the physiological condition of the nurse bees should be considered as well.

4 Compositon of larval food

To obtain a complete view of the course of changes in composition of the food of the larvae of both castes during their development, BROUWERS (in preparation) collected food samples at intervals of 12 hours and measured the content of some major components on a dry weight basis.

Among other differences he found a total sugar content of about 20 % in royal jelly and about 15 % in worker jelly collected from cells containing 1–3 day old larvae.

Very conspicuous, however, were the differences in the glucose–fructose ratio between royal jelly and worker jelly (BROUWERS, in press).

The glucose-fructose ratio of royal jelly was at the first day of larval development about 1.7, decreased slowly with increasing age of the larvae, but remained higher than 1.

In worker jelly, however, this ratio started at a level of about 1.3, decreased until the larval age of about 60 hours and remained more or less constant at a level of about 0.7.

The glucose-fructose ratio in the abundant food of the small number of larvae in the initital brood nest of a hived swarm is exceptionally low during the first few days.

It is suggested that glucose stimulates and fructose inhibits the rate of food consumption by the larva, or alternatively that glucose, as opposed to fructose, is readily metabolized.

When the glucose-fructose ratio is high, increased larval growth is only realized when large amounts of food are offered to the larva, as is the case in the queen cell.

References

Beetsma, J. (1979): The process of queen-worker differentiation in the honeybee. Bee Wld. *60* (1), 24–39.

Brouwers, E. V. M. (1982): Measurements of hypopharyngeal gland activity in the honeybee. J. apic. Res. *21* (4), 193–198.

Brouwers, E. V. M. (in press): The glucose-fructose ratio in the food of the honeybee larva. J. apic. Res.

Butler, C. G. (1954): The importance of «queen substance» in the life of a honeybee colony. Bee Wld. *35*, 169–176.

Ebert, R. (1978): Die Bedeutung der Schwerkraft bei der longitudinalen Orientierung der weiblichen Honigbienenlarve (Apis mellifera L.) und deren Beeinflussung durch hormonale Faktoren. Ph. D. Thesis, University of Würzburg.

Goewie, E. A. (1978): Regulation of caste differentiation in the honeybee (Apis mellifera L.). Meded. Landb. Hogesch. Wageningen *78* (15), 76 pp. Thesis.

Goewie, E. A. and Beetsma, J. (1976): Induction of caste differentiation in the honeybee (Apis mellifera L.) after topical application of JH-III. Proc. K. Ned. Akad. Wet. C79 (5), 466–469.

Rembold, H. and Ulrich, G. (1982): Modulation of neurosecretion during caste determination in Apis mellifera larvae. pp. 370–374 from the biology of social insects (Proc. IXth. Congr. I. U. S. S. I., Boulder, August 1982) ed. M. D. Breed, C. D. Michener and H. E. Evans. Westview Press, Boulder, Colorado.

Wilde, J. de (1975): An endorcrine view of metamorphosis, polymorphism and diapause in insects. Am. Zool. *15,* 13–27.

Wilde, J. de (1976): Juvenile hormone and caste differentiation in the honeybee (Apis mellifera L.) p. p. 5–20 from Phase and caste determination in insects ed. M. Lüscher Oxford & New York, Pergamon Press.

Wilde, J. and Beetsma, J. (1982): The physiology of caste development in social insects. Advances in Ins. Physiol. *16,* 167–246.

Wirtz, P. (1973): Differentiation in the honeybee larva. Meded. Landb. Hogesch. Wageningen *73* (5), 155 pp. Thesis.

Wirtz, P. and Beetsma, J. (1972): Induction of caste differentiation in the honeybee (Apis mellifera L.) by juvenile hormone. Ent. exp. appl. *15,* 517–520.

Fortschritte der Zoologie, Bd. 31 · Hölldobler/Lindauer (Hrsg.): Experimental Behavioral Ecology
G. Fischer Verlag · Stuttgart · New York · 1985

Individual and family recognition in subsocial arthropods, in particular in the desert isopod *Hemilepistus reaumuri*

K. E. Linsenmair

Zoologisches Institut der Universität Würzburg, F. R. Germany

Abstract

Individual recognition in the non-eusocial arthropods is, according to our present knowledge, predominantly found in the frame of permanent or temporary monogamy. In some cases, e. g. in stomatopods and possibly other marine crustaceans too, individual recognition may serve to allow identification of (i) individuals within dominance hierarchies or (ii) neighbours in territorial species thus helping to avoid the repetition of unnecessary and costly fights. Kin recognition is experimentally proven only in some isopod species (genera *Hemilepistus* and *Porcellio*) and in the primitive cockroach (termite?) *Cryptocercus*. The «signatures» or «discriminators» used in the arthropods are chemical. It is assumed that the identifying substances are mainly genetically determined and in this paper I shall discuss possible evolutionary origins. The main part of this account is devoted to the presentation of some aspects of the highly developed individual and kin identification and recognition system in the desert isopod *Hemilepistus reaumuri* – a pure monogamous species in which pairs together with their progeny form strictly exclusive family units. Amongst other things problems of (i) mate choice, (ii) learning to recognize a partner, (iii) avoiding the unadaptive familiarization with aliens are treated. Monogamy under present conditions is for both sexes the only suitable way of maximizing reproductive success; an extremely strong selection pressure must act against every attempt to abandon monogamy under the given ecological conditions. The family «badges» which are certainly always blends of different discriminator substances are extremely variable. This variability is mainly due to genetical differences and is not environmentally caused. It is to be expected that intra-family variability exists in respect of the production of discriminator substances. Since the common badge of a family is the result of exchanging and mixing individual substances, and since the chemical nature of these discriminators requires direct body contacts in order to acquire those substances which an individual does not produce itself, problems must arise with molting. These difficulties do indeed exist and they are aggravated by the fact that individuals may produce substances which do not show up in the common family badge. An efficient learning capability on the one hand and the use of inhibiting properties of newly molted isopods help to solve these problems. In the final discussion three questions are posed and – partly at least – answered; (i) why are families so strictly exclusive, (ii) how many discriminator substances have to be produced to provide a variability allowing families to remain exclusive under extreme conditions of very high population densities, (iii) what is the structure of the family badge and what does an individual have to learn apart from the badge in order not to mistake a family member for an alien or vice versa.

1 Introduction

Before animals can reproduce they have to come to numerous decisions requiring special recognition and classification performances. They have to discern self from non-self, their species from others and nearly always different categories within their species: females, males, young etc. Additionally they often have to distinguish between subcategories, e. g. females in different phases of their reproductive cycle or young of different ages. Under special conditions, still more sophisticated capabilities have been selected, those for individual and for kin recognition. These are the topic of my account, – as far as arthropods are concerned – with the exception of the eusocial species (cf. Michener 1985, this Vol.)

Only a very few species within the huge number of non-eusocial arthropod species are known to possess either or both of these abilities, and in most of these few cases our knowledge is still very limited. Therefore I shall only mention a few studies, partly to exemplify which environmental conditions may favour the selection of such capabilities and which their probable present adaptive value is. I shall mainly concentrate on some selected features of the highly developed mechanisms of identification (sensu Beecher 1982) and recognition of individuals and kin in the desert isopod *Hemilepistus reaumuri* (Linsenmair and Linsenmair 1971, Linsenmair 1972, 1975, 1979, 1984). The first aspect, that of the individual's or kin's identifying cues, will be dealt with predominantly, but not exclusively, since the second aspect, the decoding of the signals provided by individuals or kingroup members about their identity is comprehensively treated and reviewed in this volume by Shermann and Holmes.

Individual recognition requires that, independent of the site of encounter, a single individual can be sorted out of a larger homogeneous group of conspecifics by means of reliable identifying characteristics and that this individual is treated in a way which presupposes earlier experience with it. As far as kin specific recognition is concerned, an analogous definition proves to be true, omitting the final restriction: kin can in some species be recognized without previous direct contact (e. g. Blaustein and O'Hara 1981, 1982; Buckle and Greenberg 1981; Greenberg 1979; Holmes and Sherman 1982, 1983; Sherman and Holmes, this Vol.; Waldmann 1981; Wu et al 1980).

2 Individual and kin identification and recognition in non-eusocial arthropods

As far as we know, individual «signature» systems (Beecher 1982) and recognition mechanisms in the group alluded to are found mostly in connection with temporary or permanent monogamy. Following Wickler and Seibt (1981) a distinction is made between a «mate guarding» monogamy and a «shared duty» monogamy.

2.1 Individual identification and recognition of mates in monogamous species

In the *mate guarding monogamy* of the clown shrimp *Hymenocera picta* (Seibt 1974, 1980; Seibt and Wickler 1979; Wickler and Seibt 1981) males seem to have considerable difficulties in finding unpaired females. Once a male has found a suitable mate it seems to pay him to stay with his iteroparous female, most probably because leaving her and searching for a new one would on average be more costly in terms of time expenditure and/or reduction of his probability to survive. The attachment in this case is asymmetric with males being strongly attached to the partner and females much less so. The adaptive value for a male of recognizing his female individually is presumably to be seen in protection against confusion; mistaking a new female for the former one – if this female is still alive and still offers good reproductive prospects – would too often result in uncompensated time losses. The question of why females recognize their male partners too is more difficult to answer, and until now there seems to have been no satisfactory hypothesis to account for this fact. The individual recognition cues are chemical characteristics and are – contrary to conditions in many other species – effective over long distances. Nothing is known about the chemical nature of the compounds used. The cleaner shrimp *Stenopus hispidus* lives in pairs too, the partners being capable of individual recognition (JOHNSON 1977). Possibly the adaptive value of this capability is the same as in the *Hymenocera* mentioned above, but this assumption is not yet proven.

In the *shared duty monogamy* which may originate from a mate guarding monogamy (as I assume has been the case in *Hemilepistus*, cf. LINSENMAIR 1984) resources vital for reproduction cannot – under present ecological conditions – be secured or handled by one individual alone; the cooperation of at least two conspecifics is necessary. In some cases, as in the carrion beetle *Necrophorus vespilloides* (PUKOWSKI 1933), the essential resource must not be left (and has only to be defended against conspecific competitors for a short period). If under such circumstances each partner wards off only consexual rivals – a feature not only of *Neocrophorus* but also of the snapping shrimp *Alpheus ornatus* (KNOWLTON 1980) – a monogamous social and mating system may result although partners do not recognize each other individually. In cases such as that of *Hemilepistus*, where the vital burrow has to be defended for a very extended period alternately by one pair partner against competitors of both sexes (also during periods of temporary absence of the mate from the den) an anonymous relation of the partners cannot suffice. Mutual individual recognition of mates is then a necessary precondition for securing this resource reliably.

Besides *Hemilepistus*, *Cryptocercus*, which was formerly considered a primitive cockroach but is more probably a very primitive termite, seems to be the only other non-eusocial arthropod to have reached a level of social evolution which in some respects is comparable to that of the desert isopod *Hemilepistus* (SEELINGER and SEELINGER 1983). *Cryptocercus punctulatus* clearly lives in monogamous pairs or in closed family communities. The animals dwell in the rotten wood of fallen logs and by feeding on the wood they construct in time rather complicated burrows consisting of a number of chambers connected by galleries. In laboratory experiments G. and U. SEELINGER could demonstrate that these burrows are defended – mainly by the adult pair – against alien conspecifics and they assume that they possibly also deter predators (of the young). Foreign adult or juvenile roaches can be discerned from the mate and from the pair's own young – by unknown chemical identifying substances. The question of the adaptive value of the monogamous pair formation in *Cryptocercus* has to be considered still open. One adult alone could probably fulfill all tasks but its chance of surviving the extremely long time of at least 3, possibly 5, years of broodcaring could be too low.

2.2 Individual recognition and rank orders

In many vertebrates rank orders and mutually accepted territory borders are an outcome of aggressive encounters between conspecifics. By individual recognition members of a rank order or «dear neighbours» can avoid repeating already settled disputes. CALDWELL

(1979) has proved that the mantis shrimp *Gonodactylus festai* can potentially learn to recognize conspecifics individually. These highly aggressive stomatopods equipped with potentially lethal weapons fight for scarce holes. Possibly they recognize their neighbours, they certainly recognize and avoid for some time those individuals which have defeated them in a previous battle. Since there is a high risk of severe injuries in fights, the adaptive value of the identification and recognition of individuals in this respect is obvious.

2.3 Kin recognition

Kin recognition, very common in mammals, birds and the eusocial Hymenoptera (cf. e. g. the relevant articles in this Vol.), is experimentally proven in the non-eusocial arthropods besides *Hemilepistus* (see below) and one subsocial Canarian *Porcellio* species (LINSENMAIR 1984 and unpubl.) only in *Cryptocercus* (SEELINGER and SEELINGER 1983). In the latter, trophallaxis with symbiont transfer is of paramount importance for larvae during the first two instars, i. e. during the first year of their life. Which valuable advantages the family offers its members in the following 2–4 years, remains at the moment a matter of speculation; possibly it is defense of the adults against predators on the young.

2.4 Nature and origin of the individual – or kin-specific signatures

Individual and kin recognition in the group of arthropods discussed here seems always to be mediated by chemical means. (The assumption of VANNINI and GHERARDI (1981) that the crab *Potamon fluviatile* might use optical cues to identify conspecific individuals was deduced from results gained under very artificial conditions; confirmation of these results by experiments performed under more natural conditions is required.)

With the exception of *Hemilepistus* we know hardly anything about the chemical properties of the substances used, literally nothing about their degree of variability and their origin. Assuming that these substances are chiefly genetically determined (and not predominantly acquired from the surroundings) which is known to be the case in *Hemilepistus* and which is not improbable in all other species mentioned, the question arises of what the initial variability has been on which selection could have started to work. I can distinguish two main «substrates».

1) Due to variability in their enzymatic equipment, members of a population always differ to some extent in their metabolism. If a repeatedly used vital resource, e. g. a burrow, has to be left temporarily and has to be found again under difficult conditions under which external cues for (i) reorientation to the animal's own (ii) and differentiation between its own and an alien burrow are unreliable or lacking, individual variability in metabolic endproducts (showing up in secretions, faeces and urine, for instance) which had no direct previous value could gain great importance by offering themselves as cheap labels. If those substances were used we could expect a great deal of preadaptation on the decoding side of the system. Many of these compounds could already have gained sign character in connection with e. g. food choice (preference and avoidance) at earlier times. (Furthermore the normally highly developed capacity in arthropods to acquire and process chemical information will always allow the identification of many compounds with no previous biological significance.) Primitive bees and wasps (STEINMANN 1973, 1976) and isopods (*Porcellio* spec., LINSENMAIR 1979) mark their burrow entrances individually, enabling them to find their burrow even under difficult conditions of high dwelling densities. In both groups this could have been a basis for the further development of a signature system.

2) The species-specific signals could represent the other possible starting point (cf. GREENBERG 1979). In order – for example – to find the optimum balance between outbreeding and inbreeding, those individuals will be favoured which are able to use the variability always present within the species-specific features to learn what their near relatives are like, and then, in choosing a mate, to avoid conspecifics which are too similar or too different from the relatives (e. g. BATESON 1980, 1982, 1983). This does not require the ability for individual recognition but to reach it from this level should pose no very difficult evolutionary problems.

3 Individual- and family-specific recognition in the desert isopod *Hemilepistus reaumuri*

3.1 Individual recognition of the mate

3.1.1 The monogamous pair: formation and behavior

Hemilepistus family groups stay together for 9 to 10 months after the birth of the young. By the end of the two- to four-month hibernation period the previous bonding of the family members to one another and to their burrow has been dissolved. As soon as they appear above ground, most of the animals leave their birthplace forever, maintaining (by astromenotaxis) constant walking directions for considerable distances (LINSENMAIR in prep. (I)). These directions are chosen at random. This has been demonstrated by marking > 3500 isopods. The distribution of the sites of recovery revealed neither indications of a family-specific nor a population-specific directional bias. Before attempting to set up a new den, the animals travel at least several tens of meters, and sometimes more than a kilometer.

About 8 days after leaving their family burrow, at the earliest, the first emigrants settle down. Sometimes they take over an empty burrow or succeed in evicting a previous occupant, usually they construct a burrow (for details see LINSENMAIR and LINSENMAIR, 1971; LINSENMAIR, 1979, 1984). In most cases (87 %, n=63) the builders of new burrows are females; they dig a tunnel 2–4 cm long and then guard it continuously. In a procedure that usually lasts for hours, a conspecific of the other sex is allowed to enter. By marking the partners in these pairs and observing them for long periods, an observer will notice the following points:

(a) After the first 48 hours of their life together the two partners were in no cases (apart from experimental intervention) found to separate «voluntarily». If a pair does break up it is because one of the partners fails to molt properly, dies, or is driven away by a competitor.
(b) The two partners cooperate in many activities – construction, cleaning and defense of the burrow, providing food (to some extent), and later the care of the young.
(c) In the 9–13 weeks from pair formation to the birth of the young, the burrow is always guarded by one of the partners.
(d) Beginning the second day after pair formation, at the latest, the guard drives away all conspecifics other than its partner, as long as it has had direct contact with its mate within the last 3–12 hours. When the partner demands admittance – e. g. upon its return from foraging – the guard first examines it by touching it with the apical cones of the 2nd antennae, which bear chiefly contact chemoreceptive sensilla (SEELINGER 1977; KREMPIEN 1983), and then as a rule immediately allows it to enter. This special, peaceful treatment of the partner is not site- or situation-specific, and it does

not require special behavior on the part of the mate. All that is needed is that the guard is able to touch the returning isopod somewhere on its cuticular surface.

(e) The forager finds its way back to the burrow by means of allo- and idiothetic information, in some cases using an effective search behavior (HOFFMANN 1978, 1983a–c, 1984). Having returned, however, it attempts to enter only if it has identified the burrow as its own. The decisive criteria for this identification in the first days after pair formation, are provided by the individual chemical characteristics (see below) of the partner itself – perceived chemotactically.

(f) When a foreign conspecific is – e. g. following experimental manipulation of the guard – allowed direct entry (i. e. entry without the admittance routine) to a burrow, the intruder usually (in 70–90 % of cases, depending on conditions; n > 100; details in LINSENMAIR in prep. (I)) ejects the guard from the burrow, letting it back in only after a normal admittance procedure. A paired woodlouse admitted by its partner unimpeded to its burrow never behaves in such way, even if it has been experimentally separated from its mate for several days.

(g) Under the climatic conditions prevailing in the normal pair formation season, the female permits copulation only at or in her own burrow and only with her – individually known – male.

(h) After the parturial molt the gravid female accepts only her own male. Conversely, the male behaves peacefully only towards his own female during and after the parturial molt.

3.1.2 *Discussion: pair bonding in H. reaumuri*

It is evident from the above points (a–h) that *H. reaumuri* forms monogamous pairs. There are two criteria, each of which in itself is regarded as sufficient to categorize a relationship between two heterosexual conspecifics as monogamy (WICKLER and SEIBT 1981, p. 215), and both are satisfied here:

1. «... the togetherness of two heterosexual adults as a social system».
2. «... a mating system in which an individual reproduces sexually with only one partner of the opposite sex».

The partnership is not simply an association of each member with the burrow, or any other kind of an anonymous monogamy rather the pair is kept together by a mutual attachment of the two individuals – each individually recognized by the other – supplemented by a bonding of the two together to their burrow. The latter bond always proves to be the weaker in conflict experiments. The great significance of the burrow with regard to the cohesion of the pair is that it is the meeting point of the partners, which can notice and identify one another only by direct contact, for they have no mechanisms for long-distance observation of and communication with conspecifics. Without such a meeting point they would quickly lose contact with each other and they never could establish a prolonged pair bond.

The mutual attachment of the partners to one another is evident in

(a) the prolonged cohabitation in very close proximity; after pair formation each member seeks frequent direct antennal and body contact with its partner (and later with its own young), whereas alien ispods are avoided or attacked.

(b) the many behavior patterns related only to the mate (and not to other potential sexual partners), only some of which have been listed above; (see LAMPRECHT 1973; WICKLER 1976).

The partner is individually identified on the basis of extremely varied chemotactically detected characteristics («discriminators»: HÖLLDOBLER and MICHENER 1980) (see below).

3.1.3 Individual recognition: a few selected problems

Of the many questions raised by the existence of monogamy and the ability to identify the partner individually, only a very few can be considered here – and these answered only in broad outline.

3.1.3.1 Do the partners learn to recognize each other in the initial admittance ceremony?

The procedure by which the occupant of a burrow in the field first allows a sexual partner to enter the burrow lasts, depending on different conditions, from ca. 1.5 to more than 48 hours. Here we are concerned only with the question of whether the occupant and the applicant for admission get to know one another during this prolonged interaction. The intensity of the mutual touching during this procedure suggested that they do. The possibility was tested by exchanging conspecifics with the same previous histories, involved in identical parallel experiments, but belonging to different families.

Table I:

A Is admittance time prolonged by exchange of intruders (females = burrow owners)?	
Males exchanged[1] every 45 min	Males not exchanged
Time until admission[2] in min	
122 ± 94 (n = 25)	126 ± 123 (n = 40)
B Are males individually recognized 60–120 min after 1. admission	
Males replaced to original female[3]	Males crosswise exchanged[1 3]
Time until 2. admission in min	
19,5 ± 18 (n = 20)	17 ± 8 (n = 15)

[1] Exchanged males originated from different families.
[2] Admission is characterized by females allowing males (1) to assume normal guard position within the burrow without pushing them out again for at least a quarter of an hour or (ii) letting them pass into the underneath position.
[3] In experiment B males were replaced 10 min after removal.

In most of the situations tested (see Table I) individual recognition could not be demonstrated. When one of the two interacting animals was exchanged for an alien 1) no delay of the overall process was detectable and 2) at the time of admittance the intruder is not recognizably distinguished from a foreigner by the burrow owner. The only effect observed was that some of the females in the role of intruders, when shifted to another burrow began search behavior after making contact with the alien occupant (see below). The fact that individual recognition was not normally developed in the admittance process, raised the following question.

3.1.3.2 *What is the function of the prolonged admittance procedure in pair formation?*

Under natural conditions, female *H. reaumuri* are strictly semelparous. Even after the earliest possible loss of the offspring – directly after birth (n > 250), or when the eggs were removed experimentally without (n=15) or with damage to the brood pouches (n=20) – in no case was a new clutch produced. With only a single chance for reproduction, and being extremely dependent on her mate for realizing this chance (cf. p. 422), the female can be expected to choose her male with great care. I assume that it is because this choice is so important that the female usually starts to dig a burrow; thereby being in a position to decide whom to admit. How does a female choose? She could evaluate the male directly – for example, by means of the duration and intensity of its striving for admission (for details of the behavior see LINSENMAIR and LINSENMAIR 1971). However, all results contradict this presumption. First, in a direct evaluation one would have expected the male not to be admitted solely according to a time program but depending on the result of an examination of the applicant's performances. In that case, however, (i) individual recognition should be established during the admittance procedure and (ii) exchange of applicants should prolong the overall admittance time; as was shown above, neither assumption proved to be true. Second, in laboratory experiments female burrow occupants were presented with males that were (i) greatly handicapped by amputation of the first two pairs of pereiopods, (ii) obviously sick (dying a few days later), or (iii) too small to defend effectively the burrow, dug by a larger female (n ≧ 25 in each case). Choice behavior by direct evaluation should have resulted in the rejection of at least some of these males. The actual result, however, was that all the males that persisted long enough at the entrance were admitted, the times required being not prolonged greatly (Table II).

Table II: Are hampered males refused by female burrow owners?

4 legs amputated	Intact males
Admittance time (in hours)	
16,01 ± 13,2 (n = 28)	12,56 ± 16,25 (n = 30)
No admittance in 48 h	
2	2
Admittance rate	
93%	94%

The experiments were performed only 8 days after termination of hibernation, therefore overall admittance times are rather prolonged. As in all comparable experiments the experimental and the control group consisted of sisters and brothers which all had exactly the same previous history.

Field observations have shown that during the first weeks of the pair-formation phase, everywhere but in the sparsely-inhabited peripheral regions, males – being forced to wait for hours during the admittance procedure – must in this time engage in multiple fights with other males – partly vigorous battles that are very often repeated at intervals of only a few minutes, due to a large number of competitors. Ordinarily the winner is the larger

Table III: Agressive encounters in the field: weights of winners and losers

A Males			
	Heavier	Lighter	Significance
Winner	68 (284 ± 28 mg)	23 (266 ± 29)	P ≪ 0,001%
Loser	25 (290 ± 20)	73 (244 ± 39)	χ^2
B Females			
Winner	17 (246 ± 19)	9 (237 ± 10)	P = 0,03
Loser	9 (259 ± 15)	19 (204 ± 22)	χ^2

All fights took place without experimental interference at natural burrows. In most cases winners and losers – after reaching a definite decision – were put individually in small containers. They were weighed on a microbalance 1–4 hours later. In a few cases single winners started a new fight immediately after the first one observed. Therefore the number of losers in A exceeds that of winners. The fact that the average weight of those losers which are heavier is slightly higher than the weight of heavier winners is mainly caused by the participation in the aggressive encounters of old parents surviving their first breeding cycle. They are the heaviest but usually are somewhat clumsy and weakened and therefore lose nearly all fights. (They have never been observed surviving until a second pregnancy).

male (see Table III). As a result, during an admittance procedure one male applicant is often replaced by another, sometimes repeatedly. Weakened or small animals usually have no chance of keeping their places long enough to gain entry.

Conclusion. These observations gave rise to the hypothesis that the admittance procedure does indeed serve as a test of the future partner – in an indirect, very effective way that automatically takes into account whether a selection procedure is possible among applicants. In the light of this hypothesis, the results summarized in Table I are not surprising. Even if a conspecific were to become individually identifiable during an admittance procedure, the owner of the burrow should not reject an animal that replaces this already recognized potential sexual partner, because each such change can be expected to bring an improvement. It is also not surprising – indeed, it follows compellingly from the hypothesis – that any applicant (cf. Table II) will eventually be admitted if it persists long enough. After all, if there is no alternative any partner is better than none, and it would be a disadvantage to adhere to absolute criteria. Moreover, after pair formation there is another, indirect test of suitability; an animal that cannot effectively defend its burrow is very likely – if it encounters a superior opponent – to be forcibly expelled. Burrow owners, as just demonstrated, seem not to get acquainted with a potential future sexual partner during the admittance ceremony. But, as already mentioned, females in the role of applicants partly notice an exchange of the burrow owner opposite. (Table IV).

3.1.3.3 *Why do about 50% of female applicants give clear indications of recognizing the guard, whereas male applicants do not?*

Only to a very limited extent do intruders seem to be able to tell – from the intensity, form and duration of the defense behavior of the guard (and not by chemical cues) –

Table IV: Reactions after transferring intruders to a second burrow with an alien owner

Sex of intruder	Leaving burrow entrance "searching"	Uninterrupted intruding behavior	Significance
Males	1	20	P = 0,002
Females	10 (controls: 1)[1]	12 (controls: 15)	FEt[2]

After striving for admission for 1–2 hours the intruders were cautiously seized with a forceps and immediately transferred to the entrance of another burrow, identical with the first with the exception that its guard originated from another family (i.e. neither from the intruder's nor the first burrow owner's family).

[1] controls: identically treated but replaced to the first burrow
[2] FEt: Fischer's Exact test

whether the latter is paired or not. For a male intruder the problem of determining that a female burrow guard has no partner is not very difficult; when a burrow is occupied by a pair, the male is almost always present during the activity phase of intruders, and in a series of field observations during 14 activity phases in 86 % (n= 154 pairs) the male was the guard. In these cases brief touching provides the necessary information. When a female with a partner is on guard duty and cannot rapidly drive away an alien male, she will often creep backward into the burrow past her male, «handing over» the task of guard to him. For these reasons, male intruders waste little time in applying to females that already have a partner.

Once a male has found a single female and stays striving for admittance, he loses contact with the burrow only during fights with other males. If, in trying to find his way back after a fight, such a male were often in danger of encountering another single female instead, which would require him to invest more time before being admitted, it would be an advantage to the male to be able to detect his mistake. However, all field observations have indicated that such a situation is so rare and therefore this danger is so uncommon that a special adaptation is not to be expected.

Females seeking admission to a burrow have much more difficulty in finding a single sexual partner. A high percentage of their approaches are likely to be futile (30–60 % depending on how much of the pair-formation phase has elapsed, and if striving for admission for more than 15 minutes is taken as the criterion). Females that have succeeded in locating an unmated male but are distracted by a fight and must then find their way back to his burrow are in great danger of encountering the wrong burrows, occupied by paired males, for these are always much more numerous (the maximal density can reach up to 20 occupied burrows per m^2 during the mating season). Females capable of identifying males individually (and not only by behavioral criteria, which are relatively unreliable) in this situation could often save themselves time-consuming mistakes.

Discussion. In the situation just described, rapid familiarization with an unknown individual would be advantageous to a female. Otherwise, however, it would be more of a disadvantage, and should be suppressed. The social partner is distinguishable from all conspecifics comparable in sex, age, physiological state and so on solely in that it has a familiar badge. By prolonged experimentally enforced tactile contacts an isopod cannot help but get acquainted with an alien conspecific who, in most situations, is then automatically converted to a social partner. Exclusivity of the social units (see p. 431), which is necessary for various reasons, is guaranteed only as long as the learning of another animal's discriminators is limited to the situation of monogamous pair formation (and of contact with one's own young after their birth; cf. LINSENMAIR 1984 and in prep. (II)). But just during the pair-formation phase it is often impossible to avoid prolonged contact with conspecifics that should remain strangers. The simplest way to prevent such familiarization, with a general method that does not require analysis of the specific situation, is to set the frequency of contact required for familiarization so high that only the partners in a pair can achieve it. This assumption could explain why – although the partners in a newly formed pair touch one another often and extensively – 12–24 hours must elapse after admittance before both are demonstrably capable of recognizing one another reliably, whereas in other situations they learn much faster. Of course behavioral variables can play an important role in this process; a paired guard touches an intruder much more rarely than does a single guard, which is in principle prepared to allow entry. (It is not clear whether male and female intruders vary in this regard which then could explain the sex-specific differences of Table IV).

3.1.3.4 Is the pair bond endangered by unavoidable contamination with foreign discriminators?

The discriminators of the desert isopods are strongly polar, practically non-volatile compounds of low molecular weight (LINSENMAIR, SCHILDKNECHT and ESSWEIN, in prep.). They can easily be transferred from one individual to another by direct contact, the individual thereby becoming – depending on the intensity of contact – more or less strongly and long-lasting alienated (LINSENMAIR 1972, 1984, and in prep. (III); see p. 425). Paired guards cannot avoid direct body contact with foreigners during the pair formation period while defending their burrows against alien intruders. When such a guard goes out to forage, in a few cases it is rejected by its partner on its return. In this situation, and this situation alone, 15 of 20 animals responded to the defensive blows of their partners with a very effective behavior: they turned around and presented their hind ends to the guards. This part of the body always remains uncontaminated during defense of the burrow. After it had been touched by the guard's antennae, all individuals were immediately admitted.

In experiments aimed at alienating adult, mated individuals it became apparent that when one member of a pair is forced into body contact with adult foreigners – to an extent that if solely young animals had been involved the contamination would have caused long-lasting rejections by their respective families – its partner guarding the burrow sometimes reacted by no, sometimes by only brief defense. It seemed possible either that the adults are less sensitive than juveniles to foreign discriminators or that they produce less substances which are easily transferable. To test these possibilities, reciprocal contamination experiments were done with adults after hibernation and with almost full-grown (5-month-old) juveniles.

The results of these laboratory experiments are summarized in Table V. They show that the adults (i) transfer less substance than do juveniles under otherwise identical conditions, and (ii) respond less aggressively to foreign substances encountered when examining their partners than do members of families to equally intense contamination of their kin. (Other possible interpretations – that the foreign substances might be less well adsorbed by adults because of differences in the surface properties of the cuticle, and that the adults produce substances less effective in eliciting aggression – have been ruled out by experimental tests outside the scope of the present discussion.)

Conclusions: Our above question can be clearly answered: pair maintenance is not endangered by the unavoidable close contacts with aliens in burrow defense. This poten-

Table V: Alienation of adult pair partners and young family members by direct contact with alien individuals of different age

	Alienated	by	Time of contact	Admittance time ≤ 5 min	Admittance time > 5 min	Significance
I	Adults	30 Young	1 min	4	13	I : II
						$P < 0{,}02$
II	Adults	30 Adults	1 min	16	8	χ^2

				Admittance time ≤ 1 hour	Admittance time > 1 hour	
III	Young	30 Young	1 min	4	18	III : IV
						$P < 0{,}001$
IV	Adults	30 Young	1 min	14	3	χ^2

The individual to be alienated was placed into a small glass vial and then the thirty unrelated conspecifics of different age (see text) were added.
The isopods could not avoid closest contact which was additionally intensified by shaking the group up very often thus taking care that the test animal was during this procedure permanently amidst the aliens.

tial weak spot is multiply ensured against possible disturbances by suitable adjusted behavior, reduced production of exchangeable badge substances, and altered – lowered – sensitivity to contamination.

3.1.4 *What are the adaptive values of monogamy and individual recognition under present-day conditions?*

a) To acquire a burrow without undue delay and then to hold it against all competition are absolute prerequisites for the survival and reproduction of a desert isopod. Since these animals have no adaptations for efficient digging, construction of a new burrow is a laborious process which – because of climatic conditions – must begin in spring. The very strong intraspecific (to a lesser extent also interspecific) competition for this vital resource makes it necessary to keep constant guard over it. Adequate food can normally only be found outside the burrow, and a single woodlouse cannot simultaneously guard and forage. In the present situation a burrow can be permanently defended only by the cooperation of at least two adults. These should be able to tell one another apart from foreign invaders. Learning to recognize the partner's individual characteristics is certainly the most reliable solution to this problem (LINSENMAIR and LINSENMAIR 1971; LINSENMAIR 1979, 1984).

b) Even if there were no foraging problem and a desert isopod could stay indefinitely in its burrow, a female during the parturial molt and pregnancy would mostly lose its den to competitors since she cannot defend her burrow effectively enough if at all. Only if she lives together with a male during this critical time does the female have a good chance of reproduction. As a proximate cause, the individual attachment of the male to his female ensures that the pair remains together even after the parturial molt, in which the female loses her sexual attractiveness.

c) For reasons as yet unknown, single gravid females (at least under laboratory conditions) have a very small chance of survival even if all competitors are excluded. We had noticed this phenomenon in our laboratory breeds for some time; in an experiment to study it (LINSENMAIR 1984) 20 females were isolated from their males after the parturial molt, 18 of these females died, either before the birth of the young or before the time when the young would have had a chance to survive alone. In the contol group with pairs kept under otherwise identical conditions, 19 of 20 sisters of the females in the experimental group survived the critical period. It seems very unlikely that these results are pure laboratory artifacts with no relevance to the situation in the field.

d) Moreover, loss of either partner (regardless of sex) within the first two weeks after the birth of the young reduces the mean number of young that survive. In the experimental group there were 23±28 survivors (n = 38 families), 40 % (a significant difference) of the average in the control group, with 40±28 survivors (n=47); there was a particularly large increase in the incidence of total-brood mortality, with 21 % in the control group as compared with 50 % in the experimental group (LINSENMAIR 1979).

3.1.5 *Discussion: what makes monogamy the best reproductive method for both sexes?*

For females the advantages are so obvious, as the preceding section has shown, that a further discussion seems unneccessary. Males deserve closer consideration in this regard – both in general and among the Peracarida in particular, because their direct parental investment in the young is usually so much less than that of the females. In all such cases one would expect to find a polygamous mating system, relinquished only when conditions select strongly against it (cf. TRIVERS 1972; LINSENMAIR 1979, 1984; WICKLER and SEIBT 1981). Because Hemilepistus males in all our extensive experiments never seem to «play the field», it must be inferred that an attempt to do so would incur risks that would not, on average, be compensated by success with one or more additional females. As analyzed in detail elsewhere (LINSENMAIR 1979, 1984), there are only two situations worth considering in which a male could – theoretically – leave his female without a nearly 100 % risk of losing the investment he had made up to that time, which on average could never be compensated in the future. These situations are as follows:

(a)The male could leave his female immediately after the regular copulation phase, so soon before parturial molt that the female could find another partner but – because she cannot prolong the brief time until parturial molt – has no opportunity for frequent copulation with him. Results so far, though still preliminary, indicate that there is no sperm clumping (LINSENMAIR 1984), so that eggs would always be partly fertilized by sperm from the first male. A method of this sort would require (i) that the males be capable of judging the point of time within the reproductive phase of the population and hence of knowing whether there is enough time left for them to be likely to find an unmated female offering an adequate opportunity for reproduction, and (ii) that the males be very well informed as to the condition of their females, in order to choose the right time to leave them. The requirement in (ii), at least, is certainly not met; the females «withhold» all relevant information from the males (LINSENMAIR 1979, 1984). Still more important, however, is a third point; males forming a pair with a new female should be able to more than compensate the losses they incur by their infidelity towards their first female. Our first results indicate that males choosing the best time to desert their female «had to reckon» that about 50 % of their first females' eggs will be fertilized by the second male. In order to allow for a positive selection of a deserting behavior one basic requirement – besides others left out of consideration here – would be that the new pair's chances for survival (until the male reaches the end of the regular copulation phase for a second time) should exceed 50 %, but this on average is not the case (SHACHAK 1980; LINSENMAIR 1984), often their chance does not even reach 10 %.

(b) The second opportunity would arise only some weeks after the birth of the young. But because the climatic and other ecological conditions make it essential to time reproduction accurately with respect to the annual cycle, the females are too well synchronized with one another, thus a male leaving his family at that time would have no chance of finding a female still in a suitable state.

In the present-day situation the desert woodlice have good chances of reproduction only as monogamous pairs. Current conditions of selection certainly stabilize this form of a social and mating system. However, they tell us nothing about the evolution of this highly differentiated and certainly very derivative behavior. The special construction and defense of a permanent burrow in a firm substrate, the formation of pairs in which the partners individually recognize one another, brood care with family-specific recognition of the young – all are crucial adaptations to the ecological conditions of the present biotope and hence prerequisites for inhabiting it. Therefore they cannot have been formed by the selection factors prevailing at present. The early steps must have been taken under other selection pressures, under less severe abiotic ecological conditions (cf. LINSENMAIR 1984).

3.2 Kin recognition in *H. reaumuri*

The parental pair together with their young – up to 140 of them – constitute a strictly closed community. Foreign conspecifics that appear at the burrow entrance (or within the area around it bounded by the fecal wall) are always attacked by the parents and juveniles (6 weeks of age or older), and outside this boundary they are either attacked or avoided. The members of the cooperative community never behave aggressively to one another «intentionally» (i. e., after tactile examination) until the family group dissolves, unless too close contact has been made with alien conspecifics. This behavior shows that the desert woodlice are not only capable of identifying individuals, but can also distinguish kin from non-kin.

3.2.1 Interfamily variability

In the field, hundreds of experiments and very many more observations have produced no evidence of mistaken identification, despite the fact that at times *H. reaumuri* can live in very dense local populations (up to 14 families with young per m^2). The radius within which the family members forage can be as great as 6 m and is usually 1,5–4 m. Given a foraging radius of 2,5 m, with 5 families per m^2 it is to be expected that members of about a 100 families will come into the vicinity of any particular burrow. Since families in the field remain exclusive even at highest densities, it follows that the variability of the characteristics used for kin identification must be very great.

Well over 50,000 tests involving reciprocal exchange of the juveniles of two families have been carried out, and no two families have ever been found to tolerate one another. Unilateral toleration has been observed. In a systematic cross-exchange test 70 families collected in the field from an area of 50 × 50 m were examined for discriminator coincidence. Of the 4830 repetition-free combinations there were 4 in which the young from one family were not attacked by the members of a single other family. In this case a number of conditions (of animal maintenance and experimental procedure) acted to reduce aggressiveness, so that this misidentification rate of 0.1 % is certain to be an overestimate. But it does indicate that very close resemblances can exist in this frequency.

3.2.2 *Nature and origin of the family badge*

There is no doubt that the characteristics used for kin identification, the «family badge», are chemical in nature. One of the clearest demonstrations of this is that with suitable solvents extracts can be obtained that when applied to a neutral carrier such as a glass rod can elicit aggression (as foreign conspecifics do); when such extracts are applied to individuals the other members of their family treat them as aliens for a long time.

The density of family-occupied burrows and the great overlap of the foraging areas of neighboring families made it *a priori* highly unlikely that the variety and specificity of kin discriminators would reside in body odors acquired from the surroundings. All the relevant early experiments (LINSENMAIR 1972) and later extensions of them confirm this view. Only one finding need be mentioned, as follows. Hundreds of pairs and families collected in the field were kept unter constant laboratory conditions on the same substrate, and given the same food, for 5–9 months. Many hundreds of woodlouse families bred from those collected in the field have since been maintained for their whole life cycle under these conditions, some of them now in the 4th generation. Unless inbreeding occurred, none of these animals have given any sign of reduced acuity in making the distinction between kin and non-kin ($n > 500$ cross-tests). Our earlier hypothesis (LINSENMAIR 1972) – that the discriminator substances, with all their diversity and specificity, are genetically controlled secretions, very little influenced by the environment – has now been corroborated by many additional experiments (LINSENMAIR in prep. (III)) and can be regarded as firmly established.

There are several compelling reasons – (i) the degree of variability, (ii) the reproducibly graded aggressive response to members from different alien families (extending in the extreme to unilateral acceptance), and (iii) all the results of mixing experiments (see below) – to conclude that the family badge is always a blend of several components.

3.2.3 *Intrafamily variability: individual characteristics*

Because emigration of the young (see p.415) thoroughly mixes the population during the pair-formation period, we have to assume that in more than 99 % of all pairs the partners come from different families, with different badges and corresponding genetic differences. There is no evidence for sex-linkage of the genes responsible for production of the identifying substances. With normal autosomal inheritance, however, variability among the progeny of a pair is to be expected. How, then, can there be a common family badge? Does it really exist? Adoption experiments (LINSENMAIR 1972) in which parents are brought into very close contact with half of the young from another family, thereby compelling them to get acquainted with these juveniles, have shown that as a rule the foster parents will then also accept the other half of the young. This outcome refutes an alternative hypothesis to that of the family badge, namely that mutual acceptance within a family could be a matter of individual recognition alone. The results indicate that the variability of discriminator patterns among family members is somehow reduced or adjusted so that a common badge is achieved.

Given a great primary intrafamily variability in the production of individual discriminator sets, one way to achieve a common family badge could be to make a secondary adjustment, e. g. by confining the spectrum of produced substances within the family to the smallest common denominator. Then the individuals would have to match themselves to some extent to their chemical surroundings. Amputation of the apical cones, the sole structure responsible for badge identification, ought then to cause changes in production, but no such changes have been found (LINSENMAIR 1972; KREMPIEN 1983). Furthermore, week- to month-long integration of single individuals into strange families, during which such individuals soon become fully accepted (and soon «regard themselves as belonging» to the new family; LINSENMAIR in prep. (III)), ought to be accompanied by changes in production. But when such integrated individuals were isolated immediately prior to a molt and afterward tested to see how their natural and adoptive siblings would respond, the results (Table VI) gave convincing evidence against such a hypothesis. The integrated isopods after removal from their foster family and before molting – were treated in their natural families like foreigners. The alienation could have been either the result of the integrated animals' adjustment to its changed chemical environment or an effect of contamination caused through transferred substances by the foster families' members. The fact that after molting in isolation these individuals were accepted without any aggression in 84 % of all encounters by their genetical kin but only in 15 % by members of their former foster family clearly indicates the lack of an adjustment in discriminator production. It rather points to contamination as the reason for the observed alterations after the integration and before molting in isolation.

b) As the previous experiments indicate, and as we know since a long time (LINSENMAIR 1972), badge substances are exchanged between animals during body contacts. Members of a family group, living shoulder-to-shoulder within the burrow, cannot avoid this exchange and the consequent intermingling of their secretions. But if the production of the individual discriminators is under extensive genetic control and the family badge is a product of the mixing of the individual discriminators, then – in the absence of special adaptations – in the course of the (very numerous) molts communication problems must arise.

Table VI: Does integration into an alien family change the production of discriminators?

	Test animals	Where tested	Encounter releasing Attacks	No attacks	Significance
A	Not yet moulted[1]	Genetical family	192[2]	3	
			(N = 22)		A : B $P \ll 0{,}001$ (χ^2)
B	Moulted	Genetical family	164	879	
			(N = 11)[3]		B : C $P \ll 0{,}001$ (χ^2)
C	Moulted	Foster family	522	92	
			(N = 17)		

Reactions of natural and adaptive siblings towards individuals which had been integrated for 4–8 weeks into foster families and then taken out and tested before and after moulting (in complete isolation). The integrated isopods were removed from their adaptive families about 24 hours prior to the beginning of an ecdysis; they were tested 36–60 hours after terminated ecdysis.

1 Tests immediately after removal from the foster family.
2 Number of observed single reactions.
3 Number of tested isopods.

Table VII: Loss of alien substances by molting

Contaminated individuals	Reactions of uncontaminated family members		Significance
	Attacks	No attacks	
A Before molting	180	3	A : B
	(n = 15 individuals)		
B After molting (24–48 hours after termination)	36	420	B : C P ⩽ 0,001
	(n = 15 individuals)		
C Controls: contaminated like A, kept in isolation for the same period as B, but without molts	145	6	A : C not significant
	(n = 10 individuals)		χ^2

The test animals and the controls were contaminated by close direct body contacts with 10 alien conspecifics for 5 min, in A/B 24–48 hours before a molt. For molting individuals were kept in isolation in single uncontaminated containers.

3.2.4 Does molting create problems in communication because of the mixed character of the family badge?

When it molts, each family member discards along with the exuvium all the discriminators it does not itself produce (cf. Table VII; see also Table VI). Observation of undisturbed families reveals that in far fewer than 0.1 % of molts is there any evidence of the newly molted animal being killed (and even in these cases there could be reasons other than communication problems). Here, again, integration experiments and observation of artificially composed groups are useful means of claryfying the situation. We can – as the results summarized in Table VI have proved – safely assume that the substances produced by an integrated animal (which was severely attacked before its integration) depart considerably and permanently from the norm of the adoptive family. If individuals within mixed groups molt 7 days or more after integration, the group or family (if undisturbed) treats them in 96 % of all cases (see Table VIII) not differently from an unmolted mixed group or family member. But if the integrated foreigners are removed shortly before the onset of molting and replaced when molting is completed for more than 36 hours, they are predominantly (in 85 % as the results summarized in Table VI have shown) treated as aliens. Control animals, which had been integrated into families for an equally long time but which did not molt during the four to five-day-period of removal (n=50), were – after being replaced in their foster families – accepted without aggression in 98 % of cases, proving that in the test animals in fact the molting-caused changes were the reason for the altered responses of the adoptive family.

Animals molting within their foster family or mixed group after an integration time of less than 6 days have only a limited chance of surviving (see Table VIII), and all the animals (n=30), which were integrated as singles into pure foster families and which molted there within 36 hours were killed. This dependence of tolerance on integration time could indicate that learning processes are crucially involved in acceptance. But how are we to interpret the difference in response depending on whether the integrated animal molts within the family or in isolation?

Table VIII: Number of individuals surviving complete molts in mixed groups in dependence of time spent in these groups

Days after group formation	Surviving	Killed	Significance
> 2–6	56	24	P ≪ 0,001
7–35	1433	56	χ^2

Killing normally occured during or shortly after shedding the posterior half-exuvium. These animals then were cannibalized.

Preliminary answers to these questions have been given by experiments in which animals – either foreigners integrated into pure families or members of complex or highly complex mixed groups[1] – were removed for molting but returned to the group a shorter time after stripping off the anterior hemiexuvium than was the case in the experiments described above (see Table VI). Animals returned 5–16 hours after the anterior hemiexuvium is shed are attacked only rarely and they hardly ever encounter strong aggression (see Table IX). But if one waits longer, most are attacked. Evidently these early post-molt individuals enjoy temporary protection, which becomes permanent if they are returned to the family or group soon enough.

When a woodlouse is presented an unfamiliar, newly-molted individual, it can easily be seen that the newly-molted half of the body exerts, especially during the first hour after the shedding of the respective hemiexuvium, an inhibitory influence. The results summarized in Table X, however, show clearly that this protection is far from perfect at any time, for in all cases sooner or later (and always within a period of a few minutes only) the foreigner, if placed into an alien family or alien mixed group, is bitten – and the emergence of hemolymph always cancels any inhibition (even that with respect to a normal family

Table IX: Members of mixed groups replaced at different times after molting in isolation into their groups

	Time between shedding the anterior hemi-exuvium and test in hours	Reactions of group members		Significance
		Attacks	No attacks	
A	5–16	20	242 (n = 16 individuals)	P ≪ 0,001
B	22–96	180	100 (n = 16 individuals)	χ^2

The isopods were removed about 12–24 hours before molting; they were kept in isolation during the molt and the exact time of shedding the second half exuvium was determined by direct observation. After having been tested in A the animals were isolated again and used for a second time in B.

[1] In «complex mixed groups», containing between 60–120 individuals never more than 3 individuals originate from the same family, in «highly complex mixed groups», each of the 60–120 individuals was taken from another family.

Table X: Reaction towards alien conspecifics after touching their newly molted body half

	Time between shedding an hemiexuvium and test	Reactions in single encounters: Attacks	No attacks	Significance
A	1–14 min	124	397 (n = 20 individuals)	A + B : C $P \ll 0,001$
B	> 14–30 min	171	419 (n = 23 individuals)	C : D $P \ll 0,001$
C	> 30–60 min	249	331 (n = 23 individuals)	χ^2
D	> 60 min–3 h	149	40 (n = 9 individuals)	

The test animals, seized with a forceps, were presented the controls in such a way that these could touch the newly molted body half only. Often clear signs of inhibition could be observed as follows: controls quickly withdrawing their antennae, retreating some steps, etc. (cf. Linsenmair 1984).

member). Under most conditions the protection only becomes perfect when the members of the family or the mixed group have had a chance to live together with the alien animal for a prolonged period before its molt. It is to be assumed that the isopods could therewith become familiar with the special chemical properties of the deviating group member. Before proceeding to further considerations, we should answer the question of how relevant the results obtained with experimentally altered groups may be for pure families. When members of unaltered families were isolated for molting, and not replaced until 24 hours after its termination the results summarized in Table XI were obtained; they show that the same alienation problems can sometimes arise in natural families, which justifies our experimental procedure.

How can this recognition system – which apparently requires the ability to learn to recognize deviant individuals – function, given that the family badge is a mixture of all the individually produced components? Is it simply a matter of learning all the individual components of the family badge, in order to be able to tolerate then any arbitrary partial combination if it is coupled with the signal «newly molted»? The members of highly complex mixed groups with a great (about 90–95 %) but not complete overlap in the spectrum of families from which the single individuals originated, in many cases tolerate

Table XI: Reactions of pure families towards members isolated during molt and replaced 24–48 hours later

Members of 25 families	Attacks	No attacks
	27 individuals	99 individuals
family No 78/25	17 individuals	4 individuals

These families collected in the field, were kept free of any experimentally caused contaminations. A few families – family No 78/25 is an example – show an extraordinary high incidence of molting-caused alienations.

one another with no aggressiveness, even though they behave very aggressively toward pure families or differently composed mixed groups. The badges of the similar highly complex mixed groups are evidently not distinguished from one another. When members of these mixed groups are isolated for molting and afterwards returned for brief periods, alternately to their own group and to the second group that does not differ detectably in its group badge, they are much more frequently (and with high statistical significance) attacked in the latter group (Table XII). (Observations similar in principle can often be made in pure families that have been divided for long periods, i. e. > 15 days or in complex groups with complete overlap in respect to the families from which members were taken to compose the two or more experimental units.) Accordingly, two groups can have badges indistinguishable from one another and nevertheless comprise differing members (with regard to their production of individual chemical identifiers).

Here, again, experiments in which groups have been assembled experimentally provide critical evidence. A foreign juvenile brought into close contact with young of the control family can alienate between ca. 5 and 20 individuals (depending on the family combination) to such an extent that when they are returned to their uncontaminated family members they are treated as aggressively as foreigners. But when a single alien mingles with a considerably larger number of sibling juveniles from another family, alienation does not occur even after long-term exposure. A single foreigner can produce and transfer only a limited amount of badge substances. If those components which are not contained in the badge of its foster family are distributed among many adoptive siblings, these are diluted below the effectiveness threshold. That is, in the subset of individually produced substances shared by *all* members of the group, the components with respect to which this foreigner (or a single deviant family member) differs from the norm are lacking. We can now understand that two groups may have the same group-specific badge even without complete agreement between the entire sets of discriminator substances produced by each group.

How is such a deviant animal protected when it molts? The decisive point is that in fact the deviant discriminators do not show up in the common badge but – and this certainly is crucial to the deviant's survival – its individual traits are never lost. The family badge can never completely mask an individual's discriminator complex, a complete glossing over individual chemical properties is – according to many experimental results – (LINSENMAIR in prep. (III)) impossible. The physiological basis of this resistance to complete masking presumably (as other findings indicate) lies in a very firm binding of part of the animal's own secretions to the surface of its cuticle. Whereas the family badge can be perceived by touching any member of the group, even if that member itself produces only some components, the detection of individuals producing discriminator substances not present in the

Table XII: Newly molted individuals originating from one of two highly complex mixed groups with identical badges: Are they treated identically in both groups?

Newly molted tested in	Reactions of group members in single encounters		
	Aggressive	Non-aggressive	Significance
Own group	26	191	$P \ll 0{,}001$
Alien group	108	72	χ^2

7 individuals were isolated 20 hours or less before a molt and replaced within 6–20 hours after its termination.

family badge requires that they be touched directly. Within a large group it certainly does not happen that each member touches each other member in every activity period, and it is just as certain that many touches are necessary to identify a deviant conspecific reliably; in this light, the fact (documented above) that the development of tolerance requires relatively long integration times becomes rapidly understandable.

Memory plays a central role in all identification responses, and this is not the least of the reasons why it is difficult to judge the extent of intrafamily variability by using the indirect methods of behavioral experiments. It is therefore impossible to estimate the demands individual and kin recognition make on the learning abilities of the family members. We do not yet know how many components the family badges contain and how many other discriminators may be «concealed», to be learned only through direct contacts with the individual concerned. But there is considerable evidence that very complex combinations of characteristics, including many components that do not appear in the group badge, can be learned.

3.2.5 Inter- and intrafamily communication: discussion

a) Why are the family groups so strictly exclusive?

In the light of our present knowledge, being the result of intensive discussions which have received their strongest impetus from the fundamental contributions of HAMILTON (1964a, b, 1972), the basic answer is trivial: in order for the valuable and costly efforts expended within the social unit on behalf of conspecifics to have an adequate reproductive payoff, these must be confined to relatives. The society must protect itself from both intentional and accidental parasitism, by keeping foreigners out. Because a desert woodlouse family still stays together very long after the time when the juveniles pass the stage of «altricial» nestlings – though they remain thereafter for many additional weeks dependent on parental care and for months on mutual assistance among themselves – the problem of discerning foreigners from family members cannot be solved by, for example, site-specific recognition. For family members to be reliably distinguished from non-members in this case, where many families may dwell nearby, a stable but highly variable identification system allowing accurate and reliable discrimination of kin among a very large number of conspecifics is required. Acquired body odors, from the surroundings or via the food, which seem to be used exclusively (?) or at least as the essential components of the discrimination systems of some eusocial Hymenoptera (reviews: RIBBANDS 1965; MICHENER 1974; WILSON 1971; HÖLLDOBLER and MICHENER 1980), cannot meet these requirements (LINSENMAIR 1972). The signature system evolved by *Hemilepistus*, with its extremely diverse secretions – under genetic control and extremely resistant to most possible external influences – is an excellent solution to the problem of kin recognition under especially difficult conditions resulting from temporary very high dwelling densities and a more or less complete overlap of the foraging grounds of neighbouring families. But this method has one very vulnerable feature: during direct contact substances are exchanged between unrelated animals, rapidly causing alienation from their own families. Therefore such contacts must be avoided at all costs, and as long as the family badge is used for identification the community must remain exclusive. This certainly explains, at least in part, the pronounced aggressiveness exhibited toward alien conspecifics in situations in which it is comprehensible on neither ecological nor sociobiological grounds – it serves to prevent -unintended- contacts (LINSENMAIR in prep. (III)). The danger of those contacts should not be underestimated, for it is only by direct touching with the antennae that identification is possible, and in such close proximity «involuntary» contact may easily occur prior to antennal touching. (Where the density of *Hemilepistus* families was very high we always found indications of relatively large contamination problems.)

b) Interfamily variability. At a conservative estimate, in relatively densely populated regions each isopod burrow lies within the foraging radius of at least 100 families (see p. 424). An estimated upper limit, for very high population densities (10–12 families per m^2) and extensive foraging excursions (5–6 m radius) due to food shortage caused by the high numbers of individuals, would be of the order of 1000 families, the members of which could make contact with the inhabitants of a single burrow. Moreover, taking into account that within a complete family, as a rule, there exist at least three distinctly different badges: the male, the female and their progeny differ so greatly from one another that familiarity with the identifying characteristics of one of the three does not result in acceptance of the other two (a result not discussed above; see LINSENMAIR in prep. (III)). It should also be remembered that within the progeny individual variability through «personal» substances not showing up in the family badge is high. Thus the possible number of discriminator sets an individual could encounter at its burrow and mistake for kin may increase by another order, or, as an absolute maximum, another two orders of magnitude.

The estimate of 10^5 different types of identifying characteristics that a family member can encounter, none of which is to be confused with those of its siblings and parents, is probably very high and overestimated. But since we have found not a single case in which two families had the same badge, we can reasonably postulate a discriminator diversity sufficient to ensure that even in the most demanding circumstances misidentifications remain very rare exceptions. The finding of unilateral acceptance in 0.1 % of the cross-combinations in one experiment (i) may well be an overestimate (see p. 424 and LINSENMAIR in prep. (III)) and (ii) offers little threat to exclusivity. That is, a member of a family that cannot distinguish the badge of one other family from its own, so that it «feels at home» when it contacts the marked vicinity of the other family's burrow and may attempt to enter, is recognized as foreign and driven away (presumably because its family badge includes at least one extra component; LINSENMAIR in prep. (III)). Members of the family accepted by the other family never discover this fact, for to them the other family is foreign. Only in exceptional situations – for example, when deprived of their own burrow – might individuals of the more discerning family by a very improbable chance try to enter the burrow that would be «right» for them.

Let us assume, as there are good reasons to do, that badge differences are chiefly or even exclusively qualitative in nature; how many components would then be necessary to produce 10^4–10^6 distinct badges? In our chemical analysis, after a number of purification steps (ESSWEIN in prep.), 8 fractions were obtained which differ considerably from each other in the chemical properties of the compounds included. 7 of these fractions proved effective in our bioassay. If we consider these 7 as 7 independent badge dimensions we would expect each of these classes to comprise not more than between 4 and 10 (on average 8) different substances (provided they lead to phenotypic differences) in order to generate the desired number of repetition free permutations. At the genetic level, a system with 7 loci and 4–10 phenotypes associated with each locus (which, depending on gene interaction, would require 3–5 alleles per locus) would be just adequate to meet the requirement. Such a system, involving the production of only 40–70 different suitable badge substances, should be entirely realizable as far as this point of production is concerned. Nor would such a system make excessive demands on the sensory abilities of these animals, arthropods that orient to their surroundings in general chiefly by way of chemoreception.

c) Intrafamily variability and badge structure. Given normal processes of inheritance, the high interfamily variability must necessarily result in a certain amount of intrafamily variability. These differences among family members are partially masked by the intermingling of individual components that occur automatically on contact. A family badge is thereby produced, a chemical common denominator that is similar enough among at least most if its bearers (apart from the parents) that foreign conspecifics (e. g., adoptive parents) need only become familiar with some larger fraction of the individuals (50–70 % of the progeny) in order subsequently to accept the remainder – animals they have never encountered directly – as members of the group. In this experimental situation the adoptive parents employ a phenotype-matching mechanism (ALEXANDER 1979; HOLMES and SHERMAN 1982, 1983; SHERMAN and HOLMES 1984). This mechanism could also play a role in the intact family; for an animal to remind itself of the family badge and to keep track of any changes of the family badge due e. g. to losses of family members as time goes by, it needs only to touch a few of its siblings.

The family badge has gestalt character (cf. CROZIER and DIX 1979). Each member of a group learns this gestalt independently of its own discriminator production. (The interesting, multi-level problem of how the parents learn to identify their young in the first place and how they avoid eating them, in view of the fact that they always cannibalize young foreign conspecifics, is outside the scope of this discussion; see LINSENMAIR 1984, in prep. (II).) The badge, as would be expected of a gestalt model, is very vulnerable to disturbance, and is more readily made unrecognizable by the addition than by the removal of single components.

The system is greatly complicated by the fact that all members of the *Hemilepistus* families molt at more or less short intervals and in so doing, because of (i) the non-volatility of the discriminators and (ii) their intrafamily variability, alter the badge they bear, sometimes severely. Such alterations must destroy the gestalt character of the badge. A gestalt model alone, then, cannot explain the kin-recognition abilities of *Hemilepistus*.

If molting solely involved the loss of certain components of the common family badge, the family members would – in order to categorize a newly molted individual as belonging to their group or as alien – need to be able to determine whether the discriminators borne by a newly-molted animal amounted to a subset of the family pattern, with no foreign additions. To avoid acceptance of a foreigner that happens to match one of the presumably many possible subsets of the family badge, an additional safety measure would be expected – there must be some sign indicating a recent molt. (Because desert isopods can molt successfully only in burrows, owing to evaporation and other problems, no other assurance would be required.) Although the second part of this scheme does apply to *Hemilepistus* – molting is signalled – the first proposal is too simple. The reason is that the family badge contains only those components that – in a large family – are produced by several individuals. Presumably any component produced by only one or a very few individuals, when distributed among all the family members, is diluted under the threshold of effectiveness and therefore no longer influences behavior. But after such an individual has molted, the deviant substances on its surface become very conspicuous at the same time as more or less large fractions of the common pattern have been lost.

Hemilepistus can and has to learn not only the badge common to the family, but also all those chemical discriminators of deviant individuals which are not represented in the pattern borne by every family member. Such learning is possible only because these substances are bound to the producer's cuticula in a way that reliably prevents their dilution below the detection threshold. It is still not clear what the partial patterns remaining after molting must consist of in order to be acceptable – whether all the single components which are present either as personal or as family discriminators may be freely combined or whether they meet the criterion only in certain combinations, thus restricting the number of possible permutations.

The learning task that a member of a *Hemilepistus* family faces in this context cannot yet be evaluated in detail, but in any case it must be formidable. Although the memory of a

Hemilepistus is very good (LINSENMAIR and LINSENMAIR 1971; LINSENMAIR 1972), it must evidently be continually refreshed, especially in the case of the less common characteristics. Continual learning is probably also necessary because, in the course of time, changes in the badge system develop – owing, for instance, to the death of some family members.

Their permanent capacity for learning makes desert isopods especially suitable subjects for detailed analysis of a differentiated communication system. Because the signals are transmitted by contact chemoreception and not by olfaction (in which case orders of magnitude fewer molecules could be involved), chemical analysis seemed a promising approach, and one that was likely to succeed without the need to extract astronomical numbers of animals. In the event, the chemical properties of the substances have considerably hindered and delayed the analysis (HERING 1981; ESSWEIN 1982), but there are good reasons to believe that we shall soon know the general features of the spectrum of substances employed. This analysis is the most important project currently underway, for there are many questions that can be usefully pursued only after it is completed.

Acknowledgments

This research was supported by the Deutsche Forschungsgemeinschaft (Li 150/6 – 150/10 and SFB 4 «Sinnesleistungen: Anpassungen von Strukturen und Mechanismen»). I would like to express my thanks to EVI KÖPPL, Regensburg and ELISABETH BARCSAY, Würzburg for technical help, I am especially indebted to my wife, Dr. CHRISTA LINSENMAIR, and to DIETER KRAPF for their most valuable help during field work.

References

Alexander, R. D. (1979): *Darwinism and Human Affairs.* Univ. of Washington Press.

Bateson, P. P. G. (1980): Optimal outbreeding and the development of sexual preferences in Japanese quail. Z. Tierpsychol. *53,* 231–244.

– (1982): Preferences for cousins in Japanese quail. Nature *295,* 236–237.

– (1983): Optimal outbreeding. In *Mate Choice,* P. P. G. Bateson (ed), 257–277, Cambridge Univ. Press.

Beecher, M. D. (1982): Signature systems and individual recognition. Am. Zool. *22,* 477–490.

Blaustein, A. R. and R. K. O'Hara (1981): Genetic control for sibling recognition? Nature *290,* 246–248.

– (1982): Kin recognition in *Rana cascadae* tadpoles: maternal and paternal effects. Anim. Behav. *30,* 1151–1157.

Buckle, G. R. and L. Greenberg (1981): Nestmate recognition in sweat bees *(Lasioglossum zephyrum):* Does an individual recognize its own odour or only odours of its nestmates? Anim. Behav. *29,* 802–809.

Caldwell, R. L. (1979): Cavity occupation and defense behavior in the stomatopod *Gonodactylus festai:* evidence for chemically mediated individual recognition. Anim. Behav. *27,* 194–201.

Crozier, R. H. and M. W. Dix (1979): Analysis of two genetic models for the innate components of colony odors in social Hymenoptera. Behav. Ecol. Sociobiol. *4,* 217–224.

Esswein, U. (1982): Über das chemische Familienabzeichen der Wüstenassel *Hemilepistus reaumuri.* Diploma Thesis: Heidelberg University.

Greenberg, L. (1979): Genetic component of bee odor in kin recognition. Science *206,* 1095–1097.

Hamilton, W. D. (1964a): The genetical evolution of social behaviour. I. J. Theoret. Biol. *7,* 1–16.

– (1964b): The genetical evolution of social behaviour. II. J. Theoret. Biol. *7,* 17–32.
– (1972): Altruism and related phenomena, mainly in social insects. Ann. Rev. Ecol. Syst. *3,* 193–232.
Hering, W. (1981): Zur chemischen Ökologie der Tunesischen Wüstenassel *Hemilepistus reaumuri.* Ph. D. Thesis: Heidelberg University.
Hoffmann, G. (1978): Experimentelle und theoretische Analyse eines adaptiven Orientierungsverhaltens: Die «optimale» Suche der Wüstenassel *Hemilepistus reaumuri* nach ihrer Höhle. Ph. Thesis: Regensburg University.
– (1983a): The random elements in the systematic search behavior of the desert isopod *Hemilepistus reaumuri.* Behav. Ecol. Sociobiol. *13,* 81–92.
– (1983b): The search behavior of the desert isopod *Hemilepistus reaumuri* as compared with a systematic search. Behav. Ecol. Sociobiol. *13,* 93–106.
– (1983c): Optimization of brownian search strategies. Biol. Cybern. *49,* 21–31.
– (1984): Homing by systematic search. In: Varju, D., Schnitzler, H.-U. (Eds): Localization and Orientation in Biology and Engineering, 192–199. Springer Verlag Berlin-Heidelberg.
Hölldobler, B. and Michener, C. D. (1980): Mechanisms of identification and discrimination in social hymenoptera. In *Evolution of Social Behaviour:* Hypothesis and empirical tests. 35–58. Markl, H. (Ed). Dahlem Konferenzen 1980. Weinheim, Verlag Chemie GmbH.
Holmes, W. G. and P. W. Sherman (1982): The ontogeny of kin recognition in two species of ground squirrels. Amer. Zool. *22,* 491–517.
– (1983): Kin recognition in animals. Amer. Sci. *71,* 46–55.
Johnson, V. R., Jr. (1977): Individual recognition in the banded shrimp *Stenopus hispidus* (Olivier). Anim. Behav. *25,* 418–428.
Knowlton, L. (1980): Sexual selection and dimorphism in two demes of a symbiotic, pair-bonding snapping shrimp. Evolution *34,* 161–173.
Krempien, W. (1983): Die antennale Chemorezeption von *Hemilepistus reaumuri* (Audouin and Savigny) (Crustacea, Isopoda, Oniscoidea) Ph. D. Thesis: Würzburg University.
Lamprecht, J. (1973): Mechanismen des Paarzusammenhaltes beim Cichliden *Tilapia mariae* Boulanger 1899 (Cichlidae, Teleostei). Z. Tierpsych. *32,* 10–61.
Linsenmair, K. E. (1972): Die Bedeutung familienspezifischer «Abzeichen» für den Familienzusammenhalt bei der sozialen Wüstenassel *Hemilepistus reaumuri* AUDOUIN u. SAVIGNY (Crustacea, Isopoda, Oniscoidea). Z. Tierpsychol. *31,* 131–162.
– (1975): Some adaptations of the desert woodlouse *Hemilepistus reaumuri* (Isopoda, Oniscoidea) to desert environment. Verh. d. Gesell. ökol. Erlangen, 183–185.
– (1979): Untersuchungen zur Soziobiologie der Wüstenassel *Hemilepistus reaumuri* und verwandter Isopodenarten (Isopoda, Oniscoidea): Paarbindung und Evolution der Monogamie. Verh. Dtsch. Zool. Ges. Stuttgart, 60–72.
– (1984): Comparative studies on the social behaviour of the desert isopod *Hemilepistus reaumuri* and of a Canarian *Porcellio* species. Symp. zool. Soc. London. *53,* 423–453.
– I (in prep.): Family dissolving and pair formation in the desert isopod *Hemilepistus reaumuri.*
– II (in prep.): Recognition of the own progeny in the social isopod *Hemilepistus reaumuri.*
– III (in prep.): Experimental analysis of the intrafamily communication in the desert isopod *Hemilepistus reaumuri* and in some subsocial *Porcellio* species.
Linsenmair, K. E. and Linsenmair, Ch. (1971): Paarbildung und Paarzusammenhalt bei der monogamen Wüstenassel *Hemilepistus reaumuri* (Crustacea, Isopoda, Oniscoidea). Z. Tierpsych. *29,* 134–155.
Linsenmair, K. E., H. Schildknecht, and U. Esswein (in prep.): The chemistry of the family badge in the social isopod *Hemilepistus reaumuri.*
Michener, C. D. (1974): *The Social Behavior of the Bees.* Cambridge, The Belknapp Press of Harvard University Press.
Michener, C. D. (1985): From solitary to eusocial bees. *Experimental Behavioral Ecology and Sociobiology.* Symposium in memoriam K. von Frisch (Hölldobler, B. and M. Lindauer eds.) Fortschritte der Zoologie Bd. 31.
Pukowski, E. (1933): Ökologische Untersuchungen an *Necrophorus.* Z. Morph. Ökol. Tiere *27,* 518–586.

Ribbands, C. R. (1965): The role of recognition of comrades in the defence of social insect communities. Symp. zool. Soc. Lond. *14,* 159–168.

Seelinger, G. (1977): Der Antennenendzapfen der tunesischen Wüstenassel *Hemilepistus reaumuri,* ein komplexes Sinnesorgan (Crustacea, Isopoda). J. comp. Physiol. *113,* 95–103.

Seelinger, G. and U. Seelinger (1983): On the social organisation, alarm and fighting in the primitive cockroach *Cryptocercus punctulatus* Scudder. Z. Tierpsychol. *61,* 315–333.

Shachak, M. (1980): Energy allocation and life history strategy of desert isopod *Hemilepistus reaumuri.* Oecologia (Berl.) *45,* 404–413.

Sherman, P. W. and W. G. Holmes (1985): Kin recognition: issues and evidence. *Experimental Behavioral Ecology and Sociobiology,* Symposium in memoriam K. VON FRISCH (HÖLLDOBLER, B. and M. LINDAUER eds.) Fortschritte der Zoologie Bd. 31.

Seibt, U. (1974): Mechanismen und Sinnesleistungen für den Paarzusammenhalt bei der Garnele *Hymenocera picta* Dana. Z. Tierpsychol. *35,* 337–351.

– (1980): Soziometrische Analyse von Gruppen der Garnele *Hymenocera picta. 52,* 321–330.

Seibt, U. and W. Wickler (1979): The biological significance of the pair-bond in the shrimp *Hymenocera picta.* Z. Tierpsychol. *5,* 166–179.

Steinmann, E. (1973): Über die Nahorientierung der Einsiedlerbienen *Osmia bicornis* L. und *Osmia cornuta* Latr. (Hymenoptera, Apoidea). Mitt. Schweiz. Ent. Ges. *46,* 119–122.

– (1976): Über die Nahorientierung solitärer Hymenopteren: Individuelle Markierung der Nesteingänge. Mitt. Schweiz. Ent. Ges. *49,* 253–258.

Vannini, M. and F. Gherardi (1981): Dominance and individual recognition in *Potamon fluviatile* (Decapoda, Bradyura): possible role of visual cues. Mar. Behav. Physiol. *8,* 13–20.

Waldman, B. (1981): Sibling recognition in tadpoles: The role of experience. Z. Tierpsychol. *56,* 341–358.

Wickler, W. and Seibt, U. (1981): Monogamy in crustacea and man. Z. Tierpsychol. *57,* 215–234.

Wilson, E. O. (1971): *The Insect Societies.* Cambridge: The Belknapp Press of Harvard University Press.

Wu, H. M. H., W. G. Holmes, S. R. Medina, and G. P. Sackett (1980): Kin preference in infant *Macaca nemestrina.* Nature *285,* 225–227.

Fortschritte der Zoologie, Bd. 31 · Hölldobler/Lindauer (Hrsg.): Experimental Behavioral Ecology
G. Fischer Verlag · Stuttgart · New York · 1985

Kin recognition: issues and evidence

Paul W. Sherman and Warren G. Holmes

Section of Neurobiology and Behavior, Cornell University, Ithaca, New York 14853; and Department of Psychology, University of Michigan, Ann Arbor, Michigan 48109

Abstract

Investigators from a diversity of backgrounds and theoretical orientations have employed a multitude of techniques to investigate why and how animals recognize their genetic relatives. From an ultimate or evolutionary perspective, it appears that kin recognition abilities serve both to facilitate nepotism and prevent its misdirection, and to preclude close inbreeding and optimize outbreeding. From a proximate perspective, we believe that the ontogenies of all known cases of kin recognition involve learning, and that most are based on the spatial distribution of relatives or on rearing association between related individuals. In this paper we also describe an additional mechanism that facilitates recognition between relatives in the absence of spatial or associational cues. This new mechanism, phenotype matching, depends on learning about one's own phenotype or those of known relatives, and then comparing phenotypes of conspecifics whose relatedness is unknown against this template. After discussing studies of phenotype matching in insects, amphibians, birds, and mammals we summarize the results of our own field and laboratory investigations of kin recognition in Belding's ground squirrels *(Spermophilus beldingi)*. Our work along with comparative investigations of three other species of ground squirrels suggests that phenotype matching abilities are highly developed in the more nepotistic sex (females) of the most social species.

1 Introduction

In a variety of situations individuals may increase their fitness by adjusting their behavior as a function of genetic relatedness to conspecifics. Recently it has become apparent that there are several proximate mechanisms which facilitate the assessment of relatedness. These kin recognition mechanisms do not reveal relatedness *per se;* rather they provide means of assessing social or phenotypic variables that correlate with relatedness.

We define kin recognition (synonyms are kin identification and kin discrimination) as the differential treatment of conspecifics as a function of their genetic relatedness. Thus kin recognition is inferred whenever differential treatment is based on kinship, regardless of whether classes of relatives (for example, siblings as opposed to cousins) or individual relatives (sibling A versus cousin B) are identified. This definition involves no assumptions about cognition or intention; neither does it require that one specific developmental path be followed for kin-correlated behavior to emerge.

Our understanding of how kin are identified has expanded rapidly in the last five years (reviewed by HOLMES and SHERMAN 1982, 1983). Interest in this topic is largely due to two theoretical developments in evolutionary biology. The first is W. D. HAMILTON's (1964) concept of inclusive fitness. Under this hypothesis individuals can affect their reproductive success and thus their evolutionary fitness not only through the direct production of offspring and associated parental efforts, but also by assisting both descendant and collateral relatives in their reproductive efforts. Voluminous evidence supporting HAMILTON's «kin selection» hypothesis has accumulated (e. g., ALEXANDER 1979; KURLAND 1980; SHERMAN 1981b; HOOGLAND 1983) and there is no longer any doubt that genetic relatedness mediates cooperation and competition in numerous social species.

The second theoretical advancement that has focused attention on kin recognition is the idea of «optimal outbreeding» (BATESON 1978). Whereas the genetic penalties of close inbreeding have long been recognized (reviewed by HOOGLAND 1982), recent investigators have suggested that extreme outbreeding may also be disadvantageous (SHIELDS 1982). In other words, one component of mate choice may involve striking an optimal balance between inbreeding and outbreeding (BATESON 1980, 1982; COWAN 1979). Some recent empirical information seems to support this hypothesis (reviewed by BATESON 1983).

Neither the kin selection nor the optimal outbreeding hypothesis requires the existence of any particular mechanism to bring about kin-correlated behavior. Indeed, in his land-

Fig. 1: A lactating female Belding's ground squirrel *(Spermophilus beldingi)* gives an alarm call at the approach of a terrestrial predator. This female is a member of an extensively studied population at Tioga Pass, California. Research in which more than 3,000 ground squirrels have been marked individually and observed unobtrusively has shown that alarm calls are usually given by resident females with living mothers, sisters, or offspring nearby rather than by males or by females who are not resident or who have no offspring or other living kin. This pattern of calling tendencies, together with a population structure in which females are sedentary and males are nomadic, suggests that alarm calls function to alert relatives and are thus expressions of nepotism (Photo courtesy of G. D. Lepp/Bio-Tech Images).

mark paper HAMILTON (1964, pp. 21–25) described several possible kin recognition mechanisms, and added (p. 24) that «... the situations in which a species discriminates in its social behaviour tend to evolve and multiply in such a way that the coefficients of relationship involved in each situation become more nearly determinate.» In recent years theoretical discussions of potential mechanisms for identifying kin have been offered by ALEXANDER (1979), HÖLLDOBLER and MICHENER (1980), SHERMAN (1980), BEKOFF (1981), HOLMES and SHERMAN (1982), and BLAUSTEIN (1983). After reviewing these treatments, we summarize empirical information which suggests to us that at least three different kin recognition mechanisms operate in nature. Each of these mechanisms requires learning, but what is learned and how the information is used are different. In this paper, we compare and contrast these mechanisms and describe the selective circumstances, both ecological and social, that may have led to the evolution of each. Finally, we synopsize the results of our five-year laboratory and field investigation of kin recognition in Belding's ground squirrels (*Spermophilus beldingi:* Fig. 1) in light of these mechanisms.

2 Mechanisms of kin recognition

2.1 Spatial distribution

When relatives are distributed predictably in space, nepotism (favoring kin) may occur as a result of location-specific behavior. Under this mechanism, locations rather than conspecifics themselves are recognized and individuals adjust their behavior as a function of proximity to some reference point such as a nest site or burrow. Close inbreeding might also be minimized (GREENWOOD et al. 1978) or outbreeding optimized (PRICE and WASER 1979) if individuals avoided mating with conspecifics that were encountered at particular locations (e. g., too close to home). The spatial distribution mechanism depends on a close and consistent correlation between places where conspecifics are encountered and genetic relatedness. Thus it requires either viscosity of kin groups or nonrandom dispersal of kin. The reliability of this mechanism is obviously reduced to the degree that nonkin can infiltrate into groups of relatives.

Cues based on location frequently mediate parental recognition of offspring among birds, especially during the early development of young. In many cases, avian parents learn to recognize their nest or nest site rather than the chicks themselves; adults continue using location-specific cues up to the time when their chicks become mobile and broods intermix. This is exemplified by comparing species that differ in degrees of coloniality and the age at which fledging occurs (e. g., BIRKHEAD 1978). For example kittiwakes *(Rissa tridactyla),* which nest in small cracks on sheer cliff faces, often with only one nest per crack, and whose young fledge at five weeks, accept unrelated young experimentally transferred into their nest up until their own chicks are five weeks old (CULLEN 1957); thereafter, foreign chicks are rejected. Loosely colonial, marsh-nesting Franklin's gulls *(Larus pipixcan)* accept foreign chicks in their nest until their own young fledge and begin to stray, at about two weeks of age (BURGER 1974). Various species of highly colonial, ground nesting gulls, whose young are mobile within a few days of hatching, begin to identify their nestlings rather than simply the nest site when their own young are three to five days old (BEER 1970).

Studies of bank swallows *(Riparia riparia)* have provided further insights. These birds, which nest colonially in sand banks, apparently learn the spatial location of the nest hole they have excavated. Parents will feed both their own young and experimentally introduced foreign chicks, up to the time their own offspring fledge at about two weeks of age

(HOOGLAND and SHERMAN 1976; BEECHER et al. 1981a). Conversely, they ignore their own unfledged chicks when these are experimentally transferred to nearby nest burrows. Just prior to fledging, chicks begin to produce individually distinctive «signature» calls that parents learn. Parents use these calls to discriminate their own from alien young, so that once natural mixing of fledglings among nests has commenced, parents will feed only their own chicks at home and will seek out and feed misplaced offspring in other nest burrows. After fledglings have abandoned nest burrows, parents continue to rely on calls to locate and subsequently feed only their own offspring (BEECHER et al. 1981b).

2.2 Association

Association is probably the most common pathway leading to kin recognition in nature. Under this mechanism, relatives must predictably interact in unambiguous social circumstances, in which kinship is rarely confused due to the mixing of unequally related individuals. Thus, kin recognition involves learning the phenotypes of familiar associates and later recalling them; such learning may be based on the timing, rate, frequency, or duration of their social interactions. Here the term «association» refers to opportunities that individuals have to interact rather than to a particular type of learning process (e. g., associative learning). Rearing environments such as the natal nest or burrow often provide ideal settings for the operation of this mechanism because they typically separate or exclude asymmetrically related individuals.

Association is the usual mechanism of recognition between mothers and their offspring in mammals (MICHENER 1974; MICHENER and SHEPPARD 1972; HOLMES in press). For instance, mutual recognition between cows and their pups in Galapagos fur seals *(Arctocephalus galapagoensis)* and sea lions *(Zalophus californianus)* is based on familiarity with each other's vocalizations, and appears soon after pup birth (TRILLMICH 1981). Furthermore, in several colonial gull species *(Larus)*, three to five day old chicks learn to recognize as parents individuals whose vocalizations they hear most frequently at the nest (BEER 1970). Obviously, the ability of goslings to «imprint» on parents or parent-like objects (LORENZ 1935; HESS 1973) suggests learned recognition as a result of association during a sensitive period of development.

One well-studied example of the association mechanism is the work of LEON (1975, 1978) on Wistar rat pups *(Rattus norvegicus)*. LEON has shown that rat pups learn to discriminate an odor present in their mother's fecal material when they are about two weeks old. The odor is produced by microbial action in the dam's gut and its chemical composition depends on her diet. After learning their mother's odor, pups offered a choice between their own dam and an unfamiliar lactating female prefer to approach their mother; they are also attracted to her odor over that of an unfamiliar lactating female. These preferences disappear when both females are fed identical diets. Finally, pups prefer one unfamiliar lactating female over another if the diet of one of them matches that of their own dam.

Recognition of siblings also depends on learning during direct interactions in some species. Captive spiny mice *(Acomys cahirinus)* that were placed in a test arena more frequently huddled with littermates with whom they were familiar than with unfamiliar nonlittermates (PORTER et al. 1978). When *A. cahirinus* littermates were reared apart, they later behaved as if they were not siblings. Finally, when nonsiblings were reared by the same dam, they later huddled together as if they were biological siblings (PORTER et al. 1981).

Thus far the term «association» has meant direct social interactions between individuals later treated as kin. However, mediated recognition by association is also possible. If two unfamiliar relatives first interact in the presence of a third conspecific related to each and familiar to both (a «go-between»), recognition between the previously unfamiliar individuals may be facilitated. For instance, either full-siblings or maternal half-siblings born in sequential litters may sometimes learn to recognize each other as a result of mutually associating with their common parent or parents (SHERMAN 1980).

Females may mediate recognition between sire and offspring in some species (BERENSTAIN et al. 1981). For example, LABOV (1980) allowed wild-strain male and female house mice *(Mus musculus)* to mate and then separated the pairs twelve hours later. At various times before parturition, a male was placed in each female's cage and allowed to remain there. Sometimes a male was reunited with his prior mate, but in other cases the male was housed with a female inseminated by another male. LABOV reported that the probability of a male's killing his partner's pups varied inversely with the length of cohabitation prior to parturition, regardless of whether a male joined a female he had actually inseminated or one inseminated by another male. In this case, the length of male-female cohabitation apparently affected the behavior of mated males toward pups.

As was true for the spatial distribution mechanism, recognition errors – the acceptance of nonrelatives as kin – should occur under the association mechanism if unrelated individuals are inserted into kin groups at an appropriate time and place. Experimentally transferring or «cross-fostering» young so that they are reared by and among nonrelatives induces such errors in cichlid fish (NOAKES and BARLOW 1973), mice (HILL 1974; DEWSBURY 1982), domestic goats (KLOPFER and KLOPFER 1968; GUBERNICK 1980), and ground squirrels (HOLMES 1984; in press). Cross-fostering has also been used as a tool to investigate the development of parent-offspring and sibling-sibling recognition in bank swallows (HOOGLAND and SHERMAN 1976; BEECHER et al. 1981a; BEECHER and BEECHER 1983) and Belding's ground squirrels (HOLMES and SHERMAN 1982). In both these species, and in many others (e.g., MCARTHUR 1982), the appearance of differential treatment of newly cross-fostered young coincides with the time when broods first mix in nature. This suggests that the total frequency or duration of exposure to relatives (BEKOFF 1981) is not as important, under the association mechanism, as are the interactions which occur coincident with or just before the time when unequally related conspecifics first mingle (see also KAREEM 1983).

2.3 Phenotype matching: theoretical issues

Can kin identify each other even if they have not interacted in locations or social circumstances that correlate directly with their relatedness? This is an intriguing question because in many species various relatives do not have obvious locational or associational cues available to indicate kinship. Nonetheless, a growing body of empirical evidence, from insects, amphibians, birds, and mammals indicates the identification of such relatives. We believe that many of these cases involve a process of «comparing phenotypes» (ALEXANDER 1979, p. 116) or «phenotype matching» (HOLMES and SHERMAN 1982, 1983). Under this mechanism, an individual learns about its own phenotypic attributes or those of its relatives by direct association with them or cues associated with them (e.g., their odors), and acquires an image or template of these attributes. When a conspecific whose relatedness is unknown is encountered at some later time, the focal individual matches that phenotype against this learned template. In general, the degree of similarity between the observed and «expected» phenotype is the clue to relatedness. Thus phenotype matching is a perceptual process that may parallel what psychologists call «stimulus generalization» (e.g., MOSTOFSKY 1965), in which an animal's response to an unfamiliar stimulus depends on its similarity to previously learned, familiar stimuli.

Phenotype matching depends on a consistent correlation between phenotypic similarity and genotypic similarity, so that detectable traits are more alike among close relatives than among distantly related or unrelated individuals. These phenotypic traits may have genetic origins, (KUKUK et al. 1977; GREENBERG 1979; BREED 1981, 1983), environmental origins (KALMUS and RIBBANDS 1952), or be derived from a combination of both. Regardless of their source(s), if resemblances in such traits are reliably associated with genotypic similarities they may serve as cues for phenotype matching.

At this point we wish to distinguish explicitly two components that comprise kin recognition processes in general and phenotype matching in particular. First, the *production* component refers to the ways unique phenotypes develop, making individuals or groups distinctive and thus discriminable from others. Second, the *perception* component refers to the ways individuals develop abilities to discriminate among phenotypes. As pointed out by BEECHER (1982), the absence of either component precludes kin recognition.

Two different approaches have so far characterized studies of the phenotype-production component. First, mathematical models have been developed to evaluate the amount of information necessary to specify relatedness accurately (e. g., CROZIER and DIX 1979; GETZ 1981, 1982; BEECHER 1982; LACY and SHERMAN 1983). These models suggest that accuracy depends on several key factors: the number of independently assorting, detectable traits, their variation and heritability, and the algorithm used to assess kinship based on the traits. A second approach has been an empirical search for the sensory modalities that convey information about the phenotype (e. g., auditory: BEECHER et al. 1981b; MCARTHUR 1982; olfactory: KAPLAN et al. 1977; PORTER and MOORE 1981; visual: MILLER and EMLEN 1975). Terms that have been used to characterize the production component include «discriminator» or «discriminating substance» (HÖLLDOBLER and MICHENER 1980), «signatures» (BEECHER 1982), «colony odor» (especially in hymenoptera: RIBBANDS 1965; BARROWS et al. 1975; KUKUK et al. 1977; CROZIER and DIX 1979), «recognition cues» (BREED 1981, 1983), and «labeling systems» (GETZ 1981, 1982).

2.4 Phenotype matching: empirical evidence

There are three categories of participants in the phenotype matching process we envision. The *observer* is the individual attempting to assess its relationship to an *observed individual. Referents* are conspecifics whose phenotypes provide clues about the traits to be expected in various relatives. These referents, from whose phenotypes the template is learned, could be either the observer's familiar relative(s) or the observer itself.

In sweat bees *(Lasioglossum zephyrum)* referents are apparently nestmates. *L. zephyrum* females nest in groups and restrict access to their shared nest burrow by guarding the entrance. Only a single bee acts as a guard at any one time, and guards allow nestmates (usually their sisters) to enter while excluding non-nestmates. In the laboratory, GREENBERG (1979) reared guard bees with sisters and later presented guards with unfamiliar conspecifics at nest-hole entrances. Intruders were related to guards with varying degrees of closeness as a result of controlled inbreeding. GREENBERG's (1979) results show that the percentage of unfamiliar bees admitted to the nest rose as genetic relatedness between the guard and the unfamiliar intruder increased (Fig. 2a).

In a follow-up experiment, BUCKLE and GREENBERG (1981) reared some *L. zephyrum* solely with non-sisters. In nest-entrance tests these bees admitted unfamiliar intruders if they were the sisters of guards' nestmates, while rejecting their own sisters (Fig. 2b). BUCKLE and GREENBERG (1981) also reared bees in mixed groups of their sisters and non-sisters. They found that these individuals admitted both their unfamiliar sisters and unfamiliar sisters of their (unrelated) nestmates (Fig. 2b). Thus *L. zephyrum* guards discriminate among unfamiliar conspecifics on the basis of their phenotypic similarity to individuals with whom the guards were reared, rather than similarity to their own phenotype.

In the sweat bee case, phenotypic cues used to assess relatedness apparently arise from a genetic source (KUKUK et al. 1977; GREENBERG 1979). In another hymenopteran, nestmate

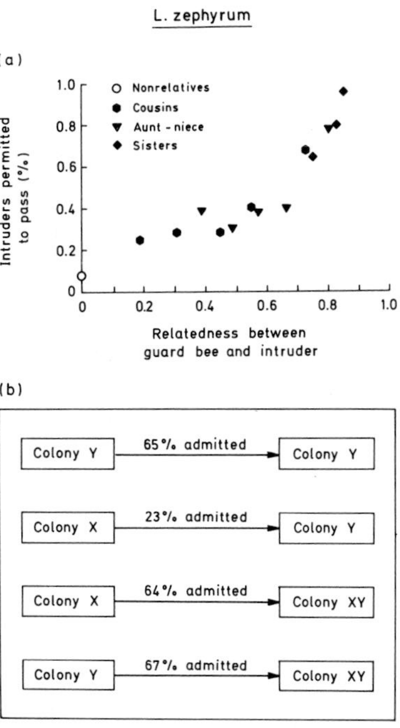

Fig. 2: Laboratory experiments on kin recognition in sweat bees *(Lasioglossum zephyrum)*. Female *L. zephyrum* guard the entrance to their shared nest, typically admitting only nestmates, which are usually their sisters.

a Guard bees were raised with sisters and later presented with intruders that they had never before encountered. The unfamiliar intruders were related to the guards with varying degrees of closeness as a result of controlled inbreeding. The likelihood that guards would admit intruders to their nest increased with the degree of genetic relatedness between guard and intruder. One possible explanation for the discrimination mechanism is that guards admit unfamilar individuals on the basis of their phenotypic similarity (probably in odor) to nestmates (Redrawn from GREENBERG 1979; copyright 1979 by the AAAS).

b Guard bees were reared in nests of six bees composed either solely of sisters (colony X and colony Y), or of three sisters from one nest and three sisters from another (colony XY). Later, bees from X, Y, and XY colonies were tested as guards as shown (arrows represent unfamiliar intruders seeking entrance to the nest). Guards reared solely with sisters were about three times as likely to admit unfamiliar sisters as unfamiliar, unrelated intruders, allowing 65 % of the unfamiliar sisters to enter as opposed to only 23 % of the unfamiliar, unrelated individuals. However, guards reared in a mixed colony were about as likely to admit their nestmates' unfamiliar sisters as their own unfamiliar sisters, allowing respectively 64 % and 67 % to enter. Guards may have matched the phenotypes of unfamiliar intruders against the phenotypes of bees with whom they were reared and admitted intruders similar to their own nestmates (Redrawn from BUCKLE and GREENBERG 1981).

recognition also appears to be based on phenotype matching but the cues may be of environmental origin. In a recent series of experiments, PFENNIG et al. (1983b) brought pupae-laden nests of the paper wasp *Polistes fuscatus* into the laboratory. As each gyne (potential queen) emerged from her natal cell she was assigned to one of four groups: (1) exposure to her natal nest and emerging nestmates for at least five days, (2) exposure to her natal nest but not to nestmates for 4–13 hours, (3) exposure to her natal nest but not to nestmates for only one hour, and (4) exposure to a «foreign nest» and «pseudo-» or «non-pseudonestmates» for 3–5 days. A foreign nest was one to which a female was transferred upon eclosion. Pseudonestmates were created when unrelated gynes were transferred to separate fragments of the same foreign nest; a non-pseudonestmate was a gyne placed on a fragment of a different foreign nest from that to which pseudonestmates had been transferred.

PFENNIG et al. (1983b) investigated nestmate recognition by assessing, in a small arena, tolerance values (tendencies of initiators of behavioral interactions to elicit retreat by a recipient) and spatial associations among triplets that were composed of combinations of nestmates, non-nestmates, or pseudonestmates. The researchers reported that gynes exposed to both their natural nest and to nestmates for at least five days (Group 1) were more tolerant of and associated more closely with unfamiliar nestmates than unfamiliar non-nestmates. Similarly, gynes exposed for 4–13 hours (Group 2) or one hour (Group 3) only to their nests showed greater tolerance for and associated more closely with unfamiliar nestmates than unfamiliar non-nestmates. Finally unrelated pseudonestmates, who were exposed to separate fragments of the same but unrelated nest (Group 4), were more tolerant of each other than they were of non-pseudonestmates, who had been exposed to fragments of yet another unrelated nest.

These results suggest that *P. fuscatus* females may absorb chemical cues from the nest soon after eclosion. In less than one hour after emergence, gynes learn these odors, probably from the nest rather than themselves or other nestmates (SHELLMAN and GAMBOA 1982; PFENNIG et al. 1983a), and use them as a template in the discrimination of nestmates from nonnestmates. This implies that phenotype matching against chemical cues learned from the natal nest underlies nestmate identification in this species. Indeed, in the most stringent test of the matching hypothesis, non-sister gynes exposed to separate fragments of the same foreign nest (Group 4) were more tolerant of each other than of another unrelated gyne that had been exposed to fragments of a different foreign nest. In the wild, *P. fuscatus* females may use their recognition abilities to discriminate between the nests of sisters, with whom they cooperate, and nonrelatives, whose nests they sometimes usurp (KLAHN and GAMBOA 1983).

The results of several investigations of discrimination in mammals (GILDER and SLATER 1978) and birds (BATESON 1978) are also suggestive of phenotype matching with nestmates as referents. In one interesting example, BATESON (1980, 1982) hatched Japanese quail *(Coturnix c. japonica)* in an incubator, reared them with siblings for thirty days, then isolated them for an additional thirty days. At sexual maturity (ca. sixty days of age), the quail were simultaneously presented with five different conspecifics: a familiar sibling, an unfamiliar sibling, an unfamiliar first cousin, an unfamiliar third cousin, and an unfamiliar, unrelated bird. Time spent in front of each stimulus bird, which is an indirect measure of relative mating preference, was recorded. The results, although quite variable, showed that both sexes preferred to stand near their (unfamiliar) first cousins over all the other choice birds. BATESON interpreted these preferences for first cousins in terms of optimal outbreeding, and suggested that the choice mechanism involves comparing phenotypes, with individuals preferring conspecifics whose appearance differs slightly but not greatly from the siblings with whom subjects grew up.

The most recent studies of captive spiny mice suggest that in addition to recognition by association (above), phenotype matching with nestmates as referents can also occur. Previous work indicated that if *A. cahirinus* juveniles were separated for eight days from littermates with whom they were reared, their tendency to huddle with these littermates was abolished (PORTER and WYRICK 1979). It now appears that if the separated littermates are each housed with another of their siblings, they retain their preference for each other (PORTER et al. 1983). In explaining these results, PORTER and his colleagues suggest that littermates' phenotypes (probably odors) «... can influence interactions among *A. cahirinus* weanlings by serving as a learned standard against which odours of conspecifics are compared» (p. 982–983).

The second general class of referents under a phenotype matching mechanism is the observer itself. Although unequivocal evidence for self-matching is not yet available, we believe that various studies point to its existence. For example, self-matching may occur in American toads *(Bufo americanus)*. WALDMAN (1981, 1982) found that in laboratory tests tadpoles reared with their siblings later preferred to associate with those individuals over

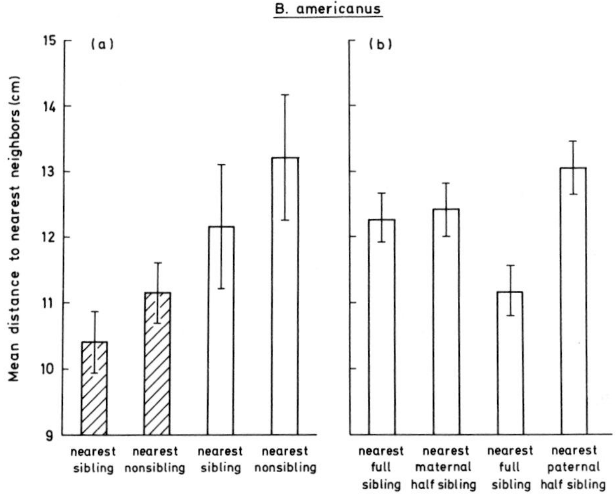

Fig. 3: Laboratory experiments on kin recognition in tadpoles of American toads *(Bufo americanus)*. In each test, two marked groups of tadpoles- either two different sets of full-siblings, full-siblings and their maternal half-siblings, or full-siblings and their paternal half-siblings – that were reared with siblings (hatched) or singly (open) were released into a laboratory pool. Later the distance from each tadpole to its nearest similarly and dissimilarly marked neighbor was recorded.

a Both for tadpoles reared with siblings and for tadpoles reared in isolation, the mean distance to the nearest full-sibling was less than the mean distance to the nearest nonsibling (bars denote 95 % confidence limits).

b Among tadpoles reared alone, there was no difference between the mean distance to the nearest full-sibling and that to the nearest maternal half-sibling; however, the mean distance to the nearest full-sibling was significantly less than that to the nearest paternal half-sibling.

Overall the findings depicted in this figure suggest that some maternal contribution to the egg (such as the jelly) served as a template against which the phenotypes of unfamiliar conspecifics were matched (Redrawn from WALDMAN 1981).

unfamiliar non-siblings. When WALDMAN (1981) divided egg masses early in development and reared each embryo individually in a petri dish, he found that as free-swimming larvae the tadpoles associated more closely with unfamilar siblings than unfamiliar non-siblings. Because in this experiment the rearing procedures eliminated social contact during development, it appears that *B. americanus* tadpoles can identify siblings by means of something learned from their own phenotype (Fig. 3a).

Indeed, WALDMAN (1981) hypothesized that substances contained in the egg or its surrounding jelly might cue recognition. He investigated this by creating paternal half-siblings (one male fertilized the eggs of two females) and maternal half-siblings (some of a female's eggs were fertilized by one male and some by another), then rearing the resulting eggs in isolation. When tested later the tadpoles showed no preference between full-siblings and maternal half-siblings, conspecifics that shared egg jelly derived from the same female. By contrast, the tadpoles preferred to associate more closely with full-siblings than with paternal half-siblings (Fig. 3b). These results are consistent with the hypothesis that sibling recognition in *B. americanus* larvae is cued by some factor associated with the egg or its surrounding matrix.

In related experiments on the Cascade frog *(Rana cascadae)*, BLAUSTEIN and O'HARA (1981) found that tadpoles reared among siblings preferred to swim near those individuals rather than unfamiliar non-siblings in a laboratory tank (also O'HARA and BLAUSTEIN 1981). The preference for siblings is apparently retained after metamorphosis (BLAUSTEIN et al. 1984). Like *B. americanus* larvae, *R. cascadae* tadpoles reared in social isolation later preferred to swim near unfamiliar siblings over unfamiliar non-siblings (BLAUSTEIN and O'HARA 1981). Moreover, in two-choice tests the isolates preferred full-siblings over either maternal or paternal half-siblings, and both sorts of half-siblings over unrelated conspecifics (BLAUSTEIN and O'HARA 1982). These results indicate that the maternal egg jelly might mediate the identification of some but certainly not all discriminable relatives in *R. cascadae*.

Phenotype matching based on self-perception is suggested by a study (WU et al. 1980) of recognition in pigtail macaques *(Macaca nemestrina)*. Laboratory matings of male monkeys with several females each produced multiple pairs of paternal half-siblings. Immediately after they were born, the infants were separated from their mothers and reared by humans. To encourage normal development each youngster was allowed to associate daily with a group of nonkin of the same age. When they were one to twelve months old, 16 of these monkeys were offered choices in a laboratory apparatus among an unfamiliar paternal half-sibling, an unfamiliar conspecific (not closely related to either the subject or its half-sibling), and an empty cage; stimulus animals were matched for sex, age, and weight. WU and her colleagues reported that subjects looked longer at their half-sibling than at either the unrelated monkey or the empty cage, and that they spent significantly more time in proximity to their half-sibling when allowed to approach the three stimuli. To account for these results, DAWKINS (1982, p. 150) has speculated «... my bet is that the monkeys recognize resemblances of relatives to perceived features of themselves.» While this hypothesis is reasonable, we caution that the macaque study has recently been repeated by FREDRICKSON and SACKETT (1984) with quite different results and interpretations (i.e., no recognition of unfamiliar paternal half-siblings was found). Thus whether or not *M. nemestrina* match phenotypes is now in dispute.

Phenotype matching based on self-perception is a plausible mechanism underlying some recently discovered kin recognition abilities of birds and small mammals. For example, male chicks *(Gallus gallus)* that are reared alone apparently «auto-imprint.» As a result they develop a social preference in adulthood for conspecifics or models similar in structure and coloration to themselves (VIDAL 1982; see also SALZEN and CORNELL 1968). In laboratory tests, house mice (*M. musculus*, C. F. L. P. strain) discriminate between unfamiliar paternal half-siblings and unfamiliar, unrelated mice, with the half-sibling pairs showing more «passive body contact» and fewer aggressive interactions in their initial contacts than the unrelated pairs (KAREEM and BARNARD 1982; also KAREEM 1983). Laboratory-reared white-footed deermice *(Peromyscus leucopus)* apparently can distinguish between siblings from parents' subsequent litters and non-siblings, despite not having associated with either prior to testing (GRAU 1982).

Finally, recognition based on phenotype matching may occur in honeybees *(Apis mellifera)*. GETZ et al. (1982) created hives composed of two groups of differently related workers by artificially inseminating one queen with the sperm of two males from phenotypically distinct strains (cordovan and Italian). The investigators discovered that when colony fission occurred the swarm that left the hive contained a slight bias toward full-sisters rather than a random assortment of full- and maternal half-sisters. However, as the authors note, since queens were inseminated with sperm from different strains, the mechanism responsible for the composition of the swarm could have been inter-strain differences in behavior (e. g., the timing of swarming) rather than kin recognition. In a follow-up study, GETZ and SMITH (1983) again used artificially inseminated *A. mellifera* queens to produce three hives containing both full- and maternal half-sisters (i. e., workers were sired by the same male or a male from a different strain). The researchers reported that in arena tests workers were more likely to bite their (different strain) half-sisters than their (same strain) full-sisters, despite having been reared with both.

With the exception of the studies of sweat bees (BUCKLE and GREENBERG 1981) and paper wasps (PFENNIG et al. 1983b), investigations of kin recognition mechanisms have rarely used the kinds of rigorously controlled rearing and testing procedures necessary to elucidate whether phenotype matching is occurring, and if so whether it is based on self-perception or on a template derived from perceptions of other relatives. However, it is clear that phenotype matching is a real possibility and one that merits detailed examination, especially when kin-correlated behavior is observed that cannot be accounted for on the basis of the spatial distribution or association mechanisms.

Table 1: Ecological and social circumstances in which phenotype matching may be favored as a mechanism of kin recognition (after HOLMES and SHERMAN 1982).

I. Multiple Mating
   (a) Maternal half-siblings reared together
   (b) Paternal half-siblings reared apart
   (c) Sire-offspring recognition

II. Inter-brood Aggregation
   (a) Grouped embryos or young
   (b) Cooperative breeding

III. Parasitism
   (a) Intra-specific parasitism
   (b) Inter-specific parasitism

IV. Dispersal or Group Size
   (a) Adult or juvenile relatives disperse
   (b) Large or widely dispersed social group

2.5 Phenotype matching: ecological and social contexts

There are at least four ecological and social circumstances in nature when a phenotype matching mechanism might be favored as a way to recognize kin (Table 1). These are conditions in which spatial or associational cues are typically unavailable or are inaccurate indicators of relatedness. It is important to note that these circumstances will not automatically lead to phenotype matching. However, if optimizing outbreeding or dispensing nepotism are advantageous, these are likely contexts for its evolution.

First, male or female polygamy may select for phenotype matching. If a male mates with several females so that paternal half-siblings are born in the same or different breeding seasons, phenotype matching may be favored when such kin do not grow up together nor live in proximity. Conversely, if a female mates with two or more males during a single receptive period so that her brood is multiply sired (e. g., HANKEN and SHERMAN 1981) or if a female's sequential broods are sired by different males, then spatial or associational cues may not distinguish full- from half-siblings. Furthermore, a male might rely on phenotype matching to identify his offspring among a mate's young (ALEXANDER 1979; DALY and WILSON 1982), especially if the male was unable to guard his mate(s) effectively after copulation or if he was only one of several males to mate with that female.

Second, when differently related young are reared together, as in a communal nest, spatial and associational cues might lead to incorrect kinship assessments. For example, females in many species of frogs, toads, and salamanders deposit eggs in a common microhabitat, and females in several bird species lay eggs in a common nest. As a result, eggs and young that differ in relatedness often develop side by side, a situation which virtually demands some type of phenotype matching if kin recognition is to occur.

Third, brood parasitism within (e. g., WELLER 1959) or between species (e. g., PAYNE 1977; ROTHSTEIN 1982) obviously reduces the reliability of spatial proximity or intra-nest associations as indicators of relatedness.

Fourth, dispersal at an early age or large group size may eliminate both spatial and associational cues about kinship. Dispersal from the natal area or group coincident with puberty may preclude older individuals from meeting younger relatives until later in life, when contextual cues about kinship are missing. For example, in many species of birds and mammals, juveniles disperse before full- or half-siblings are born in their parents' subsequent broods (GREENWOOD 1980; DOBSON 1982). Furthermore very large groups, such as those that characterize some social insects and ungulates, may contain so many individuals spread over such an extensive area that some close relatives will not meet until adulthood.

2.6 Recognition alleles

Among mechanisms of kin recognition originally described by HAMILTON (1964) was a genetic model that relied on phenotypic assessment but not «phenotype matching». He hypothesized «recognition alleles» that simultaneously caused the expression of a unique phenotypic trait, enabled their bearer to recognize the trait in others, and caused their bearer to preferentially assist conspecifics possessing the trait. DAWKINS (1976, 1982) dubbed this the «green beard effect,» metaphorically suggesting that these alleles might code for green beards and also cause their bearers to aid other green-bearded individuals. In contrast to association and phenotype matching, this mechanism postulates that the ability to recognize phenotypic attributes is conferred by the very same alleles that cause the attributes or by alleles tightly linked to them (ALEXANDER 1979).

Both HAMILTON and DAWKINS deemed the existence of recognition alleles unlikely due to their necessary complexity. Others (e. g., ALEXANDER and BORGIA 1978) have argued that such alleles are improbable as a basis for kin recognition because of the costs the behaviors they cause would impose on portions of the genome not shared by the actor and recipient. In other words, the presence of such «selfish» alleles would engender an intragenomic conflict of reproductive interests, with the result that alleles whose effects tended to nullify those of the recognition alleles would spread. Whether hypothetical recognition alleles would indeed be «outlaws» (ALEXANDER and BORGIA 1978), helping themselves at the expense of the rest of the genome, or not (RIDLEY and GRAFEN 1981) is a subject of considerable theoretical interest and debate.

Regardless of the outcome of this discussion, we think that an empirical search for recognition alleles would be difficult at best, because their existence could be inferred only after systematically eliminating all environmental and experiential cues including a subject's experience with its own phenotype. Thus it is not surprising that recognition alleles have never been identified experimentally. Although researchers might wish to heed BLAUSTEIN's (1983) urging and remain open to the existence of recognition alleles, until empirical evidence appears that cannot be understood in terms of our three aforementioned mechanisms, recognition alleles seem hypothetical.

3 Kin recognition in Belding's ground squirrels

For the past five years (1979–83) we have studied kin recognition by Belding's ground squirrels (Fig. 1). Considerable information has been accumulated about the behavioral ecology of this social rodent (SHERMAN 1977, 1980, 1981a, 1981b), allowing us to design laboratory and field experiments that closely approximate conditions under which recognition abilities develop in nature. In brief our results suggest that all three mechanisms, spatial distribution, association, and phenotype matching, are used by *S. beldingi* to identify various relatives; these mechanisms are used under different circumstances and apparently serve different evolutionary functions.

At our study site near Tioga Pass, high (3,040 m) in the central Sierra Nevada of California, Belding's ground squirrels are diurnally active from May through September; the rest of the year they hibernate (MORTON and SHERMAN 1978). Mating occurs within a week of a female's emergence in the spring, and each female typically mates with three or four different males during her single four to six hour receptive period. She rears only one litter of two to eight pups per season. The young are nursed in a solitary burrow until they are about three weeks old, when they emerge above ground. At this point we capture the juveniles, sample their blood for later paternity determinations, and permanently mark them; thus kinship is assessed before littermates mix with neighboring juveniles.

Two to five weeks after they are weaned, juvenile males begin to disperse from their natal area (SHERMAN 1977; HOLEKAMP 1983). They will rarely, if ever, associate again with their mother or sisters. Females by contrast are sedentary from birth, seldom moving very far from their natal burrow. Thus females live near and interact daily with close female relatives. After the spring mating period, some adult males move again. This post-breeding dispersal, coupled with the high mortality characteristic of males (SHERMAN and MORTON in press), virtually precludes matings between fathers and their offspring the following season. In other words males do not inbreed with close kin as a result of permanent dispersal away from places where female relatives are likely to be encountered. Thus outbreeding is enforced via the spatial distribution kin recognition mechanism.

Whereas males do not help rear offspring or assist any other kin, nepotism is important in the social behavior of female *S. beldingi*. Females assist close relatives by giving warning calls when predators appear (Fig. 1), by sharing territories and associated food resources, and by cooperating to defend nursing pups against attacks by infanticidal members of their own species. Mother-daughter and sister-sister pairs are particularly cooperative, whereas more distant kin such as nieces and cousins are treated the same as unrelated females.

To investigate the effects of sharing a nest on the development of sibling recognition in *S. beldingi,* we manipulated rearing associations in a laboratory study (HOLMES and SHERMAN 1982). Eighteen pregnant females that had mated in the field were captured near Tioga Pass and shipped to the laboratory. Pups born to these females were cross-fostered between dams within three hours of parturition, creating four kinds of rearing groups: siblings reared together by a common dam (either their own mother or a foster dam), siblings reared apart by different dams, non-siblings reared together, and non-siblings reared apart. Subsequent to the experimental transfers most nursing litters contained both biological siblings and non-siblings. Pups were weaned, fattened, and then placed in individual cages in a coldroom for a six-month hibernation period.

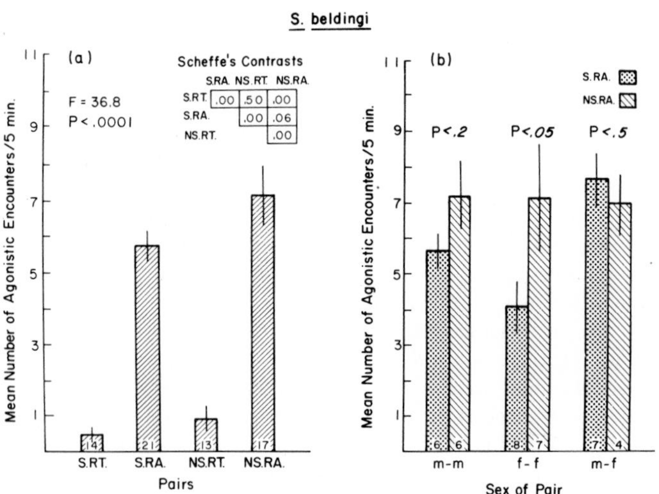

Fig. 4: Belding's ground squirrel pups born to females mated in the field were cross-fostered in the laboratory shortly after birth to create four rearing groups: siblings reared together (S. RT.), siblings reared apart (S. RA.), nonsiblings reared together (NS. RT.), and nonsiblings reared apart (NS. RA.). At the age of eight months, pairs from each group were observed in an arena for five minutes and the number of agonistic interactions was recorded.

a The mean number (± S. E.) of agonistic encounters between pairs differing in relatedness and rearing association. The results indicate that both siblings and non-siblings reared together were less aggressive than pairs reared apart. Whereas siblings and nonsiblings reared together were about equally nonaggressive, siblings reared apart were somewhat less aggressive than nonsiblings reared apart. Sample sizes and results of statistical tests are shown.

b The mean number (± S. E.) of agonistic encounters between pairs reared apart (from a), further analyzed to take into account the sex of each pair. This analysis revealed that sisters reared apart were significantly less aggressive than unrelated females reared apart; this was not true for pairs of males or males and females reared apart (From HOLMES and SHERMAN 1982).

Following hibernation, eight month old animals from the four rearing groups were observed during paired-encounter tests in an arena (1 m × 1 m in size), and the frequencies of eleven types of agonistic encounters were recorded. The results showed that young reared together, regardless of relatedness, were less aggressive than young reared apart (Fig. 4a). This was true regardless of sex or whether individuals were reared solely with siblings or with both siblings and nonsiblings. Furthermore, young reared together were about equally nonaggressive, whether or not they were related. Thus our data indicate that association during development mediates differential treatment of littermates.

Interestingly, these same tests revealed that sisters reared apart were significantly less aggressive than unrelated females reared apart (Fig. 4b). By contrast, relatedness did not seem to affect aggression between male-male or male-female pairs that had been reared apart. Thus only the females, the nepotistic sex in the field, behaved as if they could identify sisters despite not having been raised with them.

In describing the association mechanism, we argued that a sensitive period for kin recognition often occurs coincident with the time when young from different broods first mix in nature. Results of a field cross-fostering study, and a follow-up laboratory investigation, indicate such time dependence in *S. beldingi* (Holmes and Sherman 1982; Holmes in press). In the field experiment unrelated pups, matched for age with resident young, were inserted into nest burrows of lactating females at Tioga Pass. We found that foster dams accepted as offspring pups transferred into their burrows before the weaning of their own litters (about 25 days post-partum; Fig. 5). Furthermore, foster siblings treated female pups transferred before they were 25 days old as littermates the following season. By contrast, alien juveniles that were introduced into nest burrows after a female's twenty-fifth post-partum day elicited aggression from the potential foster dam, and as yearlings were treated like nonkin by foster sisters.

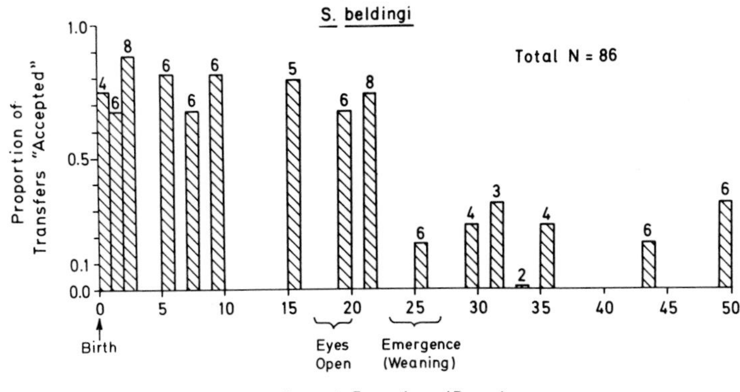

Fig. 5: To study the development of recognition between mother and offspring in *S. beldingi,* pups of different ages were cross-fostered into the burrows of lactating females in the field. For pups up to about twenty-five days of age, the percentage of acceptance was high; these pups eventually emerged above ground with the resident pups and behaved as if the foster burrowmates were nestmates. However, the percentage of acceptance dropped dramatically for pups older than twenty-five days, the age when resident pups are normally weaned and first emerge from the nest burrow. The sudden drop in acceptance at this point suggests that the onset of recognition by association coincides with the time when litters first mix in nature. Sample sizes are indicated (From Holmes and Sherman 1982).

In the follow-up laboratory study, pups were also exchanged between females (HOLMES in press). On Days 1 (i.e., 24 h post-partum), 8, 15, and 22, four pups, one from each of four groups, were presented simultaneously outside a female's nest box and the time until she retrieved each was recorded. The stimulus pups were either related-familiar, unrelated-familiar, related-unfamiliar, or unrelated-unfamiliar (related indicates whether or not a pup was the dam's own offspring, and familiar indicates whether or not the pup was reared by the dam). The time between pup introduction and retrieval by the female did not differ among the four categories of young on Days 1, 8, and 15. On Day 22 post-partum, familiar young were for the first time retrieved significantly faster than unfamiliar young. At no time did females seem to distinguish between their own offspring and the young of another female. During paired-encounter tests on Day 29, familiar dam-young pairs were significantly less agonistic than unfamiliar pairs; as in the retrieval study, there was no evidence that relatedness influenced arena encounters. Thus, both the field and laboratory

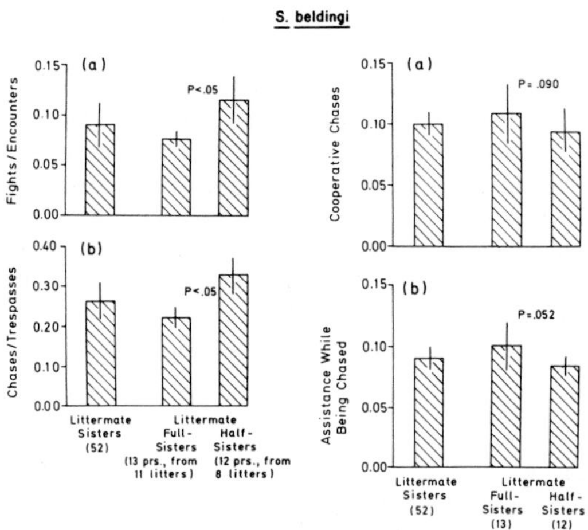

Fig. 6: Insemination of free-living female *S. beldingi* by more than one male frequently results in litters composed of full-siblings and maternal half-siblings, as shown by electrophoretic paternity exclusion analyses (HANKEN and SHERMAN 1981). Full- and half-sisters were observed in the field and their behavior was compared with that of «littermate sisters» whose exact relatedness was unknown. Numbers in parentheses indicate how many pairs were observed.

Left (a). The mean proportion (± S. E.) of times littermate sisters, full-sisters, and half-sisters fought when they encountered each other (came within 0.5 m. of one another).

Left (b). The mean proportion (± S. E.) of times littermate sisters, full-sisters, and half-sisters chased one another out of a territory following a trespass.

Right (a). The mean proportion (± S. E.) of cooperative chases in which residents were assisted by their littermate sisters, full-sisters, and half-sisters to evict trespassers who had entered their territory.

Right (b). The mean proportion (± S. E.) of chases in which females that were being chased were assisted by their littermate sisters, full-sisters, and half-sisters.

Overall the data suggest that despite being reared together, full-sisters and half-sisters are discriminable; perhaps the mechanism involves phenotype matching (From HOLMES and SHERMAN 1982).

cross-fostering data indicate a time-dependent ontogeny of dam-offspring recognition based on association, with a sensitive period that roughly coincides with the time when free-living young first emerge from their natal burrow.

Multiple mating by female *S. beldingi* typically results in multiple insemination. Electrophoretic paternity exclusion analyses, using eight different polymorphic blood proteins as phenotypic markers, revealed that 64 % to 78 % of litters have multiple sires (Hanken and Sherman 1981; Sherman unpubl.). Thus in most litters full-siblings and maternal half-siblings share a common uterine and natal burrow environment. In a field experiment, the behavior of electrophoretically identified full- and maternal half-sisters was observed in 1979–80. The data revealed that as yearlings full-sisters behaved more amicably than half-sisters, in that they fought significantly less often when they met and more frequently permitted each other unmolested access to territories (Fig. 6, left). In addition, yearling full-sisters cooperated more often in chasing distantly related and unrelated intruders from territories, and came to each other's assistance more frequently when one of the pair was chased than did half-sisters (Fig. 6, right).

Among littermates, there are no obvious differences in the kind or extent of postnatal association that would facilitate the observed discrimination between full- and half-sisters. However, it is possible that behavioral differences between littermates might result from intra-uterine associations among fetuses sired by the same male (for examples of *in utero* effects on behavior, see vom Saal 1981). Discrimination among full- and half-siblings might also indicate phenotype matching against a pup's own phenotype. The field cross-fostering data (Fig. 5) are consistent with this hypothesis because as yearlings, «accepted» foster sisters were treated like half-sisters rather than full-sisters. Thus, our observations indicate that female Belding's ground squirrels distinguish at least two categories of conspecifics, each composed of at least two groups. First, females either share a natal burrow (nestmates) or they do not (non-nestmates). Second, a female's nestmates are either relatively similar to her phenotypically (in odor, for example), in which case they are treated as full-sisters, or they are not (half-sisters and foster sisters).

4 Kin recognition in the genus Spermophilus

In the last fifteen years field studies of ground squirrels *(Spermophilus)* have provided insights into the evolution of mammalian population dynamics, life history parameters, and social organization (reviewed by Armitage 1981; Michener 1983). We highlight four species that have been studied in particular detail: *S. beldingi, S. parryii* (McLean 1982, 1983, in press), *S. richardsonii* (Michener, 1973, 1979, 1981; Michener and Michener 1977; Davis in press), and *S. tridecemlineatus* (McCarley 1966; Schwagmeyer 1979, 1980, in press; Vestal and McCarley in press). These studies have documented sexual asymmetries in dispersal similar to those we found in *S. beldingi,* and also the occurrence of nepotism among females but not males. Given the importance of kin-correlated behavior in the females' social structure, interest has recently focused on the proximal mechanisms by which females recognize their kin. A preliminary picture is emerging that relates inter-specific differences in recognition abilities and mechanisms to differences in social organization (Table 2).

For example, neonatal Belding's ground squirrels and thirteen-lined ground squirrels (Holmes and Sherman 1982; Holmes 1984) were cross-fostered intra-specifically to see how sibling recognition was affected by rearing association and genetic relatedness. Because subjects from both species were cross-fostered, reared, and studied in the same laboratory using similar procedures, inferences drawn from between-species comparisons

Table 2: Results from studies of kin recognition abilities in various ground squirrel species *(Spermophilus)*. The table indicates whether the cited recognition ability was documented in the field (F) or laboratory (L), and whether various categories of kin were (+) or were not (−) recognized when reared together (T) or apart (A).

Species	Category of Kin	Field or Lab Study	Rearing/ Recognition		Reference
S. beldingi	Dam-Offspring	F	T: +;	A: ?	HOLMES and SHERMAN 1982
		L	T: +;	A: −	HOLMES in press
	Sibling-Sibling	F	T: +;	A: ?	HOLMES and SHERMAN 1982
		L	T: +;	A: +[1]	HOLMES and SHERMAN 1982
	Maternal Half-	F	T: +;	A: ?	HOLMES and SHERMAN 1982
	Siblings	L	T: ?;	A: ?	
	Paternal Half-	F	T: ?;	A: ?	[2]
	Siblings	L	T: ?;	A: ?	[2]
S. tridecemlineatus	Sibling-Sibling	L	T: +;	A: −	HOLMES 1984
S. parryii	Sibling-Sibling	L	T: +;	A: +[1]	HOLMES and SHERMAN 1982
S. richardsonii	Dam-Offspring	L	T: +;	A: ?	MICHENER and SHEPPARD 1972; MICHENER 1974
	Sibling-Sibling	L	T: +;	A: ?	SHEPPARD and YOSHIDA 1971
		L	T: +;	A: +	DAVIS 1982

?: This rearing regimen was not used for this category of kin.
[1]: Recognition occurred between sister-sister pairs only (see Fig. 4b).
[2]: Field and laboratory studies of paternal half-sibling recognition in *S. beldingi* are currently in progress.

are strengthened. Paired-encounter tests indicate that recognition in female *S. beldingi* can be influenced by both factors (e. g., Fig. 4). In contrast, recognition in *S. tridecemlineatus* may be affected only by rearing association: whereas young that were reared together showed low levels of «exploratory» interactions, young reared apart exhibited high levels of social exploration, regardless of relatedness or sex. In other words, the only individuals thirteen-lined ground squirrel juveniles distinguished in this test were those with whom they had shared a natal nest. Although the significance of this inter-specific difference in kin recognition is currently unclear, it is worth noting that in nature dam-offspring and sister pairs interact frequently and amicably throughout life in *S. beldingi*, whereas amicable interactions in *S. tridecemlineatus* are apparently limited to dams and offspring during the first few weeks post-partum. One possible implication is that some or all of the factors favoring kin recognition by phenotype matching in *S. beldingi* – frequent multiple paternity, extensive kin networks, and widespread nepotism – may be either rare or absent in *S. tridecemlineatus*.

Because *S. parryii* and *S. richardsonii* siblings have also been cross-fostered intra-specifically (Table 2), their abilities to discriminate kin can be added to the comparative picture. Laboratory results for *S. parryii* mirror those for *S. beldingi:* both rearing association and relatedness influence the identification process among females. Like *S. beldingi,* in the field female *S. parryii* that are closely related (i. e., mother-daughter and sister pairs) share home ranges, seldom fight, and frequently interact amicably. In addition, closely related females transport their newly-emerged pups to a common burrow system, and then cooperate to protect these communal nest burrows from intra- and inter-specific predators.

It is more difficult to place the laboratory recognition data on *S. richardsonii* into our comparative perspective because DAVIS (1982) isolated his animals for 75 days immediately prior to testing them; such an isolation during the active season was not experienced by subjects in the other three species and its effects on the recognition process are unknown. This procedural difference notwithstanding, DAVIS's (1982) results suggest that association may not be the sole mechanism of sibling recognition in *S. richardsonii*. First, siblings reared apart remained closer to each other and physically contacted each other more often and for longer periods during dyadic arena tests than did non-siblings reared apart. Second, siblings reared together approached each other more often and made naso-oral contact more frequently than did non-siblings reared together. Unfortunately, small sample sizes precluded analyses by pairs' sexes. However, given the apparent similarities among the social systems and matrilineal kin group structures of *S. richardsonii*, *S. beldingi*, and *S. parryii* populations (reviewed by MICHENER 1983), we predict that phenotype matching occurs in the former species and that such abilities are especially well developed in females.

In conclusion, field and laboratory studies of four ground squirrel species suggest the existence of intriguing similarities and differences in kin recognition abilities and their ontogenies; at least some of these apparently relate to variations in dispersal, polygamy, and nepotism. We anticipate the proliferation of such investigations, and look forward to the development of the comparative picture (Table 2) in scope and detail. Particularly interesting in this regard will be comparisons among the more social *Spermophilus* species (e. g., *S. armatus, S. columbianus)* and their more solitary congeners (e. g., *S. franklinii, S. lateralis*), as well as parallel studies of prairie dogs *(Cynomys)* and marmots *(Marmota)*. Further research on the ecology and ontogeny of kin recognition abilities among invertebrates and vertebrates, particularly sciurids, seems warranted.

Acknowledgments

Many of the ideas expressed in this paper were presented in our *American Scientist* publication (HOLMES and SHERMAN 1983). For assistance with that work we thank PATRICK BATESON, MARTIN DALY, CYNTHIA KAGARISE SHERMAN, MARTIN L. MORTON, BRUCE WALDMAN, and RICHARD WRANGHAM; for help with the current paper we thank STEPHEN T. EMLEN, CYNTHIA KAGARISE SHERMAN, GEORGE J. GAMBOA, and BERT HÖLLDOBLER. RICHARD D. ALEXANDER's ideas and friendly skepticism were a consistent source of stimulation. WARREN G. HOLMES's research was supported by grants from the University of Michigan; PAUL W. SHERMAN's studies were supported by grants from the National Science Foundation.

References

Alexander, R. D. (1979): Darwinism and Human Affairs. University of Washington Press, Seattle.

Alexander, R. D. and G. Borgia (1978): Group selection, altruism, and the levels of organization of life. Ann. Rev. Ecol. Syst. *9*, 449–474.

Armitage, K. B. (1981): Sociality as a life-history tactic of ground squirrels. Oecologia *48*, 36–49.

Barrows, E. M., W. J. Bell, and C. D. Michener (1975): Individual odor differences and their social functions in insects. Proc. Natl. Acad. Sci. U.S.A. *72*, 2824–2828.

Bateson, P. (1978): Sexual imprinting and optimal outbreeding. Nature *273*, 659–660.

Bateson, P. (1980): Optimal outbreeding and the development of sexual preferences in Japanese quail. Z. Tierpsychol. *53*, 231–244.

Bateson, P. (1982): Preferences for cousins in Japanese quail. Nature *295*, 236–237.

Bateson, P. (1983): Optimal outbreeding. In Mate Choice, ed. P. Bateson, pp. 257–277. Cambridge University Press, Cambridge.

Beecher, M. D. (1982): Signature systems and individual recognition. Amer. Zool. *22*, 477–490.

Beecher, I. M. and M. D. Beecher (1983): Sibling recognition in bank swallows *(Riparia riparia)*. Z. Tierpsychol. *62*, 145–150.

Beecher, M. D., I. M. Beecher, and S. Lumpkin. (1981a): Parent-offspring recognition in bank swallows *(Riparia riparia)*: I. Natural history. Anim. Behav. *29*, 86–94.

Beecher, M. D., I. M. Beecher, and S. H. Nichols (1981b): Parent-offspring recognition in bank swallows *(Riparia riparia)*: II. Development and acoustic basis. Anim. Behav. *29*, 95–101.

Beer, C. G. (1970): Individual recognition of voice in the social behavior of birds. In Advances in the Study of Behavior, Vol. 3, eds. J. S. Rosenblatt, C. G. Beer, and R. A. Hinde, pp. 27–74. Academic Press, New York.

Bekoff, M. (1981): Mammalian sibling interactions: Genes, facilitative environments, and the coefficient of familiarity. In Parental Care in Mammals, eds. D. J. Gubernick and P. H. Klopfer, pp. 307–346. Plenum Press, New York.

Berenstain, L., P. S. Rodman, and D. G. Smith (1981): Social relations between fathers and offspring in a captive group of rhesus monkeys *(Macaca mulatta)*. Anim. Behav. *29*, 1057–1063.

Birkhead, T. R. (1978): Behavioural adaptations to high density nesting in the common guillemot *Uria aalge*. Anim. Behav. *26*, 321–331.

Blaustein, A. R. (1983): Kin recognition mechanisms: phenotypic matching or recognition alleles? Amer. Nat. *121*, 749–754.

Blaustein, A. R. and R. K. O'Hara (1981): Genetic control for sibling recognition? Nature *290*, 246–248.

Blaustein, A. R. and R. K. O'Hara (1982): Kin recognition in *Rana cascadae* tadpoles: maternal and paternal effects. Anim. Behav. *30*, 1151–1157.

Blaustein, A. R., R. K. O'Hara, and D. H. Olson (1984): Kin preference behaviour is present after metamorphosis in *Rana cascadae* frogs. Anim. Behav. *32*, 445–450.

Breed, M. D. (1981): Individual recognition and learning of queen odors by worker honeybees. Proc. Natl. Acad. Sci. *78*, 2635–2637.

Breed, M. D. (1983): Nestmate recognition in honey bees. Anim. Behav. *31*, 86–91.

Buckle, G. R. and L. Greenberg (1981): Nestmate recognition in sweat bees *(Lasioglossum zephyrum)*: Does an individual recognize its own odour or only odours of its nestmates? Anim. Behav. *29*, 802–809.

Burger, J. (1974): Breeding adaptations of Franklin's Gull *(Larus pipixcan)* to a marsh habitat. Anim. Behav. *22*, 521–567.

Cowan, D. P. (1979): Sibling matings in a hunting wasp: adaptive inbreeding? Science *205*, 1403–1405.

Crozier, R. H. and M. W. Dix (1979): Analysis of two genetic models for the innate components of colony odor in social Hymenoptera. Behav. Ecol. Sociobiol. *4*, 217–224.

Cullen, E. (1957): Adaptations in the kittiwake to cliff-nesting. Ibis *99*, 275–302.

Daly, M. and M. I. Wilson (1982): Whom are newborn babies said to resemble? Ethol. Sociobiol. *3*, 67–78.

Davis, L. S. (1982): Sibling recognition in Richardson's ground squirrels *(Spermophilus richardsonii)*. Behav. Ecol. Sociobiol. *11*, 65–70.

Davis, L. S. (in press): Behavioral interactions of Richardson's ground squirrels: asymmetries based on kinship. In Biology of Ground-Dwelling Squirrels: Annual Cycles, Behavioral Ecology, and Sociality, eds. J. O. Murie and G. R. Michener. Univ. of Nebraska Press, Lincoln.

Dawkins, R. (1976): The Selfish Gene. Oxford Univ. Press, Oxford.

Dawkins, R. (1982): The Extended Phenotype. Oxford Univ. Press, Oxford.

Dewsbury, D. A. (1982): Avoidance of incestuous breeding between siblings in two species of *Peromyscus* mice. Biol. Behav. *7*, 157–169.

Dobson, F. S. (1982): Competition for mates and predominant juvenile male dispersal in mammals. Anim. Behav. *30,* 1183–1192.

Fredrickson, W. T. and G. P. Sackett (1984): Kin preferences in primates *(Macaca nemestrina)*: Relatedness or familiarity? J. Comp. Psychol. *98,* 29–34.

Getz, W. M. (1981): Genetically based kin recognition systems. J. Theor. Biol. *92,* 209–226.

Getz, W. M. (1982): An analysis of learned kin recognition in Hymenoptera. J. Theor. Biol. *99,* 585–597.

Getz, W. M., D. Brückner, and T. R. Parisian (1982): Kin structure and the swarming behavior of the honey bee *Apis mellifera.* Behav. Ecol. Sociobiol. *10,* 265–270.

Getz, W. M. and K. B. Smith (1983): Genetic kin recognition: honey bees discriminate between full and half sisters. Nature *302,* 147–148.

Gilder, P. M. and P. J. B. Slater (1978): Interest of mice in conspecific male odours is influenced by degree of kinship. Nature *274,* 362–364.

Grau, H. J. (1982): Kin recognition in white-footed deermice *(Peromyscus leucopus).* Anim. Behav. *30,* 497–505.

Greenberg, L. (1979): Genetic component of bee odor in kin recognition. Science *206,* 1095–1097.

Greenwood, P. J. (1980): Mating systems, philopatry and dispersal in birds and mammals. Anim. Behav. *28,* 1140–1162.

Greenwood, P. J., P. H. Harvey, and C. M. Perrins (1978): Inbreeding and dispersal in the great tit. Nature *271,* 52–54.

Gubernick, D. J. (1980): Maternal «imprinting» or maternal «labelling» in goats? Anim. Behav. *28,* 124–129.

Hamilton, W. D. (1964): The genetical evolution of social behaviour, I. and II. J. Theoret. Biol. *7,* 1–52.

Hanken, J. and P. W. Sherman (1981): Multiple paternity in Belding's ground squirrel litters. Science *212,* 351–353.

Hess, E. H. (1973): Imprinting: Early Experience and the Developmental Psychobiology of Attachment. Van Nostrand Reinhold, New York.

Hill, J. L. (1974): Peromyscus: effect of early pairing on reproduction. Science *186,* 1042–1044.

Hölldobler, B. and C. D. Michener (1980): Mechanisms of identification and discrimination in social hymenoptera. In Evolution of Social Behavior: Hypotheses and Empirical Tests, Dahlem Konferenzen, ed. H. Markl, pp. 35–58. Weinheim, Verlag Chemie.

Holekamp, K. E. (1983): Proximal mechanisms of natal dispersal in Belding's ground squirrel *(Spermophilus beldingi beldingi).* Ph. D. Thesis, University of California, Berkeley.

Holmes, W. G. (1984): Sibling recognition in thirteen-lined ground squirrels: effects of genetic relatedness, rearing association, and olfaction. Behav. Ecol. Sociobiol. *14,* 225–233.

Holmes, W. G. (in press): The development of dam-young recognition in Belding's ground squirrels. J. Comp. Psych.

Holmes, W. G. and P. W. Sherman (1982): The ontogeny of kin recognition in two species of ground squirrels. Amer. Zool. *22,* 491–517.

Holmes, W. G. and P. W. Sherman (1983): Kin recognition in animals. Amer. Sci. *71,* 46–55.

Hoogland, J. L. (1982): Prairie dogs avoid extreme inbreeding. Science *215,* 1639–1641.

Hoogland, J. L. (1983): Nepotism and alarm calling in the black-tailed prairie dog *(Cynomys ludovicianus).* Anim. Behav. *31,* 472–479.

Hoogland, J. L. and P. W. Sherman (1976): Advantages and disadvantages of bank swallow *(Riparia riparia)* coloniality. Ecol. Monogr. *46,* 33–58.

Kalmus, H. and C. R. Ribbands (1952): The origin of the odours by which honeybees distinguish their companions. Proc. Roy. Soc. London (B) *140,* 50–59.

Kaplan, J. N., D. Cubicciotti III, and W. K. Redican (1977): Olfactory discrimination of squirrel monkey mothers by their infants. Develop. Psychobiol. *10,* 447–453.

Kareem, A. M. (1983): Effect of increasing periods of familiarity on social interactions between male sibling mice. Anim. Behav. *31,* 919–926.

Kareem, A. M., and C. J. Barnard (1982): The importance of kinship and familiarity in social interactions between mice. Anim. Behav. *30,* 594–601.

Klahn, J. E. and G. J. Gamboa (1983): Social wasps: Discrimination between kin an nonkin brood. Science *221,* 482–484.

Klopfer, P. H. and M. S. Klopfer (1968): Maternal imprinting in goats: Fostering of alien young. Z. Tierpsychol. *25,* 862–866.

Kukuk, P. F., M. D. Breed, A. Sobti, and W. J. Bell (1977): The contributions of kinship and conditioning to nest recognition and colony member recognition in a primitively eusocial bee, *Lasioglossum zephyrum* (Hymenoptera: Halicitidae). Behav. Ecol. Sociobiol. *2,* 319–327.

Kurland, J. (1980): Kin selection theory: A review and selective bibliography. Ethol. Sociobiol. *1,* 255–274.

Labov, J. B. (1980): Factors influencing infanticidal behavior in wild male house mice *(Mus musculus).* Behav. Ecol. Sociobiol. *6,* 297–303.

Lacy, R. C. and P. W. Sherman (1983): Kin recognition by phenotype matching. Amer. Nat. *121,* 489–512.

Leon, M. (1975): Dietary control of maternal pheromone in the lactating rat. Physiol. Behav. *14,* 311–319.

Leon, M. (1978): Filial responsiveness to olfactory cues in the laboratory rat. In Advances in the Study of Behavior, Vol. 10, eds. J. S. Rosenblatt, C. G. Beer, R. A. Hinde, and M. C. Busnel, pp. 117–153. Academic Press, New York.

Lorenz, K. (1935): Der Kumpan in der Umwelt des Vogels. J. Ornithology *83,* 137–213, 289–413.

McArthur, P. D. (1982): Mechanisms and development of parent-young vocal recognition in the piñon jay *(Gymnorhinus cyanocephalus).* Anim. Behav. *30,* 62–74.

McCarley, H. (1966): Annual cycle, population dynamics, and adaptive behavior of *Citellus tridecemlineatus.* J. Mammal. *47,* 294–316.

McLean, I. G. (1982): The association of female kin in the Arctic ground squirrel *Spermophilus parryii.* Behav. Ecol. Sociobiol. *10,* 91–99.

McLean, I. G. (1983): Paternal behaviour and killing of young in Arctic ground squirrels. Anim. Behav. *31,* 32–44.

McLean, I. G. (in press): Spacing behavior of female ground squirrels. In Biology of Ground-Dwelling Squirrels: Annual Cycles, Behavioral Ecology, and Sociality, eds. J. O. Murie and G. R. Michener. Univ. of Nebraska Press, Lincoln.

Michener, G. R. (1973): Field observations on the social relationships between adult female and juvenile Richardson's ground squirrels. Can. J. Zool. *51,* 33–38.

Michener, G. R. (1974): Development of adult-young identification in Richardson's ground squirrel. Develop. Psychobiol. *7,* 375–384.

Michener, G. R. (1979): Spatial relationships and social organization of adult Richardson's ground squirrels. Can. J. Zool. *57,* 125–139.

Michener, G. R. (1981): Ontogeny of spatial relationships and social behavior in juvenile Richardson's ground squirrels. Can. J. Zool. *59,* 1666–1676.

Michener, G. R. (1983): Kin identification, matriarchies, and the evolution of sociality in ground-dwelling sciurids. In Advances in the Study of Mammalian Behavior, eds. J. F. Eisenberg and D. G. Kleiman, pp. 528–572. Special Publication No. 7, American Society of Mammalogists.

Michener, G. R. and D. R. Michener (1977): Population structure and dispersal in Richardson's ground squirrels. Ecology *58,* 359–368.

Michener, G. R. and D. H. Sheppard (1972): Social behavior between adult female Richardson's ground squirrels *(Spermophilus richardsonii)* and their own and alien young. Can. J. Zool. *50,* 1343–1349.

Miller, D. E. and J. T. Emlen, Jr. (1975): Individual chick recognition and family integrity in the ring-billed gull. Behaviour *52,* 124–144.

Morton, M. L. and P. W. Sherman (1978): Effects of a spring snowstorm on behavior, reproduction, and survival of Belding's ground squirrels. Can. J. Zool. *56,* 2578–2590.

Mostofsky, D., ed. (1965): Stimulus Generalization. Stanford Univ. Press, Stanford.

Noakes, D. L. G. and G. W. Barlow (1973): Cross-fostering and parent-offspring responses in *Cichlasoma citrinellum* (Pisces, Cichlidae). Z. Tierpsychol. *33,* 147–152.

O'Hara, R. K. and A. R. Blaustein (1981): An investigation of sibling recognition in *Rana cascadae* tadpoles. Anim. Behav. *29,* 1121–1126.

Payne, R. B. (1977): The ecology of brood parasitism in birds. Ann. Rev. Ecol. Syst. *8,* 1–28.

Pfennig, D. W., H. K. Reeve, and J. S. Shellman (1983a): Learned component of nestmate discrimination in workers of a social wasp, *Polistes fuscatus* (Hymenoptera: Vespidae). Anim. Behav. *31*, 412–416.

Pfennig, D. W., G. J. Gamboa, H. K. Reeve, J. Shellman Reeve, and I. D. Ferguson (1983b): The mechanism of nestmate discrimination in social wasps (*Polistes*, Hymenoptera: Vespidae). Behav. Ecol. Sociobiol. *13*, 299–305.

Porter, R. H., J. A. Matochik, and J. W. Makin (1983): Evidence for phenotype matching in spiny mice *(Acomys cahirinus)*. Anim. Behav. *31*, 978–984.

Porter, R. H. and J. D. Moore (1981): Human kin recognition by olfactory cues. Physiol. Behav. *27*, 493–495.

Porter, R. H., V. J. Tepper, and D. M. White (1981): Experiential influences on the development of huddling preferences and «sibling» recognition in spiny mice. Develop. Psychobiol. *14*, 375–382.

Porter, R. H. and M. Wyrick (1979): Sibling recognition in spiny mice *(Acomys cahirinus)*: Influence of age and isolation. Anim. Behav. *27*, 761–766.

Porter, R. H., M. Wyrick, and J. Pankey (1978): Sibling recognition in spiny mice *(Acomys cahirinus)*. Behav. Ecol. Sociobiol. *3*, 61–68.

Price, M. V., and N. M. Waser (1979): Pollen dispersal and optimal outcrossing in *Delphinium nelsoni*. Nature *277*, 294–297.

Ribbands, C. R. (1965): The role of recognition of comrades in the defense of social insect communities. Symp. Zool. Soc. London *14*, 159–168.

Ridley, M. and A. Grafen (1981): Are green beard genes outlaws? Anim. Behav. *29*, 954–955.

Rothstein, S. I. (1982): Successes and failures in avian egg and nestling recognition with comments on the utility of optimality reasoning. Amer. Zool. *22*, 547–560.

Salzen, E. A. and J. M. Cornell (1968): Self-perception and species recognition in birds. Behaviour *30*, 44–65.

Schwagmeyer, P. L. (1979): The function of alarm calling behavior in *Spermophilus tridecemlineatus*, the thirteen-lined ground squirrel. Ph. D. Thesis, Univ. of Michigan, Ann Arbor.

Schwagmeyer, P. L. (1980): Alarm calling behavior of the thirteen-lined ground squirrel, *Spermophilus tridecemlineatus*. Behav. Ecol. Sociobiol. *7*, 195–200.

Schwagmeyer, P. L. (in press): Multiple mating and intersexual selection in thirteen-lined ground squirrel. In Biology of Ground-Dwelling Squirrels: Annual Cycles, Behavioral Ecology, and Sociality, eds. J. O. Murie and G. R. Michener. Univ. Nebraska Press, Lincoln.

Shellman, J. S. and G. J. Gamboa (1982): Nestmate discrimination in social wasps: the role of exposure to nest and nestmates (*Polistes fuscatus*, Hymenoptera: Vespidae). Behav. Ecol. Sociobiol. *11*, 51–53.

Sheppard, D. H. and S. M. Yoshida (1971): Social behavior in captive Richardson's ground squirrels. J. Mammal. *52*, 793–799.

Sherman, P. W. (1977): Nepotism and the evolution of alarm calls. Science *197*, 1246–1253.

Sherman, P. W. (1980): The limits of ground squirrel nepotism. In Sociobiology: Beyond Nature/Nurture?, eds. G. W. Barlow and J. Silverberg, pp. 505–544. Westview Press, Boulder.

Sherman, P. W. (1981a): Reproductive competition and infanticide in Belding's ground squirrels and other organisms. In Natural Selection and Social Behavior: Recent Research and New Theory, eds. R. D. Alexander and D. W. Tinkle, pp. 311–331. Chiron Press, New York.

Sherman, P. W. (1981b): Kinship, demography, and Belding's ground squirrel nepotism. Behav. Ecol. Sociobiol. *8*, 251–259.

Sherman, P. W. and M. L. Morton (in press): Demography of Belding's ground squirrels. Ecology.

Shields, W. M. (1982): Philopatry, Inbreeding, and the Evolution of Sex. State Univ. of New York Press, Albany.

Trillmich, F. (1981): Mutual mother-pup recognition in Galapagos fur seals and sea lions: Cues used and functional significance. Behaviour *78*, 21–42.

Vestal, B. M. and H. McCarley (in press): Spatial and social relationships among relatives in thirteen-lined ground squirrel populations. In Biology of Ground-Dwelling Squirrels: Annual Cycles, Behavioral Ecology, and Sociality, eds. J. O. Murie and G. R. Michener. Univ. of Nebraska Press, Lincoln.

Vidal, J.-M. (1982): «Auto-imprinting»: Effects of prolonged isolation on domestic cocks. J. Comp. Physiol. Psych. *96*, 256–267.

vom Saal, F. S. (1981): Variation in phenotype due to random intrauterine positioning of male and female fetuses in rodents. J. Reprod. Fert. *62*, 633–650.

Waldman, B. (1981): Sibling recognition in toad tadpoles: The role of experience. Z. Tierpsych. *56*, 341–358.

Waldman, B. (1982): Sibling association among schooling toad tadpoles: Field evidence and implications. Anim. Behav. *30*, 700–713.

Weller, M. W. (1959): Parasitic egg laying in the redhead *(Aythya americana)* and other North American Anatidae. Ecol. Monogr. *29*, 333–365.

Wu, H. M. H., W. G. Holmes, S. R. Medina, and G. P. Sackett (1980): Kin preference in infant *Macaca nemestrina*. Nature *285*, 225–227.

Fortschritte der Zoologie, Bd. 31 · Hölldobler/Lindauer (Hrsg.): Experimental Behavioral Ecology
G. Fischer Verlag · Stuttgart · New York · 1985

Coping behaviour and stress physiology in male tree shrews (*Tupaia belangeri*)

Dietrich v. Holst

Zoologisches Institut, Universität Bayreuth, F. R. Germany

Abstract

In nature, tree shrews (Tupaia belangeri) live singly or in pairs in territories which they defend vigorously against strange conspecifics.

In the laboratory adult males also immediately attack strange intruding males and normally defeat them within a few minutes. After subjugation, victors show no further sign of arousal and pay virtually no attention to the subordinate animals. The latter, however, change their behaviour completely from the moment of subjugation and may be classified in the following two distinct groups:

1. Subdominant animals: These withdraw from situations which could lead to more intense fights by active avoidance behavior. Under such conditions, they are capable of living in dominants' cages, albeit with a very reduced sphere of action, for weeks.

2. Submissive animals: These sit almost continually in a corner of their cages and hardly respond to external stimuli. Even the threats and attacks of the dominants are accepted without the animals attempting to flee or defend themselves and they normally die within less than 20 days.

Serum concentrations of adrenal, thyroidal, gonadal hormones (as well as other physiological parameters) of dominant tree shrews largely correspond to those of controls. Subordinate and submissive animals, however, show distinct stress reactions in all physiological parameters measured. In general, the physiological reactions are qualitatively the same, differing only in degree. There is, however, a clear qualitative distinction between subdominants and submissives with respect to their adrenal functions: Submissives, in relation to dominants, show no change in sympathetic-adrenomedullary activity but an increase in adrenocortical function, the opposite being true for subdominants. The relevance of these physiological reactions for the development of status and coping dependent diseases will be discussed.

1 Introduction

The dispersion of animals within their habitat is largely determined by their social behaviour. The social structure differs between species and may vary also with particular parameters (such as climatic conditions, food shortage, reproductive condition and population density) in a predictable manner. With increasing independence from physical parameters the social ‹environment›, e. g. the interaction between conspecifics, gains in importance, particularly in some insects, in birds and mammals: in these groups each individual is part of the surrounding social sphere of other individuals, which influences their behaviour and physiology constantly.

In this paper I shall present some findings of our research on tree shrews which exemplify the behavioural and physiological responses of male individuals to social conflict in order to clarify their biological significance.

Tree shrews are diurnal mammals, about the size of a squirrel, distributed throughout Southeast Asia. Their systematic position is unclear; while originally considered primates, it now seems more likely that tree-shrews provide a model of the common ancestor of all living placental mammals. They are thus classified as a separate order: Scandentia (MARTIN, 1969; STARCK, 1978). In nature, they usually live in pairs in territories defended vigorously against strange conspecifics of the same sex (KAWAMICHI and KAWAMICHI, 1979).

In large laboratory enclosures, as in nature, adult tree shrews of both sexes immediately attack strange conspecifics of their own sex and normally defeat them within a few minutes. Shortly after fighting has ended, the victor shows no further sign of arousal and pays virtually no attention to the subordinate animal. The subordinate tree-shrew, however, crouches in a corner, which it leaves only to feed and drink. It hardly moves at all, spending more than 90 % of the daily activity phase lying motionless in its hiding place and following the movements of the victor with its head.

During the following days, agonistic encounters between the two animals are very rare or do not occur at all. Nevertheless, the subordinate animal dies within less than 20 days.

2 Death as a result of constant «anxiety»

As the following results show, death does not result from the physical exertions of fighting or from wounds received, but rather from the continued presence of the victor. If an adult male is placed into the cage (floor area 120 × 70 cm, height 50 cm) of an unknown male conspecific, it is usually attacked instantly and subdued in less than 2 minutes. Following separation of the two animals by an opaque partition, the loser recovers from the fight nearly as rapidly as the winner. Under such conditions, even when the subordinate animal is subjected to an encounter every day for weeks, it hardly loses any body weight and does not die prematurely. If the two animals are separated after their first fight by a wire mesh partition, so that the loser cannot be attacked but sees the victor constantly, however, it dies within a few days. Thus death is not a (direct) adjunct of social interactions and their physiological consequences, but rather results from central nervous processes in the subordinate animal based on experience (being defeated) and learning (to recognize the victor). To put it anthropomorphously, the subordinate dies of constant «anxiety».

To determine the cause of death of subordinate tree shrews, a male was placed into the cage of an unknown conspecific of the same sex, which subjugated him completely within less than two minutes. Afterwards both animals were separated by a wire mesh partition. Every 1–2 days the figths were repeated. Experimental animals were always healthy adult tree shrews aged between 2 and 5 years (the maximum age reached by tree shrews in our institute's colony is more than 8 years). The animals were maintained under constant conditions (L : D = 12 : 12 hours; relative humidity 50–60 %) with food and water ad libitum.

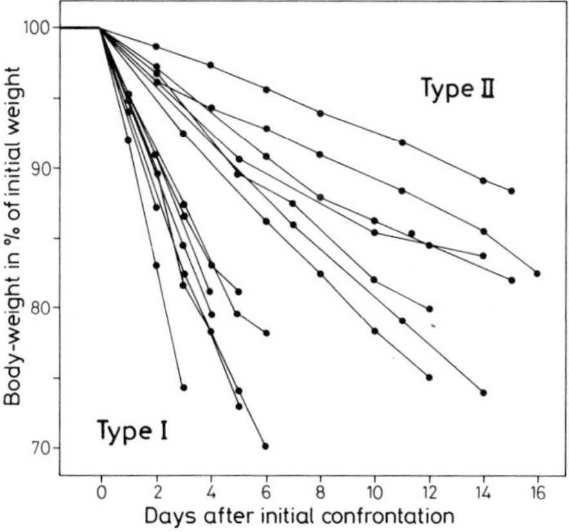

Fig. 1: Decrease of body weight of some subordinate tree shrews after initial confrontation.

From the time of first submission, all subordinate animals sat in a corner of the cage or in their sleeping boxes for practically the whole day, hardly responding to external stimuli. They did not even attempt to escape the attacks of the dominant animal, but usually suffered them without any attempt to defend themselves.

Under these conditions, all animals showed a progressive decrease in body weight and died between 2 and 20 days. The daily weight loss was more or less constant for any individual throughout the entire experiment, but differed considerably between individuals (examples: see Fig. 1).

There was a negative correlation between the survival time of subordinate animals in this situation and their daily weight losses (Fig. 2): Daily weight loss is thus a good index of an animal's stress response – at least during this form of acute stress.

On the basis of their daily body weight losses, the animals may be separated into two distinct groups: those which lived 8 days or less in this situation, having daily weight losses above 3 % (type I) and those which survived longer, having daily weight losses below 3 % (type II).

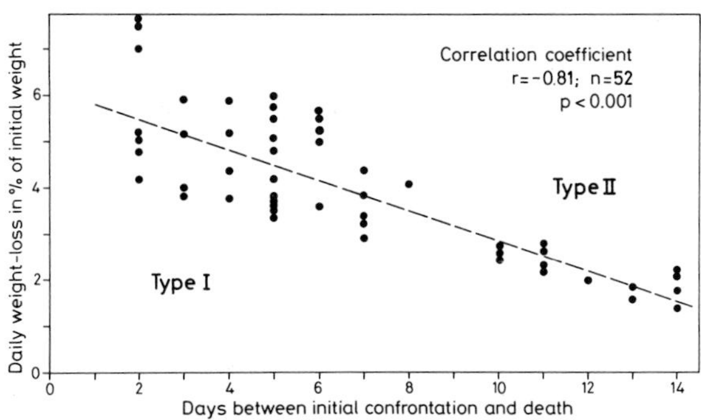

Fig. 2: Relationship between daily loss of body weight and survival time of type I and type II subordinate tree shrews under persistent social stress.

In order to obtain data on the physiological responses of the animals to this stressful situation, blood samples were taken from 48 animals before the experiments were started and every 1–3 days thereafter. Blood samples here as well as in all other experiments were always taken about one hour before onset of light (= activity) period and were used to determine various hormones (by radio-immunoassays) and other clinically important parameters. Due to their daily weight losses, 25 animals were classified as type I animals, the rest as type II.

A clear distinction was evident between type I and type II animals with respect to their kidney function. With only three exceptions the urea nitrogen (and creatinine) values in the serum of type I animals rose within 6 days to more than tenfold of their initial values, leading to death by uraemia (Fig. 3; for further details see v. Holst, 1972), while it remained within the range of controls in the longer surviving type II animals. The cause of death in the latter animals is not yet known.

With respect to all other parameters, the reactions of the individuals of both groups were qualitatively the same, differing only in degree. While all parameters changed very quickly in type I animals (see Fig. 3), the changes in the longer surviving type II animals developed more gradually. They never reached the same degree, evident in type I animals

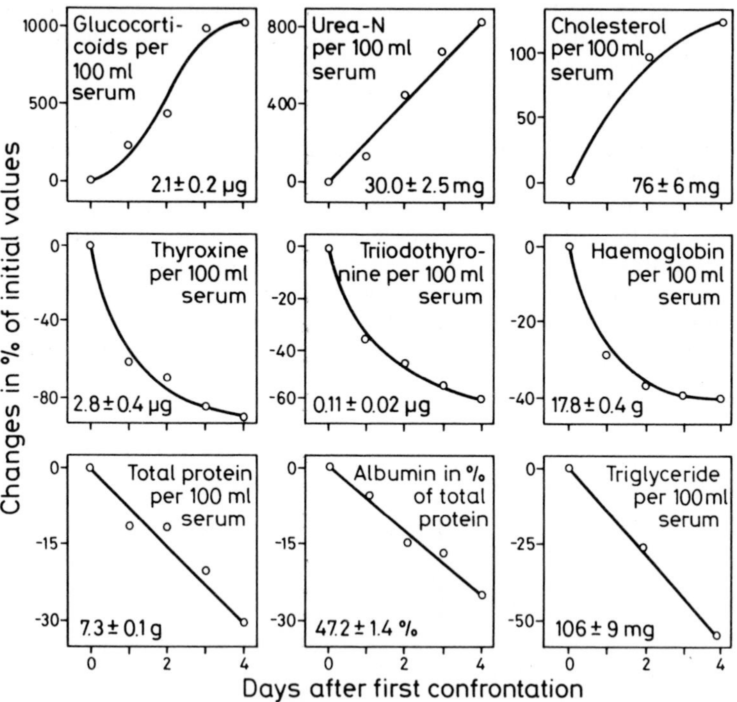

Fig. 3: Some physiological parameters in the serum (resp. blood) of type I subordinate tree shrews, and their changes (as percent of initial values) in the continued presence of dominant animals. The absolute initial values are given, along with their mean errors. All parameters differ significantly ($p < 0.001$) from the initial values within 1–2 days of confrontation (triglycerides after 4 days). For further details see text.

within less than 6 days, even shortly before their death after 10–20 days. However, after 5–8 days all parameters shown in Fig. 3 and many others, were significantly different from intitial values. In addition to the data presented in fig. 3, the weights of testes, epididymides and androgen-dependent glands declined in all animals almost linearly from the moment of subjugation; this result suggests a complete inhibition of the release of gonadotropic hormones by the adenohypophysis. Correspondingly, after less than 10 days all animals had testosterone values in their serum similar to those of castrated males, and they were sterile.

As drastic as these alterations are, we do not, as yet, have an understanding of their consequences regarding the metabolic physiology of the organism as a whole. One main reason is the inadequacy of our information about other hormones (such as epinephrine, insulin, glucagon, growth hormone), which can overlap to a great extent in their effects on metabolic processes with some hormones or may be synergistic, additive or even antagonistic to others. Nevertheless, a few physiological and patho-physiological consequences of this persistent form of stress are evident.

All animals become sterile within a few days and all show a dramatic decrease in body weight. This may be due – at least partly – to the gluconeogenetic activity of the adrenocortical hormones, which increase in all subordinate animals quite dramatically. In addition, the increased glucocorticoid concentrations in the blood result in a drastic decrease in wound healing. While insignificant abraisons or bite wounds, as they may occur in any fight, in dominant animals heal completely within 1–2 days, in subordinates larger subcutaneous wounds may develop. Even more dramatic may be the predicted effect on the immunological system: Based on data from 10 dominant and 14 subordinate individuals, the total count of leucocytes in subordinates did not increase significantly. The number of basophile granulocytes, however, decreased to about 1 %, that of eosinophilic granulocytes to 18 % and that of lymphocytes to 17 % of the values of dominant individuals (and of control animals, respectively; all differences $p < 0.001$). This implies that the healing of wounds and the immune defense against disease, parasites, or cancerous growth would be virtually completely reduced after a few days.

3 Ethological consequences of social conflict and their correlation with physiological changes

The introduction of an individual into the cage or the territory of a strange conspecific is an extremely intense form of stress in a territorial animal such as the tree shrew. To obtain physiological data on a less severe form of stress or even on adaptation to it, two male tree shrews which were unknown to each other were put together in a strange cage (floor area 120 × 70 cm, height 50 cm, with 2 separate sleeping boxes, feeding dishes, and water bottles) (see also v. Holst et al. 1983). In this situation various responses could be observed. In some cases the confrontation led to high intensity fights and the establishment of a clear dominance relationship between the two within 1–3 days. Firstly, although both animals lived together constantly, the dominant individuals, after establishment of the dominance relationship, usually paid no attention to the subordinates. Attacks against the latter were very rare or even completely lacking. Except for short periods of feeding and drinking, the so-called submissive animals sat in a corner of the cage or in their sleeping boxes for practically the whole day as was the case with the subordinate individuals observed in the first experiment. Even the very rare attacks of the dominants were tolerated, usually without any attempts to flee or to fight back. In this situation, all submissives died within a few days.

In many cases, however, this situation did not result in such immediate and intensive fighting. After less than 4 days, however, and subsequent to low intensity attacks, a dominance structure was established in which the subordinate, by active avoidance behaviour, tried to withdraw from situations which could lead to more intensive fights or – for example if it was cornered – even fought back. Under these conditions, a so-called subdominant – in contrast to a submissive – was capable of living in a dominant's cage, albeit with a reduced sphere of action, for weeks.

Finally, in some pairs low intensity fights occured daily over weeks; these did not, however, result in defined dominance relationships. For most of the time each individual appeared actively to avoid direct confrontations with the conspecific. Since these individuals correspond in their behaviour and their physiological condition to subdominant animals in the presence of the dominant they were classified as such.

So far a total of 58 animals has been investigated; of these 26 were dominant, 18 subordinate and 14 submissive. Some of these individuals were sacrificed between 4–14 days after the stabilisation of the dominance relationship (8 dominants, 10 subordinates, 10 submissives) to gain insight into changes into organ weights and histology. Furthermore, we determined the tyrosine-hydroxylase activities (TyOH) (radioenzymatically) and the norepinephrine concentrations of the left adrenals of the animals. The remaining individuals were kept up to 23 days in the experimental situation. None of the dominant individuals died within this time, but 2 subdominant individuals died on the 20th and 22nd day respectively, and all submissives were dead by the 20th day.

Before the start of the experiments and every 2–5 days thereafter, blood samples were taken from all experimental animals (and 14 control animals kept singly) to determine several physiological parameters. After the dominance relationship had been established (which was the case within the first few days), no significant time dependence with regard to the changes in physiological parameters was found and thus all values for a given parameter of an animal between days 4 and 14 were pooled and the mean considered for further analysis.

Results: The body weight of the dominant animals showed a small but significant increase after establishment of the dominance relationship. Subdominants and submissives showed a drastic decrease in body weight up to about the fourth day of the experiment.

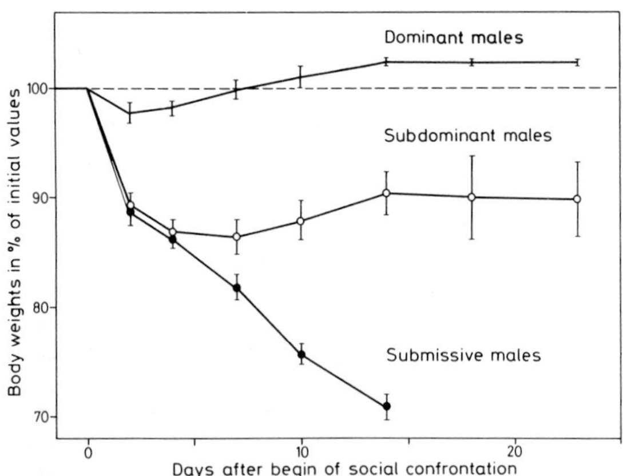

Fig. 4: Body weights (M ± SE) of dominant, subdominant and submissive tree shrews and their changes after the beginning of social confrontation.

Whereas, however, the subdominant animals stabilized at these levels, the body weight of the submissives progressively decreased; the daily weight loss was about 2.5 % of the initial weight, which is within the upper range of the values for submissives type II described in the first experiment (Fig. 4). Correspondingly, there was no apparent difference in the physiological reactions between the latter and submissives in the second experiment.

In the following I shall only consider data that originate from blood samples taken between days 4–14 of social conflict or from organs of animals sacrificed during this period. It has to be mentioned, however, that during the first 3 days even the dominants showed typical stress reactions (among other things a 50 % increase of their serum cortisol values). As can be seen from fig. 5, with the exception of testosterone, the hormone concentrations in the serum of dominant tree shrews as well as all other parameters (including organ weights) largely correspond to those of control animals without direct social contact.

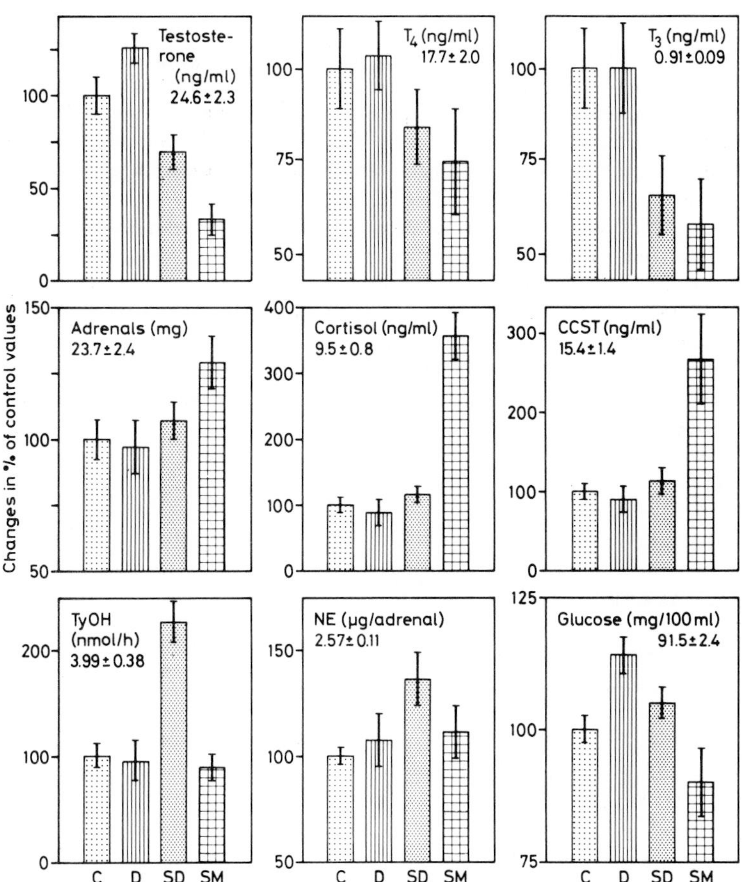

Fig. 5: Some physiological parameters of dominant (D), subdominant (SD), and submissive (SM) tree shrews in percent of those of control animals (C). The absolute values for the control animals are given, along with their mean errors. For further details see text.

There is even evidence of decreased adrenocortical function, since both glucocorticoid hormones as well as adrenal weights show a small (but not significant) decrease. The testosterone concentration in the serum of the dominants, however, is clearly increased ($p < 0.05$). In the submissives, the serum concentrations of thyroidal and gonadal hormones are lower than in controls or dominant tree shrews (testosterone: $p < 0.001$; triiodthyronine: $p \approx 0.05$). The values of these hormones in the subdominant animals are between those of the dominants and submissives.

A striking difference is evident between submissives and subdominants with respect to their adrenal function (Fig. 5). The cortisol as well as corticosterone concentrations in the serum of dominants and subdominants are within the range of those of controls, while those of submissive animals are increased. As a result, submissive animals – and only these – have very decreased lymphocyte numbers within their blood and show no wound healing (as mentioned for the subordinates in the first experiment).

In contrast, the tyrosine-hydroxylase activity of the adrenal glands in submissives does not differ from that of dominants and controls, but in subdominants it is increased by more than 100 %. The adrenal norepinephrine content in the four groups shows similar changes. This indicates a greatly increased sympathetic activity in the behaviorally active subdominant animals, but not in the passive submissive tree shrews.

The heart rates of 14 tree shrews were measured telemetrically over months; the transmitters were developed in our laboratory; weight including battery: about 1.3 g; duration of continuous registration: about 5 months (for details see Stöhr, 1982). Before social confrontation all animals showed similar marked day/night variations, with daily means about 50 % higher than night values (Fig. 6).

In all cases, social confrontation resulted in a dramatic increase in heart rates during the first day. After one to two days, however, the heart rates of dominant males ($n = 5$) returned to their original low values; the heart rates of subdominant animals ($n = 6$), however, showed dramatic increases in the mean frequencies of day values, and even more so in that of night values, thereby reaching, in some cases, night values higher than the daily mean before confrontation (e. g. Fig. 6). This was the case, independent of whether a subdominant was living together with a clear dominant male (as shown in Fig. 6) or

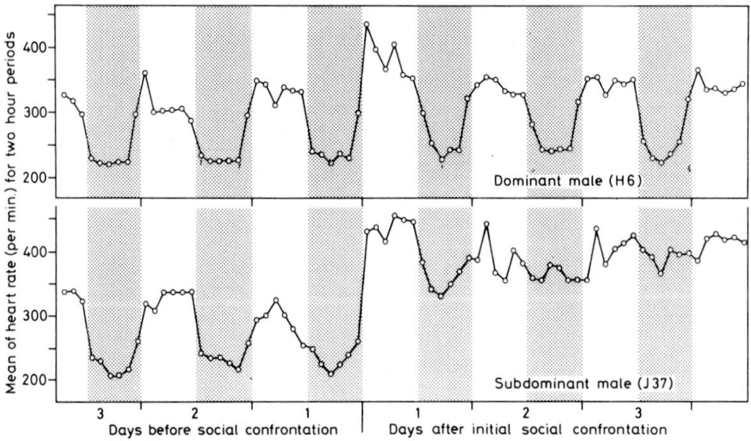

Fig. 6: Mean heart of two male tree shrews before and after begin of social confrontation.

whether both animals were living together without a dominance relationship as subdominants under constant strain.

This increased heart rate points to a constant high activation of the subordinate's sympathetic nervous system in this situation – and this not only during the day when the dominant rival is active and an attack always possible, but also during the night when the animals are completely quiet and sleeping (?) in their separate nest boxes.

It is plausible – and our preliminary histological data support this view – that this situation must lead to damage of the cardiovascular system and probably to heart failure.

Up to now we have been unable to examine submissives (n = 3), since the transmitters were always repelled after 1–2 days, but we have the impression that, in contrast to the active avoiding subdominants, they have a decreased heart rate together with a very reduced blood pressure.

4 Concluding remarks

A social confrontation leading to dominance relationships has apparently no negative effects on dominant tree shrews, even if the position is maintained by fighting. There is even an increase in testicular function and therefore probably also in an increased reproductive capacity. In addition, the data so far point to a small decrease in adrenocortical function (even if this effect is not significant). Subordinate animals, however, show distinct stress reactions in all parameters measured in our laboratory: In general, the physiological reactions are qualitatively the same, differing only in degree. There is, however, a clear distinction between active coping subdominant and passive coping submissive animals with respect to their adrenal functions: that is, the «strategy» of an animal determines (i. e. is correlated with) its physiological response: Active coping with a stressful situation has no apparent effect on adrenocortical function but leads to an increase in sympathetic-adrenomedullary activity and thus may lead to damage of the cardiovascular system, if the situation is maintained over a long period. Passive coping – comparable in human terms with depressive behaviour – has no apparent effects on the sympathetic-adrenomedullary system but leads to adrenocortical activation with its negative effects on wound healing and immune reactions.

Acknowledgements

This work would not have been possible without the cooperation of my colleagues Dr. E. Fuchs and Dr. W. Stöhr whom I would herewith like to thank. Further I would like to thank Mrs. B. Probst and Mrs. I. Zerenner-Fritzsche for their skillful technical assistance.

I am very indepted to Dr. F. Bidlingmaier (University of Munich) for providing antibodies for the determination of cortisol, corticosterone and testosterone. This work was supported by the Deutsche Forschungsgemeinschaft.

References

Holst, D. v. (1972): Renal failure as the cause of death in Tupaia belangeri exposed to persistent social stress. J. comp. Physiol. *78*, 236–273.

Holst, D. v.; Fuchs, E.; Stöhr, W. (1983): Physiological changes in male Tupaia belangeri under different types of social stress. In: Biobehavioral bases of coronary heart disease (eds.: T. M. Dembroski, T. H. Schmidt, G. Blümchen), p. 382–390. Karger, Basel.

Kawamichi, T.; Kawamichi, M. (1979): Spatial organization and territory of tree shrews (Tupaia glis). Anim. Behav. *27,* 381–393.

Martin, R. D. (1968): Reproduction and ontogeny in tree shrews (Tupaia belangeri) with reference to their general behaviour and taxonomic relationships. Z. Tierpsychol. *25,* 409–495 and 505–532.

Starck, D. (1978): Vergleichende Anatomie der Wirbeltiere auf evolutionsbiologischer Grundlage. Band 1: Theoretische Grundlagen. Stammesgeschichte und Systematik unter Berücksichtigung der niederen Chordata. Springer: Berlin, Heidelberg, New York.

Stöhr, W. (1982): Telemetrische Langzeituntersuchungen der Herzfrequenz von Tupaia belangeri: Basalwerte sowie phasische und tonische Reaktionen auf nichtsoziale und soziale Belastungen. Dissertation: Universität Bayreuth.

Fortschritte der Zoologie, Bd. 31 · Hölldobler/Lindauer (Hrsg.): Experimental Behavioral Ecology
G. Fischer Verlag · Stuttgart · New York · 1985

Epilogue:

The cognitive dimensions of animal communication

DONALD R. GRIFFIN

The Rockefeller University, 1230 York Avenue, New York, N. Y. 10021–6399 U.S.A.

KARL VON FRISCH was one of those rare scientists endowed with a «green thumb» for eye-opening discoveries about the biological mechanisms that underlie animal behavior. Again and again he showed that scientists had seriously underestimated the perceptual capabilities of so-called lower animals. I suspect that the momentum of unexpected new advances that he helped to inspire will continue far into the future and lead to discoveries that would surprise and delight him.

Around 1912 VON FRISCH upset the firmly established belief that lower vertebrates and all invertebrate animals were color blind. This opinion had been based on what seemed at the time to be sophisticated, quantitative data, chiefly similarity of spectral sensitivity curves to those of color blind human subjects. But elegantly simple experiments showed beyond doubt that fish and honeybees could indeed discriminate hues. Forty years later he effected another major extension of known visual capabilities of animals by demonstrating that honeybees make extensive use of the polarization patterns of light from the blue sky.

In the 1920s and 1930s VON FRISCH and his colleagues proved that many fish have excellent hearing. One of his papers (VON FRISCH 1923) bore the characteristically eloquent title «A catfish that comes when one whistles to him». Yet despite extensive publication of these data, not only in German journals but in English (Science 1930 and Nature 1938) the whole world of underwater sound production and hearing had to be rediscovered by those concerned with practical problems involving underwater sound; and authoritative textbooks continued for years to assert dogmatically that lower vertebrates could not possibly hear because they lack a cochlea of the mammalian form. The acoustic apparatus available to VON FRISCH was crude and uncalibrated by modern standards, but it is a sign of true genius to devise conclusive experiments with simple equipment. This fact is often overlooked by scientists who are so deeply involved in the intracacies of specialized methodology that they forget the danger of sophistication verging into sophistry.

Of course the most revolutionary of VON FRISCH's discoveries was the symbolic communication employed by honeybees. In the scientific climate of opinion prevailing forty years ago it was shocking and incredible to be told that a mere insect could communicate to its companions the direction, distance and desirability of something far away. At least in America, we were still firmly under the spell of behaviorism in psychology and of the reductionistic approach to animal behavior typified by the views of JACQUES LOEB. Therefore many remained skeptical until they could see for themselves the close correlation between the Schwänzeltänze and the location of food or other things about which bees sometimes dance. By decoding the Schwänzeltänze von Frisch provided us with the equivalent of a Rosetta Stone that opened the door to a new and previously unsuspected world.

One of the principal reasons why the symbolic communication of honeybees has revolutionized our ideas about animal behavior is that the Schwänzeltänze have the property known to linguists as displacement. One animal conveys to another information about something that is not present at the time and place where communication takes place. Nothing in the dark beehive or in the behavior of the dancer's sisters contains information about the distance or direction of the food source. This is important because it contradicts the widely held opinion that all animal communication is rigidly linked to immediate physiological conditions within the communicating animal, and that its communicative signals are incidental byproducts of its internal state of fear, hunger, fatigue or the like. At least a few minutes have elapsed since the dancer was exposed directly to stimuli at the food source, and under special conditions this interval may be several hours, as in the marathon dances described by Martin Lindauer (1955).

The Schwänzeltänze resemble human language in symbolism, displacement and versatility more than any other known type of communication employed by animals under natural conditions. But Lindauer and others have now shown how they probably evolved from simpler behavior patterns in other insects. Furthermore a few other animals are known to communicate semantically, and with some degree of displacement. Some species of ants recruit nestmates by communicative gestures conveying more than simple arousal (Hölldobler and Wilson, 1978). And, as reviewed in in this symposium by Marler, vervet monkeys use alarm calls that convey semantic information about the nature of a predator in addition to their state of fear. Thus while the Schwänzeltänze have been called an «evolutionary freak», they are not totally isolated from other known modes of animal communication. And even if only one species communicates in such a symbolic fashion, it is no longer possible to hold to the previous dogma that symbolic communication with the property of displacement is limited to human language. It would seem threatening enough to our visceral feelings of human uniqueness and superiority if one of the great apes were found to communicate symbolically under natural conditions. But when we learned from the classic experiments of Lindauer (1955) that semantically communicative exchanges are used by swarming bees to reach a consensus, the superiority of not only our species but our phylum seemed to be challenged.

One significant reaction to von Frisch's discovery that honeybees employ symbolic gestures to communicate distance and direction was that of Carl Jung (1973). Relatively late in his life he wrote that although he had believed insects were merely reflex automata, «This view has recently been challenged by the researches of Karl von Frisch ... Bees not only tell their comrades, by means of a peculiar sort of dance, that they have found a feeding-place, but ... they also indicate its direction and distance, thus enabling beginners to fly to it directly. This kind of message is no different in principle from information conveyed by a human being. In the latter case we would certainly regard such behavior as a conscious and intentional act and can hardly imagine how anyone could prove in a court of law that it had taken place unconsciously ... We are ... faced with the fact that the ganglionic system apparently achieves exactly the same result as our cerebral cortex. Nor is there any proof that bees are unconscious.»

Many experimental scientists will doubtless be reluctant to place much weight on one of the many and wide ranging comments of even so distinguished a psychiatrist as Carl Jung. Nor are we likely to accept a decisive role for law courts in settling scientific questions. But whether or not we agree with him, Jung has posed a central question. For von Frisch's Rosetta Stone has certainly narrowed the gap between human language and animal communication; and of course human language has always been considered to be closely linked to human thinking. Philosophers concerned with the nature of minds have based some of their fundamental ideas about the uniqueness of human minds on what biologists have told them about animal behavior in general, and communication in particular. Jung's reaction might be rather general among philosophers if they came to appreciate the versatility of animal communication. Therefore students of animal behavior have an important role to play in this philosophical arena, to encourage philosophers to base their views about human-animal distinctions on scientific facts rather than casual observations of their pet dogs or on elementary and perhaps outdated books on ethology.

Behavioristic psychologists have argued vehemently that mental experiences are of no importance, that they are unworthy of the attention of scientists, and that everything about both animal and human behavior can be better understood and dealt with by totally ignoring them. At least when applied to our own species, this viewpoint is so contrary to common sense that one must blink hard and be sure these colleagues really mean what they seem to be saying. But some of their considered statements are unambiguous. For example a recent rextbook on *Behaviorism, Science and Human Nature* (Schwartz and Lacey 1982) contains the following passage: «If you want to know why someone did something, do not ask. Analyze the person's immediate environment until you find the reward. If you want to change someone's actions, do not reason or persuade. Find the reward and eliminate it. The idea that people are autonomous and possess within them the power and the reasons for making decisions has no place in behavior theory.» If even human beings have no behavioral autonomy and no power to reason or make decisions, it would seem futile to search for such autonomy or decision making in animals. Although this extreme type of behavioristic view is not at all widely shared outside of relatively narrow circles in academic psychology, something quite like it has been taken over implicitly by ethologists and behavioral ecologists.

I have suggested the term «inclusive behaviorists» to include both psychologists who believe that all behavior can best be studied by considering only external contingencies of reinforcement and those biologists who confine their interests to the adaptiveness of animal behavior (Griffin 1984). The leading contemporary behaviorist has recognized for many years that evolutionary selection has influenced the behavior of animals, and presumably to some degree human behavior as well (Skinner 1966, 1981). He therefore extended his concept of contingencies of reinforcement to include evolutionary selection that has led to the more abundant survival and reproduction of animals that behaved in certain ways. Behavioral ecologists, ethologists, and sociobiologists seldom inquire whether the animals they study think or feel one way or another about what they are doing or about the situations in which they find themselves. Such questions are avoided by concentrating exclusively on the adaptive value of behavior. What inclusive behaviorists have in common is a strong aversion to considering any mental experiences that might occur in the animals they study. It is implicitly assumed that only what the animal *does* can have any effect on its survival and reproduction, so that any mental experiences would have no evolutionary consequences. But this is an assumption, not a demonstrated fact. If what animals do is guided or influenced by any thoughts or feelings they may experience, or if mental experiences themselves affect the animal's well being and reproduction, they could not escape the effects of natural selection.

This neglect of mental experiences is by no means universal among a much wider circle of thoughtful scholars who have been concerned with the nature of minds, whether those of people or animals, and the degree to which thoughts and feelings, beliefs and desires, intentions, expectations and the like do affect behavior. For example, Carles Taylor

(1964, 1970) rejects the customary argument of inclusive behaviorists that mental experiences of animals are beyond the reach of science because they are unobservable and immune from independent and objective verification: «The behaviorist view of science is a closed circle, a self-induced illusion of necessity. For there is no self evidence to the proposition that the mental is unobservable. In a perfectly valid sense I can be said to observe another man's anger, sadness, his eagerness to please, his sense of his own dignity, his uncertainty, love for his girl, or whatever. I can find out these things about another sometimes by just observing him in the common sense of that term, sometimes by listening to what he says ... What people say about themselves is never in principle uncheckable ... Teleological explanation in terms of intentional properties requires no appeal to an interactionism involving the unobservable (Taylor 1970, p. 61) ... Future generations may well look upon molar behaviorism with the wonder and awe reserved for some incomprehensible cult of a previous epoch. For the facts of insight, improvisation, goal-direction, and intentionality are so obvious in human behaviour and that of animals which are most similar to us, that it was only by special efforts that they could be overlooked or minimized to the point of appearing tractable to S-R approach.» (p. 67). Similar views have been forcefully expressed by Whiteley (1973), MacKenzie (1977) and Sober (1983). Thus biological students of animal behavior should not feel that the negative dogmatism of inclusive behaviorism is a sacred taboo.

Although inclusive behaviorists have insisted in recent years that any notions such as purpose or intention have no place in explaining animal behavior, the «crypto-phenomenologist» E. C. Tolman (1932, 1966) disagreed and developed, within academic psychology, what he called a «purposive behaviorism» that took account of expectancies and purposes. Some behaviorists have argued, mistakenly, that to assume that a goal or purpose affects the behavior of an animal requires that a future event influence what is happening here and now. It is not the future events, but the animal's thinking about them that may well influence its behavior. If we are willing to accept the reality and significance of thinking about what is likely to happen in even the short term future, it becomes reasonable to suppose that animals experiencing such thoughts will prefer one of the outcomes they can think about over others and will attempt to achieve desirable results and avoid unpleasant ones. Anticipating food, they may do those things they believe will help them get food; fearing pain, they may well do what seems likely to help them escape from predators or aggressive companions.

It is important to make clear at this point that whatever speculative extrapolations of von Frisch's discoveries I may suggest, he expressed strong doubts that bees or other insects were conscious thinkers. In *Aus dem Leben der Bienen* (English translation 1955) he wrote «There is to my knowledge not one example on record of a really intelligent action having been performed by a honeybee.» (p. 150) But he concludes his very short chapter on «The Bee's Mental Capacity» with a cautiously agnostic disclaimer: «Nobody can state with certainty whether the bees are conscious of any of their own actions.» (p. 151). In a letter written in November 1976, he expanded on this theme:

«The question where in the Animal Kingdom the limits of consciousness are located is one which has gone round and round in my head since my youth. Nevertheless I believe, then as now, that one cannot say anything definite about his question. Personally I have never doubted that animals such as apes, dogs or my parakeet who was a close friend for 15 years have some sort of consciousness and can think to a certain extent. But one can never prove this to those who do not wish to believe it ... The dances of bees are variable and are adapted to changing circumstances. But I do not believe that this is voluntary, considered behavior. They give the overwhelming impression of an instinctive performance. This seems to me to correspond to the development of their central nervous system

... (Yet) it is always better to investigate further, than not to believe in advance in the possibility of progress ... I hope for your sake that your youthful optimism will turn out to be right in the end.»

Three general reasons are commonly advanced for rejecting the possibility that any insect might think consciously about what it does or have any mental states such as beliefs, desires, intentions or the like:

(1) The arthropod nervous system, the bulk of which is located in paired segmental ganglia ventral to the digestive tract, is inherently incapable of flexible or innovative behavior.

(2) Insect behavior is rigidly instinctive, genetically programmed and never adjusted to changing circumstances.

(3) It is inherently impossible to discover what the inner life of another species is like, to them; thus even if insects do have mental experiences we can never hope to learn anything about them.

The first objection, that the arthropod nervous system is incapable of thinking, is really based on the second. Having observed countless instances of rigidity and failure to adapt behavior to changing circumstances in a wide variety of insects, it was only natural to correlate the differences in neuroanatomy of vertebrates and arthropods with these *behavioral* differences. But this correlation was accepted when biologists were limited to the gross anatomy of central nervous systems as a basis for such judgements. Modern neurobiology has disclosed nothing to suggest that the arrangements of neurons and synapses into particular shapes of ganglia, or particular locations within the body are inherently necessary for versatile behavior or thinking. But neurobiologists have not discovered anything to identify whatever properties of brains or their component cells do permit conscious thinking. Ours are not the largest brains, nor the ones with the most extreme lateralization of control over vocal communication. And there is no evidence that Wernicke's or Broca's area of the human cerebral cortex contains special types of neurons or synapses uniquely necessary for conscious thinking.

The essential properties of central nervous systems seem instead to result from interaction of excitation and inhibition at synapses. Complex patterns of excitatory and inhibitory interaction have been demonstrated in arthropod central nervous systems (Huber 1974; Hoyle 1977; Burrows 1982). Although there are relatively minor differences in neuron morphology and patterns of arrangement, the basic properties of neurons and synapses do not seem to differ in any fundamental fashion (Vowles 1961; Bullock and Horridge 1965). While only the simpler sorts of neural regulation have been analyzed in detail, no fundamental differences in basic mechanisms have been discovered. Of course insect central nervous systems are very small compared to those of the vertebrate animals with which we are most familiar, and perhaps quantitative differences in mental capabilities can be expected to correlate roughly with size or number of neurons and synapses available in a given central nervous system.

Turning to the second objection on our list, the picture of insect behavior as a set of rigid and unmodifiable instincts has grown up to a large extent because of frequently observed failures of insects to alter their behavior in what *we* can see would be an appropriate fashion. Wasps continue to lay eggs or bring food into nests after an experimenter has opened a hole through which they fall as fast as the wasp replaces them. If live and healthy worker honeybees are annointed with oleic acid, their sisters treat them as corpses. Ants and caterpillars can be induced to follow each other in endless circles. But most if not all of these instances involve conditions well outside of the normal range of circumstances to which the insects in question have become adapted in the course of evolution. Furthermore, as reviewed by Thorpe (1963) other insects do repair damaged nests and adapt their behavior to changing circumstances. Having concluded that many insects are thoughtless robots, do we reverse this conclusion when other species behave in a reasonably appropriate fashion? No, we fall back on the assumption that their DNA has been molded by natural selection to provide instructions covering the contingency in question.

Of course insects are quite capable of learning, and indeed even isolated ganglia of mollusks can show to a limited type of assocative learning (Kandel et al. 1983). The whole body of information exchanged via the symbolic dances of honeybees has been learned by the dancer. She has learned not only the location of something urgently needed by the colony but what is in fact needed. This latter type of learning is seldom considered in connection with the symbolic dances, but it is equally important. As reviewed by Lindauer (1971) worker honeybees spend a great deal of time interacting with their sisters and exchanging materials and pheromones; this provides considerable information as to the needs of the colony. Furthermore the dancer's reception by other bees and the degree to which they do exchange material with her provide information about shortages of sucrose, pollen or perhaps water to cool an overheated hive. This nexus of social communication involves chemical signals which are much more difficult to decode than the symbolic gestures, but it is important to bear in mind that it serves to elicit the searching for what is needed and the dancing about its location and desirability. And of course the dancing does not occur at all unless there is some pressing need; beehives favorably situated where all needs are readily supplied may flourish for long periods without any dancing at all.

These and numerous other examples should serve to dispel the myth that insect behavior is totally fixed and never modified by individual experience or adjusted to changing circumstances. To be sure the flexibility has limits, but undue emphasis has been placed on the experiments in which insects behave stupidly when these limits are exceeded. Within a range of circumstances for which they have been adapted in the course of evolution, insects in general, and honeybees in particular, display abundant evidence of versatile behavior, which, as Jung pointed out, would be interpreted as resulting from conscious thinking if it occurred in our own species. If learning and adjustment of behavior to changing circumstances are criteria for conscious thinking, honeybees, weaver ants, and other insects satisfy these criteria reasonably well – within the range of circumstances for which they have been adapted in the course of evolution.

Of course the scope of insect versatility seems very limited from our human perspective, but this difference may be quantitative rather than a qualitative difference in kind. Honeybees may think consciously about what is important to them and about those challenges and solutions that lie within their competence. Can we be certain that human thinking is capable of dealing with all contingencies? When viewed from a cosmic perspective, are our most highly trained and well informed leaders behaving in a wholly rational fashion in response to the newly arisen threat of nuclear warfare? Or are we following time worn behavior patterns that have been adaptive in the past under radically altered circumstances? In some future century might paleoethologists from another galaxy liken our behavior to that of wasps that continue to lay eggs in a nest where they fall through holes the animal could easily repair?

One of the clearest examples of insect versatility is provided by the behavior of swarming honeybees analyzed by the painstaking observations and experiments of Lindauer (1955). Workers live only a few weeks during the season when they are active, but swarming often occurs only at intervals of years. Yet when the swarm has emerged from its former cavity older workers that have previously been searching for food begin to search for something entirely different, namely cavities with appropriate properties to house a new colony. And they use the same symbolic gestures to communicate the direction, distance, and desirability of cavities they have discovered. If insects are genetically programmed robots, we must assume that their DNA transmits instructions to look for cavities when faced with this unprecedented situation. But such genetic instructions must also induce the bees not only to search for and dance about cavities they have visited but also to pay attention to the dances of their sisters that are reporting about other cavities. The individual robot bee sometimes changes roles from giving to receiving information about cavities when she stops dancing about the cavity she has found and follows dances of other bees that are more vigorous than her own. She may then fly out to the second cavity; and after visiting it, and assessing its qualities by the sorts of behavior described by SEELEY (1977), she may dance about its location and desirability. This process of exchanging information, and altering behavior on the basis of such information, leads after many hours or a few days to a consensus in which all or almost all the bees are dancing about the same cavity. After some hours of this, they begin to execute the Schwirrlauf or buzzing run, and after this has become very widespread the whole swarm flies off to the new cavity.

It is worthwhile to point out that honeybees do not have a monopoly on semantic communication with the property of displacement. As HÖLLDOBLER and WILSON (1978) described, weaver ants also use simple symbolic gestures when recruiting nestmates, these differing in at least two cases according to whether the recruitment is to gather food or to fight intruders. Furthermore these ants sometimes employ chain communication, in which one ant that has been stimulated by the recruiting gestures and pheromones repeats these signals to others without herself receiving the direct stimulation of the food or attacking intruders.

The versatility of symbolic communication employed by swarming honeybees to reach a consensus is one of the most significant aspects of the whole honeybee «language». Yet it has received surprisingly little attention in the thirty years since its discovery by LINDAUER, and it is not even mentioned in most reviews of honeybee communication. It certainly calls for further detailed investigation. LINDAUER's original observations required continuous watching of individually marked bees over several days, and his successors may find it difficult to summon the energy for comparable investigations. But a variety of methods could be developed to explore the extent to which bees shift roles in communicative exchanges according to the challenges they face. I often wonder whether our conviction that bees are genetically programmed robots has held back investigations that might lead to highly significant new discoveries.

As VON FRISCH and his colleagues and successors have learned of more and more complex and versatile behavior of honeybees, we are forced to postulate that their DNA encodes appropriate instructions for an exponentially expanding list of adaptive behavior patterns under a widening range of circumstances including some, such as swarming, that occur only at intervals much longer than the lifespan of the insects involved. Perhaps it would be simpler for their DNA to instruct bees how to think consciously about the problems they face and the behavior by which they try to solve them. Cognitive explanations of behavior may be more parsimonious than the customary appeal to more and more elaborate detailed instructions specifying what to do under an enormous range of conditions.

Many scientists concerned with animal behavior, such as WALKER (1983) or several of the contributors to a symposium edited by ROITBLAT, BEVER and TERRACE (1983) now recognize the probability that animal cognition is real and significant. But to speak of animal consciousness still seems too radical a departure from the inclusive behaviorism with which we have become so comfortably familiar. While we know far too little to speak with confident certainty about such questions, I suspect that when cognition becomes as complex, versatile, and multifaceted as in the behavior of honeybees, it becomes simpler and more efficient for the animal to think consciously about the many interacting factors that affect its success in life and reproduction. As scientists we tend to isolate particular behavior patterns for investigation and analysis, but the animals must often react appropriately to the interplay between several influences, and it is a special feature of conscious thinking that it can encompass a whole picture and deal with interactions and trade-offs as well as particular relationships. When we find, with the aid of von Frisch's Rosetta Stone, that socially interdependent animals communicate with each other about solutions to the challenges they face, this suggests that they may be thinking about what they are communicating. These communicated messages can be intercepted and interpreted by human observers, and this provides us with the opportunity to learn something at least about what the animals may be thinking.

The third objection listed above is in many ways the most seriously paralytic. If we remain convinced that there is no possible way to learn anything about the subjective mental experiences of another species, we will never try and therefore make no progress. While such an effort is sure to be challenging and fraught with difficulties and confusion, why must we be so pessimistic, *a priori,* when such exciting progress has been achieved in so many other areas of science? To paraphrase von Frisch, «It is always better to investigate further than to reject in advance the possibility of progress.»

It may help place this problem in appropriate perspective to remind ourselves of an elementary philosophical problem. One of the many logically impregnable philosophical positions is that of the solipsist who insists that he is the only conscious being in the universe. If he is adamant and agile in his manipulation of logical arguments, we cannot convince him that other people also experience conscious thoughts and subjective feelings, for he can explain away every piece of evidence we may offer in favor of consciousness in his companions as one or another sort of illusion. Yet no one takes this viewpoint seriously. We reject it not because of logical arguments but for reasons of plausibility and common sense. So much of what we observe in our companions' behavior, especially their nonverbal signals and what they say, indicates so strongly that they also think and feel that we are willing to make this inference with great confidence despite the logically incontrovertible arguments of the solipsist.

Yet when we consider other species, we have been so strongly swayed by the logical impossibility of proving in any absolute sense that they experience conscious thoughts and subjective feelings that we have fallen into an implicit belief in «species solipsism». Many scientific problems are so difficult that only limited progress can be expected at early stages of their investigation, but in most other areas of science this has fortunately not discouraged attempts to understand objects and processes that are not directly observable. For example behavioral ecologists and sociobiologists seek to understand how animal behavior has been influenced by natural selection. They postulate, at least implicitly, that certain behavior patterns of ancestral animals led to their more abundant reproduction. But these events cannot be observed directly, because they involved animals long dead that inhabited environments widely dispersed in space and time and quite impossible to reconstruct. Therefore we are obliged to reconstruct them in our minds, relying on inference and considerations of plausibility to an even greater degree than is necessary to develop a cognitive ethology, based upon data about events occurring in the brains of animals here and now. Furthermore animals often provide us with many types of information about what is going on in their central nervous systems. These include overt behavior and electrical correlates of brain activity such as the event related potentials that are perhaps on the threshold of providing objective indices of certain simple types of human thinking

(GALAMBOS and HILLYARD 1981; DONCHIN et al. 1983). And, last but not least, animals often seem to be communicating their thoughts and feelings to other animals.

Perhaps it is time to consider tentatively a substantial revision in the way which we think about animals and their behavior. I do not claim that the correctness of this cognitive approach to ethology can yet be established by satisfactory evidence. But I believe that it can lead to fruitful working hypotheses to be tested by scientific methods. Even if such investigations eventually disconfirm these hypotheses, or more likely fail to settle the matter conclusively, the effort promises to improve and enrich our understanding of animals and of the relationship between human and animal thinking. This cognitive perspective on ethology has been inspired by KARL VON FRISCH's discoveries, but I do not wish to imply that he would agree with all or even with any of what I am suggesting. Cognitive ethology, at this embyronic stage of its development, analogous to a gastrula or perhaps only a blastula, entails the following tentative hypotheses:

A. Mental experiences such as desires, beliefs, intentions, expectations, hopes that something nice will happen, or fears that something unpleasant may occur, are real and significant, at least in the human mind. Their occurrence in other species is a matter for empirical investigation.

B. Some mental experiences influence behavior; thus they are not inconsequential byproducts or illusions.

C. Mental experiences result from the functioning of central nervous systems; they entail no extra-physical events or processes, and they are governed by the principles of natural science. They are presumably processes regulating information flow within central nervous systems, but they also sometimes produce subjective consciousness, at least in our own species.

D. Central nervous systems function through complex excitatory and inhibitory interactions between nerve impulses, local excitation within neurons and neuroendocrine processes, primarily at synapses.

E. Neurons, synapses, neuroendocrine processes and other basic functions that underly cognition are fundamentally similar in all central nervous systems. Thus nothing now known about neurophysiology precludes cognition or consciousness in any animal with a central nervous system.

F. The existance and nature of mental experiences can be inferred to a significant though limited degree from the behavior of people and animals, and especially from their communication about such experiences. Such inferences are testable by using them to predict future behavior, including communicative behavior.

G. Suggestive evidence of the occurrence of mental experiences is supplied by versatility of adaptive behavior in response to changing circumstances, although this cannot exceed the animal's inherent repertoire of capabilities.

H. Stronger evidence of mental experiences and conscious thoughts can be gathered by intercepting and interpreting the communicative signals which two or more animals exchange; in some cases at least, these may be direct reporting of the mental experiences of the communicator.

Inclusive behaviorists can always claim that an unconscious robot could duplicate whatever behavior we may observe in any animal. But they can also apply this «robot story» to their human companions, as exemplified by the passage from SCHWARTZ and LACEY quoted above. Yet we know that our own mental experiences are real and significant, which means that this «robot story» is not a reliable and conclusive reason to deny consciousness. Therefore we can lay it to one side, just as we do the logically watertight arguments of the solipsist.

Before we can fully appreciate the significance and the potential ramifications of cognitive ethology two additional considerations require discussion. The first of these is the widely accepted view that instinctive behavior, controlled primarily by the animal's DNA, cannot under any circumstances be accompanied or influenced by conscious thinking. The second is that conscious thinking about an animal's problems or opportunities would lead to inefficient hesitation or vacillation, and that rigid and mechanical actions would be more effective. What I will call the «Frischian revolution» seems to call both into question, as I have discussed in detail elsewhere (Griffin 1978, 1981, 1984).

G. J. Romanes (1884) defined an instinct as «reflex action into which there is imported the element of consciousness.» To be sure, Romanes is no longer looked upon with favor by students of animal behavior; but, like Carl Jung, he has at least posed a significant question. If we are willing to concede that any animal might be capable of conscious thinking, why must this be limited to thoughts accompanying learned behavior? It is helpful in this context to distinguish those involuntary and reflexive patterns of human behavior such as sneezing over which we have little or no control, but which we are consciously aware of as they happen, from activities of our bodies which do not ordinarily entail conscious awareness, such as intestinal peristalisis or the constriction of the pupils in bright light. There are also intermediate categories such as breathing or postural reflexes which we do not usually think about but which can be brought to conscious attention when we wish to do so. Perhaps insects and other animals think consciously about some of their instinctive actions even though they may not perform them intentionally. During the construction and provisioning of a nest in which will develop offspring the mother wasp cannot live to see, she can scarcely think about the end result of this activity. But this does not preclude thinking about the activities in which she does engage. Perhaps she thinks about digging a burrow or constructing a nest without any concept of the eventual results. This might be roughly comparable to the early stages of human courtship in which males at least seldom give much thought to the possibility that the eventual result may be a baby.

When we dismiss the possibility of conscious thinking in insects or other animals whose behavior is largely instinctive, what we really mean is that it seems impossible for them to know, or think consciously about, the eventual results of what they are doing. But this in no way precludes conscious awareness of the immediate behavior in which they are engaged, or its short term consequences. The mother wasp might want a burrow, grasp and carry away particles of soil with the intention of forming a dark cavity into which she will crawl to lay eggs, even though she has not the remotest notion that the eggs will later hatch into larvae. We have tended to rule out the possibility of any conscious thinking at all because the animal cannot be aware of the longer term results of its behavior.

The idea that conscious thinking would be inefficient seems to have arisen in discussion of laboratory experiments with rats; for instance E. C. Tolman's purposive behaviorism was criticised for «leaving the rat lost in thought» at branch points in a maze. But when one considers the normal behavior of a wide variety of animals under natural conditions, it seems possible that at least simple thinking about alternative actions of which the animal is capable, and their likely results, would be helpful in many circumstances. The ability to think in simple «if, then» terms should permit successful solutions to many of the problems faced by animals that must make a living by finding enough food while avoiding predators. Such thoughts on the part of a hungry lion as «If I sneak slowly with my belly to the ground, those gazelles may not run away and maybe I can catch one» or «If I run into those brambles, that dog won't hurt me» on the part of a frightened rabbit seem plausible if we grant animals any capability of thinking at all. Likewise when honeybees dance about an ideal cavity they might think something like «If we all fly in this direction and about this long a time, we can get into that nice dark, dry place.» The subsequent behavior of the bees is certainly consistent with this interpretation, although such evidence is far from conclusive. But it may be possible, by much additional and enterprising investigation, to build up a body of data that will serve to increase or decrease the plausibility of such interpretations.

As scientists we strongly prefer rigorously testable hypotheses, and I wish I could suggest convincing experiments, comparable to those of KARL VON FRISCH, that would prove that a particular animal was or was not thinking a certain thought. Unfortunately, I feel we are in a state comparable to his frame of mind around 1905. But to deny that future investigators could build on what is already known and devise new and improved methods to test hypotheses about what animals may think and feel is to belittle the capabilities of human brains in general, and those of scientists in particular. Especially with the example of KARL VON FRISCH to inspire us, we should be ready, as DENNETT (1983) has put it to «cast off the straightjacket of (inclusive) behaviorism» and develop appropriate strategies and tactics for learning about animal cognition and consciousness. Perhaps these taboos will someday come to seem comparable to the conviction that bees are color blind, that fish are deaf, and that insects could not possibly communicate symbolically about something far away.

References

Bullock, T. H. and G. A. Horridge (1965): Structure and function in the nervous systems of invertebrates. Freeman, San Francisco.

Burrows, M. (1982): Interneurones co-ordinating the ventilatory movements of the thoracic spiracles in the locust. J. Exptl. Biol. *97,* 385–400.

Dennett, D. C. (1983): Intentional systems in cognitive ethology: The «Panglossian paradigm» defended. Behav. Brain Sci. *6,* 343–355. See also commentaries ibid 355–390.

Donchin, E., G. McCarthy, M. Kutas, and W. Ritter (1983): Event-related brain potentials in the study of consciousness. In: Consciousness and Self-regulation, Advances in Research and Theory, Vol. 3, ed. by R. J. Davidson, G. E. Schwartz and D. Shapiro, Ch. 3, p. 81–121, Plenum, New York.

Frisch, K. von (1923): Ein Zwergwels, der kommt, wenn man ihm pfeift. Biol. Zbl. *43,* 439–446.

– (1930): The sense of hearing in fishes. Science *71,* 515.

– (1938): The sense of hearing in fish. Nature (London) *141,* 8.

– (1955): The dancing bees, an account of the life and senses of the honey bee. Harcourt Brace, New York.

Galambos, R. and S. A. Hillyard (1981): Electrophysiological approaches to human cognitive processing. Neurosci. Res. Prog. Bull. *20,* 141–264 + vi.

Griffin, D. R. (1978/1980): Prospects for a cognitive ethology. Behav. Brain Sci. *1,* 527–538. See also commentaries ibid *1,* 555–629 and *3,* 615–623.

– (1981): The question of animal awareness, 2nd. ed. Rockefeller Univ. Press, New York. Paperback William Kaufman, Los Altos, California.

– (1984): Animal thinking. Harvard Univ. Press, Cambridge, Mass.

Hölldobler, B. and E. O. Wilson (1978): The multiple recruitment system of the African weaver and *Oecophylla longinoda* (Latreille) (Hymenoptera, Formicidae). Behav. Ecol. Sociobiol. *3,* 19–60.

Hoyle, G. ed. (1977): Identified neurons and behavior of arthropods. Plenum, New York.

Huber, F. (1974): Neural integration (central nervous system). In: The Physiology of Insecta, 2nd. ed., Vol. IV, ed. by M. Rockstein, p. 4–100. Academic Press, New York.

Jung, C. (1973): Synchronicity, a causal connecting principle. Princeton Univ. Press, Princeton, N. J.

Kandel, E. R., Abrams, T., Bernier, L., Carew, T. J., Hawkins, R. D., and Schwartz, J. A. (1983): Classical conditioning and sensitization share aspects of the same molecular cascade in *Aplesia.* Cold Spring Harbor Symposia on Quantitative Biology *48*, 821–830.

Lindauer, M. (1955): Schwarmbienen auf Wohnungssuche. Z. vergl. Physiol. *37,* 263–324.
– (1971): Communication among social bees, 2nd. ed. Harvard Univ. Press, Cambridge, Mass.
MacKenzie, B. D. (1977): Behaviourism and the limits of scientific method. Routledge and Kegan Paul, London.
Roitblat, H., T. G. Bever and H. S. Terrace eds. (1983): Animal cognition. Erlbaum, Hillsdale, N. J.
Romanes, G. J. (1969): Mental evolution in animals (1884). Reprint with posthumous essay on instinct by Charles Darwin. ABS Press, New York.
Seeley, T. (1977): Measurement of nest cavity volume by the honey bee *(Apis mellifera).* Behav. Ecol. Sociobiol. *2,* 201–227.
Schwartz, B. and H. Lacey (1982): Behaviorism, science and human nature. Norton, New York.
Skinner, B. F. (1966): The phylogeny and ontogeny of behavior. Science *153,* 1205–1213.
– (1981): Selection by consequences. Science *213,* 501–504.
Sober, E. (1983): Mentalism and behaviorism in comparative psychology. In: Comparing Behavior: Studying Man Studying Animals. ed. by D. W. Rajecki, Ch. 5 p. 113–142. Erlbaum, Hillsdale, N. J.
Taylor, C. (1964): The explanation of behaviour. Humanities Press, New York.
– (1970): The explanation of purposive behavior. In: Explanation in the Behavioral Sciences, ed. by. R. Borger and F. Cioffi, p. 49–79 Cambridge Univ. Press, London.
Thorpe, W. H. (1963): Learning and instinct in animals. 2nd. ed. Harvard Univ. Press, Cambridge, Mass.
Tolman, E. C. (1932): Purposive behavior in animals and men. Appleton-Century, New York.
– (1966): Behavior and psychological man. Univ. of California Press, Berkeley.
Vowles, D. M. (1961): Neural mechanisms in behavior. In: Current Problems in Animal Behaviour, ed. by W. H. Thorpe and O. L. Zangwill, p. 5–29. Cambridge Univ. Press, London.
Walker, S. (1983): Animal thought. Routledge and Kegan Paul, London.
Whiteley, C. H. (1973): Mind in action, an essay in philosophical psychology. Oxford Univ. Press, London.

Subject index